D1236644

MEMORY FUNCTION APPROACHES TO STOCHASTIC PROBLEMS IN CONDENSED MATTER

ADVANCES IN CHEMICAL PHYSICS

VOLUME LXII

EDITORIAL BOARD

MEMORY FUNCTION APPROACHES TO STOCHASTIC PROBLEMS IN CONDENSED MATTER

Edited by

MYRON W. EVANS

University College of North Wales

PAOLO GRIGOLINI

University of Pisa

GIUSEPPE PASTORI PARRAVICINI

University of Pisa

ADVANCES IN CHEMICAL PHYSICS
VOLUME LXII

Series Editors

Ilya Prigogine

*University of Brussels
Brussels, Belgium
and
University of Texas
Austin, Texas*

Stuart A. Rice

*Department of Chemistry
and
The James Franck Institute
University of Chicago
Chicago, Illinois*

AN INTERSCIENCE® PUBLICATION

JOHN WILEY & SONS
New York · Chichester · Brisbane · Toronto · Singapore

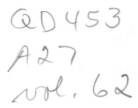

An Interscience® Publication

Library of Congress Cataloging in Publication Data:

Main entry under title:

Memory function approaches to stochastic problems in
   condensed matter.

   (Advances in chemical physics, ISSN 0065-2385; v. 62)
   "An Interscience publication."
   Includes indexes.
1.   Condensed matter.   2.   Molecular dynamics.
3.   Stochastic processes.   4.   Relaxation methods
(Mathematics).   5.   Chemistry, Physical and theoretical.
I. Evans, Myron W. (Myron Wyn), 1950–   .   II.   Grigolini,
Paolo.   III.   Pastori Parravicini, Giuseppe.   IV.   Series.

QD453.A27 vol. 62   541 s   85-5291
[QC173.4.C65]   [530.4]
ISBN 0-471-80482-7

Printed in the United States of America

10 9 8 7 6 5 4 3 2 1

INTRODUCTION

Few of us can any longer keep up with the flood of scientific literature, even in specialized subfields. Any attempt to do more and be broadly educated with respect to a large domain of science has the appearance of tilting at windmills. Yet the synthesis of ideas drawn from different subjects into new, powerful, general concepts is as valuable as ever, and the desire to remain educated persists in all scientists. This series, *Advances in Chemical Physics*, is devoted to helping the reader obtain general information about a wide variety of topics in chemical physics, which field we interpret very broadly. Our intent is to have experts present comprehensive analyses of subjects of interest and to encourage the expression of individual points of view. We hope that this approach to the presentation of an overview of a subject will both stimulate new research and serve as a personalized learning text for beginners in a field.

ILYA PRIGOGINE
STUART A. RICE

PREFACE

The main aim of this volume is to show that the projection techniques and the memory function formalism developed by Zwanzig and Mori give suitable theoretical support for a successful description of a wide variety of relaxation problems. The unifying role and the wealth of perspectives opened up by these techniques in the theory of relaxation can be compared with the ones provided by the renormalization group in field theory and cooperative phenomena. The main strategy of this volume is to show that a number of stochastic processes in apparently different fields of condensed matter (ranging from nonequilibrium statistical physics to solid state theory, molecular dynamics, dynamical properties of hydrogen bonded liquids, electron spin resonance spectra, theory of chemical rates, multiplicative noise, together with biological systems and galactic evolution in astrophysics) can be interpreted in terms of very similar mathematical equations. The theoretical foundations of this strategy are laid down in Chapter I.

Chapter II addresses another fundamental problem: under what physical conditions can the Fokker-Planck equation provide a reliable picture of fluctuation–dissipation processes? Aware as they are of the technical and conceptual difficulties involved in nonlinear statistics, the authors share Zwanzig's optimistic view that use of the Fokker-Planck equation is practicable and advantageous. Chapter II contains a brief description of rules to construct, via a suitable procedure, equations of Fokker-Planck type for the slow variables of the system under study. This theoretical method eliminates explicit analytic dependence on fast variables and thereby produces a significant simplification of the problem under discussion.

Chapter III summarizes the basic properties of the continued fractions encountered in the theory of relaxation. Continued fractions have emerged as essential for the description of correlation functions, density of states, and spectra. Although the analytical theory of continued fractions dates back to the last century, it was, for a long period of time, hardly more than mere mathematical research and speculation. The growing interest in the mathematical apparatus of continued fractions is related, on the one hand, to developments in modern projective formalism and, on the other, to the flexibility of the continued fraction techniques, especially their ability to handle non-Hermitian operators and Liouvillians.

Chapter IV applies projective techniques to solid state physics. This development has been foreshadowed by the use of projection operators in other areas of research. During the last decade, however, the situation has changed rapidly and, perhaps, the study of solids by means of continued fractions has already outdistanced more traditional lines of approach. With the advent of memory function formalism, the reduction of the huge number of degrees of freedom occurring in solid state problems is no longer based on standard group theory techniques, which hold only in ideally perfect situations. This explains why the use of the projective techniques has amounted to a major breakthrough in the physics of real, rather than ideal, solids.

The memory function formalism is advantageous not only for dealing with problems posed by solid state physics, but also to interpret the physics of the liquid state, the area explored by Chapters V through VII of this volume. Computer simulation of molecular dynamics in the liquid state is becoming of increasing importance for solving (and posing) new problems. Chapter V describes the results of an ideal, well-controlled, numerical experiment that basically consists of suddenly switching off an external electric field applied to a liquid sample of polar molecules. The decay of electric polarization (the fall transient) is gradually accelerated by increasing the intensity of the external electric field. This is a violation of linear response theory brought about with strong external fields. Can fall transient acceleration be predicted at a mesoscopic level of description?

Chapter VI illustrates what a mesoscopic level of description is meant to contain. It involves the variables of interest (usually assumed to be slow) plus a suitable set of auxiliary variables, whose role is to mimic the influence of the thermal bath on the variables of interest themselves. This level of description (reduced model theory) is less detailed than the truly microscopic one, because an overwhelming number of microscopic degrees of freedom are simulated with fluctuation–dissipation processes of standard type. The mesoscopic level, however, is still detailed enough to preserve the essential information without which the theoretical investigation becomes difficult and obscure. A new class of non-Gaussian equilibrium properties is proven to be responsible for the acceleration of the fall transient described in Chapter V. To obtain these results, use is made both of the theoretical tools already mentioned and of computer simulation (one-dimensional for translation and two-dimensional for rotation).

It is found, in particular, that when using the well-known model of the itinerant oscillator, one cannot give up the assumption that the interaction between "real" and "virtual" variables is linear without also making this interaction fluctuate randomly in time. This establishes the link with Chapter VII. This fluctuating process can be used to model the influence of hydrogen bond dynamics, the long-time effects of which are then carefully explored and compared with the results of real laboratory experiments with satisfactory agreement.

Assessing the degree of reliability of a theory may sometimes prove difficult when its application involves considerable computational difficulty. This is the case with the stochastic Liouville equation, which is usually regarded as a satisfactory approach to evaluating spectra. When applied to practical problems of electron paramagnetic resonance, however, its use is fraught with difficulties resulting from the need for diagonalizing extremely large matrices. Chapter VIII shows how this problem can be coped with efficiently using the continued fraction procedure with appropriate implementations.

The authors of Chapter IX use the theoretical methods developed in this book to illustrate the state of the art in the field of chemical reaction processes in the liquid state. The well-known Kramers theory can be properly generalized so as to deal successfully with non-Markovian effects of the liquid state. From a theoretical point of view the nonlinear interaction between reactive and nonreactive modes is still an open problem that touches on the subject of internal multiplicative fluctuations.

The subject of multiplicative fluctuations (in linear and especially nonlinear systems) is still deeply fraught with ambiguity. The authors of Chapter X set up an experiment that simulates the corresponding nonlinear stochastic equations by means of electric circuits. This allows them to shed light on several aspects of external multiplicative fluctuation. The results of Chapter X clearly illustrate the advantages resulting from the introduction of "auxiliary" variables, as recommended by the reduced model theory. It is shown that external multiplicative fluctuations keep the system in a stationary state distinct from canonical equilibrium, thereby opening new perspectives for the interpretation of phenomena that can be identified as due to the influence of multiplicative fluctuations.

Examples of multiplicative fluctuations in other disciplines are discussed in the last two chapters. The reader is guided first through biological systems, then the journey ends in the fields of stellar formation and galactic evolution. Chapter XI deals with problems of stochastic selection in population genetics. This is an illuminating example of how a mesoscopic theory based on the general principles of the reduced model theory may provide a clear answer to some basic problems while avoiding ambiguities, such as the choice of the proper stochastic algorithm.

Chapter XII is devoted to a detailed illustration of the different methods currently used to deal with stochastic problems in astrophysics, including, of course, the Langevin-Fokker-Planck method. The advantages are demonstrated by applying to astrophysics the new theoretical methods of this book, especially the continued fraction and adiabatic elimination procedures.

M. W. Evans
P. Grigolini
G. Pastori Parravicini

CONTRIBUTORS TO VOLUME LXII

DAVIDE BERTOLINI, Istituto di Fisica Atomica e Molecolare del CNR, Pisa, Italy

MARIO CASSETTARI, Istituto di Fisica Atomica e Molecolare del CNR, Pisa, Italy

MYRON W. EVANS, Department of Physics, University College of North Wales, Gwynedd, United Kingdom

SANDRO FAETTI, Dipartimento di Fisica, Università di Pisa and GNSM, Pisa, Italy

MAURO FERRARIO, Instituto di Fisica Atomica e Molecolare del CNR, Pisa, Italy

FEDERICO FERRINI, Istituto di Astronomia, Università di Pisa, Pisa, Italy

CRESCENZO FESTA, Dipartimento di Chimica e Chimica Industriale, Università di Pisa and GNSM, Pisa, Italy

TERESA FONSECA, Dipartimento di Fisica, Università di Pisa and GNSM, Pisa, Italy. Permanent address: Faculdade de Ciências do Porto, Departamento de Quìmica, Porto, Portugal

LEONE FRONZONI, Dipartimento di Fisica, Università di Pisa and GNSM, Pisa, Italy

MARCO GIORDANO, Dipartimento di Fisica, Università di Pisa and GNSM, Pisa, Italy

JOSÉ A. N. F. GOMES, Faculdade de Ciências do Porto, Departamento de Quìmica, Porto, Portugal

PAOLO GRIGOLINI, Dipartimento di Fisica, Università di Pisa and GNSM, Pisa, Italy

GIUSEPPE GROSSO, Dipartimento di Fisica, Università di Pisa and GNSM, Pisa, Italy

DINO LEPORINI, Dipartimento di Fisica, Università di Pisa and GNSM, Pisa and Scuola Normale Superiore, Pisa, Italy

FABIO MARCHESONI, Dublin Institute for Advanced Studies, Dublin, Ireland. Permanent address: Dipartimento di Fisica, Università di Perugia, Perugia, Italy

PAOLO MARIN, Dipartimento di Fisica, Università di Pisa and GNSM, Pisa and I. F. E. N. Spa, Ceparana (La Spezia), Italy

GIUSEPPE PASTORI PARRAVICINI, Dipartimento di Fisica, Università di Pisa and GNSM, Pisa, Italy

SILVANO PRESCIUTTINI, Istituto di Biochimica, Biofisica e Genetica, Università di Pisa, Pisa, Italy

GIUSEPPE SALVETTI, Istituto di Fisica Atomica e Molecolare del CNR, Pisa, Italy

STEVEN N. SHORE, Space Telescope Science Institute, Baltimore, Maryland, United States of America

ALESSANDRO TANI, Dipartimento di Chimica e Chimica Industriale, Pisa and Scuola Normale Superiore, Pisa, Italy

RENZO VALLAURI, Istituto di Elettronica Quantistica del CNR, Firenze, Italy

BRUNO ZAMBON, Dipartimento di Fisica, Università di Pisa and GNSM, Pisa, Italy

CONTENTS

MEMORY FUNCTION APPROACHES TO STOCHASTIC PROBLEMS IN CONDENSED MATTER

ADVANCES IN CHEMICAL PHYSICS

VOLUME LXII

I

THEORETICAL FOUNDATIONS

P. GRIGOLINI

CONTENTS

I. INTRODUCTION

The motivation behind this volume is to face a wide variety of problems, while strictly limiting the number of the theoretical tools to be used. Only one basic concept will be used: that of "relevant" and "irrelevant" variables,[1] which plays so fundamental a role in accounting for irreversible phenomena. Irreversibility, as currently investigated with experimental techniques, is traced back to the fact that experimental signals depend on only a limited number of variables related to the sample under study, which are therefore referred to as "relevant."[1] The "relevant" variables (usually slowly moving and, in that they directly affect the experimental signals, also termed macroscopic) interact ineluctably with the large set of the remaining variables. These degrees of freedom, usually fast moving, do not exert a direct influence on the experimental signal and thus are termed "irrelevant." The stochastic behavior of the relevant, or macroscopic, variables must be traced back to this interaction. This picture, roughly accounting for dissipation and irreversibility, must be supplemented by a rigorous mathematical treatment such as analytical continuation for well-defined dynamic quantities.[2-4] Nevertheless it is basically correct and can be regarded as a leading concept.

The simplest and most elegant theoretical technique operating in line with this leading idea is the Nakajima-Zwanzig projection method.[1,5] By using this approach we are naturally led to replace the standard master equations,

1

Markovian in nature, with a generalized equation, the most distinctive feature of which is its non-Markovian character. This projection approach applies to both rigorous Liouville equations and, so to speak, approximated Liouville equations such as the Fokker-Planck equations. The latter kind of equations represents a drastic simplification of the problem under study, since a virtually infinite number of degrees of freedom are simulated by friction and diffusion terms. Both kinds of description refer to a picture where physical variables are time-independent and their probability distributions depend on time. Borrowing a quantum-mechanical terminology, we call this the "Schrödinger representation." The well-known Mori approach[6] is usually regarded as the counterpart of the Nakajima-Zwanzig technique in the "Heisenberg representation."* In this representation, where physical variables depend on time and probability distributions do not, the time evolution of physical variables is driven by operators that are exactly the adjoints to those involved in the Schrödinger representation. The equivalence between the two procedures seems to be complete, in that the Mori approach can be extended to the non-Hermitian case,[7] so that it can also be applied to the Heisenberg picture corresponding to those Fokker-Planck equations that do not admit detailed balance and therefore cannot be brought into a self-adjoint form.

Nonlinear nonequilibrium statistical mechanics is attracting the interest of more and more physicists,[8] who are thereby involved in problems with more and more sophisticated technicalities. Prigogine and coworkers[9] have shown clearly the decisive importance of giant fluctuations leading nonlinear systems far from their stable thermodynamical equilibrium. Efforts are being made to give the phenomenology of such processes a solid and rigorous theoretical basis.[10]

This volume does not share this ambitious purpose. We have limited ourselves to showing that the theoretical background of the Zwanzig-Mori projection methods generates calculation techniques that can be applied usefully to several fields of investigation, including these exciting topics. Let us take a first illustrative example borrowed from the related field of cooperative phenomena, for which Haken[11] invented the term *synergetics*. Haken showed that spontaneous self-organization processes imply that the system under study should be divided into two subsystems: one of long-living variables and one of short-living variables. Self-organization can be understood on the basis of an adiabatic elimination of the fast variables. Now the prob-

---

*This terminology has also been adopted by Kubo.[8] A master equation (including the Fokker-Planck equation) is the "Schrödinger picture," and a Langevin equation is the "Heisenberg picture," of the same problem. We shall give our calculation techniques mainly a physical foundation. Our continued fraction procedure (CFP) stems from statistical physics explored within the framework of the Heisenberg picture.

lem arises of developing a systematic and rigorous procedure that takes into account the fact that the lifetime of the "slave" variables is finite. In this volume we shall use a procedure of this kind that stems from the Heisenberg representation. We shall refer to it as the *adiabatic elimination procedure* (AEP). Figure 1 points out that the AEP is one of the three theoretical tools that are derived from the Zwanzig-Mori projection techniques.

The AEP should allow us to derive a reliable Fokker-Planck type of equation for set **a** alone. The solution of this type of equation still involves formidable problems. In the monodimensional case (i.e., when **a** consists of one variable x), the AEP usually results in the following type of equation:

$$\frac{\partial}{\partial x}\sigma(x) = \left\{ \frac{\partial}{\partial x}\left[\frac{d\varphi}{dx}(x)\right] + Q\frac{\partial}{\partial x}g(x)\frac{\partial}{\partial x}g(x) + \varepsilon\frac{\partial^2}{\partial x^2} \right\}\sigma(x) \quad (1.1)$$

Although the equilibrium properties of this equation are well known,[12] their time behavior for $\varepsilon = 0$ is still the object of controversy.[13,14] When the additive stochastic force is present ($\varepsilon \neq 0$) and $\varphi(x)$ is a double-well potential, Eq. (1.1) describes the process of escape from a well of a fluctuating poten-

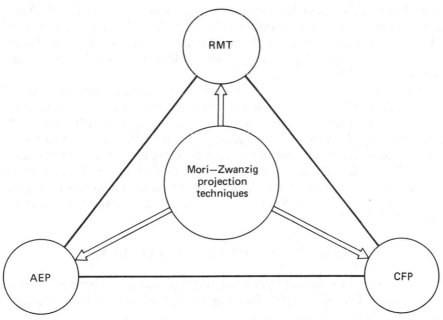

Figure 1. A scheme illustrating the common theoretical roots of the three main research tools used in this volume.

tial. This implies a generalization of the celebrated paper of Kramers.[15] The last few years have seen a tremendous proliferation of papers dealing with this problem and, as remarked in ref. 16, often ignoring earlier pioneer attempts, the simplicity of which should be retained. On the other hand, a reliable and versatile computational technique allowing a straightforward way of checking theoretical prediction would be warmly welcome, A further formidable problem is found in the field of solid state physics, where sometimes the non-Markovian properties are so significant as to prevent truncation of the Mori hierarchy of equations at any order. In both the examples above one has available the first N time derivatives of the variable of interest evaluated at the initial time. It would be quite useful if one could draw from such a precious piece of information a description valid for a wider time range, hopefully the entire time dominion of the relaxation process under study.

A technique largely satisfying these requirements is a continued fraction procedure (CFP) founded on the projection techniques applied to the Heisenberg picture. This is the second corner of the ideal triangle that symbolizes the three basic theoretical tools that will be widely used throughout this book (see Fig. 1).

The last corner of our ideal triangle symbolizes the "reduced" model theory (RMT). The last few years have heard some debates on how to build up a modeling approach to molecular dynamics in the liquid state.[17-19] These earlier attempts are based mainly on the generalized Langevin equation of Mori, which is linear in nature. In this book we shall illustrate a nonlinear version of this approach, the application of which implies wide use of both the AEP and CFP.

The major aim of this chapter is to point out the common theoretical roots of these three helpful tools. Section II will be devoted to illustrating this theoretical background in a form suitable for building up the AEP (which will be detailed by Grigolini and Marchesoni in Chapter II. Section III will project the same theoretical background in a direction favorable to establishing the CFP, which will then be illustrated accurately by Grosso and Pastori Parravicini in Chapters III and IV. A short outline of the nonlinear version of the RMT is given in Section IV. Concluding remarks are found in Section V.

II. THE ZWANZIG PROJECTION TECHNIQUE

We have now to deal with the Schrödinger picture of the physical system under study. Let us, therefore, consider the following equation of motion:

$$\frac{\partial}{\partial t}\rho(\mathbf{a},\mathbf{b};t) = \mathscr{L}\rho(\mathbf{a},\mathbf{b};t) \tag{2.1}$$

a and **b** denote sets of "relevant" and "irrelevant" variables, respectively. The operator $\mathscr{L}$ driving the probability distribution ρ is referred to as the dynamical operator. Such an operator can be either a rigorous Liouville operator,*

$$\mathscr{L} \equiv iL \qquad (2.2)$$

or a kind of effective Liouvillian. In the former case, Eq. (2.1) would be the well-known Liouville equation, the aim of which is to provide, at least in principle, a complete and detailed description of a many-body system. Rather than following this ambitious route, we shall often have recourse to a much more simplified picture, which nevertheless retains the non-Gaussian non-Markovian properties of the real physical system (see Section IV). In such a case the dynamical operator $\mathscr{L}$ will consist of a conservative part sharing the rigorous character of a Liouville operator built up according to the rules of analytical mechanics and a dissipative and diffusional one mimicking the effects of friction and stochastic forces. This will be detailed elsewhere in this volume. At a purely phenomenological level, furthermore, we can introduce stochastic forces depending on the state of the variable of interest, termed "multiplicative" by Fox.[20] Their effect on the relevant variables can be much more pronounced than in the standard additive case for reasons that will become clear later on. In this case, breakdown of the detailed balance conditions occurs, thereby preventing us from making Hermitian the corresponding dynamical operator. Throughout the present volume we shall deal frequently with this case (see for example, Chapters VII–IX, XI, and XII).

Let us assume that the dynamical operator $\mathscr{L}$ may be written as

$$\mathscr{L} = \mathscr{L}_a + \mathscr{L}_b + \mathscr{L}_1 \qquad (2.3)$$

where $\mathscr{L}_a$ and $\mathscr{L}_b$ concern only the variables **a** and **b**, respectively. $\mathscr{L}_1$ expresses the interaction between the two sets of variables. Let us assume that $\mathscr{L}_b$ admits an equilibrium distribution

$$\mathscr{L}_b \rho_{eq}(\mathbf{b}) = 0 \qquad (2.4)$$

This naturally leads to the following definition of projection operator P:

$$P\rho(\mathbf{a},\mathbf{b};t) \equiv \rho_{eq}(\mathbf{b}) \cdot \sigma(\mathbf{a};t) \equiv \rho_{eq}(\mathbf{b}) \int d\mathbf{b}\, \rho(\mathbf{a},\mathbf{b};t) \qquad (2.5)$$

If the equilibrium distribution $\rho_{eq}(\mathbf{b})$ is normalized, the operator P satisfies

---

*Note that many authors omit the $i = \sqrt{-1}$ in the definition of the Liouville operator.

the idempotency requirement

$$P^2 = P \tag{2.6}$$

To make this chapter as self-contained as possible, we shall briefly review the Zwanzig approach[1] to the generalized master equation. First of all,

$$\rho = \rho_1 + \rho_2 \tag{2.7}$$

where

$$\rho_1 \equiv P\rho \tag{2.8}$$

$$\rho_2 \equiv Q\rho \equiv (1 - P)\rho \tag{2.9}$$

Thus Eq. (2.1) can be written as

$$\frac{\partial}{\partial t}\rho = \mathscr{L}(\rho_1 + \rho_2) \tag{2.10}$$

By applying to both terms of Eq. (2.10) the operators P and Q, we obtain

$$\frac{\partial}{\partial t}\rho_1 = P\mathscr{L}(\rho_1 + \rho_2)$$
$$\frac{\partial}{\partial t}\rho_2 = Q\mathscr{L}(\rho_1 + \rho_2) \tag{2.11}$$

respectively. Formally solving the second equation, noting that this is inhomogeneous and of first order, we get

$$\rho_2(t) = e^{Q\mathscr{L}t}Q\rho(0) + \int_0^t ds\, e^{Q\mathscr{L}s}Q\mathscr{L}P\rho(t-s) \tag{2.12}$$

Substituting this result into the first of the two equations of (2.1) yields the generalized master equation[21]

$$\frac{\partial}{\partial t}\rho_1(t) = P\mathscr{L}P\rho(t) + \int_0^t ds\, P\mathscr{L}e^{Q\mathscr{L}s}Q\mathscr{L}P\rho(t-s) + P\mathscr{L}e^{Q\mathscr{L}t}Q\rho(0) \tag{2.13}$$

Note that Zwanzig[21] showed the quantum-mechanical version of this equation to be identical to that derived by Prigogine and Resibois[22] and by Montroll.[23] The most interesting feature of this master equation is its non-

Markovian form. This deserves some comment. The time evolution of the memory kernel

$$\mathcal{K}(s) \equiv P\mathcal{L}e^{Q\mathcal{L}s}Q\mathcal{L}P \qquad (2.14)$$

depends on the "irrelevant" variables **b**. When exploring the case of Eq. (2.2), the "irrelevant" system involves a very large number of degrees of freedom. This produces virtually totally random phases and also, in consequence, the decay behavior of $\mathcal{K}(t)$, a reliable description of which is then expected to be provided by proper analytical functions. When the "reduced" model theory (RMT) (see Section IV) applies, the irrelevant variables **b** of the rigorous treatment are replaced by a few variables undergoing fluctuation-dissipation processes expressed in standard Markovian terms. This allows a reliable simulation of the decay behavior of the real $\mathcal{K}(t)$ to be recovered. The lifetime of $\mathcal{K}(t)$ in general is not a vanishing value, thereby making non-Markovian the relaxation process of the system of interest.

A further interesting aspect of Eq. (2.13) is afforded by the inhomogeneous term [third term on the right-hand side of Eq. (2.13)]. Since the initial condition of the system in general is not such as to satisfy

$$Q\rho(0) = 0 \qquad (2.15)$$

this term does not vanish and exerts its influence on the dynamics of **a** for time comparable to the lifetime of $\mathcal{K}(t)$. In the non Markovian case this is, in turn, of the same order of magnitude as the relaxation time of the variables **a**. In consequence, this inhomogeneous term may have a deep influence on the relaxation process under study. The breakdown of Eq. (2.15) depends on either the fact that at the initial time the probability distribution $\rho(\mathbf{a}, \mathbf{b}; 0)$ is not factorized into a system and a thermal bath contribution or that the bath itself is far from equilibrium. To the best of our knowledge the first papers dealing with the influence of the inhomogeneous term on the decay of the system are those of Hynes and coworkers.[24a] A review on the preparation phenomena can be found in ref. 7. An alternative way is followed in ref. 24b, where the effect of preparation is incorporated by an adequate choice of the projection operator making the inhomogeneous term disappear.

Although in this volume no great attention will be devoted to this interesting topic, in Chapter II we shall establish the range of validity the AEP has when the preparation effects are neglected.

Throughout this volume we shall apply the Zwanzig approach mainly to cases where the **a** time scale is fairly well separated from the **b** time scale. In other words, the variables **b** will be assumed to be fast-relaxing, that is, precisely those termed "slave variables" by Haken.[11] A major concern of this volume, however, will be with how to correct the crudity of the standard adiabatic assumption[11] according to which the irrelevant variables are infinitely fast. An AEP avoiding this approximation will be illustrated in Chapter II. Note that, as stressed in Section I, the theoretical foundation of this technique is the Zwanzig approach, which leads from Eq. (2.1) to the master equation (2.13).

This aspect will be made more transparent by illustrating the theoretical starting point from which the AEP of Chapter II is derived.

It is convenient to express the memory kernel of the generalized master equation in terms of the correlation functions of the variable of interest. To do that, rather than using Eq. (2.1) we prefer to have recourse to the corresponding interaction picture:

$$\frac{\partial}{\partial t}\tilde{\rho}(t) = \mathscr{L}_1(t)\tilde{\rho}(t) \tag{2.16}$$

where

$$\tilde{\rho} \equiv e^{-\mathscr{L}_0 t}\rho \tag{2.17a}$$

$$\mathscr{L}_1(t) \equiv e^{-\mathscr{L}_0 t}\mathscr{L}_1 e^{\mathscr{L}_0 t} \tag{2.17b}$$

$$\mathscr{L}_0 \equiv \mathscr{L}_a + \mathscr{L}_b \tag{2.17c}$$

The approach that led us to the generalized master equation (2.13) can readily be repeated provided that the usual exponentials are replaced by time-ordered exponentials.[25] We thus obtain

$$\frac{\partial}{\partial t}P\rho(t) = P\mathscr{L}_1(t)P\tilde{\rho}(t) + \int_0^t ds\, P\mathscr{L}_1(t)\overleftarrow{\exp}\left[\int_s^t ds_1\, Q\mathscr{L}_1(s_1)\right]Q\mathscr{L}_1(s)P\tilde{\rho}(s)$$

$$+ P\mathscr{L}_1(t)\overleftarrow{\exp}\left[\int_0^t ds\, Q\mathscr{L}_1(s)\right]Q\tilde{\rho}(0) \tag{2.18}$$

The roughest approximation consists of replacing the time-ordered exponential appearing in the second term on the right-hand side of Eq. (2.18) by 1. Since we intend to explore cases where the inhomogeneous term should not play any significant role (see Chapter II), Eq. (2.18) becomes

$$\frac{\partial}{\partial t}\tilde{\rho}_1(s) = \int_0^t ds\, P\mathscr{L}_1(t)\mathscr{L}_1(s)\tilde{\rho}_1(s) \tag{2.19}$$

Note that we also assumed that $P\mathscr{L}_1(t)P = 0$. The corresponding approximated memory kernel can then be written as follows:

$$\mathscr{K}(t-s) = e^{-\mathscr{L}_s(t-s)}\mathscr{K}_B(t-s) \tag{2.20a}$$

where

$$\mathscr{K}_B(t-s) \equiv P\mathscr{L}_1 e^{(\mathscr{L}_a+\mathscr{L}_b)(t-s)}\mathscr{L}_1 P \tag{2.20b}$$

To derive Eq. (2.20b) we exploited the general properties

$$e^{\mathscr{L}_b t}P = P \tag{2.21a}$$

$$Pe^{\mathscr{L}_b t} = P \tag{2.21b}$$

(see also Chapter II). $\mathscr{K}_B(t)$ is a correlation operator providing information on the relaxation properties of the thermal bath. The AEP that will be illustrated in Chapter II basically consists of a systematic expansion of the time-ordered exponential appearing in the second term on the right-hand side of Eq. (2.18). In principle, this will afford us the way to get information from higher-order correlation functions.

To establish contact with the subject of the next section, let us note that the time evolution of a variable A belonging to the set $\mathbf{a}$ may be evaluated by the following two schemes:

$$\langle A(t) \rangle = \int A\{ e^{\mathscr{L} t}\rho(\mathbf{a},\mathbf{b};0)\}\, d\mathbf{a}\, d\mathbf{b} \tag{2.22}$$

$$\langle A(t) \rangle = \int \{ e^{\mathscr{L}^+ t}A\}\rho(\mathbf{a},\mathbf{b};0)\, d\mathbf{a}\, d\mathbf{b} \tag{2.23}$$

$\mathscr{L}^+$ is the operator adjoint to $\mathscr{L}$. We are led to Eq. (2.23) from Eq. (2.22) via integration by parts. Equation (2.22) can also be written as

$$\langle A(t) \rangle = \int A\sigma(\mathbf{a};t)\, d\mathbf{a} \tag{2.24}$$

where, according to the definition of projection operator P of Eq. (2.5),

$$\sigma(\mathbf{a};t) \equiv \frac{\rho_1(\mathbf{a},\mathbf{b};t)}{\rho_{eq}(\mathbf{b})} \tag{2.25}$$

$\rho_1(\mathbf{a},\mathbf{b};t)$ is the part of interest of $\rho(\mathbf{a},\mathbf{b};t)$ as defined by Eq. (2.8). As a consequence, $\langle A(t) \rangle$ can be determined by evaluating $\rho_1(\mathbf{a},\mathbf{b};t)$ from the generalized master equation (2.13). Note that in general $\rho(\mathbf{a}, \mathbf{b}; 0) \neq \rho_{eq}(\mathbf{b})\sigma(\mathbf{a}; t)$. This means that the correct determination of $\langle A(t) \rangle$ via Eq. (2.24) implies that the inhomogeneous term of that generalized master equation should be taken into account (see also Chapter II).

In the next section we shall focus our attention on the dynamics of A as driven by $\mathscr{L}^+$.

III. THE MORI PROJECTION TECHNIQUE

Let us consider a variable A belonging to the set $\mathbf{a}$. If we focus our attention on Eq. (2.23) rather than on Eq. (2.22), we are naturally led to replace Eq. (2.1) with

$$\frac{\partial}{\partial t}A = \Gamma A \qquad (3.1)$$

$$\Gamma \equiv \mathscr{L}^+ \qquad (3.2)$$

This is the Heisenberg representation corresponding to the Schrödinger representation of Eq. (2.1). For simplicity, in this section we shall consider only the case when $\mathbf{a}$ is monodimensional. The application of the projection technique to Eq. (3.1) will result in an equation basically equivalent to Eq. (2.13) (when related to the same monodimensional set). However, this method paves the way to a powerful computational algorithm, as widely illustrated by Grosso and Pastori-Parravicini in Chapters III and IV.

In many problems of interest dealing with relaxation processes close to equilibrium, rather than $\langle A(t) \rangle$ the relevant time evolution is that concerning the correlation function

$$\Phi(t) \equiv \frac{\langle A(0) A(t) \rangle_{eq}}{\langle A^2 \rangle_{eq}} \qquad (3.3)$$

where

$$A(t) \equiv e^{\Gamma t}A(0) \qquad (3.4)$$

$$\langle \cdots \rangle_{eq} = \int dA \, d\mathbf{b}(\cdots)\rho_{eq}(A, \mathbf{b}) \qquad (3.4a)$$

$\rho_{eq}(A, \mathbf{b})$ is the equilibrium probability distribution. If we define the scalar product $\langle B | C \rangle$ between two general variables B and C as

$$\langle B | C \rangle \equiv \int B^*(A, \mathbf{b})C(A, \mathbf{b})\rho_{eq}(A, \mathbf{b}) \, dA \, d\mathbf{b} \qquad (3.5)$$

the correlation function $\Phi(t)$ turns out to be expressed via

$$\Phi(t) \equiv \frac{\langle A|A(t)\rangle}{\langle A|A\rangle} \qquad (3.6)$$

The projection operator on the A space is then given by

$$P_A = |A\rangle \frac{1}{\langle A|A\rangle} \langle A| \qquad (3.7)$$

The well-known Mori theory[6] can be interpreted[7] as the choice of the best basis set $\{|f_0\rangle, |f_1\rangle, |f_2\rangle, \ldots\}$ for expanding the operator Γ. Since we are interested in describing the time evolution of the "state" $|A\rangle$, it is evident that this state has to be included in this basis set. Therefore,

$$|f_0\rangle \equiv |A\rangle \qquad (3.8)$$

The method of building up the remainder of the basis set is reminiscent of the rules followed by Rhodes and coworkers[26] to build up their spectroscopic basis set. The "doorway state" $|f_1\rangle$, the one carrying the entire interaction with $|f_0\rangle$,[26] is defined by

$$|f_1\rangle \equiv \Gamma_1|f_0\rangle \qquad (3.9)$$

where

$$\Gamma_1 \equiv (1 - P_0)\Gamma_0 \qquad (3.10)$$

and

$$P_0 \equiv P_A \qquad (3.11)$$

$$\Gamma_0 \equiv \Gamma \qquad (3.12)$$

The first-order state $|f_1\rangle$ is therefore obtained simply by applying Γ_0 to the zeroth-order state, with the caution of rejecting $|f_0\rangle\langle f_0| \Gamma_0 |f_0\rangle$, because the contribution coming from the state $|f_0\rangle$ is already included in the basis set. We could go on that way, as this is just how Mori proceeds in his celebrated papers.[6] However, to avoid the Hermitian assumption made by Mori, we build up a biorthogonal basis set. This means that the state $|f_1\rangle$ has to be associated with the corresponding left state (we assume $\langle \tilde{f}_0| = \langle f_0|$)

$$\langle \tilde{f}_1| \equiv \langle f_0|\Gamma_0(1 - P_0) \qquad (3.13)$$

In the Hermitian case,

$$\langle \tilde{f}_1| \approx -\langle f_1|$$ (3.14)

In general, $\langle \tilde{f}_1|$ has nothing to do with the usual dual state $\langle f_1|$ associated with $|f_1\rangle$. The projection operator P_1 will be defined as

$$P_1 = |f_1\rangle\langle \tilde{f}_1|f_1\rangle^{-1}\langle \tilde{f}_1|$$ (3.15)

The idempotent property is conserved, whereas the Hermitian one is lost. Indeed, in general, $P_1 \neq P_1^+$. The $(k+1)$th order "state" is defined by

$$|f_{k+1}\rangle \equiv \Gamma_{k+1}|f_k\rangle$$ (3.16)

where

$$\Gamma_{k+1} = (1 - P_k)\Gamma_k$$ (3.17)

and

$$P_k = |f_k\rangle\langle \tilde{f}_k|f_k\rangle^{-1}\langle \tilde{f}_k|$$ (3.18)

The corresponding left state is given by

$$\langle \tilde{f}_{k+1}| = \langle \tilde{f}_k|\Gamma_0(1 - P_0)\cdots(1 - P_k)$$ (3.19)

Note that each state is orthogonal to the preceding states. This leads to

$$P_k P_{k'} = P_{k'} P_k$$ (3.20)

and

$$\langle \tilde{f}_k|f_{k'}\rangle = \delta_{kk'}\langle \tilde{f}_k|f_k\rangle$$ (3.21)

After building up this biorthogonal basis set, we would be led naturally to expand over it the operator Γ_0 so as to introduce the following kind of time evolution:

$$|f_k(t)\rangle \equiv e^{\Gamma_0 t}|f_k\rangle$$ (3.22)

This kind of time evolution has recently been used to justify the linear version of the RMT[7] and has received some consideration from the Mori school itself.[27] The foundation of a hierarchy of Langevin equations, however, requires a more suitable definition of time evolution. Note that any state $|f_k\rangle$ divides the Mori chain into two sets: the one of the first $k-1$ states that precede $|f_k\rangle$ and that of the infinite states that follow it, the dynamics of the $|f_k\rangle$ state being driven by both groups. Although both the interactions are deterministic in nature, that with the latter group can be pictured as being stochastic, as this depends on infinite degrees of freedom. The latter interaction has therefore to be made clearly distinct from the former. This can be done via the definition

$$|f_k(t)\rangle \equiv e^{\Gamma k't}|f_k\rangle \tag{3.23}$$

As far as the state $|f_0\rangle$ is concerned, the two kinds of time evolution above coincide. In both cases Eq. (3.1) reads

$$\frac{d}{dt}|f_0(t)\rangle = \Gamma_0|f_0(t)\rangle \tag{3.24}$$

Let us now apply again to Eq. (3.24) the same projection technique as that applied to Eq. (2.1). We obtain

$$\frac{d}{dt}P_0|f_0(t)\rangle = P_0\Gamma_0P_0|f_0(t)\rangle + P_0\Gamma_0Q_0|f_0(t)\rangle \tag{3.25}$$

$$\frac{d}{dt}Q_0|f_0(t)\rangle = Q_0\Gamma_0P_0|f_0(t)\rangle + Q_0\Gamma_0Q_0|f_0(t)\rangle \tag{3.26}$$

Throughout this section we shall use the definition

$$Q_k \equiv 1 - P_k \tag{3.27}$$

which also defines the complementary projection operator already used to get (3.26). As in Section II, from Eq. (3.26) we obtain

$$Q_0|f_0(t)\rangle = e^{Q_0\Gamma_0't}Q_0|f_0\rangle + \int_0^t e^{Q_0\Gamma_0(t-\tau)}Q_0\Gamma_0P_0|f_0(\tau)\rangle\, d\tau$$

$$= \int_0^t e^{Q_0\Gamma_0(t-\tau)}Q_0\Gamma_0|f_0\rangle\Phi_0(t) \tag{3.28}$$

where

$$\Phi_0(t) \equiv \Phi(t) \tag{3.29}$$

This is a special case of the general definition

$$\Phi_k(t) \equiv \frac{\langle \tilde{f}_k | f_k(t) \rangle}{\langle \tilde{f}_k | f_k \rangle} \tag{3.30}$$

Later on we shall comment on the physics behind the fact that the first term on the right-hand side of Eq. (3.28) disappears. Note that this projection technique naturally leads us to consider time evolution as defined, via Eq. (3.23). We can thus rewrite Eq. (3.28) as

$$Q_0 | f_0(t) \rangle = \int_0^t d\tau \, | f_1(t - \tau) \rangle \Phi_0(\tau) \tag{3.31}$$

Substituting Eq. (3.31) into Eq. (3.25), we arrive at

$$\frac{d}{dt} P_0 | f_0(t) \rangle = P_0 \Gamma_0 P_0 | f_0(t) \rangle + P_0 \Gamma_0 \int_0^t \Phi_0(\tau) | f_1(t - \tau) \rangle \, d\tau \tag{3.32}$$

Multiplying Eq. (3.32) on the left by $|f_0\rangle$, we get

$$\frac{d}{dt} \Phi_0(t) = \lambda_0 \Phi_0(t) + \int_0^t \Phi_0(\tau) \frac{\langle f_0 | \Gamma_0 | f_1(t - \tau) \rangle}{\langle f_0 | f_0 \rangle} \, d\tau \tag{3.33}$$

where

$$\lambda_0 \equiv \frac{\langle f_0 | \Gamma_0 | f_0 \rangle}{\langle f_0 | f_0 \rangle} \tag{3.34}$$

which is a special case of the general definition

$$\lambda_k \equiv \frac{\langle \tilde{f}_k | \Gamma_k | f_k \rangle}{\langle \tilde{f}_k | f_k \rangle} \tag{3.35}$$

widely used throughout the remainder of this section. Note that

$$\frac{\langle f_0 | \Gamma_0 | f_1(t - \tau) \rangle}{\langle f_0 | f_0 \rangle} = \frac{\langle f_0 | \Gamma_0 (1 - P_0) | f_1(t - \tau) \rangle}{\langle f_0 | f_0 \rangle} \tag{3.36}$$

We are now in a position to illustrate the advantage coming from the introduction of the left states defined by eqs. (3.13) and (3.19). Indeed, by using Eq. (3.13) and the definition of Eq. (3.30) (for $k = 1$), we can rewrite Eq. (3.36) as

$$\frac{\langle f_0 | \Gamma_0 | f_1(t-\tau) \rangle}{\langle f_0 | f_0 \rangle} = -\Delta_1^2 \Phi_1(t-\tau) \tag{3.37}$$

where

$$\Delta_1^2 \equiv -\frac{\langle \tilde{f}_1 | f_1 \rangle}{\langle f_0 | f_0 \rangle} \tag{3.38}$$

This, in turn, is a special case of

$$\Delta_k^2 \equiv -\frac{\langle \tilde{f}_k | f_k \rangle}{\langle \tilde{f}_{k-1} | f_{k-1} \rangle} \tag{3.39}$$

By using Eq. (3.37), Eq. (3.33) can thus be rewritten as follows:

$$\frac{d}{dt}\Phi_0(t) \equiv \lambda_0 \Phi_0(t) - \Delta_1^2 \int_0^t \Phi_0(\tau) \Phi_1(t-\tau)\, d\tau \tag{3.40}$$

Applying the same approach to the general case

$$\frac{d}{dt}|f_k(t)\rangle = \Gamma_k |f_k(t)\rangle \tag{3.41}$$

which is the equation of motion corresponding to Eq. (3.23), we obtain

$$\frac{d}{dt}\Phi_k(t) \equiv \lambda_k \Phi_k(t) - \Delta_k^2 \int_0^t \Phi_k(\tau) \Phi_{k+1}(t-\tau)\, d\tau \tag{3.41'}$$

Via Laplace transforming Eq. (3.41′) we get

$$\hat{\Phi}_k(z) = \frac{1}{z - \lambda_k + \Delta_{k+1}^2 \Phi_{k+1}(z)} \tag{3.42}$$

which allows us to write $\hat{\Phi}_0(z)$ as

$$\hat{\Phi}_0(z) = \cfrac{1}{z - \lambda_0 + \cfrac{\Delta_1^2}{z - \lambda_1 + \cfrac{\Delta_2^2}{\ddots}}}$$

$$+ \cfrac{\Delta_{n-1}^2}{z - \lambda_{n-1} + \Delta_n^2 \hat{\Phi}_n(z)} \tag{3.43}$$

This result will play a central role throughout this volume, since many properties of interest can be traced back to the Laplace transform of $\Phi_0(t)$. Equation (3.43) indicates an appealing way to calculate this Laplace transform. This way is really straightforward to follow if the expansion parameters λ_k and Δ_k^2 are known. A careful inspection of the formulas of Eqs. (3.35) and (3.39) suggests that these parameters can be expressed in terms of the moments

$$s_n \equiv \langle f_0 | \Gamma_0^n | f_0 \rangle \tag{3.44}$$

We have indeed, for instance,

$$
\begin{aligned}
\Delta_1^2 &\equiv -\frac{\langle \tilde{f}_1 | f_1 \rangle}{\langle f_0 | f_0 \rangle} = -\frac{\langle f_0 | \Gamma_0 (1 - P_0) \Gamma_0 | f_0 \rangle}{\langle f_0 | f_0 \rangle} \\
&= -\frac{\langle f_0 | \Gamma_0^2 | f_0 \rangle}{\langle f_0 | f_0 \rangle} + \langle f_0 | \Gamma_0 | f_0 \rangle \frac{1}{\langle f_0 | f_0 \rangle} \langle f_0 | \Gamma_0 | f_0 \rangle \\
&= \frac{s_1^2 - s_2}{s_0}
\end{aligned} \tag{3.45}
$$

$$\lambda_0 = \langle f_0 | \Gamma_0 | f_0 \rangle = s_1 \tag{3.46}$$

$$
\begin{aligned}
\lambda_1 &= \frac{\langle \tilde{f}_1 | F_1 | f_1 \rangle}{\langle \tilde{f}_1 | f_1 \rangle} = \frac{\langle f_0 | \Gamma_0 (1 - P_0) \Gamma_0 (1 - P_0) \Gamma_0 | f_0 \rangle}{\langle f_0 | \Gamma_0 (1 - P_0) \Gamma_0 | f_0 \rangle} \\
&= \left\{ \langle f_0 | \Gamma_0^3 | f_0 \rangle - \frac{\langle f_0 | \Gamma_0 | f_0 \rangle \langle f_0 | \Gamma_0^2 | f_0 \rangle}{\langle f_0 | f_0 \rangle} \right. \\
&\quad \left. - \frac{\langle f_0 | \Gamma_0^2 | f_0 \rangle \langle f_0 | \Gamma_0 | f_0 \rangle}{\langle f_0 | f_0 \rangle} + \frac{\langle f_0 | \Gamma_0 | f_0 \rangle^3}{\langle f_0 | f_0 \rangle^2} \right\} \\
&\quad \times \left\{ \langle f_0 | \Gamma^2 | f_0 \rangle - \frac{\langle f_0 | \Gamma_0 | f_0 \rangle \langle f_0 | \Gamma_0 | f_0 \rangle}{\langle f_0 | f_0 \rangle} \right\}^{-1} \\
&= \left\{ s_3 - \frac{s_1 s_2}{s_0} - \frac{s_2 s_1}{s_0} + \frac{s_1^3}{s_0^2} \right\} \left\{ s_2 - \frac{s_1^2}{s_0} \right\}^{-1}
\end{aligned} \tag{3.47}
$$

As the order of the expansion parameters becomes higher and higher, their explicit expression in terms of the s_n becomes more and more involved. In Chapter III of this volume, Grosso and Pastori Parravicini illustrate an algorithm especially convenient from the computational point of view to evaluate these expansion parameters in terms of the s_n. The algorithm they illustrate is in turn based solely on the generalized Langevin approach of this section. Note that in all the applications considered in this volume the operator Γ will be analytical in nature. This makes it easy to evaluate the s_n's of Eq. (3.44), thereby naturally leading to the algorithm $s_n \rightarrow \lambda_i, \Delta_i$ described by Grosso and Pastori Parravicini. In some cases, however, more stable numerical results are guaranteed by generating the states $|f_i\rangle$ and $\langle \tilde{f}_i|$ and evaluating the expansion parameters directly from Eqs. (3.35) and (3.39) (see Chapter IV).

The extended version of the Mori approach here illustrated shares with the "standard" one an appealing physical meaning that is worthy of mention. Note that

$$|f_k(t)\rangle = (P_k + Q_k)|f_k(t)\rangle = \Phi_k(t)|f_k\rangle + Q_k|f_k(t)\rangle \qquad (3.48)$$

The extension of Eq. (3.31) to the kth order substituted into (3.48) leads to

$$|f_k(t)\rangle = \Phi_k(t)|f_k\rangle + \int_0^t \Phi_k(t-\tau)|f_{k+1}(\tau)\rangle \, d\tau \qquad (3.49)$$

By Laplace transforming Eq. (3.48) we get

$$|\hat{f}_k(z)\rangle = |f_k\rangle \hat{\Phi}_k(z) + |\hat{f}_{k+1}(z)\rangle \hat{\Phi}_k(z) \qquad (3.50)$$

Let us substitute Eq. (3.42) into Eq. (3.50). We then obtain

$$|\hat{f}_k(z)\rangle = \left[|f_k\rangle + |\hat{f}_{k+1}(z)\rangle \right] \left[z - \lambda_k + \Delta_{k+1}^2 \hat{\Phi}_{k+1}(z) \right]^{-1} \qquad (3.51)$$

which is the Laplace transform of

$$\frac{d}{dt}|f_k(t)\rangle = \lambda_k|f_k(t)\rangle - \int_0^t ds \, |f_k(s)\rangle \varphi_k(t-s) + |f_{k+1}(t)\rangle \qquad (3.52)$$

where

$$\varphi_k(t) = \Delta_{k+1}^2 \Phi_{k+1}(t) = -\frac{\langle \tilde{f}_{k+1}|f_{k+1}(t)\rangle}{\langle \hat{f}_k|f_k\rangle} \qquad (3.53)$$

Let us write Eq. (3.49) explicitly for A and come back from the quantumlike to the classical formalism. We obtain

$$\frac{d}{dt}A = \lambda_0 A - \int_0^t ds\, \varphi(s) A(t-s) + f(t) \qquad (3.54)$$

where*

$$\lambda_0 = \frac{\langle A(\Gamma A)\rangle_{eq}}{\langle A^2\rangle_{eq}} \qquad (3.55a)$$

$$\varphi(t) = -\frac{\langle \tilde{f}(0)f(t)\rangle_{eq}}{\langle A^2\rangle_{eq}} \qquad (3.55b)$$

$$f(t) = \exp\{(1-P_A)\Gamma t\}(1-P_A)\Gamma A \qquad (3.55c)$$

The symbol $\langle\ \rangle_{eq}$ denotes averaging over the equilibrium distribution $\rho_{eq}(A, \mathbf{b})$ [see Eq. (3.5)]. If the time scale of the variable A is well separated from that of the variables $\mathbf{b}$, we can write

$$\varphi(t) = 2\gamma\delta(t) \qquad (3.56)$$

$$-\langle \tilde{f}(0)f(t)\rangle_{eq} = 2\gamma\langle A^2\rangle_{eq}\delta(t) \qquad (3.57)$$

Let us substitute Eq. (3.56) into Eq. (3.54) and assume $\lambda_0 = 0$ while identifying A with the variable velocity v. We get

$$\dot{v} = -\gamma v + f(t) \qquad (3.58)$$

that is, the standard Langevin equation. This makes it clear why the hierarchy equations of Eq. (3.49) for $k = 0, 1, 2, \ldots$ are referred to as generalized Langevin equations. The linear version of the RMT[28] is based on making the Markovian assumption of Eq. (3.56) at the kth order with $k > 0$. This means that it is possible to extract from the thermal bath of A k suitable linear combinations of the variables $\mathbf{b}$. These linear combinations can be regarded as being variables as slow as A itself. The remainder of this set $\mathbf{b}$ is then assumed to consist of really fast relaxing variables. Sometimes, unfortunately, the Markovian truncation cannot be allowed. In Chapters III and IV, Grosso and Pastori Parravicini will face this challenging problem. Also some of the difficulties that are met in establishing a rigorous theory for molecular dynamics in the liquid state largely stem from this problem.[29]

---

*Note that when dealing with the case of the rigorous Liouvillian of Eq. (3.2), $\varphi(t)$ is defined by $\varphi(t) = \langle f(0)f(t)\rangle_{eq}/\langle A^2\rangle_{eq}$, thereby providing the more conventional form of the standard fluctuation-dissipation relationship: $\langle f(0)f(t)\rangle = 2\gamma\langle A^2\rangle_{eq}\delta(t)$.

The CFP (see Chapter III) requires the unique theoretical background behind Eq. (3.43) [which is closely related to Eq. (3.52)], whereas the AEP (see also Chapter II) is basically founded on Eq. (2.13) (written in the interaction picture). Equations (3.54) and (2.13) have been derived using the same projection technique. These two equations are related to each other still more deeply. To demonstrate this, let us note that within the quantumlike formalism of Eq. (3.5), Eq. (2.23) reads

$$\langle A(t) \rangle = \langle A(t) | F(A, \mathbf{b}) \rangle \qquad (3.59)$$

where

$$F(A, \mathbf{b}) = \frac{\rho(A, \mathbf{b}; 0)}{\rho_{eq}(A, \mathbf{b})} \qquad (3.60)$$

Let us assume that

$$\mathscr{L} F(A, \mathbf{b}) \rho(A, \mathbf{b}; t) = \{ \mathscr{L} F(A, \mathbf{b}) \} \rho(A, \mathbf{b}; t) + F(A, \mathbf{b}) \{ \mathscr{L} \rho(A, \mathbf{b}; t) \} \qquad (3.61)$$

This is certainly true when $\mathscr{L} = iL$ and L is a rigorous Liouvillian. Equation (3.58) is also satisfied when $F(A, \mathbf{b})$ depends only on A. We can thus write Eq. (3.59) as follows:

$$\langle A(t) \rangle = \langle A | e^{\mathscr{L}t} F(A, \mathbf{b}) \rangle \qquad (3.62)$$

Note also that

$$\langle A(t) \rangle = \langle A | e^{\mathscr{L}t} F(A, b) \rangle = \frac{\langle A | A \rangle \langle A | e^{\mathscr{L}t} F(A, \mathbf{b}) \rangle}{\langle A | A \rangle}$$

$$= \langle A | P_A e^{\mathscr{L}t} F(A, \mathbf{b}) \rangle \qquad (3.63)$$

The equation of motion of $P_A \exp[\mathscr{L}t] F(A, \mathbf{b})$ can be derived from Eq. (2.13) when this is thought of as being related to the new projection operator P_A. This is the basic idea of the paper of Nordholm and Zwanzig.[30a] We thus obtain

$$\frac{\partial}{\partial t} | F_1(t) \rangle = P_A \mathscr{L} | F_1(t) \rangle + \int_0^t ds \, P_A \mathscr{L} e^Q A^{\mathscr{L}s} Q_A \mathscr{L} | F_1(t-s) \rangle$$

$$+ P_A \mathscr{L} e^Q A^t | F_2(0) \rangle \qquad (3.64)$$

We used the following definitions:

$$|F_1(t)\rangle \equiv P_A e^{\mathscr{L}t}|F(A,\mathbf{b})\rangle \qquad (3.65a)$$

$$|F_2(t)\rangle \equiv Q_A e^{\mathscr{L}t}|F(A,\mathbf{b})\rangle \qquad (3.65b)$$

$$Q_A \equiv 1 - P_A \qquad (3.65c)$$

which are inspired by the corresponding ones of the preceding section. We are now in a position to determine the equation of motion of $\langle A(t)\rangle$. By substituting Eq. (3.64) into the time derivative of Eq. (3.63), we get

$$\frac{d}{dt}\langle A(t)\rangle = \left\langle A\left|\frac{\partial}{\partial t}F_1\right.\right\rangle$$

$$= \langle A|P_A\mathscr{L}F_1(t)\rangle + \int_0^t ds \left\langle A\left|P_A e^{-Q_A\mathscr{L}s}Q_A\mathscr{L}F_1(t-s)\right.\right\rangle$$

$$+ \langle A|P_A\mathscr{L}e^{-Q_A t}Q_A|F(0)\rangle \qquad (3.66)$$

Equation (3.66) can be rewritten as

$$\left[\langle A|P_A\mathscr{L}P_A e^{\mathscr{L}t} + \int_0^t ds \langle A|P_A\mathscr{L}e^{Q_A\mathscr{L}s}Q_A\mathscr{L}P_A e^{\mathscr{L}(t-s)}\right.$$

$$\left. - \langle A|P_A\mathscr{L}e^{Q_A\mathscr{L}t}Q_A - \left\langle\frac{\partial}{\partial t}A(t)\right|\right]|F(0)\rangle = 0 \qquad (3.67)$$

The vector left conjugate to that between the square brackets reads

$$\frac{d}{dt}|A(t)\rangle = \frac{|A(t)\rangle\langle A|\mathscr{L}^+|A\rangle}{\langle A|A\rangle} + \int_0^t ds \frac{|A(t-s)\rangle\langle A|\mathscr{L}^+ e^{Q_A\mathscr{L}^+ s}Q_A\mathscr{L}^+|A\rangle}{\langle A|A\rangle}$$

$$- e^{Q_A\mathscr{L}^+ t}Q_A\mathscr{L}^+ A \qquad (3.68)$$

The reader can readily check that this equation formally coincides with Eq. (3.54). [We remind the reader of Eq. (3.2).]

A still more rigorous demonstration of the complete equivalence of the Mori and Nakajima-Zwanzig approach has been derived by Grabert.[30b] For simplicity we shall not illustrate his demonstration. We shall limit ourselves to remark that this clearly establishes that the two approaches are equivalent in that they are related in the same way as the Heisenberg and Schrödinger pictures are related. Note, however, that when the scalar product implies equilibrium as Eq. (3.5) does, the Mori theory appears to be basically suitable to evaluate the equilibrium correlation function $\langle A(0)A(t)\rangle_{\mathrm{eq}}/\langle A^2\rangle_{\mathrm{eq}}$, thereby resulting in a vanishing value for the mean stochastic force. Nevertheless, when dealing with systems linear in nature, it is not difficult to use the Mori theory itself to picture the excitation of the thermal bath, which, in turn, produces a nonequilibrium stochastic force.

To shed further light on this aspect, let us consider the system of stochastic differential equations

$$\dot{v} = w$$
$$\dot{w} = -\Delta_1^2 v - \gamma w + F(t) \qquad (3.69)$$

$F(t)$ is a random variable assumed to be white and Gaussian. This means that this stochastic force is completely defined by

$$\langle f(0)f(t)\rangle = 2D\delta(t) = \gamma\langle w^2\rangle_{\mathrm{eq}}\delta(t) \qquad (3.70)$$

Let us Laplace transform Eq. (3.69):

$$z\hat{v}(z) - v(0) = \hat{w}(z)$$
$$z\hat{w}(z) - w(0) = -\Delta_1^2\hat{v}(z) - \gamma\hat{w}(z) + \hat{F}(z) \qquad (3.71)$$

By evaluating $\hat{w}(z)$ via the second equation and substituting the corresponding expression into the first one, we get

$$z\hat{v}(z) - v(0) = \frac{1}{z+\gamma}\hat{w}(0) - \frac{\Delta_1^2\hat{v}(z)}{z+\gamma} + \frac{\hat{F}(z)}{z+\gamma} \qquad (3.72)$$

Applying the inverse Laplace transformation to Eq. (3.72), we obtain

$$\frac{dv}{dt} = -\int_0^t \varphi(t-\tau)v(\tau)\,d\tau + f(t) \qquad (3.73)$$

where

$$\varphi(t) = \Delta_1^2 e^{-\gamma t} \qquad (3.74)$$

This means that Eq. (3.73) is equivalent to the generalized Langevin equation of Eq. (3.54) with $\lambda_0 = 0$, $A = v$, and the memory kernel $\varphi(t)$ defined by its Laplace transform

$$\hat{\varphi}(z) = \frac{\Delta_1^2}{z + \gamma} \qquad (3.75)$$

This agrees with the general structure allowed by the Mori theory. In other words, we are using the first two states of the Mori basis set defined by Eqs (3.8) to (3.19). Note, however, that whereas the standard Mori theory results in

$$\langle f(t) \rangle = 0 \qquad (3.76a)$$

we obtain

$$\hat{f}(z) = \frac{w(0)}{z + \gamma} + \frac{F(z)}{z + \gamma} \qquad (3.76b)$$

Its average is

$$\hat{f}(z) = \frac{\langle w(0) \rangle}{z + \gamma} \qquad (3.76c)$$

since the thermal bath of the auxiliary variable w is faster to an extreme extent than the variables v and w, thereby resulting immediately in $\langle F(z) \rangle = 0$ even if strongly excited. [This feature of the noise $F(t)$ stems from Eq. (3.70).] If the variable w is not found in its equilibrium state $\langle w \rangle_{eq} = 0$, then the stochastic force $F(t)$ is also far from its equilibrium state. As the continued fraction structure of Eq. (3.43) and, at least in principle, the corresponding expansion parameters are provided by the Mori theory, we are in a position to state that this theory can be used to provide a fairly reliable description of the thermal bath excitations. Of course, the remarks above can readily be extended to the case of an arbitrary number of auxiliary variables.

As stressed in refs. 7 and 28b, the "standard" Mori theory is deeply different in nature from the approach we have developed, which includes in the set of relevant variables both the variable f_0 (in the example under discussion, the velocity v) and a few Mori auxiliary variables (in this example only one auxiliary variable, $f_1 = w$). The auxiliary variable w plays the double role of variable of interest (being coupled to an "irrelevant" thermal bath and undergoing the influence of a standard fluctuation-dissipation process) and that of simulating the "thermal bath" of the real variable of interest. This makes it possible to study excitation and preparation processes within the framework of the Mori theory, the range of validity of which is, however, limited to the case of linear systems.

According to Zwanzig, in dealing with a system intrinsically nonlinear, it should still be possible to build up a generalized Langevin equation such as that of (3.54). However, in such a case it can be proved that the memory kernel $\varphi(t)$ is no longer independent of the state of the variable of interest.[31] This forces us to face the problem of preparation in a completely different way, as discussed elsewhere in this volume.

Although basically projected for equilibrium states, when used to build up the CFP, the Mori approach affords a way to deal also with nonequilibrium processes. Note that from a purely formal point of view the general expression of Eq. (2.23) can be recast in the form

$$\langle A(t) \rangle = \langle A | A(t) \rangle \qquad (3.77)$$

provided that use is made of the following definition of scalar product:

$$\langle B|C \rangle = \int B^*(\mathbf{a}, \mathbf{b}) C(\mathbf{a}, \mathbf{b}) w(\mathbf{a}, \mathbf{b}) \qquad (3.78)$$

where

$$w(\mathbf{a}, \mathbf{b}) = \frac{\rho(\mathbf{a}, \mathbf{b}; 0)}{A} \qquad (3.79)$$

If $w(\mathbf{a}, \mathbf{b})$ as provided by Eq. (3.79) is positive definite, we shall not find any theoretical difficulty in proposing the CFP as a calculation technique for this nonequilibrium process. Throughout this volume we shall consider stochastic variables A with vanishing mean value at equilibrium, that is, $\langle A \rangle_{eq} = 0$. The "correlation" function $\Phi(t) \equiv \langle A|A(t) \rangle / \langle A|A \rangle$ then shares the standard formal properties of the normalized equilibrium correlation functions:

$$\Phi(0) = 1, \qquad \lim_{t \to \infty} \Phi(t) = 0$$

Although the Zwanzig and Mori techniques are closely related and, from a purely formal point of view, completely equivalent, the elegant properties of the Mori theory such as the generalized fluctuation-dissipation theorem imply the physical system under study to be linear, whereas this is not necessary in the Zwanzig approach. This is the main reason we shall be able to face nonlinear problems within the context of a Fokker-Planck approach (see also the discussion of the next section). An illuminating approach of this kind can be found in a paper by Zwanzig and Bixon,[32] which has also to be considered an earlier example of the continued fraction technique applied to a "non-Hermitian" case. This method has also been fruitfully applied to the field of polymer dynamics.[33]

Basically, we shall apply the AEP, derived by the Zwanzig approach, to explore the long-time behavior of nonlinear stochastic processes, whereas the CFP derived from the Mori approach will be used as a calculation technique, which sometimes will prove useful for application to the reduced equations of motion provided by the AEP itself.

IV. MOLECULAR DYNAMICS: A CHALLENGE TO THE FOKKER-PLANCK AND LANGEVIN APPROACHES

To outline the main features of the third corner of the ideal triangle of Fig. 1, the "reduced" model theory (RMT), we shall use as an illustrative example the field of molecular dynamics in the liquid state. In this field of investigation one frequently finds results like those sketched in Fig. 2.[34] It is now clearly understood that a variable of interest such as the translational velocity v is neither Markovian nor Gaussian. A linear theory such as the generalized Langevin equation illustrated in Section III has the right ingredients to reproduce the typical nonexponential behavior of the correlation function $\langle v(0)v(t)\rangle_{eq}$ illustrated by Fig. 2, even though many Mori states are in general required for a satisfying quantitative agreement between theory and "computer experiment" to be attained.[29] Where this theory completely fails is in accounting for non-Gaussian statistical properties.

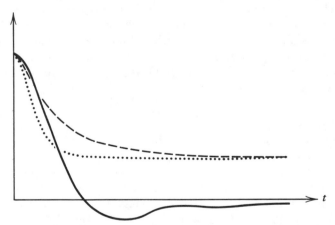

Figure 2. Schematic illustration of the non-Gaussian, non-Markovian behavior of the variable velocity v as detected via computer simulation. The solid curve denotes the correlation function $\psi(t) \equiv \langle v(0)v(t)\rangle_{eq}/\langle v^2\rangle_{eq}$ as provided by computer experiment. The dashed line denotes the correlation function $\varepsilon_2 \equiv \langle v(0)^2 v(t)^2\rangle_{eq}/\langle v^4\rangle_{eq}$ as provided by computer experiment. The dotted line denotes the Gaussian approximation to $\varepsilon_2(t)$. This is $\varepsilon_{2G}(t) = [1 + 2\psi^2(t)]/3$.

According to Fox,[20] no difficulty is met in extending the Langevin-Fokker-Planck methods to the non-Markovian processes, provided that the statistics remain Gaussian. On the other hand, van Kampen[35] exhorts us to be cautious when dealing with nonlinear (and thereby non-Gaussian) stochastic processes. According to him, in these cases the Langevin approach is dangerous and should be replaced by the master equation method. By definition, however, a master equation can be applied only to the study of Markovian stochastic processes.

The major aim of the extended version of the RMT (we use this term to keep it well distinct from the linear version of ref. 19) is to establish a compromise between these two extreme points of view. Its program is outlined as follows.

A complex model of a physical system (e.g., a Lennard-Jones fluid described in Hamiltonian terms) usually consists of a set of variables of interest $\mathbf{a}$ (e.g., the velocity v of a tagged molecule) interacting with a very large number of irrelevant variables. In principle, the dynamical properties of $\mathbf{a}$ could be described by applying the projection techniques outlined in the preceding sections. This approach would meet severe difficulties. First of all, as remarked by van Kampen,[35] the result would be merely formal, since it is difficult to evaluate a memory kernel in terms of the rigorous Liouvillian driving the complex physical system.[36] Furthermore, even in the lucky cases where the "memory kernels" admit a reliable simulation via analytical functions, the complete evaluation of an excitation-relaxation process[7] is especially difficult. Recall that in the linear case this problem finds in the RMT (linear version) a straightforward solution.[7]

The main idea of the RMT is to replace the complex model, that is, the pair of multidimensional variables $(\mathbf{a}, \mathbf{b}_R)$, where $\mathbf{b}_R$ stands for the "real" variables of the complex thermal bath, with a reduced system symbolized by the pair $(\mathbf{a}, \mathbf{b}_V)$. This is made possible by the fact that the variables $\mathbf{b}_V$ are influenced by friction and stochastic forces simulating virtually infinite degrees of freedom. The thermal baths of the variables $\mathbf{b}_V$ are assumed to be characterized by an infinitely short time scale. In consequence, the variables $\mathbf{b}_V$, when thought of as interacting with nothing but their own thermal bath, can be considered Markovian. If the variables $\mathbf{b}_V$ are also Gaussian, that is, coupled to their thermal baths via linear couplings, no conceptual difficulty is found in building up their own Fokker-Planck equations and, ultimately, the Fokker-Planck equation of the system $(\mathbf{a}, \mathbf{b}_V)$. A mere contraction of the description based on the elimination of the variables $\mathbf{b}_V$ generates the non-Markovian character of the variables $\mathbf{a}$. If the deterministic coupling between the sets $\mathbf{a}$ and $\mathbf{b}_V$ is not linear, the variables $\mathbf{a}$ are also proved to be non-Gaussian. A further way of simulating non-Gaussian properties (while keeping the set $\mathbf{a}$ non-Markovian) would be that of describing the dynamics of the variables b_V via proper master equations generating non-Gaussian statistics. In such a case the non-Gaussian statistics of the set a would be compatible with a linear a-b coupling.

The choice of the "reduced" model (a, b_V) is not totally free. Whenever the projection techniques applied to the complex system (a, b_R) provide contracted equations with well-defined formal properties, these have to be reproduced by applying the same contraction procedure to the reduced system (a, b_V). This is the main reason why this modeling approach cannot be regarded as purely phenomenological.

V. CONCLUDING REMARKS

AEP, CFP, and RMT stem from the same theoretical background. Whereas the AEP and the CFP can now be regarded as being well-tried procedures, the RMT, especially when applied to the field of molecular dynamics in the liquid state, has still to be considered as a first attempt at settling in a unitary picture the huge amount of information provided by "computer experiment." The ultimate reason for the presence in this volume of some chapters on the subject of computer experiments is to underline the program of this theoretical approach, a rigorous version of which is not yet available, so as to stimulate further research toward the attainment of a reliable and comprehensive formulation.

References

1. R. Zwanzig, *J. Chem. Phys.*, **33**, 1338 (1960); *Lectures in Theoretical Physics*, Vol. III, W. E. Brittin, B. W. Downs, and J. Downs, eds., Interscience, New York, 1961, pp. 106–141.

2. I. Prigogine, *Nature*, **246**, 67 (1973).

3. I. Prigogine, C. George, F. Henin, and L. Rosenfeld, *Chem. Scripta*, **4**, 5 (1973).

4. A. Grecos and I. Prigogine, *Physica*, **59**, 77 (1972).

5. S. Nakajima, *Prog. Theor. Phys.*, **20**, 948 (1958).

6. H. Mori, *Prog. Theor. Phys.*, **33**, 423 (1965); **34**, 399 (1965).

7. P. Grigolini, *Nuovo Cimento*, **63B**, 174 (1981).

8. R. Kubo, in "Non-Linear Nonequilibrium Statistical Mechanics," *Progr. Theoret. Phys.*, **64** (*Suppl.*), 1 (1978).

9. P. Glansdorf and I. Prigogine, *Thermodynamic Theory of Structure, Stability and Fluctuations*, Wiley, New York, 1971.

10. G. N. Bochov and Yu. E. Kuzovlev, *Physica*, **106A**, 443, 480 (1981).

11. H. Haken, *Synergetics. An Introduction. Nonequilibrium Phase Transitions and Self-Organization in Physics, Chemistry and Biology*, Springer-Verlag, Berlin, 1977.

12. A. Schenzle and H. Brand, *Phys. Rev.*, **A20**, 1628 (1979).

13. R. Graham and A. Schenzle, *Phys. Rev.*, **A25**, 1731 (1981).

14. M. Suzuki, K. Kaneko, and F. Sasagawa, *Prog. Theor. Phys.*, **65**, 828 (1981).

15. H. A. Kramers, *Physica*, **7**, 284 (1940).

16. M. Büttiker and R. Landauer, in *Nonlinear Phenomena at Phase Transition and Instabilities*, T. Riste, ed., NATO ASI Series, Plenum, New York, 1981; R. Landauer, *J. Appl. Phys.*, **33**, 2209 (1962); R. Landauer, in *Electrical Transport and Optical Properties of Inhomogeneous Media*, J. C. Garland and D. B. Tanner, eds., American Institute of Physics, New York, Sec. 9; R. Landauer and J. A. Swanson, *Phys. Rev.*, **121**, 1668 (1961); R. Landauer and J. W. F. Woo, in *Synergetics*, H. Haken, ed., B. G. Teubner, Stuttgart, 1973.

17. S. A. Adelman, *Adv. Chem. Phys.*, **XLIV**, 143 (1980).

18. S. A. Adelman and C. L. Brooks III, *J. Phys. Chem.*, **86**, 1511 (1982).

19. M. W. Evans, G. J. Evans, W. T. Coffey, and P. Grigolini, *Molecular Dynamics*, Wiley-Interscience, New York, 1982, Chapters 9 and 10.

20. R. F. Fox, *Phys. Rep.*, **48**, 179 (1978).

21. R. Zwanzig, in *Stochastic Processes in Chemical Physics: The Master Equation*, I. Oppenheim, K. E. Shuler, and G. H. Weiss, eds., MIT Press, Cambridge, 1977, p. 255.

22. P. Résibois, *Physica*, **27**, 541 (1961); I. Prigogine and P. Résibois, *Physica*, **27**, 629 (1961); P. Résibois, *Physica*, **29**, 721 (1963).

23. E. W. Montroll, *Fundamental Problems in Statistical Mechanics*, compiled by E. G. D. Cohen, North-Holland, Amsterdam, 1962, pp. 230–249; *Lecture Notes Theor. Phys. (Boulder)*, **3**, 22 (1980).

24. (a) J. H. Hynes, *Mol. Phys.*, **28**, 997 (1974); E. T. Chang, R. M. Mazo, and Y. T. Hynes, *J. Stat. Phys.*, **11**, 257 (1974). (b) H. Grabert, P. Hänggi, and P. Talkner; *J. Stat. Phys.*, **22**, 537 (1977); H. Grabert, P. Talkner, and P. Hänggi, *Z. Phys.*, **B26**, 389 (1977); H. Grabert, P. Talkner, P. Hänggi, and H. Thomas, *Z. Phys.*, **B29**, 273 (1978).

25. L. T. Muus, in *Electron Spin Relaxation in Liquids*, L. T. Muus and P. W. Atkins, eds., Plenum, New York, 1972, p. 1.

26. A. R. Ziv and W. Rhodes, *J. Chem. Phys.*, **65**, 4895 (1976), and references therein.

27. T. Karasudani, K. Nagano, H. Okamoto, and H. Mori, *Prog. Theor. Phys.*, **61**, 850 (1979).

28. (a) M. Ferrario and P. Grigolini, *J. Math. Phys.*, **20**, 2567 (1979); (b) M. Ferrario and P. Grigolini, *Chem. Phys. Lett.*, **62**, 100 (1979).

29. G. Ciccotti and J. P. Ryckaert, *Mol. Phys.*, **40**, 141 (1980).

30. (a) S. Nordholm, R. Zwanzig, *J. Stat. Phys.*, **13**, 347 (1975); (b) H. Grabert, *Z. Phys.*, **B26**, 79 (1977).

31. R. Zwanzig, *Lecture Notes Phys.*, **132**, 198 (1980).

32. M. Bixon and R. Zwanzig, *J. Stat. Phys.*, **3**, 245 (1971).

33. R. Zwanzig, *J. Chem. Phys.*, **60**, 2717 (1971); S. W. Haan and R. Zwanzig, *J. Chem. Phys.*, **68**, 1879 (1978); M. Bixon and R. Zwanzig, *ibid.*, 1896.

34. See for example, U. Balucani, R. Vallauri, V. Tognetti, P. Grigolini, and P. Marin, *Z. Phys., Condensed Matter*, **B49**, 181 (1982), and references therein.

35. N. van Kampen, *Stochastic Processes in Physics and Chemistry*, North-Holland, Amsterdam, 1981.

36. B. J. Berne, J. P. Boon, and S. A. Rice, *J. Chem. Phys.*, **45**, 1086 (1966); B. F. McCoy and S. A. Rice, *Chem. Phys. Lett.*, **35**, 431 (1975).

II

BASIC DESCRIPTION OF THE RULES LEADING TO THE ADIABATIC ELIMINATION OF FAST VARIABLES

P. GRIGOLINI and F. MARCHESONI

CONTENTS

I. INTRODUCTION

The major aim of the present volume is to use the same rigorous and generally valid theoretical structure to explain a variety of physical phenomena in a self-consistent way. A basically important part of this theory is the adiabatic elimination procedure (AEP) that can be derived from Eq. (2.18) of Chapter I.

The AEP has a long history that can be traced back to Einstein,[1a] who derived the formula

$$\left\langle |\mathbf{r} - \mathbf{r}_0|^2 \right\rangle = 6Dt \tag{1.1}$$

where D is the intensity of the stochastic force $\mathbf{F}(t)$ that drives the motion of the Brownian particle via the well-known Langevin equation

$$\dot{\mathbf{v}} = -\gamma\mathbf{v} + \mathbf{F}(t) \tag{1.2}$$

where **v** is the velocity of the Brownian particle itself. The stochastic force
F(t), which is usually assumed to be Gaussian, is defined by

$$\langle F_i(t) \rangle = 0 \tag{1.3}$$

$$\langle F_i(0) F_j(t) \rangle = 2D \delta_{ij} \delta(t) \tag{1.3'}$$

This means that this force is assumed to have a vanishingly short correlation
time. In a sense, Einstein's result provides information on the slow space
variable **r** through a contraction of the fast variable **v**. This is the end result
of any AEP.

The excellent review of Chandrasekhar[1b] provides a detailed account of
the history of the subject, to which both Smoluchowski and Einstein made
fundamental contributions. It is worth mentioning the well-known paper of
Kramers,[2] who provided a rigorous derivation of the Smoluchowski equation from the complete Fokker-Planck equation of a Brownian particle in an
external potential. This problem allows us to explain what we mean by a
systematic version of the AEP. We can state the problem as follows. Let us
consider the motion of a free Brownian particle described by the one-dimensional counterpart of Eq. (1.2),

$$\dot{v} = -\gamma v + f(t) \tag{1.4}$$

where $f(t)$ is a Gaussian white noise defined by

$$\langle f(0) f(t) \rangle = 2D \delta(t) \tag{1.5}$$

$$D \equiv \gamma \langle v^2 \rangle_{eq} \tag{1.6}$$

In Section II we shall discuss under what conditions this and the corresponding Fokker-Planck equation

$$\frac{\partial}{\partial t} \rho(v;t) = \mathcal{L}_b \rho(v;t) \equiv \gamma \left[\frac{\partial}{\partial v} v + \langle v^2 \rangle_{eq} \frac{\partial^2}{\partial v^2} \right] \rho(v;t) \tag{1.7}$$

can be considered to be rigorously equivalent to a description in terms of
Liouvillians.

When an external field is present, a coupling between velocity and space
is established, making it necessary to replace Eq. (1.4) with

$$\dot{x} = v$$
$$\dot{v} = -\frac{\partial}{\partial x} V(x) - \gamma v + f(t) \tag{1.8}$$

and Eq. (1.7) with

$$\frac{\partial}{\partial t}\rho(x,v;t) = (\mathscr{L}_1 + \mathscr{L}_b)\rho(x,v;t) \tag{1.9}$$

where

$$\mathscr{L}_1 \equiv -v\frac{\partial}{\partial x} + \left(\frac{\partial v}{\partial x}\right)\frac{\partial}{\partial v} \tag{1.10}$$

This last term is equivalent to a deterministic description of the coupling between x and v. In the presence of this coupling we must replace the probability distribution $\rho(v;t)$ with that involving both v and x, $\rho(x,v;t)$. x affords a good example of a slow variable (more generally in this chapter we shall consider a multidimensional set of slow variables **a**), and v is the fast variable to be contracted in some way (the corresponding multidimensional set of fast variables is **b**). Roughly speaking, the Smoluchowski equation is obtained by assuming that v is so fast a variable that it is valid to replace Eq. (1.8) with

$$\dot{x} = v$$
$$0 = \dot{v} = -\frac{\partial}{\partial x}V(x) - \gamma v + f(t) \tag{1.11}$$

or alternatively

$$\dot{x} = -\frac{1}{\gamma}\frac{\partial v(x)}{\partial x} + \frac{f(t)}{\gamma} \tag{1.12}$$

The probability distribution of the variable x alone, $\sigma(x;t)$, is then driven by

$$\frac{\partial}{\partial t}\sigma(x;t) = \frac{1}{\gamma}\left\{\frac{\partial}{\partial x}\left(\frac{\partial V(x)}{\partial x}\right) + \langle v^2\rangle_{eq}\frac{\partial^2}{\partial x^2}\right\}\sigma(x;t) \tag{1.13}$$

which is the usual Smoluchowski equation.

A renewal of interest in this problem began with Brinkman's work.[3] Brinkman derived an equation involving a second-order derivative with respect to time. This equation has been criticized by Hemmer,[4] who showed that for $V = 0$ and small times it does not give a better result than the Smoluchowski equation, Eq. (1.13).

The problem of eliminating fast variables has also been considered by Stratonovich,[5] who did not, however, apply his work explicitly to the case of the Brownian motion of Eq. (1.8). Stratonovich[5] noticed that many problems of interest satisfy a second-order fluctuation equation and considered the problem of establishing under which conditions that equation may be replaced by an "equivalent" one-dimensional first-order process. He studied both overdamped and underdamped processes. In the first case he developed a systematic elimination procedure and provided explicit formulas which can be recovered by the AEP illustrated in this section when fourth-order perturbation terms are considered (see Section V). In the second case he considered the energy as a slow variable, thereby making it possible to obtain simple results in a region where the assumption that the space coordinate is a slow variable is invalidated. His approach has been successfully applied by Lindenberg et al.[6] to the study of nonlinear oscillators in the presence of external noise. This is also the key factor in explaining the properties of analog electric circuits (see Faetti et al., Chapter X).

More recent investigations than Brinkman's (related to this work) are those of Landauer and Swanson[7] and Langer.[8] Lee[9] has confirmed the earlier result of Stratonovich. Equation (1.8) has also been studied by Wilemski,[10] who derived the same equation as that of Stratonovich[5] without explicit reference to the problem of the Brownian motion.

A systematic and rigorous approach to deriving equations of motion for slow variables has also been developed by Mori's school.[11-13] Their first paper[11] on this problem is of special relevance to one of the topics of the present volume—molecular dynamics in the liquid state—in that it suggests that the nonlinear nature of the underlying Hamiltonian can produce macroscopic nonlinear effects. These can find "experimental" support by means of computer simulation (see Evans, Chapter V, and Ferrario et al., Chapter VI). A further interesting aspect of this paper is that it confirms the point of view of van Kampen,[14] who urges caution in the use of the Wigner-Moyal expansion of the master equation. The result of Mori's school is a systematic approach to a Kramers-Moyal expansion in terms of a perturbation, or slowness, parameter. This approach has been generalized so as to make it possible to apply it to non-Hamiltonian systems.[12,13]

Skinner and Wolynes[15a] came back to the problem of Eq. (1.8) and solved it with a projection method in Laplace space. An interesting aspect of their work is the development of the contracted distribution function $\sigma(a; t)$ (see also Section II) inside the time-convolution integral. They pointed out that this provides perturbation terms neglected erroneously by Brinkman. This interesting feature of their approach is included in the AEP illustrated in this chapter. They explicitly evaluated correction terms up to order γ^{-5}. The projection technique has also been used by Chaturvedi and Shibata,[15b] who used a memoryless equation as the starting point of their treatment.

A rigorous analysis of the same problem has been given by San Miguel and Sancho.[16-18] They emphasized that the reduction process produces unavoidably non-Markovian statistics. Nevertheless they noted that the existence of Fokker-Planck equations does not conflict with the non-Markovian character of the stochastic process, since the corresponding solution, in harmony with ref. 19, is valid only to evaluate one-time averages and is of no use in multitime averages.

Corrections to the Smoluchowski equation (1.13) to order γ^{-5} have been evaluated by Titulaer,[20] using a Chapman-Enskog expansion.

Risken, Vollmer, and Mörsch[21] studied the Kramers equation, that is, the Fokker-Planck equation (1.9), by expanding the distribution function $\rho(x, v; t)$ in Hermitian polynomials (velocity part) and in another complete set satisfying boundary conditions (position part). The Laplace transform of the initial value problem was obtained in terms of continued fractions. An inverse friction expansion of the matrix continued fraction was then used to show that the first Hermitian expansion coefficient may be determined by a generalized Smoluchowski equation. This provides results correcting the standard Smoluchowski equation with terms of increasing power in $1/\gamma$. They evaluated explicit expressions up to order γ^{-5}.

Kaneko[22] has independently developed a systematic approach to the problem that is similar to that of ref. 21 and has applied his method to optical problems.

More recently a controversy has arisen as to whether the interaction $\mathcal{L}_1$ should be defined by Eq. (1.10) or by

$$\mathcal{L}_1 \equiv - v\frac{\partial}{\partial x} \tag{1.14}$$

and $\mathcal{L}_b$ of Eq. (1.7) by

$$\mathcal{L}_b \equiv \left(\frac{\partial v}{\partial x}\right)\frac{\partial}{\partial v} + \gamma\left[\frac{\partial}{\partial v}v + \langle v^2\rangle_{eq}\frac{\partial^2}{\partial v^2}\right] \tag{1.14'}$$

Hasegawa et al.[25] pointed out that the latter choice does not yield the correct Boltzmann distribution at equilibrium. Titulaer,[24] however, argued that this choice may well be preferable for systems far from equilibrium whereas for systems close to equilibrium it is less suitable than other schemes.

Garrido and Sancho[26] studied a multiplicative stochastic process with finite correlation time τ. By using an ordered cumulant technique they evaluated corrections up to order τ^2.

Dekker[27] has studied multiplicative stochastic processes. In his work the stochastic Liouville equation was solved explicitly through first order in an expansion in terms of correlation times of the multiplicative Gaussian "colored" noise for a general multidimensional weakly non-Markovian process. He followed the suggestions of refs. 17 and 18 and applied, Novikov's theorem. In the general multidimensional case, however, he improved the earlier work by San Miguel and Sancho.[17,18]

A cumulant expansion supplemented by a boson operator representation was used by Steiger and Fox[28] to tackle the Kramers problem of Eq. (1.9) in the multidimensional case. This allowed them to obtain a multidimensional version of their earlier results.

The problem of eliminating fast variables plays a decisive role in the subject of synergetics.[29] The concept of order parameters and slowing in non-equilibrium systems has been proved by Haken[29] to be a powerful tool for analyzing instabilities far from thermal equilibrium. Furthermore, Haken used this concept to elucidate important analogies among many-component systems from completely different disciplines. For these reasons the Haken school has also been motivated to develop a systematic procedure of adiabatic elimination.[30,31]

Finally we would like to mention the work of Lavenda and coworkers,[32] who approached the same problem by means of a generalization of the kinetic analog of Boltzmann's principle so as to include the Ornstein-Uhlenbeck process. They carried out an asymptotic analysis in the limit of high resistance. The condition for the validity of their asymptotic expansion has been proved to be identical to the modified Kramers' condition derived by Stratonovich.

The choice of AEP is somewhat subjective. However, the different AEPs available are closely related, and general agreement has been obtained on the main results. We choose to invoke the spirit of this volume, which is to obtain the widest possible range of results by using a few basic concepts, inspired by the delta-like philosophy of Chapter I. *We shall therefore emphasize the fact that the Zwanzig approach is ideally suited for deriving contracted equations of motion. The application of our AEP to nonlinear systems is in line with the general projection procedure of Zwanzig.*[33]

II.   AN ILLUSTRATIVE EXAMPLE TO MAKE READERS FAMILIAR WITH THE AEP

The AEP is a versatile theoretical tool. In this section we shall show that it can also be used to derive the Fokker-Planck equation from a rigorous classical mechanical description. Let us consider the system illustrated in

Fig. 1. Its dynamics are described by Newton's deterministic law as follows:

$$M\dot{v} = -k(x - y_1)$$
$$\dot{y}_1 = w_1 \qquad m_1\dot{w}_1 = -k(y_1 - x) - \kappa_1(y_1 - y_2)$$
$$\dot{y}_2 = w_2 \qquad m_2\dot{w}_2 = -\kappa_1(y_2 - y_1) - \kappa_2(y_2 - y_3) \qquad (2.1)$$
$$\dot{y}_3 = w_3 \qquad m_3\dot{w}_3 = -\kappa_2(y_3 - y_2) - \kappa_3(y_3 - y_4)$$

This means that a particle of mass M interacts linearly with a chain of infinitely many particles with masses m_1, m_2, m_3, and so on, which also interact with each other via a linear coupling (only nearest-neighbor interactions are considered). The dynamics of the particle of mass M is defined by its space coordinate x and velocity v. The space coordinates and velocities of the particles of mass m_i are denoted by the symbols y_i and w_i, respectively. More proper variables are the relative distances

$$\Delta_0 \equiv x - y_1$$
$$\Delta_1 \equiv y_1 - y_2$$
$$\vdots \qquad (2.2)$$
$$\Delta_i \equiv y_{i-1} - y_i$$

The set of equations (2.1) is then replaced by

$$\dot{v} = -\frac{K}{M}\Delta_0$$

$$\dot{\Delta}_0 = v - w_1 \qquad \dot{w}_1 = \frac{k}{m_1}\Delta_0 - \frac{x_1}{m_1}\Delta_1$$

$$\dot{\Delta}_1 = w_1 - w_2 \qquad \dot{w}_2 = \frac{x_1}{m_2}\Delta_1 - \frac{x_2}{m_2}\Delta_2 \qquad (2.3)$$

$$\cdots$$

$$\dot{\Delta}_i = w_i - w_{i+1} \qquad \dot{w}_{i+1} = \frac{\kappa_i}{m_{i+1}}\Delta_i - \frac{\kappa_{i+1}}{m_{i+1}}\Delta_{i+1}$$

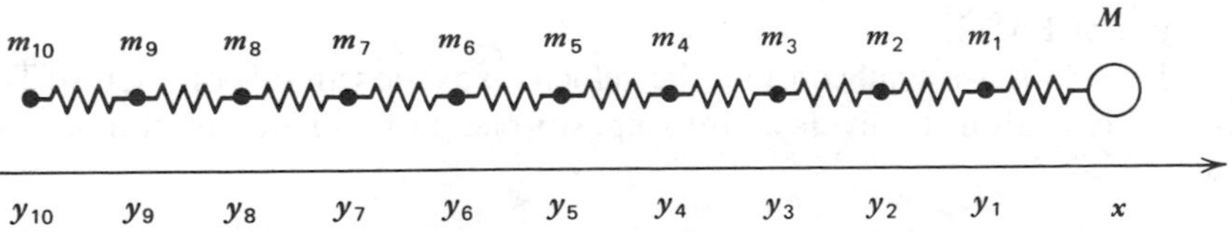

Figure 1.   Scheme of the one-dimensional system under study.

The Liouville equation corresponding to Eq. (2.3) reads[34]

$$\frac{\partial}{\partial t} \rho(x, v, w_1, w_2, \ldots, \Delta_0, \Delta_1, \ldots) = \mathscr{L}\rho(x, v, w_1, w_2, \ldots; \Delta_0, \Delta_1, \ldots)$$

(2.4)

where ρ denotes the distribution function in the phase space $x, v, w_1, w_2, \ldots,$ $\Delta_0, \Delta_1, \ldots,$ and

$$\mathscr{L} \equiv \frac{k}{M} \Delta_0 \frac{\partial}{\partial v} - (v - w_1) \frac{\partial}{\partial \Delta_0} - \left(\frac{k}{m_1} \Delta_0 - \frac{\kappa_1}{m_1} \Delta_1 \right) \frac{\partial}{\partial w_1}$$

$$- (w_1 - w_2) \frac{\partial}{\partial \Delta_1} - \left(\frac{\kappa_1}{m_2} \Delta_1 - \frac{\kappa_2}{m_2} \Delta_2 \right) \frac{\partial}{\partial w_2} \cdots \qquad (2.5)$$

Readers can convince themselves that Eq. (2.4) is completely equivalent to Eq. (2.3) as follows. Let us denote by α each variable of the set $(x, v, w_1, w_2, \ldots, \Delta_0, \Delta_1, \ldots)$. In the "Heisenberg picture" (see also Chapter I), corresponding to the "Schrödinger picture" of Eq. (2.4), the time evolution of α is driven by

$$\dot{\alpha} = \Gamma \alpha \qquad (2.6)$$

where

$$\Gamma \equiv \mathscr{L}^+ = -\mathscr{L} \qquad (2.6')$$

When $\alpha = v$ from Eq. (2.6) we recover the first equation of the set of Eqs. (2.3); when $\alpha = \Delta_0$ we obtain the second one; and so on. In other words, Eq. (2.4) and its Heisenberg picture are exactly coincident with the deterministic picture of Eq. (2.3).

Henceforth we shall consider the velocity v as the variable of interest. This naturally leads us to divide $\mathscr{L}$ into a perturbed part and an unperturbed part as follows:

$$\mathscr{L} = \mathscr{L}_1 + \mathscr{L}_b \qquad (2.7)$$

where

$$\mathcal{L}_1 \equiv \frac{k}{M} \Delta_0 \frac{\partial}{\partial v} - v \frac{\partial}{\partial \Delta_0} \tag{2.7'}$$

$$\mathcal{L}_b = w_1 \frac{\partial}{\partial \Delta_0} - \frac{k}{m_1} \Delta_0 \frac{\partial}{\partial w_1} + \frac{\kappa_1}{m_1} \Delta_1 \frac{\partial}{\partial w_1} - w_1 \frac{\partial}{\partial \Delta_1} + w_2 \frac{\partial}{\partial \Delta_1} - \frac{\kappa_1}{m_2} \Delta_1 \frac{\partial}{\partial w_2}$$

$$+ \frac{\kappa_2}{m_2} \Delta_2 \frac{\partial}{\partial w_2} - w_2 \frac{\partial}{\partial \Delta_2} + w_3 \frac{\partial}{\partial \Delta_2} - \frac{\kappa_2}{m_3} \Delta_2 \frac{\partial}{\partial w_3} \cdots \tag{2.8}$$

In this example the unperturbed part, affecting the variable of interest alone, termed $\mathcal{L}_a$ in Chapter I, vanishes.

As explained in Chapter I, to apply the Zwanzig projection method we have to establish which equilibrium distribution is attained by the heat bath, that is, the subsystem driven by $\mathcal{L}_b$. This equilibrium distribution will be denoted $\rho_{eq}(\mathbf{b})$, where $\mathbf{b} \equiv (\Delta_0, \Delta_1, \ldots, w_1, w_2, \ldots)$, and by definition satisfies

$$\mathcal{L}_b \rho_{eq}(\mathbf{b}) \equiv 0 \tag{2.9}$$

It can be shown that

$$\rho_{eq}(\mathbf{b}) = K \exp\left[- \frac{\Delta_0^2}{2\langle \Delta_0^2 \rangle_{eq}} \right] \exp\left[- \frac{\Delta_1^2}{2\langle \Delta_1^2 \rangle_{eq}} \right] \cdots \exp\left[- \frac{w_1^2}{2\langle w_1^2 \rangle_{eq}} \right]$$

$$\cdot \exp\left[- \frac{w_2^2}{2\langle w_2^2 \rangle_{eq}} \right] \cdots \tag{2.10}$$

where K is a normalization constant. It is interesting to remark that the equipartition principle is satisfied, that is,

$$m_1 \langle w_1^2 \rangle_{eq} = m_2 \langle w_2^2 \rangle_{eq} = \cdots$$
$$\kappa \langle \Delta_0^2 \rangle_{eq} = \kappa_i \langle \Delta_i^2 \rangle_{eq} = m_i \langle w_i^2 \rangle_{eq} \tag{2.11}$$

Note that the total equilibrium distribution $\rho_T(v, \mathbf{b})$ is

$$\rho_T(v, \mathbf{b}) = K' \exp\left[- \frac{v^2}{2\langle v^2 \rangle_{eq}} \right] \rho_{eq}(\mathbf{b}) \tag{2.12}$$

where K' is a normalization constant and the equipartition relationship

$$M\langle v^2 \rangle_{eq} = k\langle \Delta_0^2 \rangle_{eq} \tag{2.13}$$

holds.

From Chapter I we are led to define the projection operator P as follows:

$$\mathbf{P}\rho(v,\mathbf{b};t) = \rho_{eq}(\mathbf{b})\sigma(v;t) \equiv \rho_{eq}(\mathbf{b}) \int d\mathbf{b}\,\rho(v,\mathbf{b};t) \tag{2.14}$$

We are now in a position to apply the simple recipe described in Chapter I. From Eqs. (2.19) and (2.20) of that chapter we get

$$\frac{\partial}{\partial t}\sigma(v;t) = \frac{1}{\rho_{eq}(\mathbf{b})} \int_0^t \Phi_B(t-s)\rho_{eq}(\mathbf{b})\sigma(v;s)\,ds \tag{2.15}$$

where

$$\Phi_B(t-s) \equiv \mathbf{P}\mathscr{L}_1 e^{-\mathscr{L}_b(t-s)}\mathscr{L}_1\mathbf{P} \tag{2.16}$$

It will become clear later that at the lowest significant perturbation order, we can replace Eq. (2.15) with

$$\frac{\partial}{\partial t}\sigma(v;t) = \frac{1}{\rho_{eq}(\mathbf{b})} \left(\int_0^t \Phi_B(t-s)\,ds \right)\rho_{eq}(\mathbf{b})\sigma(v;t) \tag{2.17}$$

With increasing time, Eq. (2.17) approaches

$$\frac{\partial}{\partial t}\sigma(v;t) = \mathscr{L}_{eff}\sigma(v;t) \tag{2.18}$$

where

$$\mathscr{L}_{eff} \equiv \frac{1}{\rho_{eq}(\mathbf{b})} \left(\int_0^\infty \Phi_B(s)\,ds \right)\rho_{eq}(\mathbf{b}) \tag{2.18'}$$

the actual form of which can be derived straightforwardly by noticing that the second term on the right-hand side of Eq. (2.7'), $-v\,\partial/\partial\Delta_0$, gives a vanishing contribution when substituted in the first $\mathscr{L}_1$ appearing in the first term on the right-hand side of Eq. (2.16), since integration by parts supplemented by the requirement that $\rho_{eq}(\mathbf{b})$ properly vanish at the boundary of the system of the variables **b** leads us to

$$P\frac{\partial}{\partial\Delta_0}\cdots = 0 \tag{2.19}$$

On the other hand, $\mathscr{L}_1$ can be divided into two parts as follows:

$$\mathscr{L}_1 \equiv \frac{k}{M}\Delta_0\left(\frac{\partial}{\partial v} + \frac{Mv}{kT}\right) - v\left(\frac{\partial}{\partial\Delta_0} + \frac{k\Delta_0}{kT}\right) \tag{2.7''}$$

The first term vanishes when substituted in the second $\mathscr{L}_1$ appearing in the term on the right-hand side of eq. (2.16) [remember the property of Eq. (2.9)]. We note also that

$$P\mathscr{L}_1(t)P = Pe^{-\mathscr{L}_0 t}\mathscr{L}_1 e^{\mathscr{L}_0 t}P = P\mathscr{L}_1 P = 0 \tag{2.20}$$

This depends on the properties of Eqs. (2.9) and (2.19) supplemented by the remark that $\langle\Delta_0\rangle_{eq} = 0$ [as can be derived by inspection of Eq. (2.10)].

In consequence, we obtain

$$\mathscr{L}_{eff} = \frac{k}{M}\hat{\Phi}_{\Delta_0}(0)\left\{\langle v^2\rangle_{eq}\frac{\partial^2}{\partial v^2} + \frac{\partial}{\partial v}v\right\} \tag{2.21}$$

where $\hat{\Phi}_{\Delta_0}(z)$ is the Laplace transform of the correlation function

$$\Phi_{\Delta_0}(t) \equiv \frac{\langle\Delta_0\Delta_0(t)\rangle_{eq}}{\langle\Delta_0^2\rangle_{eq}} \tag{2.22}$$

This means that the Fokker-Planck equation can be obtained under the basic assumption alone that the time scale of the variable of interest v is much larger than that of its thermal bath, thereby recovering the well-known result of ref. 35. The AEP can thus shed light on the intimate relationship between an effective Liouvillian such as that of Eq. (2.20) and a rigorous one such as that of Eq. (2.7).

The assumption of time scale separation mentioned above implied that κ_i/m_i and k/m_1 are much larger than k/M. If the coefficients k and x_i share the same order of magnitude, the particle of interest (the Brownian particle) has to have a mass M much larger than those of the heat bath particles.

The systematic approach illustrated in the next section will make it clear that the higher-order corrections have no effect except that of renormalizing the friction

$$\gamma_v \equiv \frac{k}{M} \hat{\Phi}_{\Delta_0}(0) \tag{2.23}$$

so as to generate the new effective operator

$$\mathscr{L}_{\text{eff}} \equiv (\gamma_v + \Delta\gamma_v)\left\{ \frac{\partial}{\partial v} v + \langle v^2 \rangle_{\text{eq}} \frac{\partial^2}{\partial v^2} \right\} \tag{2.24}$$

For example, in the simple case where

$$\Phi_{\Delta_0}(t) \equiv \frac{\langle \Delta_0 \Delta_0(t) \rangle_{\text{eq}}}{\langle \Delta_0^2 \rangle_{\text{eq}}} = e^{-\gamma_c t} \tag{2.25}$$

we shall obtain

$$\gamma_v = \frac{k/M}{\gamma_c}$$
$$\Delta\gamma_v = \frac{(k/M)^2}{\gamma_c^3} + \cdots \tag{2.26}$$

which is indeed a special case of a general result in the following sections [see Eqs. (4.53), (4.54), and (4.65)].

In Chapter VI, Ferrario et al. discuss the precise conditions under which the heat bath driven by the operator $\mathscr{L}_b$ of Eq. (2.8) allows the correlation function $\Phi_{\Delta_0}(t)$ to be exponential [Eq. (2.25)]. In this case, the system of Eq. (2.1) can be replaced by the "reduced" model

$$\dot{v}(t) = w(t)$$
$$\dot{w}(t) = -\Delta_1^2 v(t) - \gamma_w w(t) + F_w(t) \tag{2.27}$$

where $F_w(t)$ is a Gaussian white noise defined by

$$\langle F_w(0) F_w(t) \rangle_{\text{eq}} = 2D_F \delta(t) \tag{2.28}$$
$$D_F = \gamma_w \langle w^2 \rangle_{\text{eq}} \tag{2.29}$$

This is the simple system described in Section III of Chapter I. The equivalence of this reduced model with the complex one of Eq. (2.1) supplemented by Eq. (2.25) can be proved straightforwardly by applying to the reduced model the same projection procedure as that which led to Eq. (2.18) of Chapter I. Interested readers can do that as a simple exercise, from which they will see that for this equivalence to be attained,

$$\gamma_w = \gamma_c \tag{2.30}$$

and

$$\Delta_1^2 = \frac{k}{M} \tag{2.31}$$

the memory kernel of both the generalized master equations can indeed be expressed, via the Gaussian statistical properties of both systems, in terms of two correlation functions $\Phi_{\Delta_0}(t)$ which turn out to be identical when Eqs. (2.30) and (2.31) are satisfied. [Note that this equivalence is shown by disregarding in both systems the inhomogeneous term, Eq. (2.18); i.e., in both cases we assume that preparation has no effect on relaxation (see also the concluding remarks).]

This poses a problem: By applying to the system of Eq. (2.27) the Laplace transformation technique illustrated in Chapter I for evaluating the correlation function

$$\Phi_v(t) \equiv \frac{\langle v(0)\dot{v}(t)\rangle_{eq}}{\langle v^2\rangle_{eq}} \tag{2.32}$$

we obtain [$\hat{\Phi}_v(z)$ is the Laplace transform of $\Phi_v(t)$]

$$\hat{\Phi}_v(z) = \frac{1}{z + (k/M)/(z + \gamma_c)} \tag{2.33}$$

The effective rate can be derived by

$$\left(\lim_{z \to 0} \hat{\Phi}_v(z)\right)^{-1} = \frac{k/M}{\gamma_c}. \tag{2.34}$$

The exact result is in apparent disagreement with the renormalized rate provided by the AEP, see Eq. (2.26).

This disagreement can be accounted for by noticing that the rate provided by the Mori theory is determined from the value of the area below the curve $\Phi_v(t)$, whereas the AEP is equivalent to fitting with an exponential decay the long-time decay behavior of $\Phi_v(t)$ (see Fig. 2). The exact time revolution of $\Phi_v(t)$ can be obtained by means of the diagonalization of the matrix $\mathbf{A}$ given

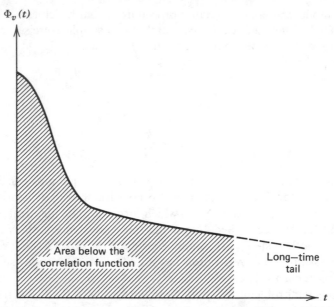

$\Phi_v(t)$

Area below the correlation function

Long–time tail

t

Figure 2. Schematic illustration of the major features of the CFP and AEP. The CFP allows the area below the correlation function $\Phi_v(t)$ to be determined. The AEP leads to the determination of the decay rate of the tail of this correlation function.

by

$$\mathbf{A} \equiv \begin{pmatrix} 0 & 1 \\ -\dfrac{k}{M} & -\gamma_c \end{pmatrix} \qquad (2.35)$$

The long-time behavior of $\Phi_v(t)$ is therefore provided by the slowest eigenvalue of $\mathbf{A}$. This can be determined through a standard perturbation technique[36] and can be proved to coincide with the result of the AEP [Eq. (2.26)].

Before ending this section, we would like to clarify the meaning of the time-scale separation between system and heat bath. If the correlation time of the variable Δ_0 is given the exponential form of Eq. (2.25), the microscopic time τ_c is immediately provided by

$$\tau_c = \frac{1}{\gamma_c} \qquad (2.36)$$

In general, the microscopic time could be defined by

$$\tau_c^{-1} = \lim_{z \to 0} \hat{\Phi}_{\Delta_0}(z) \tag{2.37}$$

where $\hat{\Phi}_{\Delta_0}(z)$ is the Laplace transform of $\Phi_{\Delta_0}(t)$ without necessarily requiring $\Phi_{\Delta_0}(t)$ to be exponential. In the remaining sections we shall show that τ_c, defined in that way, is the true perturbation parameter of the systematic approach, which we develop in Section III.

In Section III we shall provide the basic rules for a systematic expansion, which will lead us ultimately to a deeper understanding of the development just presented.

III. THE BASIC RULES FOR A SYSTEMATIC EXPANSION

The purpose of this section is to develop a systematic approach to evaluating higher-order corrections to the results discussed in Section II. Of course, this systematic approach can also be applied to physical systems different from that discussed in Section II. As a consequence of the linear nature of this system, these rules prove the exact equivalence between the Newtonian description of Eq. (2.1) and the standard Langevin equation provided that the assumption of time-scale separation is satisfied (see Section IV).

This interesting result also provides clues on how to build up the reduced model theory (RMT). As outlined in Chapter I, this approach (which will be illustrated by Ferrario et al. in Chapter VI with several applications) consists in replacing the set of variables $(\mathbf{a}, \mathbf{b}_R)$, where $\mathbf{a}$ denotes the variables of interest and $\mathbf{b}_R$ the true irrelevant variables, with a set $(\mathbf{a}, \mathbf{b}_V)$. The auxiliary variables $\mathbf{b}_V$, hopefully very few in number, should be chosen in such a way as to concentrate the nonlinearity of the system in the subset $(\mathbf{a}, \mathbf{b}_V)$. If the interaction between $(\mathbf{a}, \mathbf{b}_V)$ and the remaining (virtually infinite in number) degrees of freedom can be assumed linear, the result provided by the AEP when applied to rigorous Liouvillians guarantees that it is legitimate to simulate the latter with standard friction and noise terms. Even after replacing the system $(\mathbf{a}, \mathbf{b}_R)$ with the system $(\mathbf{a}, \mathbf{b}_V)$ supplemented by the corresponding friction and noise terms, the latter can still be complicated enough to require a further application of the AEP to eliminate the variables $\mathbf{b}_V$. In some cases it is not possible to apply the AEP directly to the $(\mathbf{a}, \mathbf{b}_R)$, whereas it is illuminating to show the AEP at work on the corresponding "reduced" system $(\mathbf{a}, \mathbf{b}_V)$ (see Chapter X).

It is therefore convenient to refer the discussion that follows to the general set $(\mathbf{a}, \mathbf{b})$, where $\mathbf{b}$ can either denote $\mathbf{b}_R$ or $\mathbf{b}_V$. The dynamics of our system are described by

$$\frac{\partial}{\partial t}\rho(\mathbf{a},\mathbf{b};t) = \mathscr{L}\rho(\mathbf{a},\mathbf{b};t) \tag{3.1}$$

where

$$\mathscr{L} = iL \qquad \text{for } \mathbf{b} = \mathbf{b}_R \tag{3.2}$$

$$\mathscr{L} = \text{Fokker-Planck operator for } \mathbf{b} = \mathbf{b}_V \tag{3.2'}$$

iL of Eq. (3.2) denotes a rigorous Liouvillian operator such as that of Section II.

We assume that the operator $\mathscr{L}$ consists of an unperturbed part $\mathscr{L}_0$ and a perturbation part $\mathscr{L}_1$, that is,

$$\mathscr{L} = \mathscr{L}_0 + \mathscr{L}_1 \tag{3.3}$$

In general $\mathscr{L}_0$ drives both $\mathbf{a}$ and $\mathbf{b}$, but it does not have any interaction term between sets $\mathbf{a}$ and $\mathbf{b}$, that is,

$$\mathscr{L}_0 = \mathscr{L}_a + \mathscr{L}_b \tag{3.4}$$

This property is satisfied by the system illustrated in Section II and every other system explored with the same method in the present volume.

As pointed out in Chapter I, we have to apply the Zwanzig projection method to the interaction picture. For clarity let us rewrite the corresponding result (see Chapter I):

$$\frac{\partial}{\partial t}\tilde{\rho}_1 = P\mathscr{L}_1 P\tilde{\rho}(t)$$

$$+ \int_0^t ds\, P\mathscr{L}_1(t)\overleftarrow{\exp}\left[\int_s^t ds_1\, Q\mathscr{L}_1(s_1)\right]Q\mathscr{L}_1(s)P\tilde{\rho}(s)$$

$$+ P\mathscr{L}_1(t)\overleftarrow{\exp}\left[\int_0^t ds\, Q\mathscr{L}_1(s)\right]Q\tilde{\rho}(0) \tag{3.5}$$

where

$$Q \equiv 1 - P$$

Throughout this chapter we shall use the projection operator P defined as

$$P\rho(\mathbf{a},\mathbf{b};t) = \rho_{\text{eq}}(\mathbf{b})\int d\mathbf{b}\,\rho(\mathbf{a},\mathbf{b};t) \equiv \rho_{\text{eq}}(\mathbf{b})\sigma(\mathbf{a};t) \tag{3.6}$$

which reduces to Eq. (2.14) when $\mathbf{a} = v$. Recall that

$$\mathscr{L}_b \rho_{eq}(\mathbf{b}) = 0 \qquad (3.7)$$

that is, $\rho_{eq}(\mathbf{b})$ is the equilibrium distribution of the heat bath.

For simplicity we shall disregard the inhomogeneous term appearing in Eq. (3.5) by assuming that

$$\rho(\mathbf{a}, \mathbf{b}; 0) = \sigma(\mathbf{a}; 0) \cdot \rho_{eq}(\mathbf{b}) \qquad (3.8)$$

In Section VII we shall justify our more general neglect of the inhomogeneous term.

Some useful properties are:

$$[P, \mathscr{L}_a] = 0 \qquad (3.9)$$

$$P e^{\mathscr{L}_b t} = P \qquad (3.10)$$

$$e^{\mathscr{L}_b t} P = P \qquad (3.11)$$

The first one is a trivial consequence of the fact that P, as defined in Eq. (3.6), does not act on the variables $\mathbf{a}$. The second [(3.10)] results from

$$P \mathscr{L}_b \cdots = 0 \qquad (3.12)$$

which in turn can be proved by means of integration by parts supplemented by the property that the heat bath equilibrium distribution properly vanishes in the boundary region. The third property [(3.11)] is a direct consequence of Eq. (3.7).

Equations (3.9)–(3.11) allow us to write

$$\frac{\partial}{\partial t} \rho_1(t) = \frac{\partial}{\partial t} P e^{\mathscr{L}_a t} e^{\mathscr{L}_b t} \tilde{\rho}_1(t)$$

$$= \mathscr{L}_a \rho_1(t) + P e^{\mathscr{L}_a t} \frac{\partial}{\partial t} \tilde{\rho}_1(t) \qquad (3.13)$$

Note that the time-independent nature of the projection operator P [already exploited to derive Eq. (3.5)] allows us to use the property

$$\left[\frac{\partial}{\partial t}, P \right] = 0 \qquad (3.14)$$

By substituting Eq. (3.5) (without the inhomogeneous term) into Eq. (3.13), we get

$$\frac{\partial}{\partial t}\sigma(\mathbf{a};t) = \mathscr{L}_a\sigma(\mathbf{a};t) + \rho_{eq}^{-1}(\mathbf{b})\,P\mathscr{L}_1Pe^{\mathscr{L}_a t}\tilde{\rho}(t)$$
$$+ \int_0^t K(t-\tau)\sigma(\mathbf{a};\tau)\,d\tau \qquad (3.15)$$

where

$$K(t-\tau) \equiv \rho_{eq}^{-1}(\mathbf{b})\,e^{\mathscr{L}_a t}P\mathscr{L}_1(t)\overleftarrow{\exp}\left[\int_\tau^t dt'(1-P)\mathscr{L}_1(t')\right]$$
$$\cdot(1-P)\mathscr{L}_1(\tau)e^{-\mathscr{L}_a\tau}P\rho_{eq}(\mathbf{b}) \qquad (3.16)$$

To obtain this result we have used both Eq. (3.9) and Eq. (3.10).

Let us focus on the third term on the right-hand side of Eq. (3.15). By developing $\sigma(t)$ into a Taylor power series around $\tau = t$, we get

$$\int_0^t K(t-\tau)\sigma(\mathbf{a};\tau)\,d\tau = \sum_{r=0}^\infty \int_0^t d\tau\,K(t-\tau)\frac{(t-\tau)^r}{r!}(-)^r\left(\frac{\partial^r\sigma}{\partial\tau}(\mathbf{a};t)\right)_{\tau=t}$$
$$= \sum_{r=0}^\infty\left(\int_0^t K(s)\frac{s^r}{r!}(-)^r\right)\left(\frac{\partial^r}{\partial\tau^r}\sigma(\mathbf{a};\tau)\right)_{\tau=t} \qquad (3.17)$$

For times t of the same order of magnitude as the relaxation times of $\mathbf{a}$ [i.e., much larger than the lifetime of $K(t)$], Eq. (3.17) can be rewritten as

$$\int_0^t K(t-\tau)\sigma(\mathbf{a};\tau)\,d\tau = \sum_{r=0}^\infty\left(\int_0^\infty K(s)\frac{s^r}{r!}(-)^r\right)\left(\frac{\partial^r}{\partial\tau^r}\sigma(\mathbf{a};t)\right)_{\tau=t}$$
$$= \sum_{r=0}^\infty\frac{1}{r!}\frac{d^r}{dz^r}\hat{K}(z)\bigg|_{z=0}\frac{\partial^r\sigma}{\partial t^r}(\mathbf{a};t) \qquad (3.18)$$

where

$$\hat{K}(z) = \int_0^\infty e^{-zs}K(s)\,ds \qquad (3.19)$$

The time expansion of $\sigma(\tau)$ around $t=\tau$ will be proven to be responsible for corrections of order higher than the result discussed in Section II. Let us expand into a power series the time-ordered exponential of Eq. (3.16). We get

$$K(t-\tau) \equiv \rho_{eq}^{-1}(\mathbf{b})\exp(\mathscr{L}_a t)P\mathscr{L}_1(t)(1-P)\mathscr{L}_1(\tau)P\rho_{eq}(\mathbf{b})$$

$$+ \sum_{K=3}^{\infty} \rho_{eq}^{-1}(\mathbf{b})\exp(\mathscr{L}_a t)\int_{\tau}^{t} ds_1\, P\mathscr{L}_1(t)(1-P)\mathscr{L}_1(s_1)$$

$$\cdot \int_{\tau}^{s_1} ds_2(1-P)\mathscr{L}_1(s_2) \cdots \int_{\tau}^{s_{k-3}} ds_{k-2}(1-P)$$

$$\cdot \mathscr{L}_1(s_{k-2})(1-P)\mathscr{L}_1(\tau)\exp(-\mathscr{L}_a\tau)P\rho_{eq}(\mathbf{b}) \qquad (3.20)$$

By making the following change of variables:

$$\tau_0 \equiv t - \tau$$

$$\tau_n \equiv s_n - \tau, \qquad n > 0$$

$$K(\tau_0) = \sum_{k=2}^{\infty} K^{(k)}(\tau_0)$$

$$\equiv \rho_{eq}^{-1}(\mathbf{b})\exp\{\mathscr{L}_a(\tau_0+\tau)\}P\mathscr{L}_1(\tau_0+\tau)(1-P)\mathscr{L}_1(\tau)\exp(-\mathscr{L}_a\tau)P\rho_{eq}(\mathbf{b})$$

$$+ \sum_{k=3}^{\infty} \rho_{eq}^{-1}(\mathbf{b})e^{\mathscr{L}_a\tau_0}\int_0^{\tau_0} d\tau_1 \int_0^{\tau_1} d\tau_2 \cdots$$

$$\cdot \int_0^{\tau_{k-3}} d\tau_{k-2}\, P\mathscr{L}_1(\tau_0+\tau)(1-P)\mathscr{L}_1(\tau_1+\tau)$$

$$\cdots (1-P)\mathscr{L}_1(\tau_{k-2}+\tau)(1-P)\mathscr{L}_1(\tau)\exp(-\mathscr{L}_a\tau)P\rho_{eq}(\mathbf{b}) \qquad (3.21)$$

which implies [as the right-hand side of the definition of Eq. (3.21) turns out to be independent of τ]

$$K^{(2)}(\tau_0) = \rho_{eq}^{-1}(\mathbf{b})e^{\mathscr{L}_a\tau_0}P\mathscr{L}_1(\tau_0+\tau)(1-P)\mathscr{L}_1(\tau)e^{-\mathscr{L}_a\tau}P\rho_{eq}(\mathbf{b}) \qquad (3.21')$$

$$K^{(k)}(\tau_0) = \rho_{eq}^{-1}(\mathbf{b})e^{\mathscr{L}_a\tau_0}\int_0^{\tau_0} d\tau_1 \int_0^{\tau_1} d\tau_2 \cdots \int_0^{\tau_{k-3}} P\mathscr{L}_1(\tau_0+\tau)(1-P)$$

$$\cdot \mathscr{L}_1(\tau_1) \cdots (1-P)\mathscr{L}_1(\tau_{k-2})(1-P)\mathscr{L}_1(\tau)e^{-\mathscr{L}_a\tau}P\rho_{eq}(\mathbf{b}),$$

$$k \geq 3 \qquad (3.21'')$$

and by substituting Eq. (3.21) into Eq. (3.18) we get the complete expansion

$$\int_0^t K(t - \tau)\sigma(\tau)\,d\tau = \sum_{r=0}^{\infty} \sum_{k=2}^{\infty} H_r^{(k)} \frac{\partial^r}{\partial t^r}\sigma(\mathbf{a}; t) \qquad (3.22)$$

where

$$H_r^{(k)} = \frac{1}{r!}\frac{d^r}{dz^r}\hat{K}^{(k)}(z)\Big|_{z=0} \qquad (3.23)$$

$\hat{K}^{(k)}(z)$ being the Laplace transform of $K^{(k)}(t)$. Equation (3.15) can thus be written as

$$\frac{\partial}{\partial t}\sigma(\mathbf{a}; t) = \mathscr{D}\sigma(\mathbf{a}; t) + \sum_{r=0}^{\infty} D_r \frac{\partial^r}{\partial t^r}\sigma(\mathbf{a}; t) \qquad (3.24)$$

where

$$\mathscr{D} \equiv \mathscr{L}_a + \rho_{eq}^{-1}(\mathbf{b})\,P\mathscr{L}_1 P\rho_{eq}(\mathbf{b}) \qquad (3.25)$$

$$D_r \equiv \sum_{k=2}^{\infty} H_r^{(k)} \qquad (3.26)$$

$\mathscr{D}$ is a zero-order perturbation order, whereas $H_r^{(k)}$ is $(k + r)$th perturbation order. The lowest order term contributing to D_r is $(r + 2)$th order.

Note that little attention has been given in the literature to the case $\mathscr{L}_a \neq 0$ except in recent papers.[37,38] Let us now consider this case explicitly. First of all, let us rewrite $K^{(k)}$ given by Eqs. (3.21′) and (3.21″) as follows:

$$K^{(2)}(\tau_0) = \rho_{eq}^{-1}(\mathbf{b})\,P\mathscr{L}_1 \exp\{\mathscr{L}_0 \tau_0\}\,Q\mathscr{L}_1 P\rho_{eq}(\mathbf{b}) \qquad (3.27′)$$

$$K^{(k)}(\tau_0) = \rho_{eq}^{-1}(\mathbf{b})\int_0^{\tau_0} d\tau_1 \int_0^{\tau_1} d\tau_2 \cdots \int_0^{\tau_{k-3}} d\tau_{k-2} P\mathscr{L}_1 e^{\mathscr{L}_0(\tau_0 - \tau_1)}Q$$

$$\cdots Qe^{\mathscr{L}_0(\tau_{k-3} - \tau_{k-2})}Q\mathscr{L}_1 e^{\mathscr{L}_0 \tau_{k-2}}Q\mathscr{L}_1 P\rho_{eq}(\mathbf{b}) \qquad (3.27″)$$

This was made possible by using Eqs. (3.9)–(3.11). The main problem we shall have to face is to provide explicit expressions for the commutator

$$\left[e^{\mathscr{L}_0 s}, \mathscr{L}_1\right] \qquad (3.28)$$

In the explicit case considered, this is possible if $\mathscr{L}_a = 0$. In order to take the case $\mathscr{L}_a \neq 0$ into account we shall develop $K^{(k)}$ into power as follows:

$$K^{(k)}(\tau_0) = \sum_{n=0}^{\infty} K^{(k,n)}(\tau_0)$$

$$K^{(2,n)}(\tau_0) \equiv \sum_{n=0}^{\infty} \frac{1}{n!} P_{\text{eq}}^{-1}(\mathbf{b}) P\mathscr{L}_1 e^{\mathscr{L}_b \tau_0} \mathscr{L}_a^n \tau_0^n Q\mathscr{L}_1 P P_{\text{eq}}(\mathbf{b}) \tag{3.29}$$

$$K^{(k,n)}(\tau_0) \equiv P_{\text{eq}}^{-1} \sum_{n_1+n_2+\cdots+n_{k-1}=n} \frac{1}{(n_1!n_2!\cdots n_{k-1}!)}$$

$$\cdot \int_0^{\tau_0} d\tau_1 \int_0^{\tau_1} d\tau_2 \cdots \int_0^{\tau_{k-3}} d\tau_{k-2} \, P\mathscr{L}_1$$

$$\cdot e^{\mathscr{L}_b(\tau_0-\tau_1)} \mathscr{L}_a^{n_1}(\tau_0-\tau_1)^{n_1} Q e^{\mathscr{L}_b(\tau_1-\tau_2)} \mathscr{L}_a^{n_2}(\tau_1-\tau_2)^{n_2} Q \cdots$$

$$\cdot e^{\mathscr{L}_b(\tau_{k-3}-\tau_{k-2})} \mathscr{L}_a^{n_{k-2}}(\tau_{k-3}-\tau_{k-2})^{n_{k-2}} Q e^{\mathscr{L}_b \tau_{k-2}}$$

$$\cdot \mathscr{L}_a^{n_{k-1}} \tau_{k-2}^{n_{k-1}} Q\mathscr{L}_1 P P_{\text{eq}}(\mathbf{b}) \tag{3.30}$$

We are therefore naturally led to define

$$H_r^{(k,n)} \equiv \frac{1}{r!} \frac{d^r}{dz^r} \hat{K}^{(k,n)}(z)\Big|_{z=0} \tag{3.31}$$

and recast Eq. (3.26) as follows:

$$D_r \equiv \sum_{r=2}^{\infty} H_r^{(k,n)}. \tag{3.32}$$

The present version of our expansion is useful only if we can find a criterion for truncating the series of Eqs. (3.24) and (3.32). Throughout this section it is understood that $\mathscr{L}_1$ is small in comparison with $\mathscr{L}_0$, and $\mathscr{L}_a$ with respect to $\mathscr{L}_b$, Eq. (3.32). In Section IV we shall give a more refined definition of this concept of "smallness." Here we limit ourselves to assuming that $\mathscr{L}_1, \mathscr{L}_a, \mathscr{L}_b$ are associated with the time scale parameters $\tau_1, \tau_a,$ and τ_b, respectively. Let $\tau_a, \tau_1 \gg \tau_b$. In Section IV we shall prove that $H_r^{(k,n)}$ is a perturbation term of $(k+r-1)$th order in τ_1 and nth order in τ_a. If τ_a and τ_1 are of the same order, we can conclude that $H_r^{(k,n)}$ is a perturbation term of the $(k+n+r-1)$th order. When this is taken into account properly, Eq. (3.24) can be rewritten as

$$\frac{\partial}{\partial t} \sigma(a;t) = \mathbf{D}\sigma(a;t) \tag{3.33}$$

where **D** is given by (up to the τ_1^4 order)

$$\mathbf{D} = \mathscr{D} + D_0 + D_1\mathscr{D} + D_1D_0 + D_1^2\mathscr{D} + D_2\mathscr{D}^2 + D_2D_0\mathscr{D}$$
$$+ D_2\mathscr{D}D_0 + D_0\mathscr{D}^3 + D_1^2D_0 + D_2D_0^2 + 0\left(\tau_1^5\right). \qquad (3.34)$$

The minimum perturbation order of each term of Eq. (3.34) can be evaluated by summing indices and exponents factor by factor. By substituting the explicit expression for D_i, we finally get the perturbation expansion of Eq. (3.24) at any required order in τ_1 and in τ_a. In Sections IV and V, we shall do this explicitly for some examples of interest.

IV. MORE ON THE ILLUSTRATIVE EXAMPLE OF SECTION II

We are now in a position to answer the question left unsolved in Section II. The systematic expansion of Section III can be used straightforwardly when supplemented by a new ingredient—creation and destruction operators. Let us define

$$a_- \equiv \frac{v}{\left\langle v^2\right\rangle_{\text{eq}}^{1/2}} + \left\langle v^2\right\rangle_{\text{eq}}^{1/2} \frac{d}{dv} \qquad (4.1')$$

$$a_+ \equiv -\left\langle v^2\right\rangle_{\text{eq}}^{1/2} \frac{d}{dv} \qquad (4.1'')$$

These adimensional operators satisfy the commutation rule

$$[a_-, a_+] = \mathbf{I} \qquad (4.1''')$$

where **I** denotes the identity operator.

Let us consider the operator

$$\Gamma_R \equiv -a_+a_- \qquad (4.2)$$

Since

$$a_- e^{-v^2/2\langle v^2\rangle_{\text{eq}}^{1/2}} = 0 \qquad (4.3)$$

the operator Γ_R has an equilibrium solution of canonical form.

When defining the nondimensional variable

$$y \equiv \frac{v}{\langle v^2 \rangle_{eq}^{1/2}} \qquad (4.4)$$

the operator reads

$$\Gamma_R = \frac{d}{dy}y + \frac{d^2}{dy^2} \qquad (4.5)$$

Let us look for the eigenvalues $-\varepsilon_n$ and the corresponding eigenvectors f_n of this operator; that is, let us solve

$$\Gamma_R f_n(y) = -\varepsilon_n f_n(y) \qquad (4.6)$$

It is convenient to assume that the eigenvectors $f_n(y)$ can be written as

$$f_n(y) = c^{-y^2/2} \varphi_n(y) \qquad (4.7)$$

After some algebra we obtain from Eq. (4.6) the equation

$$\varphi_n''(y) - y\varphi_n'(y) + \varepsilon_n \varphi_n(y) = 0. \qquad (4.8)$$

As is well known,[39] Eq. (4.8) results in

$$\varphi_n(y) = He_n(y) \qquad (4.9)$$

$$\varepsilon_n = n \qquad (4.10)$$

The symbol $He_n(y)$ denotes the Hermitian polynomials, which are known[40] to be orthogonal to each other when weighted with the function $\exp[-y^2/2]$, that is,

$$N_n \int_{-\infty}^{+\infty} \varphi_n(y)\varphi_{n'}(y)\exp\left[-\frac{y^2}{2}\right] dy = \delta_{nn'}, \qquad (4.11)$$

with

$$N_n^{-1} = (2\pi)^{1/2} n! \qquad (4.12)$$

Note that the operator Γ_R will usually be involved in actual calculations through expressions such as

$$\int_{-\infty}^{+\infty} g_1(y)[\Gamma_R g_2(y)] \, dy.$$

Integration by parts, supplemented by the assumption that $g_2(y)$ tends to vanish for $|y| \to \infty$, allows us to rewrite this quantity as

$$\int_{-\infty}^{+\infty} g_2(y)(\Gamma_R^+ g_1(y))\, dy,$$

where

$$\Gamma_R^+ \equiv -y\frac{d}{dy} + \frac{d^2}{dy^2}. \tag{4.13}$$

In consequence, the Hermitian polynomials $\mathrm{He}_n(y)$ can also be identified with the eigenstate $\tilde{f}_n(y)$ of Γ_R^+. Note that

$$\Gamma_R^+ \tilde{f}_n(y) = -\varepsilon_n \tilde{f}_n(y) \tag{4.14}$$

has precisely the same analytical form as Eq. (4.8). In other words, the Hermitian polynomials can be identified with the left eigenstates of Γ_R. This naturally leads us to define the following biorthogonal basis set (we return to the original variable v):

$$|n\rangle = \sqrt{N_n}\, e^{-v^2/2\langle v^2\rangle_{\mathrm{eq}}} \mathrm{He}_n\left(\frac{v}{\langle v^2\rangle_{\mathrm{eq}}^{1/2}}\right) \tag{4.15}$$

$$\langle \hat{n}| = \sqrt{N_n}\, \mathrm{He}_n\left(\frac{v}{\langle v^2\rangle_{\mathrm{eq}}^{1/2}}\right) \tag{4.16}$$

the weighting function of the corresponding scalar product now being 1.

Well-known recursion formulas concerning the form $H_n(x)$ of the Hermitian polynomials,[41] after some algebra, lead us to

$$\frac{v}{\langle v^2\rangle_{\mathrm{eq}}}|n\rangle = (n+1)^{1/2}|n+1\rangle + (n)^{1/2}|n-1\rangle \tag{4.17}$$

$$\langle v^2\rangle_{\mathrm{eq}}^{1/2}\frac{d}{dv}|n\rangle = -(n+1)^{1/2}|n+1\rangle \tag{4.18}$$

which can also be written

$$\frac{v}{\langle v^2\rangle_{\mathrm{eq}}^{1/2}} = \sum_{n=0}^{\infty} |n+1\rangle(n+1)^{1/2}\langle \tilde{n}| + \sum_{n=0}^{\infty} |n-1\rangle n^{1/2}\langle \tilde{n}| \tag{4.17'}$$

$$\langle v^2\rangle_{\mathrm{eq}}^{1/2}\frac{d}{dv} = -\sum_{n=0}^{\infty} |n+1\rangle(n+1)^{1/2}\langle \tilde{n}| \tag{4.18'}$$

By substituting Eq. (4.17′) into (4.1′) and Eq. (4.18′) into both (4.1′) and (4.1″), we get

$$a_- = \sum_{n=0}^{\infty} |n-1\rangle n^{1/2}\langle \tilde{n}| \qquad (4.19)$$

$$a_+ = \sum_{n=0}^{\infty} |n+1\rangle (n+1)^{1/2}\langle \tilde{n}| \qquad (4.20)$$

which justify the terms "destruction" and "creation" operator for a_- and a_+, respectively. Equations (4.17′) and (4.18′) can then be written

$$v = \langle v^2\rangle_{eq}^{1/2}(a_+ + a_-) \qquad (4.21)$$

$$\frac{d}{dv} = -\frac{1}{\langle v^2\rangle_{eq}}a_+ \qquad (4.22)$$

When Eqs. (4.21) and (4.22) are used, the interaction operator of Eq. (2.7′) turns out to be

$$\mathscr{L}_1 = \Omega(a_- A_+ - a_+ A_-) \qquad (4.23)$$

provided that the following definitions are accepted:

$$\Omega \equiv \left(\frac{k}{M}\right)^{1/2} \qquad (4.24)$$

$$A_- \equiv \frac{\Delta_0}{\langle \Delta_0^2\rangle_{eq}^{1/2}} + \langle \Delta_0^2\rangle_{eq}^{1/2}\frac{d}{d\Delta_0} \qquad (4.25)$$

$$A_+ \equiv -\langle \Delta_0^2\rangle_{eq}^{1/2}\frac{d}{d\Delta_0} \qquad (4.26)$$

Note that A_- and A_+ share the same properties as the creation and destruction operators a_- and a_+. Proceeding in this way, we can write $\mathscr{L}_b$ of Eq. (2.8) as

$$\mathscr{L}_b = \sum_{i=1}^{\infty} \left(A_+^{(i)}b^{(i)}_- - A_-^{(i)}b^{(i)}_+\right)\Omega_i - \sum_{i=0}^{\infty} \left(A_+^{(i)}b^{(i+1)}_- - A_-^{(i)}b^{(i+1)}_+\right)\omega_{i+1}$$

$$(4.27)$$

where

$$b_+^{(i)} \equiv -\left\langle w_i^2 \right\rangle_{eq}^{1/2} \frac{d}{dw_i} \tag{4.28a}$$

$$b_-^{(i)} \equiv \frac{w_i}{\left\langle w_i^2 \right\rangle_{eq}^{1/2}} + \left\langle w_i^2 \right\rangle_{eq}^{1/2} \frac{d}{dw_i} \tag{4.28b}$$

$$A_+^{(i)} \equiv -\left\langle \Delta_i^2 \right\rangle_{eq}^{1/2} \frac{d}{d\Delta_i} \tag{4.28c}$$

$$A_-^{(i)} \equiv \frac{\Delta_i}{\left\langle \Delta_i^2 \right\rangle_{eq}^{1/2}} + \left\langle \Delta_i^2 \right\rangle_{eq}^{1/2} \frac{d}{d\Delta_i} \tag{4.28d}$$

and $(x_0 \equiv k)$

$$\Omega_i \equiv \left(\frac{\kappa_{i-1}}{m_{i-1}} \right)^{1/2} \tag{4.29a}$$

$$\omega_i \equiv \left(\frac{\kappa_{i-1}}{m_i} \right)^{1/2} \tag{4.29b}$$

Note that the equilibrium distribution of Eq. (2.10) can be written by adopting Fock's notation as

$$\rho_{eq}(\mathbf{b}) = | \overbrace{0,0,0,\dots}^{\Delta_i} ; \overbrace{0,\dots,0}^{w_i} \rangle \tag{4.30}$$

Application of $\mathcal{L}_1$ to $\rho_{eq}(\mathbf{b})$ generates the excited state $|1,0,0,\dots;0,\dots,0\rangle$. The time evolution of this excited state, as driven by $\mathcal{L}_b$, appears to be a linear combination of states $|n_0, n_1, n_2, \dots; \nu_1, \nu_2, \dots\rangle$ satisfying the constraint

$$n_{exc} \equiv \sum_{n=0}^{\infty} n_i + \sum_{i=1}^{\infty} \nu_i = 1 \tag{4.31}$$

This depends on the fact that $\mathcal{L}_b$ as provided by Eq. (4.27) keeps unchanged the excitation number n_{exc}.

We shall apply the systematic approach of Section III to evaluate

$$\mathbf{D} = D_0 + D_1 D_0$$

that is, we are interested in evaluating corrections up to $\bar{\gamma}^{-3}$. A major advantage of using creation and destruction operators is that we are led to a reduced equation with the following structure:

$$\frac{\partial}{\partial t}\sigma(v;t) = \left\{ \alpha a_+ a_- + \beta a_+ a_-^3 + \eta a_+ a_- a_+ a_- + \delta a_+^2 a_-^2 + \varepsilon a_+^3 a_- \right\}\sigma(v;t)$$

(4.32)

This structure is found immediately by noticing that

$$PA_+ \cdots = 0$$
$$\cdots A_- P = 0$$

This prevents A_- from appearing on the extreme right and A_+ on the extreme left of the terms $K^{(k)}(t)$, Eqs. (3.21') and (3.21''), thereby obliging each term of the reduced Fokker-Planck operator to begin acting via an a_- operator and end its action via an a_+ operator. This guarantees that for any perturbation the canonical equilibrium is attained [see Eq. (4.3)].

According to the general rules of Section III, the parameters α, β, η, δ, and ε are defined by

$$\alpha \equiv -\Omega^2 \left[A_-(\tau_0)A_+ \right]^{(0)}$$

(4.32a)

$$\beta \equiv -\Omega^4 \left[A_-(\tau_0)A_+(\tau_1)QA_+(\tau_2)A_+ \right]^{(0)}$$

(4.32b)

$$\eta \equiv \Omega^4 \left\{ \left[A_-(\tau_0)A_+(\tau_1)QA_-(\tau_2)A_+ \right]^{(0)} \right.$$
$$\left. + \left[A_-(\tau_0)A_+ \right]^{(0)} \left[A_-(\tau_0)A_+ \right]^{(1)} \right\}$$

(4.32c)

$$\delta \equiv \Omega^4 \left[A_-(\tau_0)A_-(\tau_1)QA_+(\tau_2)A_+ \right]^{(0)}$$

(4.32d)

$$\varepsilon \equiv -\Omega^4 \left[A_-(\tau_0)A_-(\tau_1)QA_-(\tau_2)A_+ \right]^{(0)}$$

(4.32e)

We have adopted the suggestive definition

$$\left[B(\tau_0)QC(\tau_1)QD(\tau_2)\cdots F(\tau_n)QG \right]^{(m)}$$
$$\equiv \rho_{eq}(\mathbf{b})^{-1}\frac{d^m}{d\tau_0^m}\int_0^\infty d\tau_0 e^{-z\tau_0}\int_0^{\tau_0} d\tau_1 \cdots$$
$$\cdot \int_0^{\tau_{n-1}} d\tau_n\, PB(\tau_0)QC(\tau_1)QD(\tau_2)\cdots F(\tau_n)QGP\rho_{eq}(\mathbf{b})$$

(4.33)

Note that

$$PA_-(\tau_0)P = 0 \qquad (4.34a)$$

$$PA_+P = 0 \qquad (4.34b)$$

These properties are of general validity in that they also hold when the first of the set of Eqs. (2.3) is replaced by

$$\dot{v} = \frac{k}{M}\varphi'(\Delta_0) \qquad (4.35)$$

where φ is a potential not necessarily harmonic in nature. For Eq. (4.34a) to be satisfied, we need only that $\varphi(\Delta_0) = \varphi(-\Delta_0)$. Equations (4.34) allowed us to write β, η, δ, and ε with only one projection operator Q (in the middle). We have, for instance,

$$\left[A_-(\tau_0)QA_+(\tau_1)QA_+(\tau_2)QA_+ \right]^{(0)} = \left[A_-(\tau_0)A_+(\tau_1)QA_+(\tau_2)A_+ \right]^{(0)}$$

$$(4.36)$$

In the linear case we are dealing with,

$$\beta = 0 \qquad (4.37)$$

$$\varepsilon = 0 \qquad (4.38)$$

This is so because, as remarked above, the time evolution $e^{\mathscr{L}_b t}$ leaves unchanged the excitation number n_{exc} defined by Eq. (4.31). In consequence, applying, for instance, A_+ three times, we obtain a combination of states $|n_0, n_1, n_2, \ldots; \nu_1, \nu_2, \ldots\rangle$ with $n_{exc} = 3$. The application of a A_- alone is not enough to reestablish equilibrium. For the same reasons we have

$$QA_-(\tau_2)A_+P = 0 \qquad (4.39)$$

$e^{\mathscr{L}_b \tau_2}A_+P$ is a linear combination of states with excitation number $n_{exc} = 1$. Application of A_- to a linear combination of an infinite number of such states means that $|1, 0, \ldots, 0, 0, \ldots\rangle$ is chosen, thereby implying

$$A_-(\tau_2)A_+P = PA_-(\tau_2)A_+P \qquad (4.40)$$

which in turn leads to Eq. (4.39).

If Eqs. (4.37) and (4.38) are taken into account, Eq. (4.32) reads

$$\frac{\partial}{\partial t}\sigma(v;t) = \{(\alpha+\eta)a_+a_- + (\delta+\eta)a_+^2 a_-^2\}\sigma(v;t) \qquad (4.41)$$

To obtain this equation we also used the commutation rules of Eq. (4.1'''), which means

$$a_+a_-a_+a_- = a_+a_- + a_+^2 a_-^2 \qquad (4.42)$$

We are therefore left with the problem of evaluating α, η, and δ. Note that, due to Eq. (4.39), η is given by

$$\eta = [A_-(\tau_0)A_+]^{(1)}[A_-(\tau_0)A_+]^{(0)} \qquad (4.43)$$

To effect this calculation, we have to introduce a further theoretical ingredient: the normal mode technique. Let us denote by $\{A_j\}$ the set of variables $\{\Delta_0, \Delta_1, \Delta_2,\ldots; w_1, w_2, w_3,\ldots\}$. It is possible to make a unitary transformation to the new set of variables $\{\tilde{A}_j\}$ satisfying the requirement

$$\frac{d}{dt}\tilde{A}_j = \Gamma_b\tilde{A}_j = -\Lambda_j\tilde{A}_j \qquad (4.44)$$

where Γ_b is the operator adjoint of $\mathscr{L}_b$. We can write Δ_0 in terms of the new variables $\tilde{A}_j$ as

$$\Delta_0 = \sum_j U_{0j}\tilde{A}_j \qquad (4.45)$$

More generally,

$$A_j = \sum_{j'} U_{jj'}A_{j'} \qquad (4.46)$$

By using the reverse transformation, we have

$$\tilde{A}_j = \sum_{j'} U_{jj'}^{-1}A_{j'} \qquad (4.47)$$

We also have

$$\frac{\partial}{\partial \Delta_0} = \sum_j \left(\frac{d\tilde{A}_j}{d\Delta_0}\right)\frac{\partial}{\partial\tilde{A}_j} \qquad (4.48)$$

From Eq. (4.47),

$$\frac{d\tilde{A}_j}{d\Delta_0} = U_{j0}^{-1} \tag{4.49}$$

which permits Eq. (4.48) to be written as

$$\frac{\partial}{\partial\Delta_0} = \sum_j U_{j0}^{-1} \frac{\partial}{\partial\tilde{A}_j} \tag{4.50}$$

We can thus write

$$PA_-(\tau_0)A_+P = P\left\{ \frac{\Delta_0}{\langle\Delta_0^2\rangle^{1/2}} + \langle\Delta_0^2\rangle^{1/2}\frac{d}{d\Delta_0} \right\} e^{\mathscr{L}_b\tau_0}\left\{ -\langle\Delta_0^2\rangle^{1/2}\frac{d}{d\Delta_0} \right\}P$$

$$= -P\left\{ \Delta_0 e^{\mathscr{L}_b\tau_0}\frac{\partial}{\partial\Delta_0} \right\}P = -P\left\{ \left(e^{\Gamma_b\tau_0}\Delta_0\right)\frac{\partial}{\partial\Delta_0} \right\}P \tag{4.51}$$

By using Eqs. (4.45), (4.44), and (4.50) we also have

$$PA_-(\tau_0)A_+P = -P\left\{ \left(e^{\Gamma_b\tau_0}\Delta_0\right)\frac{\partial}{\partial\Delta_0} \right\}P$$

$$= P\sum U_{0j}U_{j0}^{-1}e^{-\Lambda_j\tau_0}P \tag{4.52}$$

As usual, we integrate by parts. Replacing Eq. (4.52) into Eq. (4.32a), we get

$$\alpha = -\Omega^2\sum_j \frac{U_{0j}U_{j0}^{-1}}{\Lambda_j} \tag{4.53}$$

From Eq. (4.43) we immediately derive

$$\eta = -\Omega^4\sum_{jj'} \frac{U_{0j}U_{j0}^{-1}U_{0j'}U_{j'0}^{-1}}{\Lambda_{j'}^2\Lambda_j} \tag{4.54}$$

We believe that it is useful to show the details of the calculations leading to the value of δ. First of all, we note that

$$PA_-(\tau_0)A_-(\tau_1)QA_+(\tau_2)A_+P = PA_-e^{\mathscr{L}_b(\tau_0-\tau_1)}A_-e^{\mathscr{L}_b(\tau_1-\tau_2)}A_+e^{\mathscr{L}_b\tau_2}A_+P \tag{4.55}$$

Substituting the explicit expressions of Eqs. (4.25) and (4.26) into Eq. (4.55), we get

$$PA_-(\tau_0)A_-(\tau_1)QA_+(\tau_2)A_+P$$

$$= P\frac{\Delta_0}{\left\langle\Delta_0^2\right\rangle_{eq}^{1/2}}e^{\mathscr{L}_b(\tau_0-\tau_1)}\left\{\frac{\Delta_0}{\left\langle\Delta_0^2\right\rangle_{eq}^{1/2}}+\left\langle\Delta_0^2\right\rangle_{eq}^{1/2}\frac{d}{d\Delta_0}\right\}e^{\mathscr{L}_b(\tau_1-\tau_2)}$$

$$\cdot\left\langle\Delta_0^2\right\rangle_{eq}\frac{d}{d\Delta_0}e^{\mathscr{L}_b\tau_2}\frac{d}{d\Delta_0}P \tag{4.56}$$

Use was made of the property $P(\partial/\partial\Delta_0)\cdots=0$ [Eq. (2.19)].

To calculate explicitly the term on the right-hand side of Eq. (4.56), we need to introduce a further property. First of all, note that

$$e^{\mathscr{L}_b t}\frac{d}{d\Delta_0}e^{-\mathscr{L}_b t}=e^{\mathscr{L}_b^x t}\frac{d}{d\Delta_0} \tag{4.57}$$

where

$$\mathscr{L}_b^x\frac{d}{d\Delta_0}\equiv\mathscr{L}_b\frac{d}{d\Delta_0}-\frac{d}{d\Delta_0}\mathscr{L}_b \tag{4.58}$$

This general relationship can be proved by expanding into a Taylor series both sides of Eq. (4.57).[42] Using Eq. (4.50) and noticing that

$$\mathscr{L}_b^x\frac{d}{dA_j}=\Lambda_j\left\{\frac{d}{dA_j}A_j\frac{d}{dA_j}-\frac{d}{dA_j}\frac{d}{dA_j}A_j\right\}$$

$$=\Lambda_j\left\{\frac{d}{dA_j}A_j\frac{d}{dA_j}-\frac{d}{dA_j}A_j\frac{d}{dA_j}-\frac{d}{dA_j}\right\}$$

$$=-\Lambda_j\frac{d}{dA_j} \tag{4.59}$$

from Eq. (4.57) [by expanding its right-hand side into a Taylor series and applying Eq. (4.59)], we get

$$e^{\mathscr{L}_b t}\frac{d}{d\Delta_0}e^{-\mathscr{L}_b t}=\sum_j e^{-\Lambda_j t}U_{j0}^{-1}\frac{d}{d\tilde{A}_j} \tag{4.60}$$

This is equivalent to [let us multiply both sides of Eq. (4.60) by $e^{\mathscr{L}_b t}$]

$$e^{\mathscr{L}_b t}\frac{d}{d\Delta_0} = \sum_j e^{-\Lambda_j t}U_{j0}^{-1}\frac{\partial}{\partial\tilde{A}_j}e^{\mathscr{L}_b t} \tag{4.61}$$

By applying this property two times to Eq. (4.56) we can shift toward the right $\exp\{\mathscr{L}_b(\tau_1 - \tau_2)\}$ and $\exp\{\mathscr{L}_b\tau^2\}$ so as to make them disappear through application of Eq. (3.11). Thus we obtain

$$PA_-(\tau_0)A_-(\tau_1)QA_+(\tau_2)A_+P$$

$$= P\frac{\Delta_0}{\langle\Delta_0^2\rangle_{eq}^{1/2}}e^{\mathscr{L}_b(\tau_0-\tau_1)}\left\{\frac{\Delta_0}{\langle\Delta_0^2\rangle_{eq}^{1/2}} + \langle\Delta_0^2\rangle^{1/2}\frac{d}{d\Delta_0}\right\}\langle\Delta_0^2\rangle_{eq}$$

$$\cdot\langle\Delta_0^2\rangle_{eq}\sum_{jj'}U_{j0}^{-1}U_{j'0}^{-1}\frac{\partial}{\partial\tilde{A}_j}\frac{\partial}{\partial\tilde{A}_{j'}}e^{-\Lambda_j(\tau_1-\tau_2)}e^{-\Lambda_{j'}\tau_1}P$$

$$= P\sum_{\substack{ii'\\jj'}}U_{0i}e^{-\Lambda_i(\tau_0-\tau_1)}\tilde{A}_iU_{0i'}\tilde{A}_{i'}U_{j0}^{-1}U_{j'0}^{-1}e^{-\Lambda_j(\tau_1-\tau_2)}e^{-\Lambda_{j'}\tau_1}\frac{\partial}{\partial\tilde{A}_j}\frac{\partial}{\partial\tilde{A}_{j'}}P$$

$$= -\sum_{i'jj'}U_{0j}e^{-\Lambda_j(\tau_0-\tau_1)}U_{0i'}\tilde{A}_{i'}U_{j0}^{-1}U_{j'0}^{-1}e^{-\Lambda_j(\tau_1-\tau_2)}e^{-\Lambda_{j'}\tau_1}\frac{\partial}{\partial\tilde{A}_{j'}}P$$

$$\quad -\sum_{ijj'}U_{0i}e^{-\Lambda_i(\tau_0-\tau_1)}\tilde{A}_iU_{0j}U_{j0}^{-1}U_{j'0}\frac{\partial}{\partial\tilde{A}_{j'}}e^{-\Lambda_j(\tau_1-\tau_2)}e^{-\Lambda_{j'}\tau_1}P$$

$$= P\sum_{jj'}U_{0j}U_{j0}^{-1}U_{0j'}U_{j'0}^{-1}e^{-\Lambda_j(\tau_0-\tau_1)}e^{-\Lambda_j(\tau_1-\tau_2)}e^{-\Lambda_{j'}\tau_1}P$$

$$\quad + P\sum_{jj'}U_{0j}U_{j0}^{-1}U_{0j'}U_{j'0}e^{-\Lambda_{j'}(\tau_0-\tau_1)}e^{-\Lambda_j(\tau_1-\tau_2)}e^{-\Lambda_{j'}\tau_1}P \tag{4.62}$$

To get Eq. (4.62), we made $\exp\{\mathscr{L}_b(\tau_0 - \tau_1)\}$ disappear by applying this operator to its left side and using the property

$$e^{\mathscr{L}_b^+(\tau_0-\tau_1)}\Delta_0 = \sum_j U_{0j}\tilde{A}_je^{-\Lambda_j(\tau_0-\tau_1)} \tag{4.45'}$$

which, in turn, immediately results from Eqs. (4.44) and (4.45). To get the actual value of δ, we need to substitute Eq. (4.62) into Eq. (4.32d) (and make the corresponding time integration). To get the final result we have to exploit another relationship:

$$\int_0^\infty d\tau_0 \int_0^{\tau_1} d\tau_2 \int_0^{\tau_2} d\tau_3 \cdots \int_0^{\tau_{n-1}} d\tau_n e^{-(\lambda_0 \tau_0 + \lambda_1 \tau_1 + \lambda_2 \tau_2 + \cdots \lambda_n \tau_n)}$$

$$= [\lambda_0(\lambda_0 + \lambda_1) \cdots (\lambda_0 + \lambda_1 + \cdots + \lambda_n)]^{-1}. \tag{4.63}$$

This useful formula can be proved by an iterative procedure. By using Eqs. (4.63), (4.32d), and (4.62), we derive

$$\delta = \Omega^4 \sum_{jj'} U_{0j} U_{j0}^{-1} U_{0j'} U_{j'0}^{-1} \left\{ \frac{1}{\Lambda_j (\Lambda_j + \Lambda_{j'}) \Lambda_{j'}} + \frac{1}{\Lambda_{j'}^2 (\Lambda_j + \Lambda_{j'})} \right\}$$

$$= \Omega^4 \sum_{jj'} U_{0j} U_{j0}^{-1} U_{0j'} U_{j'0}^{-1} \frac{1}{\Lambda_{j'}^2 \Lambda_j} \tag{4.64}$$

We note that $\eta = -\delta$. From Eq. (4.41) we derive, therefore,

$$\frac{\partial}{\partial t} \sigma(v; t) = (\alpha + \eta) a_+ a_- \sigma(v; t) \tag{4.65}$$

We have proved, therefore, that the effect of higher-order corrections is to renormalize the friction coefficient, while the standard Fokker-Planck form is left completely unchanged. In the linear case this is, therefore, an exact equation. It will be shown later on (Ferrario et al., Chapter VI) that nonlinearity destroys this standard form. This will lead us to the following conclusions:

1. The Fokker-Planck equation is based on a clear-cut time-scale separation between the variable of interest and its heat bath. In the linear case this is an exact equation even when finite correlation times are considered.
2. In the nonlinear case the standard Fokker-Planck equation can be regarded as a good approximation only when for a given nonlinearity the heat bath relaxation time is small enough to make the effects of higher-order corrections negligible.

V. APPLICATIONS TO EFFECTIVE LIOUVILLIANS

This section is devoted to illustrating the AEP of Section III at work in the case of effective Liouville operators. As pointed out in Section IV, as well as in Chapter I, these effective Liouvillians are constructed for the purpose of simulating real systems.

As an explanatory example, we shall refer ourselves to a model of activation in a chemical reaction process. A fairly satisfactory form is

$$\dot{x} = v$$

$$\dot{v} = -\frac{\partial V}{\partial x} - \psi(x)\xi - \int_0^t \varphi(t-\tau)v(\tau)\,d\tau + \chi(t) \tag{5.1}$$

$$\dot{\xi} = -\lambda\xi(t) + F(t)$$

where x is the reaction coordinate. If no external potential were present, the velocity v would be driven by the generalized Langevin equation

$$\dot{v} = -\int_0^t \varphi(t-\tau)v(\tau)\,d\tau + \chi(t) \tag{5.2}$$

where the stochastic force $\chi(t)$ and the memory kernel $\varphi(t)$ are related to each other via the generalized fluctuation-dissipation relationship

$$\varphi(t) = \frac{\langle \chi(t)\chi(0)\rangle_{eq}}{\langle v^2 \rangle_{eq}} \tag{5.3}$$

It is well known (see, for instance, Chapter I) that this is a satisfying description of the influence of the molecules of a liquid sample on a tagged molecule. We believe, therefore, that this may be a realistic description of a chemical reaction taking place in the liquid state.

The variable ξ can be interpreted as being a chaotic radiation field with a bandwidth λ. This radiation field can also be supplemented by a proper frequency (see, e.g., ref. 43). This involves an additional "virtual" variable without implying a conceptual improvement of the problem under study.

A degree of simplification can be attained by assuming

$$\varphi(t) = 2\gamma\delta(t) \tag{5.4}$$

This allows us to replace Eq. (5.1) with

$$\dot{x} = v$$

$$\dot{v} = -\frac{\partial V(x)}{\partial x} - \psi(x)\xi - \gamma v + \chi(t) \tag{5.5}$$

$$\dot{\xi} = -\lambda\xi(t) + F(t)$$

If we also assume γ to be extremely large, then

$$\dot{x} = -\frac{1}{\gamma}\frac{\partial V(x)}{\partial x} - \frac{\psi(x)\xi}{\gamma} + \frac{\chi(t)}{\gamma}$$

$$\dot{\xi} = -\lambda\xi(t) + F(t)$$

$$(5.6)$$

If we also assume that the temperature of the thermal source providing the fluctuation $\chi(t)$ is vanishingly small, we recover

$$\dot{x} = f(x) + g(x)\xi$$

$$\dot{\xi} = -\lambda\xi(t) + F(t)$$

$$(5.7)$$

which is the stochastic multiplicative process studied in depth by the Barcelona group.[16-18,26]

A. One-Dimensional Stochastic Differential Equation of the Langevin Type

The phenomenological Langevin equations studied in this section are not considered legitimate by the RMT. This will be discussed further by Faetti et al. in Chapter X. However, this kind of equation has been studied exhaustively. It is useful, therefore, to show the AEP in this case so as to compare our approach with the different ones currently being applied by other groups.

The system described by Eqs. (5.7) is equivalent to the stochastic equation in the variable x alone,

$$\dot{x} = f(x) + g(x)\xi$$

$$(5.8)$$

where $\xi(t)$ is a Gaussian noise with

$$\langle \xi(t) \rangle = 0$$

$$\langle \xi(t)\xi(s) \rangle = D\lambda \ \exp[-\lambda|t-s|]$$

$$(5.9)$$

provided[16]: (i) $F(t)$ is a white Gaussian noise with zero mean value and

$$\langle F(t)F(0) \rangle = 2D\lambda^2\delta(t)$$

and (ii) a fluctuation-dissipation relationship for the auxiliary variable ξ is understood and it is initially prepared at its Gaussian equilibrium with

$$\langle \xi^2(0) \rangle = \langle \xi^2 \rangle_{eq} \equiv \lambda D$$

$$(5.10)$$

In this case the terms $\mathscr{L}_a$, $\mathscr{L}_b$, and $\mathscr{L}_1$ of the dynamical operator $\mathscr{L}$ of Eqs. (3.3) and (3.4) read

$$\mathscr{L}_a = -\frac{\partial}{\partial x}f(x)$$

$$\mathscr{L}_b = \lambda\left\{\frac{\partial}{\partial\xi}\xi + \langle\xi^2\rangle_{eq}\frac{\partial^2}{\partial\xi^2}\right\} \qquad (5.11)$$

$$\mathscr{L}_1 = -\frac{\partial}{\partial x}g(x)\xi$$

We have therefore to face a specific case where $\mathscr{L}_a \neq 0$. This model belongs to a class of problems studied by means of an ordered-cumulant technique by Garrido and Sancho.[26] Their results provide a reliable check on the present approach.

Following the notation of Section III we can identify ξ as a fast-relaxing variable and, by solving Eq. (3.7), determine:

$$\rho_{eq}(\xi) = \mathscr{N}\exp\left(-\frac{\xi^2}{2\langle\xi^2\rangle_{eq}}\right) \qquad (5.12)$$

where $\mathscr{N}$ is a normalization constant. Let us adopt the preparation condition of Eq. (3.8), which allows us to neglect the inhomogeneous term in the Nakajima-Zwanzig expansion, Eq. (3.5). In Section VII we shall discuss how it can affect the explicit form of the reduced Fokker-Planck equation, in spite of the fact that the fate of the inhomogeneous term is to rapidly become vanishingly small.

When truncating the expansion of Eqs. (3.33) and (3.34) at the third order, we find

$$\mathbf{D} = D_0 + D_1 + D_2 D_0 + D_2 D_1 + D_3 D_0^2 \qquad (5.13)$$

where $D_0 = \mathscr{L}_a$. By expanding the D's of Eq. (5.13) [see Eq. (3.32)] and reordering the terms order by order up to order 3, we obtain ($\tau \equiv \lambda^{-1}$)

$$\mathbf{D} = D(\tau^0) + D(\tau^1) + D(\tau^2) + D(\tau^3) \qquad (5.14)$$

with

$$\begin{aligned}
D(\tau^0) &= \mathscr{L}_a \\
D(\tau^1) &= H_1^{(0,0)} \\
D(\tau^2) &= H_1^{(0,1)} + H_2^{(0,0)}\mathscr{L}_a \\
D(\tau^3) &= H_1^{(0,2)} + H_1^{(2,0)} + H_2^{(0,0)}H_1^{(0,0)} + H_2^{(0,1)}\mathscr{L}_a + H_3^{(0,0)}\mathscr{L}_a^2
\end{aligned} \qquad (5.15)$$

The basic rules of Section IV help us in explicitly calculating these operators; our final result is

$$D(\tau^0) = -\frac{\partial}{\partial x}f(x)$$

$$D(\tau^1) = \tau\langle\xi^2\rangle_{eq}\frac{\partial}{\partial x}g(x)\frac{\partial}{\partial x}g(x)$$

$$D(\tau^2) = -\tau^2\langle\xi^2\rangle_{eq}\frac{\partial}{\partial x}g(x)\frac{\partial}{\partial x}M(x)$$

$$D(\tau^3) = \tau^3\langle\xi^2\rangle_{eq}\frac{\partial}{\partial x}g(x)\frac{\partial}{\partial x}[M'(x)f(x) - M(x)f'(x)]$$

$$(5.16)$$

where

$$M(x) = f(x)g'(x) - f'(x)g(x) \qquad (5.17)$$

and the prime denotes the first derivative with respect to x.

The expansion of Eqs. (5.16) agrees exactly with the result of ref. 26. A comparison between the two approaches at higher order is quite cumbersome, but no discrepancy is expected.

We can now make some concluding remarks:

1. As pointed out in Section III we have to give the conditions of applicability of our AEP to the case under study. The first condition should be $\tau \ll \tau_f$, where the "mechanical" time scale τ_f is of the order $x/f(x)$. This constraint, however, can be bypassed, by transferring to the interaction picture, that is, by taking the integral of motion of (5.8) as a new variable. In the present form we must assume $\tau \ll \tau_f$. The second condition is essential: $\tau \ll \tau_g$, where τ_g is a time scale somehow related to $g(x)$ and of the order $x/g(x)\langle\xi^2\rangle_{eq}^{-1/2}$. If for simplicity we choose $g(x) = x$, then $\tau_g \cong \langle\xi^2\rangle_{eq}^{-1/2}$. When these conditions are applied to the system of Eq. (5.7) with $\langle\xi^2\rangle_{eq}$ fixed, Eq. (5.16) becomes an expansion with perturbation parameter τ.

Equations (5.7) were introduced so as to treat the non-Markovian process of Eq. (5.8) in the frame of the time-independent Fokker-Planck formalism. The equivalence has been shown to require that the fluctuation-dissipation relationship (5.10) holds: the white noise limit can then be recovered by making τ vanish for a fixed value of D. If we substitute Eq. (5.10) into Eqs. (5.16), we realize that $D(\tau^3)$ is actually a second-order term with respect to the autocorrelation time τ. Further terms proportional to $D^2\tau^2$ come from higher-order terms of our expansion, Eqs. (3.33) and (3.34). In contrast, Eqs. (5.16) contain *all* the terms in $D\tau$.

2.   Equations (5.16) account for non-Markovian corrections up to the first order and can be rewritten as

$$\frac{\partial}{\partial t}\sigma(x;t) = -\frac{\partial}{\partial x}f(x)\sigma(x;t) + D\frac{\partial}{\partial x}g(x)\frac{\partial}{\partial x}[g(x) - \tau M(x)]\sigma(x;t)$$

(5.18)

As pointed out by the authors of refs. 16–18 and 26 for the class of models discussed in this section, the first-order correction to the white noise only amounts to a renormalization of the diffusion term. In ref. 26 it has been shown explicitly that such a property breaks down only when $D(\tau^4)$ terms are added. In the following (see Ferrario et al., Chapter VI) we shall introduce systems where this behavior is broken even at the lower orders.

3.   By exploiting the diffusional form of Eq. (5.18), which allows an analytical expression for $\sigma_{st}(x) = \sigma(x; t = \infty)$, or by applying a numerical algorithm based on a continued fraction expansion (Grosso and Pastori Parravicini, Chapter III), we can compute the correlation function $\langle x(t)x(0)\rangle_0$. Here $\langle \cdots \rangle_0$ means an average over the equilibrium realizations accounted by $\sigma_{st}(x)$. In this case the assumption we made to get rid of the inhomogeneous term of Eq. (3.5) should now read:

$$\rho(x,\xi;0) = \sigma_{st}(x)\rho_{eq}(\xi)$$

(5.19)

This corresponds to a physical situation when two insulated systems at the same temperature are put in contact with each other. The factorization on the right-hand side of Eq. (5.19) is generally valid when the interaction between x and ξ is not taken into account in the initial conditions. In Section VII, however, we shall show that the choice of projection operator corresponding to different preparations should not affect the final reduced equation of motion if one can exactly resummate the corresponding perturbative expansions.

B.   Two-Dimensional Stochastic Differential Equations of the Langevin Type

We shall explore the case where $\gamma < \infty$ and the temperature of the thermal source is not zero. The corresponding SDE of the Langevin type, Eq. (5.5), can be rewritten as

$$\begin{aligned}
\dot{x} &= v \\
\dot{v} &= f(x) - \gamma v + g(x)\xi + \chi(t) \\
\dot{\xi} &= -\lambda\xi + \eta(t)
\end{aligned}$$

(5.20)

where the auxiliary variable ξ accounts for the finite autocorrelation time τ $\equiv \lambda^{-1}$ of the external noise (see Section V.A) and $f(x) = -\partial V(x)/\partial x$. $\chi(t)$ is a white Gaussian noise with vanishing mean value, and

$$\langle \chi(t)\chi(0) \rangle = 2\gamma \langle v^2 \rangle_{eq} \delta(t) \tag{5.21}$$

the related fluctuation-dissipation relationship. The system of Eq. (5.20) can depict a Brownian particle bounded in an external potential and coupled to a heat bath $\chi(t)$ in the presence of an external noisy source.[43] For that reason $\chi(t)$ and $\xi(t)$ are supposed to be statistically independent.

In this problem, in addition to λ^{-1}, discussed in Section IV, we have another time scale, γ^{-1}. Let us assume that our experimental apparatus only allows us to observe long-time regions corresponding to $t \sim \gamma \tau_f^2$, where τ_f is the mechanical time scale introduced in provision (i) of Section V.A. In consequence, the dynamics induced by the interplay of inertia and external noise belongs to the short-time region. In such a limit $\tau, \gamma^{-1} \ll \tau_f$, both ξ and v play the role of fast-relaxing variable while x is our variable of interest. By applying our AEP, we then recover the corrections to the diffusional approximation due to the non-white external noise.

In the notation of Section V.A, the Fokker-Planck operator driving the the probability function $p(x, v, \xi; t)$ can be split into the sum of an unperturbed part

$$\mathcal{L}_0 = \lambda \left\{ \frac{\partial}{\partial \xi}\xi + \langle \xi^2 \rangle_{eq} \frac{\partial^2}{\partial \xi^2} \right\} + \gamma \left\{ \frac{\partial}{\partial v}v + \langle v^2 \rangle_{eq} \frac{\partial^2}{\partial v^2} \right\} \tag{5.22}$$

and a perturbation

$$\mathcal{L}_1 = v \frac{\partial}{\partial x} - [f(x) + g(x)\xi] \frac{\partial}{\partial v} \tag{5.22'}$$

In the present case $\mathcal{L}_a = 0$. Equation (3.7) can be easily solved:

$$\rho_{eq}(x, v) = \mathcal{N}_b \exp\left(-\frac{v^2}{2\langle v^2 \rangle_{eq}} \right) \exp\left(-\frac{\xi^2}{2\langle \xi^2 \rangle_{eq}} \right) \equiv \rho_{eq}(v)\rho_{eq}(\xi) \tag{5.23}$$

where $\mathcal{N}_b$ is the normalization constant. According to Eqs. (3.6) and (3.8), we can choose a projection operator P and initial conditions

$$\rho(x, v, \xi; 0) = \sigma(x; 0)\rho_{eq}(x, v) \tag{5.24}$$

such that the inhomogeneous term of the expansion (3.5) turns out to be identically zero.

By expressing v, d/dv, ξ, and $d/d\xi$ is terms of creation and destruction operators, Eqs. (4.21) and (4.22), we can give $\mathscr{L}_1$ the following form:

$$\mathscr{L}_1 = -\frac{a_+}{\langle v^2 \rangle_{eq}} j(x) - \langle v^2 \rangle_{eq} a_- \frac{\partial}{\partial x} + \frac{g(x)}{\langle v^2 \rangle_{eq}} \left(\langle \xi^2 \rangle q_- + q_+ \right) a_+ \tag{5.25}$$

where

$$j(x) = \langle v^2 \rangle_{eq} \frac{\partial}{\partial x} - f(x) \tag{5.26}$$

We shall use the basis set of the direct product $\{ f_n(v) \} \otimes \{ f_n(\xi) \}$:

$$\langle \tilde{n}_1 \tilde{n}_2 | = \tilde{f}_{n_1}(v) \tilde{f}_{n_2}(\xi)$$
$$|n_1 n_2 \rangle = f_{n_1}(v) f_{n_2}(\xi) \tag{5.27}$$

where $\langle \tilde{n}_1 \tilde{n}_2 | m_1 m_2 \rangle = \delta_{n_1 m_1} \delta_{n_2 m_2}$ and f_n are the Hermitian polynomials defined in Section IV, and the property

$$\mathscr{L}_0 | n_1 n_2 \rangle = -(n_1 \gamma + n_2 \lambda) | n_1 n_2 \rangle \tag{5.28}$$

The projection operator is then defined by $P\rho(x, v, \xi; t) = |00\rangle \sigma(x; t)$ and can be expressed as $P = |00\rangle \langle \tilde{0}\tilde{0}|$.

By following this way, after a tedious calculation, we evaluate $\mathbf{D}$ up to the fifth order:

$$\mathbf{D} = D_0 + D_1 D_0 + D_1^2 D_0 + D_2 D_0^2 \tag{5.29}$$

with

$$D_0 = H_0^{(2)} + H_0^{(4)}$$
$$D_1 = H_1^{(2)} \tag{5.30}$$
$$D_2 = H_2^{(2)}$$

since $\mathcal{D}$ and the terms $H_r^{(2s+1)}$ vanish. Our final result reads: [44]

$$
\frac{\partial}{\partial t}\sigma(x;t) = \frac{1}{\gamma}\frac{\partial}{\partial x}j\sigma(x;t) - \frac{1}{\gamma^3}\frac{\partial}{\partial x}j\frac{\partial}{\partial x}j\sigma(x;t) + \frac{1}{\gamma^3}\frac{\partial^2}{\partial x^2}j^2\sigma(x;t)
$$

$$
+ \left\langle \xi^2 \right\rangle_{eq}\left\{ \frac{\partial^2}{\partial x^2}g^2\frac{1}{\gamma^2(\gamma+\lambda)} + \frac{\partial}{\partial x}g\frac{\partial}{\partial x}g\frac{1}{\lambda(\gamma+\lambda)\gamma} \right\}\sigma(x;t)
$$

$$
+ \left\{ \frac{1}{2\gamma^5}\frac{\partial^3}{\partial x^3}j^3 + \frac{1}{\gamma^5}\frac{\partial^2}{\partial x^2}j\frac{\partial}{\partial x}j^2 - \frac{5}{2}\frac{1}{\gamma^5}\frac{\partial^2}{\partial x^2}j^2\frac{\partial}{\partial x}j \right.
$$

$$
\left. - \frac{1}{\gamma^5 x}\frac{\partial}{\partial x}j\frac{\partial^2}{\partial x^2}j^2 + \frac{2}{\gamma^5}\frac{\partial}{\partial x}j\frac{\partial}{\partial x}j\frac{\partial}{\partial x}j \right\}\sigma(x;t)
$$

$$
+ \left\langle \xi^2 \right\rangle_{eq}\left\{ \frac{\partial^3}{\partial x^3}jg^2\frac{1}{2\gamma^4(\gamma+\lambda)} + \frac{\partial^3}{\partial x^3}g^2j\frac{1}{\gamma^4(2\gamma+\lambda)} \right.
$$

$$
+ \frac{\partial^3}{\partial x^3}gjg\frac{1}{\gamma^3(2\gamma+\lambda)(\gamma+\lambda)} + \frac{\partial^2}{\partial x^2}j\frac{\partial}{\partial x}g^2\frac{1}{\gamma^4(\gamma+\lambda)}
$$

$$
+ \frac{\partial^2}{\partial x^2}jg\frac{\partial}{\partial x}g\frac{1}{\gamma^3\lambda(\gamma+\lambda)} + 2\frac{\partial^2}{\partial x^2}g\frac{\partial}{\partial x}gj\frac{1}{\gamma^3(\gamma+\lambda)(2\gamma+\lambda)}
$$

$$
+ 2\frac{\partial^2}{\partial x^2}g\frac{\partial}{\partial x}jg\frac{1}{\gamma^2(\gamma+\lambda)^2(2\gamma+\lambda)} + \frac{\partial^2}{\partial x^2}gj\frac{\partial}{\partial x}g\frac{1}{\gamma^2\lambda(\gamma+\lambda)^2}
$$

$$
+ 2\frac{\partial}{\partial x}g\frac{\partial^2}{\partial x^2}gj\frac{1}{\gamma^2\lambda(2\gamma+\lambda)(\gamma+\lambda)}
$$

$$
+ 2\frac{\partial}{\partial x}g\frac{\partial^2}{\partial x^2}jg\frac{1}{\gamma(2\gamma+\lambda)2(\gamma+\lambda)^2} + \frac{\partial}{\partial x}g\frac{\partial}{\partial x}j\frac{\partial}{\partial x}\frac{1}{\gamma\lambda^2(\gamma+\lambda)^2}
$$

$$
- \frac{\partial}{\partial x}j\frac{\partial^2}{\partial x^2}g^2\frac{1}{\gamma^4(\gamma+\lambda)} - \frac{\partial}{\partial x}j\frac{\partial}{\partial x}g\frac{\partial}{\partial x}g\frac{1}{\gamma^3\lambda(\gamma+\lambda)}
$$

$$
- \frac{\partial}{\partial x}g\frac{\partial}{\partial x}g\frac{\partial}{\partial x}j\left(\frac{1}{\gamma^3\lambda^2} + \frac{1}{\gamma^2\lambda(\gamma+\lambda)^2} \right)
$$

$$
- \frac{\partial^2}{\partial x^2}g^2\frac{\partial}{\partial x}j\left(\frac{3}{2\gamma^4(\gamma+\lambda)} + \frac{1}{\gamma^3(\gamma+\lambda)^2} \right)\right\}\sigma(x;t) \qquad (5.31)
$$

This expansion exhibits several interesting properties. Some of them will be the subject of a more detailed investigation elsewhere in the present volume (Fonseca et al., Chapter IX). Here we limit ourselves to the following remarks:

(i)   If we put $g(x) = 0$, Eqs. (5.20) represent the well-known Kramers problem of a Brownian particle in an external potential $V(x)$. Equation (5.31) provides the corrections to the Smoluchowski equation evaluated by other authors.[15,20]

(ii)   In case (i) each contribution to **D** ends with $j(x)$. This means that the corresponding equilibrium distribution $p_{eq}(x)$ obeys the equation $j(x)p_{eq}(x) = 0$; that is,

$$\sigma_{eq}(x) = \mathcal{N}_x \exp\left(-\frac{V(x)}{\langle v^2 \rangle_{eq}}\right) \tag{5.32}$$

where $\mathcal{N}_x$ is the normalization constant. Furthermore, Eqs. (5.22) and (5.22′) guarantee that

$$\rho_{eq}(xv) = \sigma_{eq}(x)\rho_{eq}(v) \tag{5.33}$$

is the true equilibrium distribution. In the general case, when $g(x) \neq 0$, the stationary equilibrium distribution loses the canonical form. Faetti et al. (Chapter X) will show that this case means a continuous flux of energy into the system, which is therefore kept far from the canonical equilibrium distribution.

VI.   ROTATIONAL BROWNIAN MOTION

A further interesting application of the AEP is to obtain a diffusion-like equation from a more complete equation of motion—one that involves the molecular angular velocity. If the molecular principal axis system is chosen to describe the molecular motion, we obtain

$$\dot{\theta}_i = \omega_i$$

$$\dot{\omega}_i = r_i \omega_j \omega_k - \gamma_i \omega_i + f_i(t) - \frac{1}{I_i}\frac{\partial V}{\partial \theta_i}(\Omega) \tag{6.1}$$

where

$$r_i = \frac{I_j - I_k}{I_i} \tag{6.2}$$

i, j, k denotes a circular permutation; I_i, I_j, I_k are the molecular moments of inertia; and V is an external potential depending on molecular orientation and $f_i(t)$ white stochastic forces defined by

$$\langle f_i(0) f_j(t) \rangle = 2 D_i \delta_{ij} \delta(t)$$

Using the algebra of rotation generators, that is,

$$iJ_i \equiv \frac{\partial}{\partial \theta_i} \tag{6.3}$$

the Fokker-Planck operator of the system under study

$$\frac{\partial}{\partial t} \rho(\omega, \Omega; t) = \mathscr{L} \rho(\omega, \Omega; t) \tag{6.4}$$

is driven by a dynamical operator $\mathscr{L}$ defined as follows:

$$\mathscr{L} \equiv -i \omega_i J_i - \frac{\partial}{\partial \omega_i} (r_i \omega_j \omega_k - \gamma_i \omega_i) + i \frac{1}{I_i} \frac{\partial}{\partial \omega_i} (J_i V) + D_i \frac{\partial^2}{\partial \omega_i^2} \tag{6.5}$$

Note that the unperturbed part $\mathscr{L}_b$ in this case should be defined as follows:

$$\mathscr{L}_b \equiv -\frac{\partial}{\partial \omega_i} (r_i \omega_j \omega_k - \gamma_i \omega_i) + D_i \frac{\partial^2}{\partial \omega_i^2} \tag{6.6}$$

which has the equilibrium distribution

$$\rho_{eq}(b) \equiv \exp\left(-\sum_i \frac{I_i \omega_i^2}{2kT} \right). \tag{6.7}$$

$\mathcal{L}_b$ of Eq. (6.6), unfortunately, is nonlinear. However, the linear approximation

$$\mathcal{L}_b \equiv \frac{\partial}{\partial \omega_i} \omega_i + D_i \frac{\partial^2}{\partial \omega_i^2} \tag{6.8}$$

also has $\rho_{eq}(\mathbf{b})$ as its equilibrium distribution. This allows us to use as thermal bath operator $\mathcal{L}_b$ of Eq. (6.8) without contravening canonical constraints. (In general, to replace the true heat bath Hamiltonian with a part of it has the bad effect of destroying the canonical nature of the equilibrium distribution—see Ferrario et al., Chapter VI).

Therefore we are allowed to adopt the definition

$$\mathcal{L}_0 \equiv \gamma_i \frac{\partial}{\partial \omega_i} \omega_i + D_i \frac{\partial^2}{\partial \omega_i^2} \tag{6.9}$$

$$\mathcal{L}_1 \equiv -i\omega_i J_i + i\frac{1}{I_i} \frac{\partial}{\partial \omega_i} (J_i V) - \frac{\partial}{\partial \omega_i} r_i \omega_j \omega_k \tag{6.10}$$

By adopting the projection operator corresponding to the equilibrium distribution of Eq. (6.7), we get

$$P\mathcal{L}_1(t)P = -iJ_i P\omega_i e^{\mathcal{L}_0 t} \tag{6.11}$$

$$P\mathcal{L}_1(t)Q\mathcal{L}_1(s) = -iJ_i P\omega_i e^{\mathcal{L}_0(t-s)}$$

$$\times \left\{ -i\omega_{i'} J_{i'} + i\frac{1}{I_{i'}} \frac{\partial}{\partial \omega_{i'}} (J_j V) - r_{j'} \frac{\partial}{\partial \omega_{j'}} \omega_{j'} \omega_k \right\} e^{-\mathcal{L}_0 s_1}. \tag{6.12}$$

This leads us to

$$P\mathcal{L}_1(t)Q\mathcal{L}_1(s)P = -J_i J_{i'} P\omega_i \omega_{i'} Pe^{-\gamma_i(t-s)} - \frac{1}{I_i} J_i (J_i V) e^{-\gamma_i(t-s)}P \tag{6.13}$$

Equation (6.13) leads to

$$\frac{\partial}{\partial t} \sigma(\Omega; t) = \left\{ -J_i^2 \frac{\langle \omega_i^2 \rangle}{\gamma_i} - \frac{1}{I_i} \frac{J_i(J_i V)}{\gamma_i} \right\} \sigma(\Omega; t) \tag{6.14}$$

This is the well-known Favro[45] equation. The generators J_i can be defined in terms of the Euler angles θ, φ, and ψ as follows:

$$-iJ_x = \cos\psi\frac{\partial}{\partial\theta} + \frac{\sin\psi}{\sin\theta}\frac{\partial}{\partial\varphi} - \cot\theta\sin\psi\frac{\partial}{\partial\psi}$$

$$-iJ_y = -\sin\psi\frac{\partial}{\partial\theta} + \frac{\cos\psi}{\sin\theta}\frac{\partial}{\partial\varphi} - \cot\theta\cos\psi\frac{\partial}{\partial\psi} \qquad (6.15)$$

$$-iJ_z = \frac{\partial}{\partial\varphi}$$

If the potential V depends only on the angle θ, it is straightforward to show that Eq. (6.14) is exactly equivalent to

$$\frac{\partial}{\partial t}\sigma(\theta;t) = \frac{1}{\zeta}\frac{1}{\sin\theta}\frac{\partial}{\partial\theta}\sin\theta\left[kT\frac{\partial}{\partial\theta} - \left(\frac{\partial}{\partial\theta}V\right)\right]\sigma(\theta;t) \qquad (6.16)$$

Equation (6.16) can be obtained via integration on the irrelevant variables ψ and φ, assuming that

$$\frac{\left\langle\omega_x^2\right\rangle}{\gamma_x} = \frac{\left\langle\omega_y^2\right\rangle}{\gamma_y}. \qquad (6.17)$$

It is not necessary to invoke a complete molecular isotropy if V depends only on θ. When $\partial V/\partial\theta = \mu E\sin\theta$, we obtain the well-known Debye equation.[46]

It is very interesting to remark that at this perturbation order there is no sign of whether the rotator is a symmetric or a spherical top.

The influence of the peculiar nature of the top under investigation is certainly present in the higher order terms. By applying the systematic procedure developed in Section III, it is possible to show that these depend on the asymmetry parameters r_i. For simplicity we omit writing the corresponding explicit expressions.

VII.   ON THE CHOICE OF THE BEST PROJECTION OPERATOR

The theory developed in the preceding sections is to some extent arbitrary. This mainly depends on the fact that the division of the dynamical operator into an unperturbed and a perturbation part is not unique. It has also been stressed that certain choices cannot describe an approach toward the correct equilibrium distribution.[25] Titulaer,[24] on the other hand, argued that, though failing in describing systems close to equilibrium, these choices may well be preferable far from equilibrium. A related problem concerns the influence of the inhomogeneous term [third term on the right-hand side of Eq. (3.5)] on decay. The division of $\mathscr{L}$ into a perturbed and an unperturbed part would seem dictated by the need for making this term vanish.

To shed light on that, let us consider the system

$$
\begin{aligned}
\dot{x} &= -\gamma x + y \\
\dot{y} &= -\Gamma y + f(t)
\end{aligned}
\tag{7.1}
$$

where $f(t)$ is a white Gaussian noise defined by

$$
\langle f(0)f(t)\rangle = 2\Gamma\langle y^2\rangle_{\mathrm{eq}}\delta(t)
\tag{7.1'}
$$

The variable y is driven by a standard fluctuation-dissipation process and is not influenced by the variable x. The dynamical operator driving the pair of variables x and y reads

$$
\mathscr{L} = \gamma\frac{\partial}{\partial x}x - y\frac{\partial}{\partial x} + \Gamma\frac{\partial}{\partial y}y + \Gamma\langle y^2\rangle_{\mathrm{eq}}\frac{\partial^2}{\partial y^2}
\tag{7.2}
$$

Since the motion of the variable y is unaffected by that of the variable x, we are naturally led to dividing the operator $\mathscr{L}$ as follows:

$$
\begin{aligned}
\mathscr{L} &= \mathscr{L}_a + \mathscr{L}_1 + \mathscr{L}_b \\
\mathscr{L}_a &= \gamma\frac{\partial}{\partial x}x \\
\mathscr{L}_b &= \Gamma\frac{\partial}{\partial y}y + \Gamma\langle y^2\rangle_{\mathrm{eq}}\frac{\partial^2}{\partial y^2} \\
\mathscr{L}_1 &= -y\frac{\partial}{\partial x}
\end{aligned}
\tag{7.3}
$$

thereby imposing the projection operator

$$
P\rho(x, y; t) \equiv \rho_{\mathrm{eq}}(y)\int dy\rho(x, y; t)
\tag{7.4}
$$

where

$$
\rho_{\mathrm{eq}}(y) \propto \exp\left\{-\frac{y^2}{2\langle y^2\rangle_{\mathrm{eq}}}\right\}
\tag{7.4'}
$$

As illustrated in Section V.A, $\mathscr{L}_a$ itself has to be assumed "small" so as to make possible an expansion in the corresponding perturbation parameter. This can be bypassed via the following division:

$$\mathscr{L}_a \equiv 0$$

$$\mathscr{L}_b \equiv (\Gamma + \gamma)\frac{\partial}{\partial y}(y - \gamma x) + \Gamma\langle y^2\rangle_{eq}\frac{\partial^2}{\partial y^2} \tag{7.5}$$

$$\mathscr{L}_1 \equiv -\frac{\partial}{\partial x}(y - \gamma x) + \Omega^2 x\frac{\partial}{\partial y} - \gamma\frac{\partial}{\partial y}(y - \gamma x)$$

where

$$\Omega^2 \equiv \gamma\Gamma \tag{7.6}$$

This leads to the projection operator

$$P\rho(x, y; t) \equiv \rho_{eq}(y|x)\int dy\rho(x, y; t) \tag{7.7}$$

where $\rho_{eq}(y|x)$ denotes the equilibrium distribution of y conditioned by the slow variable being given the value x. In the case of standard Brownian motion in an external force field, the division of $\mathscr{L}$ proposed by Haake[23] leads to a projection operator of the same kind as that of Eq. (7.7), that is, with the equilibrium distribution of the irrelevant variable conditioned by the state of the relevant one. This has the unwanted effect of producing incorrect equilibrium distributions. In the present case, on the contrary, the division of Eq. (7.5) will be shown to have the beneficial effect of describing correctly equilibrium at any perturbation order.

The technical difficulty deriving from the projection operator of Eq. (7.7) is bypassed by the following change of variables:

$$\tilde{x} = x \tag{7.8}$$
$$\tilde{y} = y - \gamma x$$

from which

$$\frac{\partial}{\partial x} = -\gamma\frac{\partial}{\partial\tilde{y}} + \frac{\partial}{\partial\tilde{x}}$$

$$\frac{\partial}{\partial y} = \frac{\partial}{\partial\tilde{y}} \tag{7.9}$$

Substituting Eq. (7.9) into Eq. (7.5), we obtain

$$\mathcal{L}_a = 0$$

$$\mathcal{L}_b = (\gamma + \Gamma)\left(\frac{\partial}{\partial \tilde{y}}\tilde{y} + \langle y^2 \rangle_{eq}\frac{\partial^2}{\partial \tilde{y}^2}\right) \tag{7.10}$$

$$\mathcal{L}_1 = \Omega^2 \tilde{x}\frac{\partial}{\partial \tilde{y}} - \frac{\partial}{\partial \tilde{x}}\tilde{y}$$

with the "standard" projection operator

$$P\rho(\tilde{x}, \tilde{y}; t) = \rho_{eq}(\tilde{y}) \int d\tilde{y}\rho(\tilde{x}, \tilde{y}; t) \tag{7.11}$$

Note that the dynamical operator defined by Eq. (7.10) is the same as that concerning a linear oscillator with frequency Ω and damping $\gamma + \Gamma$. This means that the corresponding systematic expansion can be derived from that of Eq. (5.31) by assuming $g(x) = 0$ and $f(x) = -\Omega^2 x$. As to the division of Eq. (7.3), on the contrary, since $\mathcal{L}_a \neq 0$, we can have recourse to the approach detailed in Section V.A. Thus the division of Eq. (7.5) leads us to

$$\frac{\partial}{\partial t}\sigma(x; t) = \frac{\gamma\Gamma}{\gamma + \Gamma}\left[1 + \frac{\gamma\Gamma}{(\gamma + \Gamma)^2} + \frac{2(\gamma\Gamma)^2}{(\gamma + \Gamma)^4} + \cdots\right]$$

$$\cdot \left[\frac{\partial}{\partial x}x + \frac{\langle y^2 \rangle}{\gamma(\gamma + \Gamma)}\frac{\partial^2}{\partial x^2}\right]\sigma(x; t) \tag{7.12}$$

and that of Eq. (7.3) produces

$$\frac{\partial}{\partial t}\sigma(x; t) = \gamma\left[\frac{\partial}{\partial x}x + \frac{\langle y^2 \rangle_{eq}}{\gamma\Gamma}\frac{\partial^2}{\partial x^2}\left(1 - \frac{\gamma}{\Gamma} + \left(\frac{\gamma}{\Gamma}\right)^2 \cdots\right)\right]\sigma(x; t) \tag{7.13}$$

where $\sigma(x; t)$ denotes the distribution of the variable x at time t.

It can be shown that the friction factor of Eq. (7.12) is nothing but the expansion of

$$\varepsilon_{slow} = \frac{(\gamma + \Gamma) - [(\gamma + \Gamma)^2 - 4\Omega^2]^{1/2}}{2} \tag{7.14}$$

in terms of the perturbation parameter $2\Omega/(\gamma + \Gamma)$. $-\varepsilon_{slow}$, in turn, is the slowest eigenvalue of the matrix

$$\mathbf{A} \equiv \begin{pmatrix} 0 & 1 \\ -\Omega^2 & -(\gamma + \Gamma) \end{pmatrix} \qquad (7.15)$$

that is, the counterpart of that of Eq. (2.35). As argued in Section II, the correlation function $\langle x(0)x(t) \rangle_{eq} \langle x^2 \rangle_{eq}$ can be expressed in terms of the eigenstates and eigenvalues of this matrix, that is, as the sum of two exponential time decay terms. For $t \to \infty$, only the term with the slowest damping, $\varepsilon_{slow} = \gamma$, survives. The perturbation expansion in the large parentheses in Eq. (7.13) is the expansion of $(1 + \gamma/\Gamma)^{-1}$ in terms of the perturbation parameter γ/Γ.

In consequence, both Eq. (7.12) and Eq. (7.13) prove to coincide with

$$\frac{\partial}{\partial t} \sigma(x;t) = \gamma \left[\frac{\partial}{\partial x} x + \frac{\langle y^2 \rangle}{\gamma(\gamma + \Gamma)} \frac{\partial^2}{\partial x^2} \right] \sigma(x;t) \qquad (7.16)$$

Note that the initial distributions making the inhomogeneous term of Eq. (3.5) vanish are

$$\rho(x, y) \propto \varphi(x) \exp\left\{ -\frac{y^2}{2\langle y^2 \rangle_{eq}} \right\} \qquad (7.17)$$

for the division of Eq. (7.3), and

$$\rho(x, y) \propto \varphi'(x) \exp\left\{ -\frac{(y - \gamma x)^2}{2\langle y^2 \rangle_{eq}} \right\} \qquad (7.18)$$

for the division of Eq. (7.5). $\varphi(x)$ and $\varphi'(x)$ denote distribution of x of arbitrary form. This means that the long-time regime is completely independent of the initial conditions, thereby justifying our neglecting the inhomogeneous term of Eq. (3.5).

We see, however, that if the expansion series are truncated, Eq. (7.12) provides incorrect information on dynamical properties and Eq. (7.13) results in an incorrect equilibrium distribution. This means that some caution must be exerted when the correct equilibrium distribution is not known.

VIII. CONCLUDING REMARKS

When the different approaches to adiabatic elimination are compared, substantial general agreement is found. The approach of the Barcelona group[26] based on a generalization of the cumulant method, the Chapman-

Ensog method applied by Titulaer,[20] and the approach described in this chapter provide the same result up to the perturbation orders so far considered. The only reason for controversy seems to be the division of the operator $\mathscr{L}$ into perturbed and perturbation parts.[23-25] However, any choice of a perturbation part which cannot be given the form

$$a_- A_+ - a_+ A_-$$

does not allow canonical equilibrium distribution to be retained even though in some cases it can produce a more efficient description of the process far from equilibrium.[24]

The investigation of ref. 47, carried out by using the AEP of this Chapter, shows that certain preliminary divisions of the operator $\mathscr{L}$ into a perturbation and an unperturbed part result in a summation at infinite order of the perturbation series coming from different choices. This provides a solution to the intriguing problem of summation at infinite order of perturbative expansions, alternative to that of Lugiato and coworkers.[48] With this caution in mind, however, the AEP can be regarded as a powerful technique for assessing under what conditions a Fokker-Planck equation can be considered a rigorous description of the problem under study. As shown in this chapter, a linear system is rigorously represented by a Fokker-Planck equation under the sole assumption that the time scale of the heat bath is much shorter than that of the system of interest. Ferrario et al. (Chapter VI) will show that the microscopic nonlinearity of a real system is to a great extent averaged out so that the reliability of a Fokker-Planck equation depends on a delicate interplay between nonlinearity and the random mechanisms destroying memory. This result will also be based on the AEP of this chapter.

In consequence, the AEP allows us to establish to what degree of accuracy a Fokker-Planck equation simulates a nonlinear system. This is a basic requirement for a rigorous foundation for the RMT. As outlined in Chapter I, the RMT is based on the introduction of suitable auxiliary variables so that a description in terms of the standard Fokker-Planck equation can be recovered. It is then of fundamental importance to establish general rules for governing the equations under which the interaction between auxiliary variables and heat bath (and sometimes also variables of interest and heat bath) admits a description in terms of Fokker-Planck equations.

References

1. (a) A. Einstein, *Ann. Phys.*, **17**, 549 (1905); **19**, 371 (1906); (b) S. Chandrasekhar, in *Selected Topics on Noise and Stochastic Processes*, N. Wax, ed., Dover, New York, 1954, p. 3.

2. H. A. Kramers, *Physica*, **7** 284 (1940).

3. H. C. Brinkman, *Physica*, **22**, 149 (1956).

4. P. C. Hemmer, *Physica*, **27**, 79 (1961).

5. R. L. Stratonovich, *Topics in the Theory of Random Noise*, Vol. 1, Gordon and Breach, New York, 1963.

6. V. Seshadri, B. J. West, and K. Lindenberg, *Physica*, **102A**, 470 (1980).

7. R. Landauer and J. A. Swanson, *Phys. Rev.*, **121**, 1668 (1961).

8. J. S. Langer, *Phys. Rev. Lett.*, **21**, 973 (1968); *Ann. Phys. (NY)*, **54**, 258 (1969).

9. P. A. Lee, *J. Appl. Phys.*, **42**, 325 (1971).

10. G. Wilemski, *J. Stat. Phys.*, **14**, 153 (1976).

11. H. Mori, H. Fujisaka, and H. Shigematsu, *Prog. Theor. Phys.*, **51**, 109 (1974).

12. H. Mori, T. Morita, and K. T. Mashiyama, *Prog. Theor. Phys.*, **63**, 1865 (1980).

13. T. Morita, H. Mori, and K. T. Mashiyama, *Prog. Theor. Phys.*, **64**, 500 (1980).

14. N. G. van Kampen, *Can. J. Phys.*, **39**, 5516 (1961); *Prog. Theor. Phys. Suppl.*, **64** 381 (1964).

15. (a) J. L. Skinner and P. G. Wolynes, *Physica*, **96A**, 561 (1979). (b) S. Chaturvedi and F. Shibata, *Z. Phys.—Condensed Matter*, **35**, 297 (1979).

16. M. San Miguel and J. M. Sancho, *J. Stat. Phys.*, **22**, 605 (1980).

17. M. San Miguel and J. M. Sancho, *Phys. Lett.*, **76A**, 97 (1980).

18. M. San Miguel and J. M. Sancho, *Stochastic Non-linear Systems in Physics, Chemistry and Biology*, Synergetics Ser. Vol. 8, L. Arnold and R. Lefever, eds., Springer, Berlin, 1981, p. 137.

19. P. Hänggi, H. Thomas, H. Grabert, and P. Talkner, *J. Stat. Phys.*, **18**, 155 (1978).

20. U. M. Titulaer, *Physica*, **100A**, 234 (1980).

21. H. Risken, H. D. Vollmer, and M. Mörsch, *Z. Phys.—Condensed Matter*, **B40**, 343 (1981).

22. K. Kancko, *Progr. Theor. Phys.*, **66**, 129 (1981).

23. F. Haake, *Z. Phys.—Condensed Matter*, **B48**, 31 (1982).

24. U. M. Titulaer, *Z. Phys.—Condensed Matter*, **B50**, 71 (1983).

25. H. Hasegawa, M. Mizuno, and M. Mabuchi, *Prog. Theor. Phys.*, **67**, 98 (1982).

26. L. Garrido and J. M. Sancho, *Physica*, **115A**, 479 (1982).

27. H. Dekker, *Phys. Lett.*, **90A**, 26 (1982).

28. U. R. Steiger and R. F. Fox, *J. Math. Phys.*, **23**, 1678 (1982).

29. H. Haken, *Synergetics, An Introduction*, Springer-Verlag, Berlin, Heidelberg, New York, 1978.

30. H. Haken, *Z. Phys.—Condensed Matter*, **B29**, 61 (1978).

31. A. Wunderlin and H. Haken, *Z. Phys.—Condensed Matter*, **B44**, 135 (1981).

32. B. H. Lavenda and R. Serra, *J. Math. Phys.*, (in press); B. H. Lavenda and E. Santamato, *ibid.*, **22**, 2926 (1981); E. Santamato and B. H. Lavenda, *ibid.*, **23**, 2452 (1981).

33. R. Zwanzig, *Phys. Rev.*, **124**, 983 (1961).

34. D. N. Zubarev, *Nonequilibrium Statistical Thermodynamics*, Consultants Bureau, New York, 1974.

35. R. J. Rubin, *J. Math. Phys.*, **1**, 309 (1960); **2**, 373 (1961); J. L. Lebowitz and R. J. Rubin, *Phys. Rev.*, **131**, 2381 (1963); P. Mazur and E. Montroll, *J. Math. Phys.*, **1**, 70 (1960); M. Maekawa and K. Wada, *Phys. Lett.*, **80A**, 293 (1980); R. I. Cukier, K. E. Shuler, and J. D. Weeks, *J. Stat. Phys.*, **5**, 99 (1972).

36. L. D. Landau and E. M. Lifshitz, *Quantum Mechanics — Nonrelativistic Theory*, Pergamon Press, Oxford, 1958.

37. S. Chaturvedi, *Z. Phys.—Condensed Matter*, **B51**, 271 (1983).

38. F. Marchesoni and P. Grigolini. *Z. Phys.—Condensed Matter*, **55**, 257 (1984).
39. M. Abramowitz and J. A. Stegun, *Handbook of Mathematical Functions*, Dover, 1965, p. 781.
40. Ref. 39, p. 774.
41. R. S. Gradsteyn and I. M. Ryshik, Table of Integrals Series and Products Academic Press, New York, 1963, p. 1033.
42. L. T. Muus, in *Electron Spin Relaxation in Liquids*, P. W. Atkins, ed., Plenum Press, New York, 1972, p. 1.
43. S. Faetti, P. Grigolini, and F. Marchesoni, *Z. Phys.—Condensed Matter*, **B47**, 353 (1982).
44. F. Marchesoni and P. Grigolini, *Physica*, **A121**, 269 (1983).
45. L. D. Favro, Fluctuation phenomena in solids, ed. R. E. Burger (Academic Press, New York, 1965) p. 79.
46. Y. McConnell, Rotational Brownian Motion and Dielectric Theory, Academic Press 1980, London.
47. T. Fonseca, P. Grigolini, D. Pareo, J. Chem. Phys (in press); S. Faetti, L. Fronzoni, P. Grigolini, to be published in Phys. Rev. A.
48. L. A. Lugiato, Physica, **81A** 565 (1976); F. Casagrande, E. Eschenazi, and L. A. Lugiato, Phys. Rev. **29A**, 239 (1984); L. A. Lugiato, P. Mandel, L. M. Narducci, Phys. Rev. **29A**, 1438 (1984).

III

CONTINUED FRACTIONS IN THE THEORY OF RELAXATION

G. GROSSO and G. PASTORI PARRAVICINI

CONTENTS

I. INTRODUCTION

In these notes we describe some relevant aspects of continued fractions as they are encountered in the memory function approaches to the theory of relaxation. A thorough treatment of all properties, theorems, and subtleties of this field of mathematical research is not within the scope of the present volume. The subject is fairly old in origin[1] and has a wide and celebrated literature. One can consult, for instance, the classical textbooks and reviews of Wall,[2] Akhiezer,[3] and Shohat and Tamarkin,[4] or the modern treatments of the subject and related problems by Brezinski,[5] Jones and Thron,[6] and Draux.[7] Our purpose is rather to guide the reader within the mathematical apparatus of continued fractions and take advantage of its beautiful and exact results in treating those physical problems described in the theory of relaxation by appropriate biorthogonal sets representation.

The mathematics taught to researchers in science is almost exclusively focused on the Hilbert theory of infinite matrices; an object such as a continued fraction may thus be unfamiliar to them. Historically mathematical discoveries have followed a different pattern. The analytic theory of continued fractions of Stieltjes[8] inspired and preceded by almost a decade the pioneering work of Hilbert and his school. The development of quantum mechanics in the early part of this century has granted to the latter a leading role.[9]

In the early 1960s, with the creation of the modern projective and memory function approaches to nonequilibrium statistical mechanics,[10,11] continued fractions have undergone a renaissance as a fundamental tool in calculating correlation functions, density of states, and spectra. The prime motivation for this chapter is to present the algebraic and analytic properties of continued fractions relevant to the theory of relaxation. An effort has been made to make this presentation self-contained, to demonstrate the statements or at least to make them plausible, and to avoid a flood of theorems and lemmas. The elegant subjects of the moment problem, orthogonal polynomials and error bounds, are described by focusing on those aspects pertinent to the theory of relaxation. Technical but interesting topics often overlooked in the literature, such as product-difference algorithms and modified moments, are also discussed.

II. CONTINUED FRACTIONS: DEFINITIONS AND BASIC ALGEBRAIC PROPERTIES

By a continued fraction we denote an expression of the type

$$F = \cfrac{a_1}{b_1 + \cfrac{a_2}{b_2 + \cfrac{a_3}{b_3 + \cdots}}} \qquad (2.1)$$

the elements a_n and b_n may be any complex numbers and are called *partial numerators* and *partial denominators*, respectively. Unless specifically stated, we assume that all divisions can be performed, that is, no divisor vanishes. Other common notations for F are

$$F = K\left(\frac{a_n}{b_n}\right) = \frac{a_1}{b_1 +} \frac{a_2}{b_2 +} \frac{a_3}{b_3 +} \cdots$$

$$= \frac{a_1|}{|b_1} + \frac{a_2|}{|b_2} + \frac{a_3|}{|b_3} + \cdots \qquad (2.2)$$

The nth *approximant* (sometimes also called the nth convergent) of a continued fraction is defined by truncating it at the nth level, that is, by setting the partial numerator a_{n+1} equal to 0. The value of an infinite continued fraction is defined as the limit of its sequence of approximants if this limit exists and is finite.

The truncation at successive steps also defines the nth *numerator* A_n and the nth *denominator* B_n of the continued fraction. The first few approximants are

$$\frac{A_1}{B_1} = \frac{a_1}{b_1} \qquad \frac{A_2}{B_2} = \frac{a_1 b_2}{b_1 b_2 + a_2} \qquad \frac{A_3}{B_3} = \frac{a_1 b_2 b_3 + a_1 a_3}{b_1 b_2 b_3 + b_1 a_3 + a_2 b_3}$$

The quantities A_n and B_n may be obtained by means of the *fundamental recurrence formulas*

$$
\begin{aligned}
A_{n+1} &= b_{n+1} A_n + a_{n+1} A_{n-1} \\
B_{n+1} &= b_{n+1} B_n + a_{n+1} B_{n-1} \qquad n = 0,1,2,\dots
\end{aligned}
\tag{2.3}
$$

with the initial conditions

$$
\begin{aligned}
A_{-1} &= 1 & A_0 &= 0 \\
B_{-1} &= 0 & B_0 &= 1
\end{aligned}
\tag{2.4}
$$

as can be easily demonstrated by induction. Both numerators A_n and denominators B_n obey the same three-term recurrence relation (2.3), but with the different initial conditions specified by (2.4).

We also notice that

$$
\begin{aligned}
A_{n-1} B_n - B_{n-1} A_n &= A_{n-1}(b_n B_{n-1} + a_n B_{n-2}) - B_{n-1}(b_n A_{n-1} + a_n A_{n-2}) \\
&= -a_n (A_{n-2} B_{n-1} - B_{n-2} A_{n-1})
\end{aligned}
$$

By successive applications, we obtain (with $a_0 \equiv 1$)

$$A_{n-1} B_n - B_{n-1} A_n = (-1)^n a_1 a_2 \cdots a_n \qquad n = 0,1,2,\dots \tag{2.5}$$

which is called the *determinant formula*.

From the structure of the continued fraction (2.1), we see that its value remains unchanged if a_n, b_n, a_{n+1} (for all $n \geq 1$) are multiplied by the same constant (other than zero). Operations of this kind do not affect the sequence of approximants and are called *equivalence transformations*.

By use of equivalence transformations we can always transform a continued fraction $K(a_n/b_n)$ into an equivalent one whose partial numerators equal unity, or into an equivalent one whose partial denominators equal unity. We have, in fact,

$$K\left(\frac{a_n}{b_n}\right) = K\left(\frac{1}{\beta_n}\right) \tag{2.6}$$

where

$$\beta_1 = \frac{b_1}{a_1} \qquad \beta_2 = \frac{a_1}{a_2} b_2 \qquad \cdots$$

and in general (by induction)

$$\beta_{2n+1} = \frac{a_2 a_4 \cdots a_{2n}}{a_1 a_3 \cdots a_{2n+1}} b_{2n+1} \qquad \beta_{2n} = \frac{a_1 a_3 \cdots a_{2n-1}}{a_2 a_4 \cdots a_{2n}} b_{2n} \qquad n \geq 1$$

We have also

$$K\left(\frac{a_n}{b_n}\right) = K\left(\frac{\alpha_n}{1}\right) \tag{2.7}$$

where

$$\alpha_1 = \frac{a_1}{b_1} \qquad \alpha_n = \frac{a_n}{b_n b_{n-1}} \qquad n \geq 2$$

A continued fraction is called a *contraction* of an assigned continued fraction if the sequence of approximants of the former matches a subset of approximants of the latter. The opposite situation is called an *extension*.

Of particular interest are the *even part* and the *odd part* of a continued fraction. The even (odd) part of a continued fraction is defined as the continued fraction whose sequence of approximants is given by the even (odd) approximants of the original fraction.

As an example consider the continued fraction of the type

$$F = \cfrac{a_1}{1 + \cfrac{a_2}{1 + \cfrac{a_3}{1 + \cdots}}} \tag{2.8}$$

The first few even approximants are

$$\frac{A_2}{B_2} = \frac{a_1}{1+a_2}$$

$$\frac{A_4}{B_4} = \frac{a_1}{1+\cfrac{a_2}{1+\cfrac{a_3}{1+a_4}}} = \frac{a_1}{1+a_2-\cfrac{a_2 a_3}{1+a_3+a_4}}$$

and in general

$$F_e = \frac{a_1}{1+a_2-\cfrac{a_2 a_3}{1+a_3+a_4-\cfrac{a_4 a_5}{1+a_5+a_6-\cdots}}} \qquad (2.9)$$

In a similar way, one can obtain the odd part of the continued fraction (2.8) as

$$F_o = a_1 - \frac{a_1 a_2}{1+a_2+a_3-\cfrac{a_3 a_4}{1+a_4+a_5-\cfrac{a_5 a_6}{1+a_6+a_7-\cdots}}} \qquad (2.10)$$

III.   ELEMENTARY ANALYTIC PROPERTIES OF CONTINUED FRACTIONS

A continued fraction of the type

$$F(x) = \frac{a_1 x}{1+\cfrac{a_2 x}{1+\cfrac{a_3 x}{1+\cdots}}} \qquad a_i \neq 0 \qquad (3.1)$$

is referred to as a *Stieltjes-type* continued fraction (in general a_i are complex numbers $\neq 0$, and x is complex variable). Any fraction of type (3.1), or equivalent to it, is called an *S-fraction*.

Let the formal Taylor expansion of (3.1) be

$$F(x) = \sum_{n=1}^{\infty} c_n x^n \qquad (3.2)$$

The first few terms are

$$F(x) = a_1x - a_1a_2x^2 + \left(a_1a_2^2 + a_1a_2a_3\right)x^3 - \cdots$$

It is seen that a_2 affects the coefficients of x^2 and higher powers; a_3 affects the coefficients of x^3 and higher powers; and so on. In general, a_n affects the coefficients of x^n and higher powers. *Hence the series expansion of the nth approximant matches exactly the series expansion of the Stieltjes fraction up to powers of order n*, while higher power coefficients in general will be different. We can thus write

$$\frac{A_n(x)}{B_n(x)} = c_1x + \cdots + c_nx^n + d_{n+1}x^{n+1} + d_{n+2}x^{n+2} + \cdots \quad (3.3)$$

The first few approximants of the S-fraction (3.1) are

$$\frac{A_1(x)}{B_1(x)} = \frac{a_1x}{1}$$

$$\frac{A_2(x)}{B_2(x)} = \frac{a_1x}{1 + a_2x}$$

$$\frac{A_3(x)}{B_3(x)} = \frac{a_1x + a_1a_3x^2}{1 + (a_2 + a_3)x}$$

$$\frac{A_4(x)}{B_4(x)} = \frac{a_1x + (a_1a_3 + a_1a_4)x^2}{1 + (a_2 + a_3 + a_4)x + a_2a_4x^2}$$

(3.4)

In general the approximants can be obtained using the fundamental recurrence formulas (2.3), which in our specific case are

$$A_{n+1}(x) = A_n(x) + xa_{n+1}A_{n-1}(x)$$
$$B_{n+1}(x) = B_n(x) + xa_{n+1}B_{n-1}(x) \quad n = 0,1,2,\ldots$$

(3.5)

with the initial conditions $A_{-1} = 1$; $A_0 = 0$; $B_{-1} = 0$; $B_0 = 1$. The determinant formula (2.5) has the aspect of a Wronskian relation (with $a_0 \equiv 1$):

$$A_{n-1}(x)B_n(x) - B_{n-1}(x)A_n(x) = (-1)^n a_1a_2\cdots a_n\cdot x^n \quad n = 0,1,2\ldots$$

From Eqs. (3.4) and (3.5) it can be easily seen that the degree of $A_{2n}(x)$, $B_{2n}(x)$, and $A_{2n-1}(x)$ is n and the degree of $B_{2n-1}(x)$ is $n-1$. These polynomials can be written as

$$A_{2n}(x) = \alpha_1^{(n)}x + \alpha_2^{(n)}x^2 + \cdots + \alpha_n^{(n)}x^n$$

$$B_{2n}(x) = 1 + \beta_1^{(n)}x + \beta_2^{(n)}x^2 + \cdots + \beta_n^{(n)}x^n$$

$$A_{2n-1}(x) = \gamma_1^{(n)}x + \gamma_2^{(n)}x^2 + \cdots + \gamma_n^{(n)}x^n \tag{3.6}$$

$$B_{2n-1}(x) = 1 + \delta_1^{(n)}x + \delta_2^{(n)}x^2 + \cdots + \delta_{n-1}^{(n)}x^{n-1}$$

The expressions for the coefficients α, β, γ, δ can be worked out explicitly from the fundamental recurrence formulas (3.5). We have need in particular for a few of them as follows:

$$\beta_n^{(n)} = a_2 a_4 \cdots a_{2n} \qquad \gamma_n^{(n)} = a_1 a_3 \cdots a_{2n-1}, \qquad n = 1,2,\ldots \tag{3.7}$$

From Eqs. (3.4) and (3.5) we also have

$$\beta_1^{(1)} = a_2 \qquad \beta_1^{(n)} - \beta_1^{(n-1)} = a_{2n-1} + a_{2n} \qquad n = 2,3,\ldots \tag{3.8}$$

From Eqs. (3.4) and (3.6), we see that the sequence of approximants of a Stieltjes-type continued fraction fill the stairlike sequence [1|0], [1|1], [2|1], [2|2],... of the Padé table.[5,12]

If we consider the Stieltjes fraction (3.1), write $x = 1/z$, and perform an equivalence transformation by multiplying by z the composing fractions, we obtain the S-fraction of the type

$$F(z) = \cfrac{a_1}{z + \cfrac{a_2}{1 + \cfrac{a_3}{z + \cfrac{a_4}{1 + \cdots}}}} \tag{3.9}$$

whose peculiar structure is to have as partial denominators z and 1 alternately. The sequence of approximants of the S-fraction (3.9) fill the doubly repeated sequence [0|1], [0|1], [1|2], [1|2], [2|3], [2|3],... of the Padé table in the variable z.

A continued fraction of the type

$$F(x) = \cfrac{x}{1 - a_0 x - \cfrac{b_1^2 x^2}{1 - a_1 x - \cfrac{b_2^2 x^2}{1 - a_2 x - \cdots}}} \tag{3.10}$$

is referred to as a *Jacobi-type* continued fraction (in general a_n and b_n are complex numbers, $b_n \neq 0$, and x is a complex variable). Any fraction of type (3.10), or equivalent to it, is called a *J-fraction*. We write b_n^2 instead of b_n to avoid in the following the use of the exponent $1/2$. The sequence of approximants to the J-fraction (3.10) fill the diagonal sequence [1|1], [2|2], [3|3]... of the Padé table.

If we consider the J-fraction (3.10), write $x = 1/z$, and perform an equivalence transformation, multiplying by z the composing fractions we obtain a J-fraction of the type

$$F(z) = \cfrac{1}{z - a_0 - \cfrac{b_1^2}{z - a_1 - \cfrac{b_2^2}{z - a_2 - \cdots}}} \tag{3.11}$$

The sequence of approximants of the J-fraction (3.11) fill the sequence [0|1], [1|2], [2|3]... of the Padé table.

The J-fractions of type (3.11) are the most general continued fractions encountered in the memory function approaches and are thus of central importance in the theory of relaxation. In fact, Eq. (3.11) has the familiar structure encountered on performing the Laplace transform of the hierarchy of coupled integro-differential equations used in the calculation of correlation functions (see Chapter I and refs. 11 and 13).

Given the Stieltjes-type continued fraction (3.1), it is interesting to consider its even part. From Eqs. (2.8) and (2.9) we have

$$F_e(x) = \cfrac{a_1 x}{1 + a_2 x - \cfrac{a_2 a_3 x^2}{1 + (a_3 + a_4)x - \cfrac{a_4 a_5 x^2}{1 + (a_5 + a_6)x - \cdots}}} \tag{3.12}$$

Similarly, the even part of the S-fraction (3.9) is

$$F_e(z) = \cfrac{a_1}{z + a_2 - \cfrac{a_2 a_3}{z + (a_3 + a_4) - \cfrac{a_4 a_5}{z + (a_5 + a_6) - \cdots}}} \qquad (3.13)$$

Thus the even part of an S-fraction is always a J-fraction.

However, it must be noted explicitly that the converse is not true and that the structure of the J-fractions (3.10) and (3.11) is *similar but more general* than Eqs. (3.12) and (3.13); thus caution must be used in applying mathematical theorems valid for S-fractions or their even contractions to the memory J-fractions. Note, for instance, that (3.11) allows the possibility $a_0 = 0$, $b_1^2 \neq 0$, while (3.13) does not allow $a_2 = 0$, $a_2 a_3 \neq 0$.

However, the close formal relationship between S-fractions and J-fractions allows one in a number of situations to consider properties and theorems for some particular form and to transfer *appropriately* the results to other situations.

We have not yet considered the convergence properties of continued fractions. Generally in physics we solve this problem through physical arguments, rather than resorting to rigorous mathematical theorems, because the number of really accessible parameters of the continued fraction is limited. However, it is instructive to mention the following important convergence theorem of E. B. Van Vleck,[14] because it is linked to the physical concept of the constant chain model.[15]

Consider an S-fraction

$$F(x) = \frac{a_1 x}{1+} \frac{a_2 x}{1+} \cdots \qquad (3.14)$$

whose coefficients a_n have the property

$$\lim_{n \to \infty} a_n = a \neq 0 \qquad (3.15)$$

Let L denote the rectilinear cut from $-1/4a$ to ∞ (and lined up with the origin, see Fig. 1). Consider an arbitrary finite closed domain, whose distance from L is finite. In this domain the S-fraction is a regular analytic function (except for the possibility of poles) and coincides in the region around the origin with the corresponding Taylor series.

Instead of reporting a rigorous mathematical proof,[14] we present the following argument, which clarifies the essentials of the theorem and provides a guide to the treatment of tails of continued fractions in solid state physics (see also Chapter IV and references therein).

Consider the "tail" of order n of the S-fraction (3.14):

$$t_n(x) = \frac{a_n x}{1+} \frac{a_{n+1}x}{1+} \cdots \qquad (3.16)$$

We have exactly

$$t_n(x) = \frac{a_n x}{1 + t_{n+1}(x)} \qquad (3.17)$$

Because of relation (3.15) we expect for n sufficiently large $t_n(x) \simeq t_{n+1}(x) \simeq t(x)$. Such an approximation for the tails gives

$$t^2 + t - ax = 0$$

so that

$$t = \frac{1}{2}\left(-1 \mp \sqrt{1 + 4ax}\right) \qquad (3.18)$$

Suppose for simplicity that a is real and so is x; then if $1 + 4ax < 0$, we have from (3.18) that t has an imaginary part, which is not compatible with the definition (3.16) for large n. We can overcome this difficulty by cutting away the offending part of the real axis, that is, the line $[-\infty, -1/4a]$ if $a > 0$

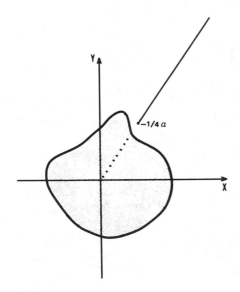

Figure 1.   Schematic illustration of the convergence theorem of E. B. Van Vleck (see ref. 14).

and the line $[-1/4a, \infty]$ if $a < 0$. In the case where a is a generical complex number, we have to perform an appropriate transformation of the cut from the real axis to the complex plane. The possible poles are connected to the first linear transformations before the asymptotic region is reached and would be absent in the constant chain model ($a_n = a$, $n = 1, 2, \ldots$).

IV. RELATION BETWEEN POWER SERIES AND CONTINUED FRACTIONS

In this section we consider the problem of expanding a given power series into an S-fraction (later we consider the relation between power series and J-fractions). The correspondence between power series and S-fractions is determined by the requirement that the expansion of the nth approximant of the S-fraction coincides term by term with the power series up to order n. We discuss when a one-to-one correspondence is possible and how to determine the parameters of the continued fraction. We consider this problem from a pure *algebraic point of view*. The occurrence of convergence or divergence of the power series *does not imply* that the same property holds for the corresponding continued fraction. It is well known that certain continued fractions associated with power series can converge outside the domain of convergence of the series. In general it is possible to find examples in which a power series *converges* (*diverges*) and the corresponding fraction either *converges* or *diverges*.

Consider an arbitrary formal power series of the form

$$F(x) = \mu_0 x + \mu_1 x^2 + \mu_2 x^3 + \cdots + \mu_{n-1} x^n + \cdots \qquad (4.1)$$

where $\{\mu_0, \mu_1, \mu_2 \ldots\}$ is a given sequence of complex constants.

It is convenient to define the Hankel determinants formed with the sequence of complex numbers $\{\mu_0, \mu_1, \mu_2, \ldots\}$ as

$$D_n = \begin{vmatrix} \mu_0 & \mu_1 & \cdots & \mu_n \\ \mu_1 & \mu_2 & \cdots & \mu_{n+1} \\ \cdots & \cdots & \cdots & \cdots \\ \mu_n & \mu_{n+1} & \cdots & \mu_{2n} \end{vmatrix} \qquad n = 0, 1, 2, \ldots \qquad (4.2)$$

We will denote by Ω_n the Hankel determinants formed with the sequence of constants:

$$\Omega_n = \begin{vmatrix} \mu_1 & \mu_2 & \cdots & \mu_{n+1} \\ \mu_2 & \mu_3 & \cdots & \mu_{n+2} \\ \cdots & \cdots & \cdots & \cdots \\ \mu_{n+1} & \mu_{n+2} & \cdots & \mu_{2n+1} \end{vmatrix} \qquad n = 0, 1, 2, \ldots \qquad (4.3)$$

The elements of Ω_n are obtained from the elements of D_n by advancing the subscripts by unity. In the following we will need also the modified Hankel determinants of the type

$$
R_n = \begin{vmatrix} \mu_0 & \mu_1 & \cdots & \mu_n \\ \mu_1 & \mu_2 & \cdots & \mu_{n+1} \\ \cdots & \cdots & \cdots & \cdots \\ \mu_{n-1} & \mu_n & \cdots & \mu_{2n-1} \\ \mu_{n+1} & \mu_{n+2} & \cdots & \mu_{2n+1} \end{vmatrix} \qquad n = 0,1,2,\ldots \qquad (4.4)
$$

It can be seen that the elements of R_n are obtained from the elements of D_n by advancing the subscripts of the last row by unity.

We give now a method for expanding a power series into an S-fraction. It is summarized by the following theorem.

Theorem 4.1. There is a one-to-one correspondence between a power series

$$
F(x) = \mu_0 x + \mu_1 x^2 + \mu_2 x^3 + \cdots \qquad (4.5)
$$

for which $D_n \neq 0$, $\Omega_n \neq 0$ and an S-fraction

$$
F(x) = \frac{a_1 x}{1+} \; \frac{a_2 x}{1+} \; \frac{a_3 x}{1+} \cdots \qquad (4.6)
$$

in which $a_n \neq 0$. The expression of the partial numerators is

$$
a_{2n} = -\frac{\Omega_{n-1} D_{n-2}}{\Omega_{n-2} D_{n-1}} \qquad n = 1,2,3,\ldots \qquad (4.7)
$$

$$
a_{2n+1} = -\frac{\Omega_{n-2} D_n}{\Omega_{n-1} D_{n-1}} \qquad n = 1,2,3,\ldots \qquad (4.8)
$$

$(D_{-1} = \Omega_{-1} = 1; \; a_1 = D_0)$.

Proof. We give a proof of this theorem, because of its important implications in the calculation of correlation functions; we will follow the algebraic procedure of Perron.[16]

We have already seen that the approximants of an S-fraction have a Taylor expansion of the type

$$
\frac{A_{2n}(x)}{B_{2n}(x)} = \frac{\alpha_1 x + \cdots + \alpha_n x^n}{1 + \beta_1 x + \cdots + \beta_n x^n} = \mu_0 x + \cdots + \mu_{2n-1} x^{2n} + \cdots \qquad (4.9)
$$

$$
\frac{A_{2n-1}(x)}{B_{2n-1}(x)} = \frac{\gamma_1 x + \cdots + \gamma_n x^n}{1 + \delta_1 x + \cdots + \delta_{n-1} x^{n-1}} = \mu_0 x + \cdots + \mu_{2n-2} x^{2n-1} + \cdots
$$

$$
(4.10)
$$

where we have indicated up to the powers which are *exactly* reproduced by the approximants. For simplicity we have dropped the superscript (n) to the α, β, γ, and δ given by Eqs. (3.6), (3.7), and (3.8).

We consider now Eq. (4.9), multiply by the denominator, and obtain

$$\alpha_1 x + \cdots + \alpha_n x^n \equiv (1 + \beta_1 x + \cdots + \beta_n x^n)(\mu_0 x + \cdots + \mu_{2n-1} x^{2n} + \cdots)$$

If we require that the coefficients of x^ν for $\nu = n+1, n+2, \ldots, 2n$ of the first and second members are equal, we have

$$
\begin{cases}
0 = \mu_0 \beta_n + \mu_1 \beta_{n-1} + \cdots + \mu_{n-1} \beta_1 + \mu_n \\
0 = \mu_1 \beta_n + \mu_2 \beta_{n-1} + \cdots + \mu_n \beta_1 + \mu_{n+1} \\
\cdots\cdots\cdots\cdots\cdots\cdots\cdots\cdots\cdots\cdots\cdots\cdots\cdots\cdots \\
0 = \mu_{n-1} \beta_n + \mu_n \beta_{n-1} + \cdots + \mu_{2n-2} \beta_1 + \mu_{2n-1}
\end{cases}
\tag{4.11}
$$

From Kramer's rule we obtain

$$
\beta_n = \frac{1}{D_{n-1}}
\begin{vmatrix}
-\mu_n & \mu_1 & \mu_2 & \cdots & \mu_{n-1} \\
-\mu_{n+1} & \mu_2 & \mu_3 & \cdots & \mu_n \\
\cdots\cdots\cdots\cdots\cdots\cdots\cdots\cdots\cdots \\
-\mu_{2n-1} & \mu_n & \mu_{n+1} & \cdots & \mu_{2n-2}
\end{vmatrix}
= (-1)^n \frac{\Omega_{n-1}}{D_{n-1}}
$$

$$\tag{4.12}$$

From Eqs. (4.12) and (3.7), Eq. (4.7) is demonstrated.

To prove Eq. (4.8) we can use essentially the same procedure, starting now from Eq. (4.10). We multiply by the denominator and obtain

$$\gamma_1 x + \cdots + \gamma_n x^n \equiv (1 + \delta_1 x + \cdots + \delta_{n-1} x^{n-1})$$
$$\cdot (\mu_0 x + \cdots + \mu_{2n-2} x^{2n-1} + \cdots)$$

If we require that the coefficients of x^ν for $\nu = n, n+1, \ldots, 2n-1$ of the first and second members are equal, we have

$$
\begin{cases}
\gamma_n = \mu_0 \delta_{n-1} + \mu_1 \delta_{n-2} + \cdots + \mu_{n-2} \delta_1 + \mu_{n-1} \\
0 = \mu_1 \delta_{n-1} + \mu_2 \delta_{n-2} + \cdots + \mu_{n-1} \delta_1 + \mu_n \\
\cdots\cdots\cdots\cdots\cdots\cdots\cdots\cdots\cdots\cdots\cdots\cdots\cdots\cdots \\
0 = \mu_{n+1} \delta_{n-1} + \mu_n \delta_{n-2} + \cdots + \mu_{2n-3} \delta_1 + \mu_{2n-2}
\end{cases}
$$

This system of n linear nonhomogeneous equations in $n-1$ unknown quantities $\delta_1, \delta_2, \ldots, \delta_{n-1}$ has solution only if

$$
\begin{vmatrix}
\mu_0 & \mu_1 & \cdots & \mu_{n-2} & \mu_{n-1} - \gamma_n \\
\mu_1 & \mu_2 & & \mu_{n-1} & \mu_n \\
\cdots & \cdots & \cdots & \cdots & \cdots \\
\mu_{n-1} & \mu_n & & \mu_{2n-3} & \mu_{2n-2}
\end{vmatrix} = 0
$$

If we expand the above determinant according to the elements of the last column and the corresponding minors, we have

$$
D_{n-1} + (-1)^{n+1}(-\gamma_n)\Omega_{n-2} = 0
$$

Thus

$$
\gamma_n = (-1)^{n+1}\frac{D_{n-1}}{\Omega_{n-2}} \tag{4.13}
$$

From Eqs. (4.13) and (3.7), Eq. (4.8) follows.

The above results with straightforward implementations lead to the following theorem for a J-fraction, obtained as an *even* contraction of an S-fraction:

Theorem 4.2. There is a one-to-one correspondence between a power series

$$
F(x) = \mu_0 x + \mu_1 x^2 + \mu_2 x^3 + \cdots
$$

for which $D_n \neq 0$, $\Omega_n \neq 0$, and a J-fraction (even contraction of an S-fraction)

$$
F(x) = \cfrac{a_1 x}{1 + a_2 x -} \quad \cfrac{a_2 a_3 x^2}{1 + (a_3 + a_4)x -} \quad \cfrac{a_4 a_5 x^2}{1 + (a_5 + a_6)x -} \cdots
$$

in which $a_n \neq 0$, $n = 1, 2, \ldots$. The expression for the products $a_{2n}a_{2n+1}$ and the sums $a_{2n+1} + a_{2n+2}$ are

$$
a_{2n}a_{2n+1} = \frac{D_n D_{n-2}}{D_{n-1}^2} \qquad n = 1, 2, \ldots \tag{4.14}
$$

$$
a_{2n+1} + a_{2n+2} = -\frac{1}{\Omega_{n-1}}\left[\frac{D_{n-1}\Omega_n}{D_n} + \frac{D_n \Omega_{n-2}}{D_{n-1}}\right] \qquad n = 1, 2, \ldots \tag{4.15}
$$

$(\Omega_{-1} = 1; \ a_1 = D_0)$, or equivalently

$$a_{2n+1} + a_{2n+2} = -\left(\frac{R_n}{D_n} - \frac{R_{n-1}}{D_{n-1}}\right), \qquad n = 1,2,\dots \qquad (4.16)$$

Proof. The results (4.14) and (4.15) are straightforward algebraic manipulations of the results (4.7) and (4.8). What is new in the theorem is Eq. (4.16). Consider again the system of linear homogeneous equations (4.11), and let us apply the standard Kramer's rule to obtain $\beta_1^{(n)}$ [we have placed here the superscript (n) to β_1]. We have

$$\beta_1^{(n)} = \frac{1}{D_{n-1}} \begin{vmatrix} \mu_0 & \mu_1 & \cdots & \mu_{n-2} & -\mu_n \\ \mu_1 & \mu_2 & & \mu_{n-1} & -\mu_{n+1} \\ \vdots & & & & \vdots \\ \mu_{n-1} & \mu_n & & \mu_{2n-3} & -\mu_{2n-1} \end{vmatrix} = -\frac{R_{n-1}}{D_{n-1}} \qquad (4.17)$$

From Eqs. (4.17) and (3.8), Eq. (4.16) follows.

If in Theorem 4.2, we use expressions (4.14) and (4.16), the restriction $\Omega_n \neq 0$ appears to be superfluous. The condition that partial numerators of the J-fraction are $\neq 0$ is guaranteed by $D_n \neq 0$. This is a plausible argument for the following theorem.

Theorem 4.3. There is a one-to-one correspondence between a power series

$$F(x) = \mu_0 x + \mu_1 x^2 + \mu_2 x^3 + \cdots$$

for which $D_n \neq 0$ and a J-fraction (not necessarily an even contraction of an S-fraction)

$$F(x) = \frac{b_0^2 x}{1 - a_0 x -} \ \frac{b_1^2 x^2}{1 - a_1 x -} \ \frac{b_2^2 x^2}{1 - a_2 x -} \ \cdots$$

in which the partial numerators $b_n^2 \neq 0$. The expression for the parameters of the J-fraction are

$$b_n^2 = \frac{D_n D_{n-2}}{D_{n-1}^2} \qquad n = 0,1,2,\dots \qquad (4.18)$$

$$a_n = \frac{R_n}{D_n} - \frac{R_{n-1}}{D_{n-1}} \qquad n = 0,1,2,\dots \qquad (4.19)$$

$(D_{-1} = D_{-2} = 1; \ R_{-1} = 0)$.

The rigorous proof of Theorem 4.3, along lines similar to those exploited for Theorem 4.1, can be found in the book of Perron.[16] An alternative demonstration using the formal tool of orthogonal polynomials is given later in this chapter (see Section VII.A).

V.   RELATION BETWEEN GREEN'S FUNCTION AND CONTINUED FRACTIONS

A.   Introductory Remarks

A large number of physical problems are described by operators (or superoperators) which are too complicated for a direct diagonalization but are still manageable enough to allow a reasonably simple calculation of moments. In this case one can obtain precious information on the physical system of interest via a *continued fraction expansion of the Green's function of the given operator (or superoperator)*. The mathematical procedures of using the moments to arrive at the continued fraction expansion of the resolvent are known as the *moment problem*. These mathematical procedures are the same regardless of the fact that moments refer to operators or superoperators. For simplicity and without loss of generality, we will illustrate such procedures using notations and terminology appropriate to the former. However, we wish to stress that everything we say is automatically applicable to superoperators, such as those encountered, for instance, in nonequilibrium statistical mechanics, as they are the central purpose of the present volume.

We have not yet specified if the operator to be handled is Hermitian (real eigenvalues) or whether it is a relaxation operator (eigenvalues either real or in the lower half of the complex plane). The moment problem related to a *Hermitian operator* is addressed as the *classical moment problem*, while by *relaxation moment problem* we mean the treatment of *relaxation operators*.

In the classical moment problem, one has to distinguish (not only for historical reasons) the case in which the eigenvalues have a lower bound and extend, say, along the positive real axis (*Stieltjes moment problem*), and the case in which the eigenvalues may extend along the whole real axis (*Hamburger moment problem*). While the Hamburger moment problem as well as the relaxation moment problem are solved in general via J-fractions, a Stieltjes moment problem can be solved also via S-fractions.

Consider a Hermitian or relaxation operator H and a state $|f_0\rangle$. The moments of H with respect to the state of interest $|f_0\rangle$ are defined as the diagonal matrix elements of H^n

$$\mu_n = \frac{\langle f_0|H^n|f_0\rangle}{\langle f_0|f_0\rangle} \tag{5.1}$$

When H is Hermitian, the power moments μ_n are real quantities, and even moments are positive; both properties are lost in the case of a general relaxation operator.

We wish now, having only moments as ingredients, to express the Green's function or resolvent

$$G_{00}(E) = \langle f_0 | \frac{1}{E - H} | f_0 \rangle \tag{5.2}$$

in the form of a continued fraction

$$G_{00}(E) = \cfrac{b_0^2}{E - a_0 - \cfrac{b_1^2}{E - a_1 - \cfrac{b_2^2}{E - a_2 - \cdots}}} \tag{5.3}$$

where E is defined in the usual way by adding an (infinitesimal) positive imaginary part; b_0^2 is the normalization factor of the state of interest, and without loss of generality we assume $b_0^2 = 1$. The parameters a_n are real and b_n^2 are positive if H is Hermitian. If in addition H has positive eigenvalues, the parameters a_n are positive.

The possibility of expressing *diagonal matrix elements $G_{00}(E)$ of the resolvent* as a continued fraction of type (5.3) is quaranteed by the following physical argument.*

Consider the standard dynamical equation

$$\frac{d}{dt}|f_0(t)\rangle = -iH|f_0(t)\rangle$$

and the autocorrelation function

$$\Phi_0(t) = \frac{\langle f_0 | f_0(t) \rangle}{\langle f_0 | f_0 \rangle} = \frac{\langle f_0 | e^{-iHt} | f_0 \rangle}{\langle f_0 | f_0 \rangle} \tag{5.4}$$

The Laplace transform of $\Phi_0(t)$ is

$$\hat{\Phi}_0(z) = \int_0^\infty e^{-zt} \Phi_0(t)\, dt$$

$$= \langle f_0 | \frac{1}{z + iH} | f_0 \rangle \qquad \mathrm{Re}\, z > 0 \tag{5.5}$$

---

*For a demonstration of the results mentioned in this section, see Chapter IV with explicit reference to operators and Chapter I with explicit reference to superoperators. The procedures followed in these chapters are formally different, although of course equivalent.

From Eqs. (5.2) and (5.5) we see that the diagonal matrix elements of the resolvent and the Laplace transform of an autocorrelation functions are related by

$$G_{00}(E) \equiv (-i)\hat{\Phi}_0(-iE) \qquad \text{Im } E > 0 \qquad (5.6)$$

The memory function approaches (Chapters I and IV) guarantee than an autocorrelation function satisfies the integro-differential equation

$$\frac{d\Phi_0(t)}{dt} = -ia_0\Phi_0(t) - b_1^2\int_0^t \Phi_1(t-\tau)\Phi_0(\tau)\,d\tau \qquad (5.7)$$

where the kernel $\Phi_1(t-\tau)$ is the memory function for $\Phi_0(t)$. If we Laplace transform Eq. (5.7), we obtain

$$z\hat{\Phi}_0(z) - 1 = -ia_0\hat{\Phi}_0(z) - b_1^2\hat{\Phi}_1(z)\hat{\Phi}_0(z)$$

that is,

$$\hat{\Phi}_0(z) = \frac{1}{z + ia_0 + b_1^2\hat{\Phi}_1(z)}$$

or

$$G_{00}(E) = (-i)\hat{\Phi}_0(-iE) = \frac{1}{E - a_0 - b_1^2 G_{11}(E)} \qquad (5.8)$$

The procedure can be iterated for $G_{11}(E)$ itself, and expansion (5.3) follows.

We remark explicitly that the above reasoning does not apply to cross-correlation functions; hence in general it is *not* possible to express *off-diagonal* matrix elements of the resolvent into a continued fraction of type (5.3). Whenever necessary, the off-diagonal matrix elements can be obtained by the following trick[15]:

$$G_{ab}(E) = \langle f_a| \frac{1}{E-H} |f_b\rangle$$

$$\equiv \frac{1}{4}\langle f_a + f_b| \frac{1}{E-H} |f_a + f_b\rangle - \frac{1}{4}\langle f_a - f_b| \frac{1}{E-H} |f_a - f_b\rangle$$

$$- \frac{i}{4}\langle f_a + if_b| \frac{1}{E-H} |f_a + if_b\rangle + \frac{i}{4}\langle f_a - if_b| \frac{1}{E-H} |f_a - if_b\rangle$$

$$(5.9)$$

where each of the four terms on the right-hand side can be expanded into a continued fraction. Because of result (5.9), from now on we can confine our attention only to diagonal matrix elements of the resolvent.

The expansion of $G_{00}(E)$ into a J-fraction is assured by the dynamical equations of motion, and we wish to obtain the explicit expression of the parameters in terms of moments. We start from the Dyson identity

$$\frac{1}{E-H} \equiv \frac{1}{E} + \frac{1}{E} H \frac{1}{E-H} \tag{5.10}$$

Exploiting the above identity repeatedly, using Eqs. (5.2) and (5.1), we obtain the formal series expansion

$$G_{00}(E) = \sum_{n=0}^{\infty} \frac{\mu_n}{E^{n+1}} \tag{5.11}$$

According to Theorem 4.3, we can associate with (5.11) the J-fraction

$$G_{00}(E) = \frac{1}{E-a_0-} \frac{b_1^2}{E-a_1-} \frac{b_2^2}{E-a_2-} \cdots \tag{5.12}$$

provided the Hankel determinants D_n formed with the sequence of moments $\{\mu_0, \mu_1, \ldots, \mu_n, \ldots\}$ are different from zero. The conditions

$$D_n = \begin{vmatrix} \mu_0 & \mu_1 & \cdots & \mu_n \\ \mu_1 & \mu_2 & \cdots & \mu_{n+1} \\ \cdots & \cdots & \cdots & \cdots \\ \mu_n & \mu_{n+1} & \cdots & \mu_{2n} \end{vmatrix} \neq 0 \qquad n = 0,1,2,\ldots \tag{5.13}$$

are equivalent to saying that the states $|f_0\rangle, H|f_0\rangle, \ldots, H^n|f_0\rangle, \ldots$ are linearly independent. According to Theorem 4.3 the parameters b_n^2 and a_n are given by the formulas

$$b_n^2 = \frac{D_n D_{n-2}}{D_{n-1}^2} \qquad n = 0,1,2,\ldots \tag{5.14}$$

$$a_n = \frac{R_n}{D_n} - \frac{R_{n-1}}{D_{n-1}} \qquad n = 0,1,2\ldots \tag{5.15}$$

$(D_{-1} = D_{-2} = 1; R_{-1} = 0)$.

A simple implementation of Theorem 4.3 shows also that if $D_n \neq 0$ for $n \leq \bar{n}$ and $D_n = 0$ for any $n > \bar{n}$, then $G_{00}(E)$ truncates exactly at the $\bar{n}$ approximant.

If, in addition to the conditions $D_n \neq 0$, we also have the conditions $\Omega_n \neq 0$, we can use, according to Theorem 4.2, the alternative form for the parameters

$$a_n = \frac{1}{\Omega_{n-1}} \left[\frac{D_{n-1}\Omega_n}{D_n} + \frac{D_n\Omega_{n-2}}{D_{n-1}} \right] \qquad n = 0,1,2,\ldots \qquad (5.16)$$

($\Omega_{-1} = 1$; $\Omega_{-2} = 0$). In this circumstance we can expand $G_{00}(E)$ also with an S-fraction.

We have already noticed that the b_n^2 are positive if H is Hermitian; b_n^2 can, in fact, be interpreted as the normalization factors of a hierarchy of states obtained using projection operators. The positiveness of b_n^2 implies the positiveness of D_n, and vice versa. The parameters a_n are positive if H is Hermitian with positive eigenvalues; a_n can in fact be interpreted as the expectation values of H on a hierarchy of states obtained using the memory function approaches. The positiveness of a_n implies that of Ω_n and vice versa.

B. The Classical Moment Problem

The moment problem has been almost exclusively studied in the literature having (implicitly) in mind Hermitian operators (classical moment problem). With the progress of the modern projective methods of statistical mechanics and the description of relaxation phenomena via effective non-Hermitian Hamiltonians or Liouvillians, it is important to consider the moment problem also in its generalized form. In this section we consider some specific aspects of the classical moment problem, and in Section V.C we focus on peculiar aspects of the relaxation moment problem.

It is possible to make evident some properties of the resolvent of a Hermitian operator, by using the identity

$$1 = \sum_\alpha |\psi_\alpha\rangle\langle\psi_\alpha| \qquad (5.17)$$

where ψ_α denotes the eigenstates of H with (real) energy E_α. Inserting the above identity in Eq. (5.2), we obtain

$$G_{00}(E + i\varepsilon) = \sum_\alpha \frac{1}{E + i\varepsilon - E_\alpha} |\langle\psi_\alpha|f_0\rangle|^2$$

$$= \sum_\alpha \frac{E - E_\alpha - i\varepsilon}{(E - E_\alpha)^2 + \varepsilon^2} |\langle\psi_\alpha|f_0\rangle|^2 \qquad \varepsilon > 0 \qquad (5.18)$$

From Eq. (5.18) we see the *Herglotz property* of $G_{00}(E)$: for $\mathrm{Im}\, E > 0$,

$G_{00}(E)$ is analytic and $\mathrm{Im}\,G_{00}(E) < 0$. We also see that any approximant of (5.3), where a_n are real and b_n^2 are positive, satisfy the Herglotz property since it *represents a half linear chain of variables of energies* a_n *and nearest-neighbor interaction* b_n.

A physical quantity of remarkable interest is the *projected density of states*, defined for Hermitian operators as

$$n(E) = \sum_{\alpha} |\langle \psi_\alpha | f_0 \rangle|^2 \delta(E - E_\alpha) \qquad (5.19)$$

From Eq. (5.18), keeping the limit $\varepsilon \to 0+$ and remembering that

$$\lim_{\varepsilon \to 0+} \frac{1}{\pi} \frac{\varepsilon}{(E - E_\alpha)^2 + \varepsilon^2} = \delta(E - E_\alpha)$$

we obtain the standard *spectral density theorem*[17]

$$n(E) = -\frac{1}{\pi} \lim_{\varepsilon \to 0+} \mathrm{Im}\,G_{00}(E + i\varepsilon) \qquad (5.20)$$

The Herglotz property of $G_{00}(E)$ and of the approximants $G_{00}^{(n)}(E)$ guarantees that $n(E)$ given by (5.20) or any approximate expression of the type $(-1/\pi)\mathrm{Im}\,G_{00}^{(n)}(E + i\varepsilon)$, $\varepsilon > 0$ (often used in practice) are definite positive.

The spectral density theorem, using Eq. (5.6), can also be written in terms of the correlation function $\hat{\Phi}_0(z)$ as

$$n(E) = -\frac{1}{\pi} \mathrm{Im}(-i)\hat{\Phi}_0(-iE) \qquad \mathrm{Im}\,E \to 0+$$

or equivalently

$$n(E) = \frac{1}{\pi} \lim_{\varepsilon \to 0+} \mathrm{Re}\,\hat{\Phi}_0(-iE + \varepsilon) \qquad (5.21)$$

which is the standard stationary response result used in statistical mechanics.[18]

Inserting the identity (5.17) into (5.1), the moments μ_n can be rewritten in the form

$$\mu_n = \sum_{\alpha} E_\alpha^n |\langle \psi_\alpha | f_0 \rangle|^2 \qquad (5.22)$$

In the case of a continuous distribution of eigenvalues, we have

$$\mu_n = \int_{-\infty}^{+\infty} E^n n(E) \, dE \qquad (5.23)$$

and μ_n are the familiar moments of nth order of the spectral density of states $n(E)$.

It is possible to rewrite $G_{00}(E)$ in a more effective form by inserting the unit operator (5.17) into (5.2); we have

$$G_{00}(E) = \int_{-\infty}^{+\infty} \frac{n(E')}{E - E'} \, dE' \qquad (5.24)$$

This integral, to be taken in the Stieltjes-Lebesgue sense, is known as *the Stieltjes transform* of the spectral density of states $n(E)$. In general the Stieltjes transform defines two analytic functions, one in the upper complex plane and the other in the lower complex plane, and we refer to the former consistently with the Herglotz properties of $G_{00}(E)$.

The sequence of real numbers $\{\mu_0, \mu_1, \mu_2, \dots\}$ associated by means of (5.23) with distribution function $n(E)$, are expected to have peculiar properties that we wish to investigate. The function $n(E)$ may reduce to a sum of δ-like functions (with positive coefficients), or may be different from zero in one or more finite connected regions (case of a model solid with one or more groups of energy bands), or may cover the positive or the entire real axis: *the only basic requirements on $n(E)$ are the positive definite property, normalization to 1, and existence of power moments.*

Given a sequence of real numbers $\{\mu_0, \mu_1, \dots\}$, we say that they constitute a *Hamburger sequence* if there is a definite positive function $n(E)$ satisfying

$$\mu_n = \int_{-\infty}^{+\infty} E^n n(E) \, dE \qquad n = 0, 1, 2, \dots \qquad (5.25)$$

A necessary and sufficient condition for the existence of a solution of the Hamburger moment problem[19] (5.25) is that the Hankel determinants $D_n > 0$, $n = 0, 1, 2, \dots$. Furthermore, if and only if $D_n > 0$ for $n < \bar{n}$ and $D_n = 0$ for any $n \geq \bar{n}$, then $n(E)$ reduces to the sum of $\bar{n}$ Dirac δ distributions (with positive coefficients).

In the case $D_n > 0$ for $n = 0, 1, 2, \dots$, the solution of the Hamburger moment problem may be *unique*, in which case we speak of a *determined moment problem*; or there may be infinitely many solutions and the moment problem is called *undetermined*. Notice that the possibility of an indeterminate case is peculiar to infinite intervals; in the case of a finite interval the solution of the moment problem, if it exists, is also unique.

As far as uniqueness is concerned, a number of sufficient conditions are known in the literature. A useful sufficient condition (see, e.g., ref. 4, p. 20) is that $n(E)$ decreases exponentially or faster for large E; this implies that μ_n grow no faster than $n!$ for large n.

It should be noticed that the original Stieltjes formulation of the moment problem considered the less general case in which the distribution function $n(E)$ could be different from zero only for positive values of E.

Given a sequence of real numbers $\{\mu_0, \mu_1, \dots\}$ we say that they constitute a *Stieltjes sequence* if there is a definite positive function $n(E)$ satisfying the equations

$$\mu_n = \int_0^\infty E^n n(E)\, dE \tag{5.26}$$

A necessary and sufficient condition for the existence of a solution of the Stieltjes moment problem (5.26) is that the Hankel determinants $D_n > 0$ and, in addition, $\Omega_n > 0$ ($n = 0, 1, 2, \dots$).

C. The Relaxation Moment Problem

It is not possible to extend right away the results of the classical moment problem to the relaxation moment problem.[20] However, our survey of Section V.B has been done in such a way that it is possible to select which relations maintain their validity in the relaxation moment problem and which are to be disregarded. Thus little remains to be said except for a few comments.

The moments μ_n are now in general complex numbers, and even moments are not necessarily positive real quantities. The Hankel determinants D_n are also in general complex numbers, and relations such as $D_n > 0$ or $\Omega_n > 0$ of the classical moment problem have no counterpart in the relaxation moment problem. Also, any relation of the classical moment problem involving integrals on the real axis E has no counterpart in the relaxation moment problem, where eigenvalues E can be anywhere in the lower half of the complex plane. These differences with respect to the classical situation find an easily understandable framework when interpreted within the projection operator techniques. In the relaxation case, in fact, one has to introduce a biorthogonal basis set for arriving at the continued fraction expansion of the resolvent (see Chapters I and IV), since H and $H^\dagger$ are now different from each other. In consequence, the parameters b_n^2 cannot be interpreted as ordinary normalization factors and can thus be complex numbers or negative numbers, although in general different from zero.

An important property which is preserved in the relaxation case is the Herglotz property of the Green's function $G_{00}(E)$. It can be expanded in the form

$$G_{00}(E) = \cfrac{1}{E - a_0 - \cfrac{b_1^2}{E - a_1 - \cfrac{b_2^2}{E - a_2 - \cdots}}} \qquad (5.27)$$

where a_n and b_n^2 are in general complex numbers, but still for $G_{00}(E)$ the Herglotz property holds.

As a final remark before closing this section, we emphasize that everything that has been said for Hermitian and relaxation operators also applies to Hermitian or relaxation superoperators (see also Chapters I and IV). The formal changes to be performed are trivial: the state of interest $|f_0\rangle$ is to be replaced by the operator of interest $|A_0\rangle$, the operator H by the superoperator $(-L)$ where $L = [H,\ldots]$, and the scalar product by a suitable average on an appropriate equilibrium distribution. The moments now have the form

$$\mu_n = \frac{\langle A_0|(-L)^n|A_0\rangle}{\langle A_0|A_0\rangle} \qquad (5.28)$$

All the mathematical apparatus of Hankel determinants and continued fractions expansion apply also to Hermitian or relaxation superoperators.

VI. PRODUCT-DIFFERENCE RECURSION ALGORITHMS

In Section V, we have formally provided simple expressions [Eqs. (5.14), (5.15), and (5.16)] that allow passing from the moments to the parameters of the continued fractions. From a purely algebraic point of view the situation is satisfactory, but not from an operative point of view, an aspect which has often been overlooked in the literature. Indeed, formulas based on Hankel determinants could hardly be used for steps up to $n \simeq 10$, because of numerical instabilities inherent in the moment problem.[20-22] On the other hand, in a variety of physical problems (typical are those encountered in solid state physics[15,21]), the number of moments practically accessible may be several tens up to 100 or so; the same happens in a number of "simulated models" of remarkable interest in determining the asymptotic behavior of continued fractions. In these cases, more convenient algorithms for the economical evaluation of Hankel determinants must be considered. But the point to be stressed is that in any case one must know the moments with a desired accuracy and, if necessary, in extended arithmetic precision[20]; otherwise one must resort to stabilizing procedures based on modified moments or/and recursion methods.

A Hankel determinant D_n is a function of $2n + 1$ independent parameters (the moments); yet when constructed explicitly it requires a matrix with $(n + 1)^2$ elements. The problem of finding efficient algorithms, which take into account the peculiar "persymmetric" structure of the Hankel matrices [left diagonals of (5.13) are formed with the same element], has been considered in the literature by several authors. We discuss here in detail a recent satisfactory solution[23] of this problem, obtained within the memory function formalism, and then compare it with other algorithms.

Consider the "correlation function"

$$\Phi_0(t) = \frac{\langle f_0 | e^{-iHt} | f_0 \rangle}{\langle f_0 | f_0 \rangle} = \sum_{n=0}^{\infty} \frac{\mu_n}{n!} (-it)^n \qquad \mu_0 = 1 \qquad (6.1)$$

Using the standard memory function techniques (Chapters I and IV), we have that $\Phi_0(t)$ satisfies the Volterra integro-differential equation

$$\frac{d\Phi_0(t)}{dt} = -ia_0\Phi_0(t) - b_1^2 \int_0^t \Phi_1(t - \tau)\Phi_0(\tau)\,d\tau \qquad (6.2)$$

It is convenient to expand the memory function $\Phi_1(t)$ into powers of t;

$$\Phi_1(t) = \sum_{n=0}^{\infty} \frac{\sigma_n}{n!} (-it)^n \qquad (6.3)$$

where σ_n are the yet unknown moments of the memory function $\Phi_1(t)$.

An early idea of expanding both the correlation function and its memory function in Taylor series is given in refs. 24 and 25. Notice that the convolution integral in (6.2) takes the form

$$\int_0^t \Phi_1(t - \tau)\Phi_0(\tau)\,d\tau = \sum_{n=0}^{\infty} \sum_{m=0}^{\infty} \sigma_m \mu_n \frac{i}{(m + n + 1)!} (-it)^{m+n+1} \qquad (6.4)$$

Substituting the power expansion (6.1), (6.3), and (6.4) into (6.2) and comparing term by term the coefficients with the same power of t, we obtain the following simple expression[23] of a_0, b_1^2 and σ_n:

1. The first pair of coefficients a_0 and b_1^2 are trivially given by

$$a_0 = \mu_1 \qquad b_1^2 = \mu_2 - \mu_1^2 \qquad (6.5)$$

2. In the case $b_1^2 \neq 0$, the moments σ_n of the memory function are related to the moments μ_n of the correlation function via the simple PD recursion relation

$$\sigma_n = \frac{\mu_{n+2} - \mu_1 \mu_{n+1}}{\mu_2 - \mu_1^2} - \sum_{m=0}^{n-1} \sigma_m \mu_{n-m} \qquad (6.6)$$

3. In the particular case $b_1^2 = 0$, the numerators in the fraction must be zero, that is,

$$\mu_n = \mu_1^n \qquad (6.7)$$

In the case of no truncation ($b_1^2 \neq 0$), the procedure can be repeated on the memory function itself, and we obtain for the next pair of coefficients

$$a_1 = \sigma_1 \qquad b_2^2 = \sigma_2 - \sigma_1^2$$

and we can proceed iteratively.

The memory function PD method can be recast in the following operative way.

One begins to construct an auxiliary M matrix, whose first row is filled with the moments ($\mu_0 = 1$ for convenience)

$$M_{0,n} = \mu_n \qquad (6.8)$$

The other rows are filled up using only *one* immediate predecessor row and the recursion formulas

$$M_{i,j} = \frac{M_{i-1,j+2} - M_{i-1,1} M_{i-1,j+1}}{M_{i-1,2} - M_{i-1,1}^2} - \sum_{m=0}^{j-1} M_{i,m} M_{i-1,j-m}$$

$$i \geq 1; \; j = 0,1,2,\ldots \quad (6.9)$$

From the elements of the second and third columns, one evaluates

$$a_i = M_{i,1} \qquad b_i^2 = M_{i,2} - M_{i,1}^2 \qquad (6.10)$$

With respect to the use of Hankel determinants, the memory function PD algorithm is very convenient because it needs only two arrays of $2n$ storage locations, instead of $\simeq n^2$ storage locations, to construct n steps of the continued fraction. It is thus possible to use multiple-precision arithmetic, when necessary, and to overcome round-off errors and numerical instabilities.

It is interesting to compare the memory function procedure with the well known Gordon algorithm.[26] In order to facilitate comparison with our method, we report it with an inessential change of rows and columns. The Gordon algorithm can be summarized as follows.

One begins to construct an auxiliary M matrix whose first row is filled with zeros except for the first element taken as unity, while the second row contains the moments

$$M_{0,0} = 1 \qquad M_{0,n} = 0 \qquad (n \geq 1)$$

$$M_{1,n} = \mu_n \qquad (n \geq 0) \tag{6.11}$$

The other rows are filled up using only *two* immediate predecessors and the PD recursion formulas

$$M_{i,j} = M_{i-1,1}M_{i-2,j+1} - M_{i-2,1}M_{i-1,j+1} \qquad i \geq 2; \, j \geq 0 \tag{6.12}$$

From the elements of the first column one evaluates

$$c_i = \frac{M_{i+1,0}^2}{M_{i+1,0}M_{i-1,0}} \qquad i \geq 1 \tag{6.13}$$

and obtains

$$h_0^2 = 1 \qquad b_n^2 - c_{2n}c_{2n+1}$$

$$a_0 = -c_2 \qquad a_n - -(c_{2n+1} + c_{2n+2}) \qquad n \geq 1 \tag{6.14}$$

The Gordon algorithm becomes inapplicable if some c_i vanishes.

As an illustrative example of the procedures described, we consider the parabolic spectral density, normalized to 1, given by

$$n(E) = \frac{1}{2\pi B^2} \sqrt{E(4B - E)} \tag{6.15}$$

for $0 \leq E \leq 4B$, which was originally introduced by Hubbard[27] to represent in solid state physics the essential features of a single connected band of width $W = 4B$. In this case the moments are

$$\mu_n = \int_{-\infty}^{+\infty} E^n n(E) \, dE$$

$$= 2^{n+3}(2B)^n \frac{(2n+1)!!}{(2n+4)!!} \qquad n = 0, 1, \ldots \tag{6.16}$$

TABLE I

Construction of the Auxiliary M Matrix for Evaluation of the Parameters of the Continued Fraction Starting from the Moments of the Hubbard Density of States. The Memory-Function PD Method[23] is applied.

		0	1	2	3	4	5	6	7	8	9	10	11
$a_0=2$	$b_1^2=1$	1	2	5	14	42	132	429	1430	4862	16796	58786	208012
$a_1=2$	$b_2^2=1$	1	2	5	14	42	132	429	1430	4862	16796		
$a_2=2$	$b_3^2=1$	1	2	5	14	42	132	429	1430				
$a_3=2$	$b_4^2=1$	1	2	5	14	42	132						
$a_4=2$	$b_5^2=1$	1	2	5	14								
$a_5=2$		1	2										

The moments μ_n up to 11 are reported in Table I, in units of B^n. Equation (6.5) gives $a_0=2$, $b_1^2=1$. We now apply the PD algorithm (6.9), and we see that in this particular case $\sigma_n \equiv \mu_n$ for all n. We immediately have that the continued fraction corresponding to the moment sequence (6.16) is the constant chain model

$$G_{00}(E) = \cfrac{1}{E-2B-\cfrac{B^2}{E-2B-\cfrac{B^2}{E-2B-\cdots}}} \qquad (6.17)$$

In Table II we report the same illustrative example following the Gordon procedure. The convenience of the memory function method is apparent. Furthermore, the Gordon method fails unless $\Omega_n \neq 0$, a restriction which is overcome in the memory function method. Another quite interesting P-algorithm has been provided by Hänggi et al., and we refer to the original papers[28] for illustration and discussion of stability aspects.

A glance at Tables I and II shows the ill-conditioned nature of the moment problem. We can easily see the internal difficulty[29] which appears in handling power moments if we note that in the simple example of a given connected density of states different from zero in the interval $[0,1]$ the power moments μ_n reproduce well the density $n(E)$ near 1, while information of its behavior near 0 is very poorly accounted, due to the limited weight of E^n for $E \to 0$ and increasing values of n. In the diagrammatic expansion of μ_n on a lattice, one sees that the moment μ_n contains large contributions due to low-order moments with respect to contributions which are properly of order n in μ_n. It would be desirable to use "modified moments," in which

TABLE II

Construction of the Auxiliary Matrix for Evaluation of the Parameters of the Continued Fraction Starting from the Moments of the Hubbard Density of States. The Gordon PD Method[26] is Applied.

0	1	2	3	4	5	6	7	8	9	10	11
1	0	0	0	0	0	0	0	0	0	0	0
1	2	5	14	42	132	429	1,430	4,862	16,796	58,786	
-2	-5	-14	-42	-132	-429	-1,430	-4,862	-16,796	-58,786		
1	4	14	48	165	572	2,002	7,072	25,194			
3	14	54	198	715	2,574	9,282	33,592				
-2	-12	-54	-220	-858	-3,276	-12,376					
8	54	264	1,144	4,680	18,564						
12	96	528	2,496	10,920							
-120	-1,056	-6,240	-31,200								
1,152	11,520	74,880									
165,888	1,797,120										
-159,252,480											

nth powers of E are replaced by appropriate polynomials of degree n, more "balanced" in the interval of interest. The techniques to deal with modified moments are based on the properties of orthogonal polynomials, which are the subject of the next section.

VII.   PROPERTIES OF ORTHOGONAL POLYNOMIALS

A.   Definitions and basic properties

It is well known that orthogonal polynomials originate from the theory of continued fractions (see, for instance, the books 5, 7, 30, and 31 and references therein). We briefly review now the definitions and the basic properties of these polynomials and their connection with continued fractions, with the main purpose of emphasizing the aspects which are at the basis of the recursion methods.

Let $dN(E) = n(E)\,dE$ be a given distribution in the interval $[E_a, E_b]$ with $n(E)$ a nonnegative function measurable in the Lebesgue's sense; moreover, let $N(E)$ be absolutely continuous and well behaved at $\pm\infty$.[32] The Stieltjes-Lebesgue integral

$$\langle f|g\rangle = \int_{E_a}^{E_b} f(E)g(E)\,dN(E) \tag{7.1}$$

defines the *scalar product* between any two real functions $f(E)$ and $g(E)$; if $\langle f|g\rangle = 0$, we say that f and g are orthogonal with respect to the distribution $dN(E)$.

Starting from a sequence of independent real functions of the class $L_N^2(a, b)$:

$$f_0(E), f_1(E), \ldots, f_n(E)$$

n finite or infinite, we can easily obtain an orthonormal set

$$\phi_0(E), \phi_1(E), \ldots, \phi_n(E)$$

as

$$\phi_n(E) = \frac{1}{\sqrt{\Delta_n \Delta_{n-1}}}
\begin{vmatrix}
\langle f_0|f_0\rangle & \langle f_0|f_1\rangle & \cdots & \langle f_0|f_n\rangle \\
\vdots & & & \vdots \\
\langle f_{n-1}|f_0\rangle & \langle f_{n-1}|f_1\rangle & \cdots & \langle f_{n-1}|f_n\rangle \\
f_0(E) & f_1(E) & \cdots & f_n(E)
\end{vmatrix}$$

$$n = 0, 1, 2, \ldots \tag{7.2}$$

where $\Delta_{-1} = 1$ and

$$\Delta_n = \begin{vmatrix} \langle f_0|f_0\rangle & \cdots & \langle f_0|f_n\rangle \\ \vdots & & \vdots \\ \langle f_n|f_0\rangle & \cdots & \langle f_n|f_n\rangle \end{vmatrix} > 0 \qquad n = 0,1,2,\ldots \qquad (7.3)$$

Notice that $\phi_n(E) = \sqrt{\Delta_{n-1}/\Delta_n}\, f_n(E) + \cdots$, that is, the coefficient by which $f_n(E)$ enters in $\phi_n(E)$ is certainly positive and different from zero.

Let us consider now a finite or infinite interval $[E_a, E_b]$ and suppose that the moments

$$\mu_n = \int_{E_a}^{E_b} E^n\, dN(E) \qquad n = 0,1,2,\ldots$$

exist. Following the procedure just presented, we can now obtain from the sequence of *powers*

$$1, E, E^2, E^3, \ldots, E^n, \ldots$$

a set of *orthonormal polynomials*

$$P_0(E), P_1(E), \ldots, P_n(E), \ldots$$

which are given as before by

$$P_n(E) = \frac{1}{\sqrt{D_n D_{n-1}}} \begin{vmatrix} \mu_0 & \mu_1 & \mu_2 & \cdots & \mu_n \\ \cdots & \cdots & \cdots & \cdots & \cdots \\ \mu_{n-1} & \mu_n & \mu_{n+1} & \cdots & \mu_{2n-1} \\ 1 & E & E^2 & \cdots & E^n \end{vmatrix} \qquad n \geq 1 \quad (7.4)$$

and

$$D_n = \begin{vmatrix} \mu_0 & \mu_1 & \cdots & \mu_n \\ \mu_1 & \mu_2 & \cdots & \mu_{n+1} \\ \cdots & \cdots & \cdots & \cdots \\ \mu_n & \mu_{n+1} & \cdots & \mu_{2n} \end{vmatrix} > 0 \qquad n \geq 0 \qquad (7.5)$$

$[D_{-1} = 1;\ P_0(E) = 1]$. The expression of D_n can be recognized as the standard expression of Hankel determinants, which are known to be essentially positive quantities (in the classical moment problem).

From the definition (7.4), we can write the orthogonal polynomials $P_n(E)$ in the form

$$P_n(E) = \sqrt{\frac{D_{n-1}}{D_n}}\, E^n - \frac{R_{n-1}}{\sqrt{D_n D_{n-1}}}\, E^{n-1} + \cdots \qquad (7.6)$$

where R_n are the modified Hankel determinants, already encountered in previous sections and defined by

$$
R_n = \begin{vmatrix} \mu_0 & \mu_1 & \cdots & \mu_n \\ \mu_{n-1} & \mu_n & \cdots & \mu_{2n-1} \\ \mu_{n+1} & \mu_{n+2} & \cdots & \mu_{2n+1} \end{vmatrix} \tag{7.7}
$$

From (7.6) we see that $P_n(E)$ is a polynomial of degree n with positive coefficients multiplying E^n.

From the definition (7.4) and the standard property of a determinant to be zero when two rows (or columns) are equal, we easily obtain

$$
\int E^m P_n(E)\, dN(E) = \begin{cases} 0 & \text{if } m < n \\ \sqrt{\dfrac{D_n}{D_{n-1}}} & \text{if } m = n \\ \dfrac{R_n}{\sqrt{D_n D_{n-1}}} & \text{if } m = n+1 \\ \cdots\cdots\cdots\cdots \end{cases} \tag{7.8}
$$

Using Eqs. (7.8) it is easily verified that the orthogonal polynomials $\{P_n(E)\}$ satisfy the orthonormality condition

$$
\int P_m(E)P_n(E)\, dN(E) = \delta_{mn} \qquad m,n = 0,1,2,\ldots \tag{7.9}
$$

In dealing with orthogonal polynomials we have encountered Hankel determinants and modified Hankel determinants, which were an important tool in discussing power series and continued fractions. Thus we may expect a close analogy between the theory of orthogonal polynomials and that of continued fractions. This expectation is corroborated by the following theorem.

Theorem. For any three consecutive orthonormal polynomials defined by Eq. (7.4), we have the recurrence formula

$$
b_{n+1}P_{n+1}(E) = (E - a_n)P_n(E) - b_n P_{n-1}(E) \qquad n = 1,2,3,\ldots \tag{7.10}
$$

where a_n and b_n are real constants and the latter are positive. Furthermore, the values of a_n and b_n^2 are given by

$$b_n^2 = \frac{D_n D_{n-2}}{D_{n-1}^2} \qquad n = 1, 2, 3, \dots \qquad (7.11)$$

$$a_n = \frac{R_n}{D_n} - \frac{R_{n-1}}{D_{n-1}} \qquad n = 1, 2, \dots \qquad (7.12)$$

$(b_0^2 = 1; \ a_0 = \mu_1)$.

Proof. We use only the definitions and assumptions of this section; we obtain thus an elegant and alternative demonstration of the by now familiar expressions (7.11) and (7.12).

To prove the important general property that orthogonal polynomials are connected by a three-term recurrence relation, we consider the polynomial $b_{n+1}P_{n+1}(E) - EP_n(E)$, whose degree is n provided the b_n^2 are defined by (7.11). Thus it can be expanded as a linear combination of polynomials up to $P_n(E)$:

$$b_{n+1}P_{n+1}(E) - EP_n(E) = \lambda_0 P_0(E) + \lambda_1 P_1(E) + \cdots + \lambda_n P_n(E)$$

$$(7.13)$$

The values of λ_i $(i = 0, 1, \dots, n)$ can be obtained by multiplying both members of Eq. (7.13) by $P_i(E)$ and integrating on the distribution function $dN(E)$. We easily see that $\lambda_i = 0$ if $i \leq n - 2$, because of orthogonality. Thus the three-term recurrence formula (7.10) follows.

The second part of the theorem is a straightforward consequence of the first part.

To verify that b_n^2 do satisfy Eq. (7.11), let us multiply both members of Eq. (7.10) by $P_{n-1}(E)$ and integrate on the distribution function $dN(E)$; we have

$$b_n = \int P_n(E) E P_{n-1}(E) \, dN(E)$$

$$= \int P_n(E) E \left\{ \sqrt{\frac{D_{n-2}}{D_{n-1}}} E^{n-1} + \cdots \right\} dN(E) = \sqrt{\frac{D_n D_{n-2}}{D_{n-1}^2}}$$

because of Eq. (7.8).

To verify that a_n satisfy Eq. (7.12), let us multiply both members of Eq. (7.10) by $P_n(E)$ and integrate on the distribution function $dN(E)$; we have

$$a_n = \int P_n(E) E P_n(E) \, dN(E)$$

$$= \int P_n(E) E \left\{ \sqrt{\frac{D_{n-1}}{D_n}} E^n - \frac{R_{n-1}}{\sqrt{D_n D_{n-1}}} E^{n-1} + \cdots \right\} dN(E) = \frac{R_n}{D_n} - \frac{R_{n-1}}{D_{n-1}}$$

and Eq. (7.12) is thus proved too. Notice the simplicity and elegance of using arguments based on orthogonal polynomials.

Often in physics one encounters a sequence of polynomials related by three-term recurrence relations; one could ask if there is a definite positive function $n(E)$ for which the set of polynomials are orthogonal and which are the parameters a_n and b_n^2 of the recurrence relations. It is easily verified that most of the polynomials of common usage in physics (Legendre polynomials, Laguerre polynomials, Tchebycheff polynomials, etc.) constitute an orthogonal set with respect to a definite positive weight function. For convenience, we report in Table III a few classes of orthogonal polynomials, together with the weight function and the parameters a_n and b_n^2, and we refer to the literature[31,33] for a more extensive list.

We wish to add now another important property of orthogonal polynomials.[5-7,30-34] From the recurrence formula (7.10) the *Christoffel-Darboux* relation can be easily deduced:

$$\sum_{i=0}^{n} P_i(E_1) P_i(E_2) = b_n \frac{P_{n+1}(E_1) P_n(E_2) - P_n(E_1) P_{n+1}(E_2)}{E_1 - E_2} \quad (7.14)$$

which for $E_1 = E_2$ can be expressed in terms of the derivatives of the polynomials with respect to E:

$$\sum_{i=0}^{n} P_i^2(E) = b_n \left\{ P_{n+1}'(E) P_n(E) - P_n'(E) P_{n+1}(E) \right\} \quad (7.15)$$

Thus for real E, $P_n(E)$ and $P_{n+1}(E)$ cannot have common zeros.

Peculiar properties of the zeros of orthogonal polynomials come from the fact that they are zeros of a *Sturm sequence* $\{ P_n(E) \}$; in fact;

1. $P_0(E)$ is a constant $\neq 0$, and $P_n(E)$ is of "precise degree n."
2. If $P_n(E_0) = 0$ for some $n \geq 1$, it follows from the recurrence relation that $P_{n-1}(E_0) P_{n+1}(E_0) < 0$.
3. If in E_0, $P_n(E_0) = 0$, we have from (7.15) that $P_n'(E_0) P_{n-1}(E_0) > 0$.

So we can say that the zeros of $P_n(E)$ are real and distinct in the interval

TABLE III

Illustrative List of Some Orthogonal Polynomials with the Interval of Definition, the Corresponding Weights, and Continued Fraction Parameters

Name of Polynomials and Notations	Interval	Weight Function $n(E)$	Parameters of the Recurrence Relations for the Monic Polynomials: $f_{n+1}(E) = (E - a_n)f_n(E) - b_n^2 f_{n-1}(E)$ $\quad(b_0^2 \equiv 1)$
Tchebycheff of the first kind, $T_n(E)$	$[-1,1]$	$(1-E^2)^{-1/2}$	$a_n = 0$, $\quad n=0,1,2,\ldots$ $b_1^2 = \frac{1}{2}$ $b_n^2 = \frac{1}{4}$, $\quad n=2,3,4,\ldots$
Tchebycheff of the second kind, $U_n(E)$	$[-1,1]$	$(1-E^2)^{1/2}$	$a_n = 0$, $\quad n=0,1,2,\ldots$ $b_n^2 = \frac{1}{4}$, $\quad n=1,2,3,\ldots$
Shifted Tchebycheff of the first kind, $T_n^*(E)$	$[0,1]$	$(E-E^2)^{-1/2}$	$a_n = \frac{1}{2}$, $\quad n=0,1,2,\ldots$ $b_1^2 = \frac{1}{8}$ $b_n^2 = \frac{1}{16}$, $\quad n=2,3,4,\ldots$
Shifted Tchebycheff of the second kind, $U_n^*(E)$	$[0,1]$	$(E-E^2)^{1/2}$	$a_n = \frac{1}{2}$, $\quad n=0,1,2,\ldots$ $b_n^2 = \frac{1}{16}$, $\quad n=1,2,3,\ldots$
Legendre, $P_n(E)$	$[-1,1]$	1	$a_n = 0$, $\quad n=0,1,2,\ldots$ $b_n^2 = n^2/(4n^2-1)$, $\quad n=1,2,3,\ldots$
Shifted Legendre, $P_n^*(E)$	$[0,1]$	1	$a_n = \frac{1}{2}$, $\quad n=0,1,2,\ldots$ $b_n^2 = n^2/4(4n^2-1)$, $\quad n=1,2,3,\ldots$
Laguerre, $L_n(E)$	$[0,\infty]$	e^{-E}	$a_n = 2n+1$, $\quad n=0,1,2,\ldots$ $b_n^2 = n^2$, $\quad n=1,2,3,\ldots$
Hermite, $H_n(E)$	$[-\infty,+\infty]$	e^{-E^2}	$a_n = 0$, $\quad n=0,1,2,\ldots$ $b_n^2 = n/2$, $\quad n=1,2,3,\ldots$
Hermite, $He_n(E)$	$[-\infty,+\infty]$	$e^{-E^2/2}$	$a_n = 0$, $\quad n=0,1,2,\ldots$ $b_n^2 = n$, $\quad n=1,2,3,\ldots$

$[E_a, E_b]$. A further useful property is expressed by the *separation theorem*, which guarantees that between two successive zeros of $P_n(E)$ there is only one zero of $P_{n+1}(E)$.

Before closing this section, we wish briefly to comment on how to deal with the relaxation moment problem; in this case the usual definition of moments,

$$\mu_n = \int E^n \, dN(E) \qquad (7.16)$$

has no meaning, and we have to consider the more general form

$$\mu_n = \langle f_0 | H^n | f_0 \rangle \qquad (7.17)$$

where H is a non-Hermitian operator of the relaxation type.[20] Now the scalar product between two polynomials cannot be defined as the integration on the real axis E of the product of the two polynomials with an appropriate distribution function, as the eigenvalues of a non-Hermitian relaxation operator are spread in the whole lower half of the complex plane. We can define a *formal scalar product*[2,35] as the operator which, acting on polynomials, replaces powers of E of order n with the corresponding moments μ_n defined by

$$\mathscr{F}\left(a_0 + a_1 E + a_2 E^2 + \cdots + a_n E^n + \cdots\right)$$
$$= a_0 \mu_0 + a_1 \mu_1 + a_2 \mu_2 + \cdots + a_n \mu_n + \cdots$$

Most of the properties so far described remain valid, as the formal scalar product reduces to standard integration on a distribution function in the Hermitian case. For instance, the theorem of this section remains valid, dropping the restriction that a_n and b_n are real. Of course, the peculiar properties of the zeros of orthogonal polynomials remain valid only for real polynomials. From now on we consider explicitly only the latter situation.

B. Relation between Orthogonal Polynomials and Continued Fractions

The theorem of Section VII.A already suggests the equivalence of the mathematical tools of continued fractions and orthogonal polynomials. We further clarify this connection.

The orthonormalized polynomials $\{P_n(E)\}$ contain the powers of E^n with coefficient $\sqrt{D_{n-1}/D_n}$. It may be convenient to consider the set of orthogonal (but not normalized) polynomials $\{\Pi_n(E)\}$ defined by

$$\Pi_n(E) = \sqrt{\frac{D_n}{D_{n-1}}} \, P_n(E) \tag{7.18}$$

These polynomials $\{\Pi_n(E)\}$ contain E^n with coefficient $+1$ and are called *monic*. The three-term recurrence relation (7.10), using (7.18), becomes

$$\Pi_{n+1}(E) = (E - a_n)\Pi_n(E) - b_n^2\Pi_{n-1}(E) \tag{7.19}$$

Consider now the continued fraction of the type

$$F(E) = \cfrac{1}{E - a_0 - \cfrac{b_1^2}{E - a_1 - \cfrac{b_2^2}{E - a_2 - \cdots}}} \tag{7.20}$$

The recurrence formulas (2.3) specified for the numerators $A_n(E)$ and denominators $B_n(E)$ of the nth approximant of the continued fraction (7.20) are

$$A_{n+1}(E) = (E - a_n)A_n(E) \quad b_n^2 A_{n-1}(E)$$
$$B_{n+1}(E) = (E - a_n)B_n(E) - b_n^2 B_{n-1}(E) \qquad n = 0,1,2,\ldots \tag{7.21}$$

with the initial conditions

$$A_{-1} = -1 \qquad A_0 = 0 \tag{7.22a}$$
$$B_{-1} = 0 \qquad B_0 = 1 \tag{7.22b}$$

Thus the denominators $B_n(E)$ coincide with the orthogonal polynomials $\Pi_n(E)$.

Both $A_n(E)$ and $B_n(E)$ obey Eq. (7.21), but with the different initial conditions specified in (7.22). The $\{B_n(E)\}$ are said to be regular solutions of (7.19), while $\{A_n(E)\}$ are said to be irregular solutions.

Let us consider once more the fundamental recurrence relation for the orthonormalized polynomials $\{P_n(E)\}$:

$$b_{n+1}P_{n+1}(E) = (E - a_n)P_n(E) - b_n P_{n-1}(E) \tag{7.23}$$

Equations (7.23) can be considered as a set of linear equations

$$
\left.\begin{array}{r}
(a_0 - E)P_0(E) + b_1 P_1(E) = 0 \\
b_1 P_0(E) + (a_1 - E)P_1(E) + b_2 P_2(E) = 0 \\
b_2 P_1(E) + (a_2 - E)P_2(E) + b_3 P_3(E) = 0 \\
\cdots\cdots\cdots\cdots\cdots\cdots\cdots\cdots\cdots\cdots
\end{array}\right\} \tag{7.24}
$$

The infinite sequence (7.24) terminates if we consider the values $E = E_i$ at which $P_n(E)$ vanishes. Then the solution of the homogeneous system of the remaining n linear equations is connected with the familiar determinantal relation

$$
\begin{vmatrix}
a_0 - E & b_1 & & \vdots & \\
b_1 & a_1 - E & b_2 & \vdots & \\
& b_2 & a_2 - E & \vdots & \\
\cdots & \cdots & \cdots & \ddots & \cdots \\
& & & & b_{n-1} \\
& & \vdots & b_{n-1} & a_{n-1} - E
\end{vmatrix} = 0 \tag{7.25}
$$

The roots E_i of $P_n(E)$ are thus the eigenvalues of the symmetric matrix of order n (appearing in 7.25), and the $P_n(E_i)$ play the role of eigenvectors

$$
\mathbf{v}_i = \frac{1}{\sqrt{\displaystyle\sum_{k=0}^{n-1} P_k^2(E_i)}} \{ P_0(E_i), P_1(E_i), \ldots, P_{n-1}(E_i) \} \tag{7.26}
$$

So if we consider the value of orthonormal polynomials $P_n(E)$ at the energies E_i at which the first neglected polynomial vanishes, a new orthogonality relation is achieved from the orthogonality property of the matrix of eigenvectors:

$$
\sum_{i=0}^{n-1} \rho(E_i) P_s(E_i) P_t(E_i) = \delta_{st} \tag{7.27}
$$

where now normalization plays the role of a weight function

$$
\rho(E_i) = \frac{1}{\displaystyle\sum_{k=0}^{n-1} P_k^2(E_i)}
$$

It is important to note that these weight functions are exactly the *weights of the Gaussian quadrature*[34]; this fact will be useful in the discussion of error bounds.

As is well known to numerical analysts, the approach of diagonalizing a tridiagonal matrix is better conditioned than the problem of finding the roots of polynomials. If the latter procedure is followed, one must be prepared to use extended precision arithmetic in the computer programs.

We wish to add a few more comments on the Gaussian quadrature procedure. Consider an integral of the type

$$I = \int f(E)n(E)\,dE \qquad (7.28)$$

where $n(E)$ is a positive definite function (leading to a determinate moment problem). Notice that (7.28) includes integrals of the type $\int_a^b f(E)\,dE$, which correspond to $n(E)$ unity in the finite interval $[a, b]$ and zero otherwise.

From the knowledge of $2N+1$ moments of $n(E)$, we can construct the continued fraction approximant

$$F_N(E) = \cfrac{1}{E - a_0 - \cfrac{b_1^2}{\ddots - \cfrac{b_N^2}{E - a_N}}} = \left\langle \phi_0 \left| \frac{1}{E - H_N} \right| \phi_0 \right\rangle \qquad (7.29)$$

The model Hamiltonian H_N can be written as

$$H_N = a_0|\phi_0\rangle\langle\phi_0| + b_1|\phi_0\rangle\langle\phi_1| + b_1|\phi_1\rangle\langle\phi_0| + \cdots + a_N|\phi_N\rangle\langle\phi_N|$$

and can be indicated graphically as in Fig. 2.

We can diagonalize the Hamiltonian H_N and obtain the eigenvalues $E_0^{(N)}, E_1^{(N)}, \ldots E_N^{(N)}$ [which are the roots of $P_{N+1}(E)$] and the eigenvectors

$$|\psi_i\rangle = \sum_{m=0}^{N} c_{im}|\phi_m\rangle \quad \text{with } c_{im} = P_m\big(E_i^{(N)}\big) \frac{1}{\sqrt{\displaystyle\sum_{m'=0}^{N} P_{m'}\big(E_i^{(N)}\big)}}$$

Expressing in Eq. (7.29) $|\phi_0\rangle$ in terms of $|\psi_i\rangle$, we have

$$|\phi_0\rangle = \sum_{i=0}^{N} |\psi_i\rangle\langle\psi_i|\phi_0\rangle = \sum_{i=0}^{N} c_{i0}|\psi_i\rangle$$

Figure 2.   Graphical representation of the chain model Hamiltonian H_N.

and

$$\langle \phi_0 | \frac{1}{E - H_N} | \phi_0 \rangle = \sum_{i=0}^{N} \rho(E_i^{(N)}) \frac{1}{E - E_i^{(N)}}$$

where

$$\rho(E_i^{(N)}) = |c_{i0}|^2 = \frac{1}{\sum\limits_{m=0}^{N} P_m^2(E_i^{(N)})} \tag{7.30}$$

Then

$$n^{(N)}(E) = -\frac{1}{\pi} \lim_{\varepsilon \to 0+} \langle \phi_0 | \frac{1}{E + i\varepsilon - H_N} | \phi_0 \rangle = \sum_{i=0}^{N} \rho(E_i^{(N)}) \delta(E - E_i^{(N)})$$

Remembering that $n(E) = \lim\limits_{N \to \infty} n^{(N)}(E)$, we obtain the standard expression for Gaussian quadrature:

$$I = \int f(E) n(E) \, dE \simeq \int f(E) n^{(N)}(E) \, dE = \sum_{i=0}^{N} f(E_i^{(N)}) \rho(E_i^{(N)}) \tag{7.31}$$

The integrals of the form (7.28) can thus be expressed in terms of the values of $f(E)$ at selected points, which are the zeros of $P_{N+1}(E)$ with suitable weights.

The approximate Gaussian quadrature (7.31) becomes *exact if $f(E)$ is any polynomial of degree $m \leq 2N + 1$*. In fact, in this case we construct the

Lagrange interpolation polynomial of degree N,

$$F_N(E) = \sum_{k=0}^{N} f(E_k^{(N)}) \frac{P_{N+1}(E)}{P'_{N+1}(E_k^{(N)})(E - E_k^{(N)})} \qquad (7.32)$$

We can easily write

$$f(E) = F_N(E) + P_{N+1}(E) \cdot r_N(E)$$

where $r_N(E)$ is a polynomial of degree N [and is thus orthogonal to $P_{N+1}(E)$ because of relations (7.9)]. We thus have

$$I = \int f(E) n(E) \, dE = \int F_N(E) n(E) \, dE$$

$$= \sum_{k=0}^{N} f(E_k^{(N)}) w(E_k^{(N)})$$

where the weights w are defined as

$$w(E_k^{(N)}) = \int \frac{P_{N+1}(E)}{P'_{N+1}(E_k^{(N)})(E - E_k^{(N)})} n(E) \, dE \qquad (7.33)$$

The weights are completely independent of the type of function $f(E)$.

We want now to show that the weights defined in Eq. (7.33) are exactly the same as those of Eq. (7.30). Let us start from the Christoffel-Darboux identity (7.14): if we choose E_2 as a zero of $P_{N+1}(E)$ and call $E_1 = E$, we have

$$\sum_{i=0}^{N} P_i(E) P_i(E_2) = b_N \frac{P_{N+1}(E) P_N(E_2)}{E - E_2}$$

We have then

$$w(E_2^{(N)}) = \int \frac{P_{N+1}(E)}{P'_{N+1}(E_2^{(N)})(E - E_2^{(N)})} n(E) \, dE$$

$$= \frac{1}{P'_{N+1}(E_2^{(N)}) P_N(E_2^{(N)})} \int \frac{P_{N+1}(E) P_N(E_2^{(N)})}{E - E_2^{(N)}} n(E) \, dE$$

$$= \frac{1}{P'_{N+1}(E_2^{(N)}) P_N(E_2^{(N)})} \frac{1}{b_N} \sum_{i=0}^{N} P_i(E_2^{(N)}) \int P_i(E) n(E) \, dE$$

Using again orthogonality relations for the sequence $\{P_N(E)\}$, we have in the above integral a contribution only for $i = 0$,

$$w\left(E_2^{(N)}\right) = \frac{1}{b_N P'_{N+1}\left(E_2^{(N)}\right) P_N\left(E_2^{(N)}\right)}$$

We use then the particular form (7.15) of the Christoffel-Darboux identity, taking into account that $E_2^{(N)}$ is a zero of $P_{N+1}(E)$,

$$\sum_{i=0}^{N} P_i^2\left(E_2^{(N)}\right) = b_N P'_{N+1}\left(E_2^{(N)}\right) P_N\left(E_2^{(N)}\right)$$

from which we have the result

$$w\left(E_2^{(N)}\right) = \frac{1}{\displaystyle\sum_{i=0}^{N} P_i^2\left(E_2^{(N)}\right)}$$

which is exactly Eq. (7.30).

Finally we note that if $f(E)$ is a polynomial of degree $m > 2N + 1$, the integration (7.31) is no longer exact, and to retain its validity we must introduce an error term on the right-hand side.

C.   Orthogonal Polynomials and Modified Moments

It is well known that the determination of abscissas and weights of the Gaussian quadrature from power moments is an exponentially ill-conditioned problem due to the presence of rounding errors.[36]

As suggested by Sack and Donovan[37] and Gautschi,[38] the introduction of *modified moments* has considerably contributed to a better conditioning of this problem. Let us examine this problem following the paper of Blumstein and Wheeler.[29]

Consider the standard power moments

$$\mu_k = \int E^k n(E)\, dE$$

of a given density of states $n(E)$, not yet known. Let us indicate by $\{P_n(E)\}$ the monic orthogonal polynomials associated with $n(E)$ that satisfy the recurrence relations

$$P_{n+1}(E) = (E - a_n) P_n(E) - b_n^2 P_{n-1}(E)$$

If we have some overall information on $n(E)$, it is convenient to consider a weight function $\bar{n}(E)$, different from zero in the same interval as $n(E)$, whose orthogonal polynomials $\{\bar{P}_n(E)\}$ and parameters $\bar{a}_n$, $\bar{b}_n^2$ are known. It is then convenient to introduce the modified moments

$$\nu_k = \int \bar{P}_k(E)n(E)\,dE \qquad (7.34)$$

The transformation from μ_k to ν_k can be easily performed by introducing the matrix M defined as

$$M_{i,j} = \int \bar{P}_i(E)E^j n(E)\,dE \qquad i,j = 0,1,2,\ldots$$

By construction, the first row of M contains the power moments, and the first column the modified moments.

To work out the M matrix, let us introduce the recursion relations

$$\bar{P}_{i+1}(E) = (E - \bar{a}_i)\bar{P}_i(E) - \bar{b}_i^2 \bar{P}_{i-1}(E)$$

in

$$M_{i+1,j} = \int \bar{P}_{i+1}(E)E^j n(E)\,dE$$

We have

$$M_{i+1,j} = M_{i,j+1} - \bar{a}_i M_{ij} - \bar{b}_i^2 M_{i-1,j} \qquad (7.35)$$

($M_{-1,j} = 0$, $M_{0,j} = \mu_j$, and $\bar{b}_0^2 = 1$). Then from the first row of matrix M we can generate successive rows using (7.35), and the modified moments are so obtained.

From the knowledge of the modified moments, we can now obtain the desired parameters a_n and b_n^2 corresponding to $n(E)$, as follows.

Consider the matrix N defined as

$$N_{ij} = \int P_i(E)\bar{P}_j(E)n(E)\,dE$$

The modified moments fill the first row of matrix N. We have

$$N_{i+1,j} = \int (E - a_i) P_i(E) \bar{P}_j(E) n(E) \, dE - b_i^2 \int P_{i-1}(E) \bar{P}_j(E) n(E) \, dE$$

$$= N_{i,j+1} - (a_i - \bar{a}_i) N_{i,j} - b_i^2 N_{i-1,j} + \bar{b}_i^2 N_{i,j-1}$$

The use of the orthogonality conditions (see Section VII.A) assure that

$$N_{i,j} = 0 \qquad \text{if } i > j$$

and allows a determination of a_i and b_i^2. In fact, from

$$N_{i+1,i-1} \equiv 0 \qquad \text{and} \qquad N_{i+1,i} \equiv 0$$

one easily obtains

$$b_i^2 = \frac{N_{ii}}{N_{i-1,i-1}}$$

and

$$a_i = \bar{a}_i + \frac{N_{i,i+1}}{N_{ii}} - \frac{N_{i-1,i}}{N_{i-1,i-1}}$$

The determination of the sequence $\{a_i\}$ and $\{b_i^2\}$ from modified moments is well conditioned if $\bar{n}(E)$ is reasonably near to $n(E)$.

The modified moments are useful whenever it is possible to guess an auxiliary distribution $\bar{n}(E)$ closely simulating the actual one. However, in the construction of the N matrix, the principal ingredients are just the modified moments, whose determination from the power moment is still a notoriously ill-conditioned problem.

Recently Lambin and Gaspard[39] presented an implemented version of the modified moments method, which calculates *directly* the modified moments for Hamiltonians of the tight binding form. We will discuss this issue in connection with solid state physics problems (see Chapter IV).

We want to note finally that, as pointed out by Blumstein and Wheeler[29] and by Magnus,[40] the results of this section can also be obtained within the original procedure of Gautshi,[38] starting from the knowledge of the Gram matrix corresponding to the sequence $\{\bar{P}_k(E)\}$ with density function $n(E)$.

VIII. ERROR BOUNDS AND CONTINUED FRACTIONS

The use of a continued fraction representation is always connected with the difficult problem of an appropriate termination. As we shall see, in a number of physical problems one can exploit physical arguments to obtain a satisfactory, and in some cases exact, description of asymptotic behavior. A remarkable case is the situation of crystals, where the presence of translational symmetry leads to allowed energy bands, and this gives precious information on the tail behavior of fractions. In other situations, Markovian truncation is possible, which means that the tail of order n can be approximated with the square root of the nth partial numerator. In other situations again, the parameters of the continued fraction are complex numbers, which means that any stage actually takes into account a large number of degrees of freedom.

In addition to this, it is often important to have also exact mathematical criteria of the truncation errors performed when an infinite continued fraction is replaced by its approximant of order n. In a number of situations these criteria do exist and are also economical.[41-44] Keeping in line with the general spirit of our review, we focus on error bounds in continued fractions, describing specific (though sufficiently general) physical problems. We do not dwell on the interesting theory of error bounds in more abstract situations.

A first aspect of interest concerns the determination of error bounds when truncating an infinite continued fraction. A second aspect concerns averages over nonnegative distribution functions. We give significant examples of both kinds.

The first situation we wish to consider concerns the error bounds in the spectral density of a pure dissipation operator[43] Γ, whose eigenvalues $\gamma_m = -i\varepsilon_m$ ($\varepsilon_m \geq 0$) lie on the negative imaginary axis. Consider the equation of motion of the type

$$\frac{d\phi_0(t)}{dt} = -i\Gamma\phi_0(t) \tag{8.1}$$

and the autocorrelation function

$$c(t) = \langle \phi_0 | e^{-i\Gamma t} | \phi_0 \rangle \tag{8.2}$$

($|\phi_0\rangle$ is normalized to 1). The corresponding spectral density is given by the usual expression

$$I(E) = -\frac{1}{\pi}\operatorname{Im}\langle \phi_0 | \frac{1}{E - \Gamma} | \phi_0 \rangle \qquad \operatorname{Im} E \to 0+ \tag{8.3}$$

We can also write

$$I(E) = -\frac{1}{\pi}\operatorname{Im}\langle \phi_0 | \frac{i}{iE - i\Gamma} | \phi_0 \rangle$$

$$= -\frac{1}{\pi}\operatorname{Re}\langle \phi_0 | \frac{1}{iE - H} | \phi_0 \rangle \tag{8.4}$$

where the operator H ($\equiv i\Gamma$) is a Hermitian operator with positive eigenvalues. Because of this the continued fraction

$$S(z) = \left\langle \phi_0 \middle| \frac{1}{z - H} \middle| \phi_0 \right\rangle \tag{8.5}$$

can be expanded into an S-type continued fraction [see Eq. (3.9)]:

$$S(z) = \cfrac{1}{z - \cfrac{a_2}{1 - \cfrac{a_3}{z - \cfrac{a_4}{1 - \cdots}}}} \qquad a_n > 0, \forall n \tag{8.6}$$

For any S-type continued fraction such as (8.6), it is possible[44] to obtain simple a posteriori bounds for the truncation errors valid for all $n \geq 1$ and z not on the positive real axis. The bounds are

$$|S(z) - S_n(z)| \leq \begin{cases} |S_n(z) - S_{n-1}(z)| & \text{if } \dfrac{\pi}{2} \leq |\arg z| \leq \pi \\[2ex] \cot g\left(\tfrac{1}{2}|\arg z|\right)|S_n(z) - S_{n-1}(z)| & \text{if } 0 \leq |\arg z| \leq \dfrac{\pi}{2} \end{cases} \tag{8.7}$$

While Eq. (8.7) represents a powerful result to handle Stieltjes operators, no similar result exists for a Hamburger operator or a general relaxation operator. In Fig. 3 we give a schematic representation of the theorem summarized by Eq. (8.7).

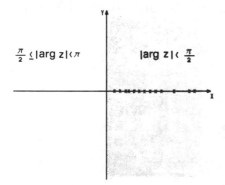

Figure 3. Schematic illustration of the theorem of Henrici et al.[44] for error bounds. Eigenvalues of the operator must be on the positive real axis.

We specify theorem (8.7) for $z = iE$ (E real); we have

$$|S(iE) - S_n(iE)| \leq |S_n(iE) - S_{n-1}(iE)| \tag{8.8}$$

Hence we have for the spectral density the following bounds:

$$|I(E) - I_n(E)| = \frac{1}{\pi}|\operatorname{Re} S(iE) - \operatorname{Re} S_n(iE)|$$

$$\leq \frac{1}{\pi}|S_n(iE) - S_{n-1}(iE)| \tag{8.9}$$

which constitutes an important result first discussed by Hänggi.[43]

A second significant example occurs in the calculation of the canonical partition function

$$\int_0^\infty f(E)n(E)\,dE \tag{8.10}$$

where $f(E) = e^{-\beta E}$, $n(E)$ is a positive definite density of states, and the zero of energy has been taken to be the energy of the ground state of the system. The determination of error bounds for integrals of type (8.10) is connected with the theory of Gaussian quadrature, which allows us to express integrals of the form (8.10) in terms of sums of the values $f(E)$ at selected points with suitable weights as indicated in Section VII.

Since we can associate with the moments of $n(E)$ a Stieltjes-type continued fraction, it is convenient to express the integral (8.10) in terms of even and odd approximants of the S-fraction corresponding to $n(E)$. Following Shohat and Tamarkin (see, e.g., ref. 4, p. 119), Gordon[26] presents an expression of the remainder when truncation at order n is performed:

$$\int_0^\infty f(E)n(E)\,dE = \begin{cases} \displaystyle\sum_{j=1}^N f(E_j^{eN})\rho^{eN}(E_j^{eN}) + \frac{\alpha_{2N}\alpha_{2N+1}f^{(2N)}(\bar{E})}{(2N)!} \\[2ex] \displaystyle\sum_{j=1}^N f(E_j^{oN})\rho^{oN}(E_j^{oN}) + \frac{\alpha_{2N-1}^1\alpha_{2N-2}^1 f^{(2N-1)}(\bar{\bar{E}})}{(2N-1)!} \end{cases}$$

$$\tag{8.11}$$

where $\bar{E}, \bar{\bar{E}}$ are somewhere in the interval $[0, \ldots, \infty]$, α are the coefficients of the continued fraction for the moment problem; α^1 are the coefficients of the continued fraction for the moment problem, where μ_{j+1} replaces μ_j.

Since $\alpha_{2N}\alpha_{2N+1} > 0$, as well as $\alpha_{2N-1}^1\alpha_{2N-2}^1 > 0$, we see that the remainder terms have the sign of derivatives of $f(E)$ of order $2N$ and $2N-1$. In our case $f(E) = e^{-\beta E}$: the even derivatives are positive for any value of $E > 0$, while the odd derivatives are negative; thus, even and odd partial sums in Eq. (8.11) give lower and upper bounds, respectively.

Another situation in which error bounds can be provided is in the calculation of a correlation function of a purely dissipative operator (eigenvalues $\gamma_m = -i\varepsilon_m$, $\varepsilon_m > 0$, on the negative imaginary axis), earlier discussed in this section with regard to the spectral density bounds. Consider a correlation function

$$c(t) = \langle \phi_0 | e^{-i\Gamma t} | \phi_0 \rangle \qquad (8.12)$$

($|\phi_0\rangle$ is normalized to 1). Insert the unit operator

$$1 = \sum_m |\psi_m\rangle\langle\psi_m|$$

where $|\psi_m\rangle$ are the eigenvectors of Γ with energy $-i\varepsilon_m$; defining the projected density of states

$$n(\varepsilon) = \sum_m |\langle\phi_0|\psi_m\rangle|^2 \delta(\varepsilon - \varepsilon_m)$$

we obtain for the correlation function (8.12) the expression

$$c(t) = \int_0^\infty e^{-\varepsilon t} n(\varepsilon)\, d\varepsilon \qquad (8.13)$$

Thus for the calculation of error bounds in the correlation function associated to a pure dissipative operator, we can exploit the properties summarized by Eq. (8.11).

With a slightly different approach but within the spirit of Gaussian quadrature, one can also find realistic bounds for the integral

$$F(E_0) = \int_{-\infty}^{E_0} f(E)n(E)\, dE \qquad (8.14)$$

where $f(E)$ is a monotonically decreasing function over the energies where $n(E)$ is nonzero. In the case $f(E) = 1$, Eq. (8.14) gives the integrated density of states. Let us suppose we have $2n$ moments of the density $n(E)$: from our knowledge of them we have the coefficients $\{a_0, a_1, \ldots, a_{n-1}\}$ and $\{b_1, b_2, \ldots, b_n\}$ for the recursion problem. We can thus determine the sequence of orthonormal polynomials $\{P_0(E), P_1(E), \ldots, P_n(E)\}$.

Let us construct the $P_{n+1}(E)$ polynomial

$$P_{n+1}(E) = (E - a_n)P_n(E) - b_n P_{n-1}(E) \qquad (8.15)$$

(with a_n not yet specified) in such a way that $P_{n+1}(E_0) = 0$; this is possible by choosing

$$a_n = E_0 - b_n \frac{P_{n-1}(E_0)}{P_n(E_0)} \qquad (8.16)$$

We now consider the $n+1$ roots of $P_{n+1}(E)$ indicated by $\{E_0, E_1, \ldots, E_n\}$.
We can now construct two polynomials $P_+(E)$ and $P_-(E)$ of degree $2n$ which bound the given function $f(E)$ as indicated below:

$$P_+(E_m) = \begin{cases} f(E_m) & E_m \leq E_0 \\ 0 & E_m > E_0 \end{cases}$$

$$ \qquad (8.17)$$

$$P_-(E_m) = \begin{cases} f(E_m) & E_m < E_0 \\ 0 & E_m \geq E_0 \end{cases}$$

Moreover, $P'_\pm(E_m) = f'(E_m)$ for $E_m \neq E_0$. The polynomials $P_+(E)$ and $P_-(E)$ bound the function above and below (see Fig. 4); moreover, they can

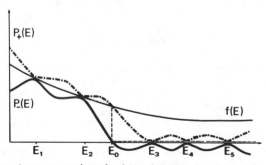

Figure 4.   Schematic representation of polynomials, upper and lower bounds of a monotonic decreasing function.

be integrated *exactly* with a Gaussian quadrature:

$$F_-(E_0) = \int_{-\infty}^{E_0} P_-(E) n(E) \, dE = \sum_{E_m \langle E_0} \rho(E_m) f(E_m)$$

$$F_+(E_0) = \int_{-\infty}^{E_0} P_+(E) n(E) \, dE = F_-(E_0) + \rho(E_0) f(E_0)$$

where, as seen in Section VII,

$$\rho(E_m) = \frac{1}{\displaystyle\sum_{k=0}^{N} P_k^2(E_m)}$$

The integral value $F(E_0)$ is so bounded by $F_+(E_0)$ and $F_-(E_0)$, and the difference

$$F_+(E_0) - F_-(E_0) = \rho(E_0) f(E_0)$$

goes to zero as $N \to \infty$.

From the discussion of this section, we see that the knowledge of a number of moments of a definite positive function poses constraints on the remainder of the corresponding continued fraction, and error bounds can in principle be determined (and in a number of situations they are simple indeed).

Out of the detailed mathematical aspects, some of them summarized in this section, there is a more general physical concept at the heart of the theory of error bounds. It is a fact that the memory function formalism provides in a natural manner a framework by which the short-time behavior, via the kernel of the integral equations, makes its effects felt in the long-time tail. The mathematical apparatus of continued fractions can adequately describe memory effects, and this explains the central role of this tool in the theory of relaxation.

References

1. For a quite interesting survey of the more than two millenium history of continued fractions, since Euclid to our times, we recommend the article by C. Brezinski in *Padé Approximation and its Applications, Lecture Notes in Mathematics*, 888, M. G. de Bruin and H. van Rossum, eds., Springer, Berlin, 1981, p. 1.

2. H. S. Wall, *Analytic Theory of Continued Fractions*, Van Nostrand, Princeton, 1948; Chelsea, New York, 1973.

3. N. I. Akhiezer, *The Classical Moment Problem*, Oliver and Boyd, Edinburgh, 1965 (original edition in Russian published in Moscow in 1961).

4. J. A. Shohat and J. D. Tamarkin, *The Problem of Moments* (*Mathematical Surveys* I),

American Mathematical Society, New York 1943; 2nd ed., 1950.

5. C. Brezinski, *Padé-Type Approximation and General Orthogonal Polynomials*, Birkhäuser, Basel, 1980.

6. W. B. Jones and W. J. Thron, *Continued Fractions* (Encyclopedia of Mathematics and its Applications, Vol. 11, G. C. Rota, ed.), Addison-Wesley, London, 1980.

7. A. Draux, *Polynômes Orthogonaux Formel. Applications, Lecture Notes in Mathematics*, A. Dold and B. Eckmann, eds., Springer, Berlin, 1983.

8. T. J. Stieltjes, *Ann. Fac. Sci. Univ. Toulouse*, **3**, H1 (1889); **8**, J1 (1894); **9**, A5 (1895).

9. R. Courant and D. Hilbert, *Methoden der Mathematischen Physik*, Springer, Berlin, Vol. 1, 1924; Vol. 2, 1937. English translation: *Methods of Mathematical Physics*, Interscience, New York, Vol. 1, 1953; Vol. 2, 1962.

10. R. Zwanzig, in *Lectures in Theoretical Physics*, Vol. 3, W. E. Brittin, B. W. Downs, and J. Downs, eds., Interscience, New York, 1961, p. 106.

11. H. Mori, *Prog. Theor. Phys.*, **33**, 423 (1965); **34**, 399 (1965).

12. See, for instance, G. A. Baker, Jr., *Essentials of Padé Approximants*, Academic Press, New York, 1975; G. A. Baker, Jr., and P. Graves-Morris, *Padé Approximants*. Part I: *Basic Theory*; Part II: *Extensions and Applications* (Vols. 13 and 14 of the Encyclopedia of Mathematics and its Applications, G. C. Rota, ed.), Addison-Wesley, London, 1981.

13. P. Grigolini, *Nuovo Cimento*, **B63**, 174 (1981); *J. Stat. Phys.*, **27**, 283 (1982); M. W. Evans, G. J. Evans, W. T. Coffey, and P. Grigolini, *Molecular Dynamics*, John Wiley, New York, 1982, and references quoted therein.

14. E. B. Van Vleck, *Trans. Amer. Math. Soc.*, **5**, 253 (1904). See also, for instance, H. S. Wall, ref. 2, pp. 138–140; 209–210.

15. R. Haydock, in *Solid State Physics*, **35**, 216 (1980). For a review of the quite important findings and developments of continued fractions in solid state theory, see also the other contributing papers of D. W. Bullett, V. Heine, and M. J. Kelly in the same volume.

16. O. Perron, *Die Lehre von den Kettenbrüchen*, Teubner, Leipzig, 1913; 2nd ed., 1929.

17. See for instance J. M. Ziman, *Models of Disorder*, Cambridge University Press, 1979.

18. R. Kubo, *Adv. Chem. Phys.*, **15**, 101 (1969).

19. H. Hamburger, *Math. Ann.*, **81**, 235 (1920); **82**, 120, 168 (1921).

20. P. Giannozzi, G. Grosso, and G. Pastori Parravicini, *Phys. Stat. Sol.* (b) **128**, 643 (1985).

21. See, for instance, J. P. Gaspard and F. Cyrot-Lackmann, *J. Phys.*, **C6**, 3077 (1973).

22. See, for instance, P. Turchi, F. Ducastelle, and G. Tréglia, *J. Phys.*, **C15**, 2891 (1982).

23. P. Grigolini, G. Grosso, G. Pastori Parravicini, and M. Sparpaglione, *Phys. Rev.*, **B27**, 7342 (1983).

24. M. Bixon and R. Zwanzig, *J. Stat. Phys.*, **3**, 245 (1971).

25. M. W. Evans, *Chem. Phys. Lett.*, **39**, 601 (1976).

26. R. G. Gordon, *J. Math. Phys.*, **9**, 655 (1968).

27. J. Hubbard, *Proc. Roy. Soc.*, **A281**, 401 (1964).

28. P. Hänggi, F. Rösel, and D. Trautmann, *Z. Naturforsch.*, **33A**, 402 (1978); *J. Computational Phys.*, **37**, 242 (1980).

29. C. Blumstein and J. C. Wheeler, *Phys. Rev.*, **B8**, 1764 (1973). See also J. C. Wheeler and C. Blumstein, *Phys. Rev.*, **B6**, 4380 (1972); J. C. Wheeler, M. G. Prais, and C. Blumstein, *Phys. Rev.*, **B10**, 2429 (1974); J. C. Wheeler, *J. Chem. Phys.*, **80**, 472 (1984).

30. G. Szegö, *Orthogonal Polynomials*, Colloquium Publications, Vol. 23, American Mathe-

matical Society, New York, 1939; 4th ed., 1975.

31.   T. S. Chihara, *An Introduction to Orthogonal Polynomials*, Gordon and Breach, New York, 1978.

32.   For a discussion of the form of the distribution $N(E)$, see, for instance, P. G. Nevai, *Orthogonal Polynomials*, Memoires American Mathematical Society, Providence, RI, 1979. In essence, in handling $dN(E) = n(E)\,dE$, one can consider $n(E)$ as the (projected) density of states of a Hermitian operator and $N(E)$ as the corresponding integrated density of states.

33.   U. W. Hochstrasser, in *Handbook of Mathematical Functions*, M. Abramcvitz and I. Stegun, eds., Dover, New York, 1964, p. 771.

34.   C. Lanczos, *Applied Analysis*, Prentice-Hall, Englewood Cliffs, NJ, 1956.

35.   M. Dupuis, *Prog. Theor. Phys.*, **37**, 502 (1967).

36   W. Gautschi, *Math. Comp.*, **22**, 251 (1968); see also the discussion in ref. 29.

37.   R. A. Sack and A. F. Donovan, *Numer. Math.*, **18**, 465 (1972).

38.   W. Gautschi, *Math. Comp.*, **24**, 245 (1970).

39.   P. Lambin and J. P. Gaspard, *Phys. Rev.*, **B26**, 4356 (1982).

40.   A. Magnus, in *Padé Approximation and its Applications, Lecture Notes in Mathematics*, 765, L. Wuytack, ed., Springer, Berlin, 1979, p. 150.

41.   C. M. M. Nex, *Phys. Rev.*, **A11**, 653 (1978).

42.   R. G. Gordon, *Adv. Chem. Phys.*, **15**, 79 (1969); C. T. Corcoran and P. W. Langhoff, *J. Math. Phys.*, **18**, 651 (1977).

43.   This beautiful example of error bounds in spectral density is provided by P. Hänggi, *Z. Naturforsch.*, **33A**, 1380 (1978).

44.   P. Henrici and P. Pfluger, *Numer. Math.*, **9**, 120 (1966); W. B. Jones and W. J. Thron, *SIAM J. Num. Anal.*, **8**, 693 (1971); W. B. Jones, *Rocky Mountain J. Math.*, **4**, 241 (1974).

IV

MEMORY FUNCTION METHODS IN SOLID STATE PHYSICS

G. GROSSO and G. PASTORI PARRAVICINI

CONTENTS

I. INTRODUCTION

Periodicity is often considered the central tenet of solid state physics. In the presence of perfect order and translational symmetry, one can exploit the beautiful and exact results of group theory embodied in the Bloch theorem; via the construction of coherent phase states, the crystal problem greatly simplifies, the order of difficulty being dictated by the small number of degrees of freedom of the unit cell rather than the huge number of degrees of freedom of the whole crystal. The most significant consequences are the possibility of using powerful k-space methods and quasi-particle concepts. It seems almost superfluous to mention the remarkable success achieved in a number of traditional subjects such as electron band structure, Fermi surfaces, lattice dynamics, exciton resonances, transport and optical properties, etc. The theoretical study of this discipline is well established and codified in textbooks.[1,2]

Since its formulation, solid state theory has been concerned also with non-strictly-periodic systems, due principally to the theoretical and technological importance of defects (point impurities, color centers, dislocations, surfaces, etc.). However, most of these theoretical studies and approaches exploit the results of the ideal periodic crystal as the basic ingredient on which to include impurity effects.

The situation becomes more complicated and challenging in a number of problems where the fundamental tenet of periodicity is not applicable and is useless as an initial starting point.

In the last decade an abundant literature has focused more and more on the properties of low-symmetry systems having large unit cells which render unwieldy the traditional description in terms of the Bloch theorem. Low-symmetry systems include complicated ternary or quaternary compounds, man-made superlattices, intercalated materials, etc. The k-space picture becomes totally useless for higher degrees of disorder as exhibited by amorphous materials, microcrystallites, random alloys, phonon-induced disorder, surfaces, adsorbed atoms, chemisorption effects, and so on.

In the mentioned situations one may encounter Hamiltonians that cannot be handled by direct diagonalization but yet are sufficiently manageable to allow calculation of moments and Green's-function matrix elements. In these cases one gives up the idea of extracting the whole of the information embodied in the Hamiltonian; rather, one confines oneself to the aim of determining some average predictive value from the moments (or appropriate manipulations of them). This is the basic philosophy behind *the moment method*,[3,4] *the recursion method*,[5,6] and other formally equivalent lines of approach, hereafter referred to as *memory function methods* because of the "memory aspects" exhibited by the corresponding dynamic equations. To give an idea of the expanse of solid state problems which have been considered within one or the other of the memory function methods, we can mention the tight-binding model for electrons (in the presence of both diagonal and off-diagonal disorder), the dynamical matrix description of the vibrational modes, the spin Hamiltonian for magnetic systems, the vibronic models of impurities, the dynamic Jahn-Teller effect, the exciton-phonon interaction effects on the line shape of exciton resonances, and so on.

A major novelty of this chapter is to point out the unifying framework behind seemingly different methods of solid state physics and the connection with the *projective techniques*[7,8] routinely adopted in statistical mechanics (see, for instance, Chapter I and references quoted therein). In the physics of crystals, however, significant new physical concepts appear, because of topological aspects, possible translational symmetry or short-range order, bonding formation, and so on. For instance, correlation functions exhibit peculiar asymptotic behavior in the presence of energy gaps and van Hove singularities in the quasi-particle density of states. As we shall see, these and other typical solid state effects can be embodied naturally in the memory function formalism and in the mathematical apparatus of continued fractions (see, e.g., Chapter III and references quoted therein), which turns out to be a versatile and invaluable tool in this field of research as in many others.

II. REPRESENTATION OF MODEL HAMILTONIANS IN A LOCAL BASIS

In this section we briefly survey a number of physical problems whose Hamiltonians can be conveniently described in a local basis. Hamiltonians of this kind allow a reasonably simple calculation of moments and are thus natural candidates for the memory function methods we are going to describe.

Consider, for instance, the electron structure of solids (periodic or aperiodic) and the Schrödinger equation

$$\left[\frac{\mathbf{p}^2}{2m_e} + V(\mathbf{r}) \right] \psi(\mathbf{r}) = E \psi(\mathbf{r}) \tag{2.1}$$

where $V(\mathbf{r})$ is an appropriate effective one-electron potential. The wavefunctions $\psi(\mathbf{r})$ are of the Bloch type only in the case of a periodic crystalline potential, but we wish here to avoid such an assumption in order to describe more general situations. Equation (2.1) tacitly contains a number of assumptions, such as the adiabatic approximation, the one-electron approximation, and other more or less standard simplifications to the original complicated many-body problem.

It is well known that one of the standard approaches to Eq. (2.1) is the linear combination of atomic orbitals (LCAO) method; it consists in expanding the states of the solid in linear combination of atomic (or molecular) orbitals of the composing atoms (or molecules). This method, when not applied in oversimplified form, provides an accurate description of core and valence bands in any type of crystal (metals, semiconductors, and insulators). Applied with some caution, the method also provides precious information on lowest lying conduction states, replacing whenever necessary atomic orbitals with appropriate localized orbitals.[2]

Let us denote by $\phi_{\mu m}(\mathbf{r} - \mathbf{d}_\mu - \boldsymbol{\tau}_m)$ the μth atomic-like orbital of interest, centered in position $\mathbf{d}_\mu$ within the unit cell $\boldsymbol{\tau}_m$. For simplicity, let us suppose that $\{\phi_{\mu m}\}$ are orthonormal; that is, the overlap matrix

$$S_{\mu m, \nu n} = \langle \phi_{\mu m} | \phi_{\nu n} \rangle = \delta_{\mu \nu} \delta_{mn} \tag{2.2}$$

is unity.

The projection operator in the manifold of the orthonormal states $\{\phi_{\mu m}\}$ is given by

$$P = \sum_{\mu m} |\phi_{\mu m}\rangle \langle \phi_{\mu m}| \tag{2.3}$$

The Hamiltonian

$$\mathcal{H} = \frac{\mathbf{p}^2}{2m_e} + V(\mathbf{r})$$

appearing in Eq. (2.1), within the manifold $\{\phi_{\mu m}\}$ takes the form

$$H = P\mathcal{H}P = \sum_{\mu m} |\phi_{\mu m}\rangle \langle \phi_{\mu m}| \frac{\mathbf{p}^2}{2m_e} + V(\mathbf{r}) | \sum_{\nu n} |\phi_{\nu n}\rangle \langle \phi_{\nu n}|$$

We arrive thus at the tight-binding representation

$$H = \sum_{\mu m} E_{\mu m} |\phi_{\mu m}\rangle \langle \phi_{\mu m}| + \sum_{\mu m \neq \nu n} T_{\mu m, \nu n} |\phi_{\mu m}\rangle \langle \phi_{\nu n}| \tag{2.4}$$

In Eq. (2.4), $\phi_{\mu m}$ define the orthonormal basis set, $E_{\mu m}$ are the diagonal matrix elements corresponding to the atomic energy levels, and $T_{\mu m, \nu n}$ are the off-diagonal (hopping) matrix elements representing the interaction between pairs of neighbors (often nearest neighbors only, or in some cases a few more orders of neighbors). After expanding ψ in the orthonormal set $\{\phi_{\mu m}\}$, Eq. (2.1) becomes

$$\left(E_{\mu m} - E \right) A_{\mu m} + \sum_{\substack{\nu n \\ (\neq \mu m)}} T_{\mu m, \nu n} A_{\nu n} = 0 \tag{2.5}$$

For notational simplicity, we assume whenever possible a single orbital per site; the model tight-binding Hamiltonian then takes the form

$$H = \sum_m E_m |\phi_m\rangle \langle \phi_m| + \sum_{m \neq n} T_{mn} |\phi_m\rangle \langle \phi_n| \tag{2.6}$$

If the functions $\{\phi_{\mu m}\}$ are not orthonormal, the overlap matrix S is given by

$$S_{\mu m,\nu n} = \langle \phi_{\mu m}|\phi_{\nu n}\rangle \neq \delta_{\mu\nu}\delta_{mn} \qquad (2.7)$$

In the manifold of the nonorthonormal set $\{\phi_{\mu m}\}$ the projection operator is now expressed as

$$P = \sum_{\mu m}\sum_{\nu n} |\phi_{\mu m}\rangle (S^{-1})_{\mu m,\nu n}\langle \phi_{\nu n}| \qquad (2.8)$$

The representation of the crystal Hamiltonian in the form (2.4) is still valid, but the explicit expressions for $E_{\mu m}$ and $T_{\mu m,\nu n}$ are more complicated. To eliminate inessential detail, we consider only orthonormal basis sets (nonorthogonal sets can be handled by similar procedures, but additional labor is required). The Anderson concept of localization,[9] as well as recent successful applications of the overlap-reduced semiempirical tight-binding method[10] amply justify this assumption in a number of band structure calculations.

An eigenvalue problem of type (2.5) is also encountered in an apparently different problem: the lattice dynamics of a generical three-dimensional solid. In the harmonic approximation, the potential energy of the system is expanded to second order in the displacements $\mathbf{u}_n$ by

$$V = V_0 + \frac{1}{2}\sum_{nn'} \mathsf{G}_{nn'}\mathbf{u}_n\mathbf{u}_{n'} + \cdots$$

where the second-rank tensor $\mathsf{G}_{nn'}$ is given by

$$\mathsf{G}_{nn'} = \left(\frac{\partial^2 V}{\partial \mathbf{u}_n\partial \mathbf{u}_{n'}}\right)_0$$

The classical equations of motion are

$$M_n\ddot{\mathbf{u}}_n(t) = -\sum_{n'}\mathsf{G}_{nn'}\cdot\mathbf{u}_{n'}(t)$$

Looking for solutions in the form of traveling waves of frequency ω, one obtains

$$(\mathsf{G}_{nn} - M_n\omega^2)\mathbf{u}_n + \sum_{n'(\neq n)}\mathsf{G}_{nn'}\cdot\mathbf{u}_{n'} = 0 \qquad (2.9)$$

A formal analogy is evident between Eq. (2.5), describing electron propagation, and Eq. (2.9), which is the basic equation of lattice dynamics.

Hamiltonians formally similar to Eq. (2.4) are encountered not only in the central problems of lattice dynamics and electron propagation, but also in a large variety of other problems. Among them we mention the Frenkel theory of excitons, the coupled electron-lattice impurities in the entire range of coupling, the Jahn-Teller (or pseudo-Jahn-Teller) systems, interacting spins, and so on.

For a periodic system, a model Hamiltonian of type (2.4) or (2.6) can be easily diagonalized. Consider, for instance, a crystal with one orbital per unit cell; we have

$$H = \sum_{\tau_n} E_0 |\phi_n\rangle\langle\phi_n| + \sum_{\tau_m \neq \tau_n} T_{mn} |\phi_m\rangle\langle\phi_n| \qquad (2.10)$$

where E_0 is independent of the translation vector τ_n, and T_{mn} depends only on the relative value $\tau_m - \tau_n$. We can use the standard k-space procedures and form the Bloch sum of vector $\mathbf{k}$

$$|\Phi_{\mathbf{k}}\rangle = \frac{1}{\sqrt{N}} \sum_{\tau_m} e^{i\mathbf{k}\cdot\tau_m} |\phi_m\rangle \qquad (2.11)$$

From Eqs. (2.10) and (2.11) it follows that the allowed energy levels are given by

$$E_{\mathbf{k}} = E_0 + \sum_{\tau_m \neq 0} T_{0m} e^{i\mathbf{k}\cdot\tau_m} \qquad (2.12)$$

and are grouped in a single energy band.

In the absence of periodicity (because of impurities, surfaces, adatoms, etc.) and also in the instance of stochastic disorder (binary alloys, orbital energy fluctuations because of phonons, etc.), k-space procedures are useless, and one must resort to the calculation of the Green's function or configurationally averaged Green's function. In this chapter we describe the more commonly adopted techniques for calculating Green's function stressing the central role of the memory function theoretical approach. We will illustrate procedures and concepts with specific reference to the electronic structure of solids, but we wish to stress that memory function techniques are quite general in nature and have deeply influenced not only solid state physics, statistical mechanics, or the physics of liquids, but also several other areas of investigation of interdisciplinary nature. This can be seen from the other review articles of this volume.

III. MEMORY FUNCTION METHODS

A. The Moment Method

By the moment method we mean the technique of directly using *power moments* to determine the Green's function and to reconstruct the spectral density. From a mathematical point of view, this problem goes back to the last century (see Chapter III), but the applications in several branches of physics have more recent origin.

The usefulness of moments in electronic state calculations is evident in the pioneering paper of Cyrot-Lackmann.[3] In early attempts[11] at describing the electronic structure of transition metals, a conjecture about its features was made. For instance, a common procedure consisted of approximating the unknown density of states by the product of a Gaussian and a polynomial, with coefficients fitted to the first moments.

The preconception about the results was the most unsatisfactory aspect of this procedure, which was refined in the basic article of Gaspard and Cyrot-Lackmann[4]; the use of Stieltjes transform and continued fraction analysis poses a sound mathematical framework to the use of moments.

In Chapter III we amply illustrated the relationships between Green's functions, continued fractions, and moments; here we need only summarize a few relevant results and focus on some peculiar aspects (such as topological aspects) inherent in the structure of the Hamiltonians on a lattice.

Given a state of interest $|f_0\rangle$ (normalized to 1) and a model crystal Hamiltonian H, we consider the Green's function

$$G_{00}(E) = \langle f_0 | \frac{1}{E - H} | f_0 \rangle \qquad (\text{Im } E \to 0+). \qquad (3.1)$$

When the spectral moments

$$\mu_n = \langle f_0 | H^n | f_0 \rangle \qquad (3.2)$$

exist and the moment problem is determined (which is the situation commonly met in physics), we consider the continued fraction expansion of Eq. (3.1):

$$G_{00}(E) = \cfrac{1}{E - a_0 - \cfrac{b_1^2}{E - a_1 - \cfrac{b_2^2}{E - a_2 - \cdots}}} \qquad (3.3)$$

The parameters a_n and b_n^2 can be obtained from the moments by evaluating Hankel determinants or working out product-difference algorithms. The projected density of states $n(E)$ is then given by

$$n(E) = -\frac{1}{\pi} \lim_{\varepsilon \to 0+} \text{Im} \, G_{00}(E + i\varepsilon) \qquad (3.4)$$

For preliminary empirical estimations, one often adopts Eq. (3.4), maintaining for E a finite imaginary part and truncating the fraction (3.3) at a given level.

A much more satisfactory technique is to add an analytic tail to the continued fraction (3.3) on the basis of physical considerations, as discussed in Section IV.

Alternatively, one can use error bound theory (see Chapter III) if any kind of explicit termination of the J-fraction (3.3) is avoided. This approach has been followed extensively in the literature by the different schools.[4,6,12] The approximate local density of states is obtained by differentiation of the two continuous curves $N_+(E)$ and $N_-(E)$, which are upper and lower bounds of the integrated density of states

$$n(E) \simeq \frac{1}{2} \frac{d}{dE} [N_+(E) + N_-(E)] \qquad (3.5)$$

The general and exact bounds $N_+(E)$ and $N_-(E)$ become more and more stringent as the number of available moments increases.

In solid state theory the power moments of order n are related with *closed paths*[13] of length n and depend thus on the topology and composition of the lattice. They can be evaluated by a walk-counting technique or by manageable computational procedures.

In the case of a solid with one orbital per site, we can write the moments as

$$\mu_n = \langle \phi_0 | H^n | \phi_0 \rangle$$

$$= \sum_{i_1, i_2, \ldots, i_{n-1}} \langle \phi_0 | H | \phi_{i_1} \rangle \langle \phi_{i_1} | H | \phi_{i_2} \rangle \cdots \langle \phi_{i_{n-1}} | H | \phi_0 \rangle$$

For instance, the moment of order 4 on a square lattice with nearest neighbor interactions only can be written diagrammatically as:

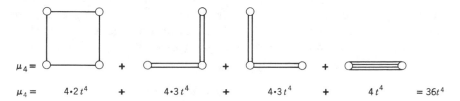

$$\mu_4 = \qquad 4 \cdot 2 t^4 \qquad + \qquad 4 \cdot 3 t^4 \qquad + \qquad 4 \cdot 3 t^4 \qquad + \qquad 4 t^4 \qquad = 36 t^4$$

where t is the hopping integral. The first diagram involves atoms in the first and second shell; the second diagram, atoms in the first, second, and third shell; the last two diagrams involve only nearest-neighbor atoms (already included in the lowest-order moment μ_2). In the case of a simple cube,

$$\mu_4 = 6 \cdot 4 t^4 + 6 \cdot 5 t^4 + 6 \cdot 5 t^4 + 6 t^4 = 90 t^4$$

It is clear that the evaluation of moments from closed path counting becomes cumbersome as n increases; the higher moments are evidently dominated by walks weaving around the central atom. This is the basic reason for the well-known ill-conditioned nature of the problem.

The evaluation of moments can be performed using a computer and a cluster method. One can consider the expressions

$$\mu_{2n} = \sum_i \langle \phi_0 | H^n | \phi_i \rangle \langle \phi_i | H^n | \psi_0 \rangle$$

$$\mu_{2n+1} = \sum_i \langle \phi_0 | H^n | \phi_i \rangle \langle \phi_i | H^{n+1} | \phi_0 \rangle$$

Successive powers $H^n | \phi_0 \rangle$ can then be generated iteratively from a knowledge of $H^{n-1} | \phi_0 \rangle$; the required storage locations equal the total number of orbitals in the cluster. However, power moments evaluated using standard arithmetic precision are hardly significant for n higher than a few tens or so because of the instability aspects.

In Chapter III we discussed the opportunity of replacing power moments with "balanced" modified moments.[14] The determination of the modified moments from the power moments is still an ill-conditioned problem. Recently, however, Lambin and Gaspard[15] have shown that in the case of tight-binding Hamiltonians it is possible to carry out a *direct* calculation of the modified moments. The technique is reasonably stable if some overall information (for instance, the bound edges) on the density of states is available, so that the auxiliary chosen orthogonal polynomials are near the "true" ones. A similar procedure could be advantageous for disordered systems, because of the convenience in averaging moments rather than continued fractions; but the stabilizing procedure, if successful, would be in any case an ad hoc procedure. The other memory function methods we are going to describe are systematically free from these unpleasant numerical instabilities, since they are implemented either by a process of "minimized iterations" or by "projection techniques."

B. The Lanczos Method

The Lanczos method[16,17] is a powerful tool originally introduced to detect the *latent roots and the principal axes* of a matrix by a process of *minimized iterations* with least squares. The algorithm is particularly convenient for matrices of very large dimensions, which at the same time are in *sparse* form or with a large spread of eigenvalues. A sparse matrix M can be stored economically by leaving out all the zeros; all we need is a matrix-vector multiplication to obtain $M|x_0\rangle$ from a vector $|x_0\rangle$. The Lanczos method generates from a given matrix M and a given test state $|x_0\rangle$ a quite convenient tridiagonal matrix. The eigenvalues of the tridiagonal matrix converge to the eigenvalues of the original matrix that are selected out by the starting vector. In addition, the weights of the starting vector $|x_0\rangle$ on the eigenvectors of M can be monitored, as schematically represented in Fig. 1. By appropriately changing the initial vector, it is possible to explore other energy regions of the spectrum of M without being forced to direct diagonalization, a practically impossible task for large dimensions (a thousand or more).

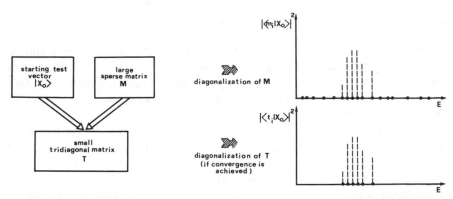

Figure 1.   A schematic pictorial representation of the Lanczos method. The eigenvectors of M (here supposed to be Hermitian) are indicated by $|m_i\rangle$, and the corresponding eigenvalues are reported on the real E axis. The eigenvectors of the tridiagonal matrix generated by $|x_0\rangle$ are denoted by $|t_i\rangle$; the eigenvalues selected out by the test state are also indicated for convenience.

Following Lanczos,* we consider first the case of a Hermitian matrix M and operate with it on a given starting vector $|x_0\rangle$. We choose then the linear combination of the *iterated* vector $M|x_0\rangle$ with $|x_0\rangle$ itself which minimizes the length of the new vector

$$|x_1\rangle = M|x_0\rangle - a_0 e^{i\alpha}|x_0\rangle \qquad (3.6)$$

where $a_0 \exp(i\alpha)$ is a generic complex number (with a_0 real and α a phase). Exploiting the hermiticity of M, we obtain

$$\langle x_1|x_1\rangle = \langle x_0|M^2|x_0\rangle - a_0(e^{i\alpha} + e^{-i\alpha})\langle x_0|M|x_0\rangle + a_0^2\langle x_0|x_0\rangle \quad (3.7)$$

The minimization of $\langle x_1|x_1\rangle$ with respect to the variational parameter α gives $\alpha = 0$ (mod π). As far as a_0 is concerned, we obtain from Eq. (3.7)

$$0 \equiv \frac{\partial\langle x_1|x_1\rangle}{\partial a_0} = -2\langle x_0|M|x_0\rangle + 2a_0\langle x_0|x_0\rangle$$

and hence the variational parameter a_0 is

$$a_0 = \frac{\langle x_0|M|x_0\rangle}{\langle x_0|x_0\rangle} \qquad (3.8)$$

It is easy to see that the new vector

$$|x_1\rangle = M|x_0\rangle - a_0|x_0\rangle \qquad (3.9)$$

is orthogonal to the original vector $|x_0\rangle$ because of Eq. (3.8).

The process can be iterated by operating on $|x_1\rangle$ and finding the minimum conditions for the amplitude of the new vector

$$|x_2\rangle = M|x_1\rangle - a_1|x_1\rangle - b_1^2|x_0\rangle \qquad (3.10)$$

---

*We follow basically the original paper of Lanczos, cited in ref. 16. However, we prefer the standard quantum-mechanical symbolism to indicate vectors and row-by-column matrix-vector multiplication. Furthermore, we focus on Hermitian (or non-Hermitian) matrices, rather than the symmetric (or unsymmetric) matrices specifically considered by Lanczos. These facts imply some straightforward adjustments in reporting the results of ref. 16. Finally, notice that the definition of the biorthogonal set as given by our Eq. (3.18) holds not only for non-Hermitian operators with real eigenvalues, but also for non-Hermitian operators with complex eigenvalues, such as relaxation operators.

where a_1 and b_1^2 are now used as variational parameters. Since M is Hermitian, we can prove that a_1 and b_1^2 are real numbers as shown earlier for a_0 (b_1^2 is used instead of b_1 to avoid a fractional exponent later on and to conform with the notation of ref. 6. The latter notation is adopted throughout this chapter.)

The amplitude of the vector $|x_2\rangle$ is

$$\langle x_2|x_2\rangle = \langle x_1|M^2|x_1\rangle - 2a_1\langle x_1|M|x_1\rangle - 2b_1^2\langle x_1|M|x_0\rangle$$
$$+ a_1^2\langle x_1|x_1\rangle + \left(b_1^2\right)^2\langle x_0|x_0\rangle$$

From the minimum conditions

$$\frac{\partial\langle x_2|x_2\rangle}{\partial a_1} = 0 \quad \text{and} \quad \frac{\partial\langle x_2|x_2\rangle}{\partial\left(b_1^2\right)} = 0$$

we obtain, respectively,

$$a_1 = \frac{\langle x_1|M|x_1\rangle}{\langle x_1|x_1\rangle} \tag{3.11}$$

$$b_1^2 = \frac{\langle x_0|M|x_1\rangle}{\langle x_0|x_0\rangle} = \frac{\langle x_1|x_1\rangle}{\langle x_0|x_0\rangle} \tag{3.12}$$

The process can be continued, and one would think that for the next step the best linear combination is of the type

$$|x_3\rangle = M|x_2\rangle - a_2|x_2\rangle - b_2^2|x_1\rangle - \gamma|x_0\rangle \tag{3.13}$$

But we can easily show that $\gamma = 0$; in fact, from the condition

$$0 \equiv \frac{\partial\langle x_3|x_3\rangle}{\partial\gamma} = \langle x_0|M|x_2\rangle + \langle x_2|M|x_0\rangle - 2\gamma\langle x_0|x_0\rangle$$

and the orthogonality of $|x_2\rangle$ to the previous vectors, we see that γ vanishes. As aptly noted by Lanczos (ref. 16, p. 266), "*the most remarkable feature of this successive minimization process is that the best linear combination never includes more than three terms....*"

We can summarize the Lanczos method by writing down the hierarchy of the best linear combinations generated starting from $|x_0\rangle$:

$$|x_{n+1}\rangle = M|x_n\rangle - a_n|x_n\rangle - b_n^2|x_{n-1}\rangle \tag{3.14}$$

$(n \geq 0; \ |x_{-1}\rangle = 0)$ and

$$a_n = \frac{\langle x_n | M | x_n \rangle}{\langle x_n | x_n \rangle} \qquad n = 0, 1, 2, \ldots \tag{3.15}$$

$$b_n^2 = \frac{\langle x_{n-1} | M | x_n \rangle}{\langle x_{n-1} | x_{n-1} \rangle} = \frac{\langle x_n | x_n \rangle}{\langle x_{n-1} | x_{n-1} \rangle} \qquad n = 1, 2, \ldots \tag{3.16}$$

If the set of orthogonal vectors $\{|x_0\rangle, |x_1\rangle, \ldots\}$ is then normalized and used as a basis for the matrix M, we obtain the standard tridiagonal form with diagonal elements $\{a_0, a_1, \ldots\}$ and off-diagonal elements $\{b_1, b_2, \ldots\}$.

A slightly different procedure must be followed if the matrix M is non-Hermitian: the main difference is that we must now operate with M and $M^\dagger$, which are no longer equal. We have thus to consider the biorthogonal sets of vectors[16]

$$\{|x_0\rangle, |x_1\rangle, |x_2\rangle, \ldots\} \qquad \text{and} \qquad \{|\tilde{x}_0\rangle, |\tilde{x}_1\rangle, |\tilde{x}_2\rangle, \ldots\} \tag{3.17}$$

whose defining relations are appropriate generalizations of Eq. (3.14). Namely, we have

$$|x_0\rangle = |\tilde{x}_0\rangle$$

$$|x_1\rangle = M|x_0\rangle - a_0|x_0\rangle \qquad\qquad |\tilde{x}_1\rangle = M^\dagger|\tilde{x}_0\rangle - a_0^*|\tilde{x}_0\rangle$$

$$|x_{n+1}\rangle = M|x_n\rangle - a_n|x_n\rangle - b_n^2|x_{n-1}\rangle \quad |\tilde{x}_{n+1}\rangle = M^\dagger|\tilde{x}_n\rangle - a_n^*|\tilde{x}_n\rangle - b_n^{2*}|\tilde{x}_{n-1}\rangle$$

$$\tag{3.18}$$

With the further prescription of performing scalar products between opposing vectors, we consider the amplitude $\langle \tilde{x}_n | x_n \rangle$ and search for extrema. We obtain

$$a_n = \frac{\langle \tilde{x}_n | M | x_n \rangle}{\langle \tilde{x}_n | x_n \rangle} \tag{3.19}$$

$$b_n^2 = \frac{\langle \tilde{x}_n | x_n \rangle}{\langle \tilde{x}_{n-1} | x_{n-1} \rangle} \tag{3.20}$$

The Lanczos algorithm, in either the one-sided or two-sided formulation, is extremely useful because it avoids an accumulation of rounding errors. For more details and criteria to establish truncation errors and convergence of the process, we refer to the literature.[18]

Once a tridiagonal representation of M is obtained, one can use standard methods to diagonalize the tridiagonal matrix and obtain the eigenvalues (in this particular aspect, the Lanczos algorithm, as commonly used in the literature, differs from the other memory function methods where the Green's function is constructed). In general, the roots that appear early and converge better are those with the highest overlap with the starting state. If, for instance, a vector $|x_0\rangle$ is known from physical considerations to involve the lowest eigenvalues of M, we can choose as starting state $|x_0\rangle$ and determine reliably the lowest-lying eigenvalues of M. The Lanczos method has found wide application in solid state physics, mainly in relation to vibronic models of color centers and Jahn-Teller systems[19] when a large number of phonons must be taken into account.

C. The Recursion Method

The recursion method, formulated by Haydock, Heine, and Kelly in the early 1970s,[5] has had a major impact in opening new perspectives and developments in solid state physics. The method is essentially a stable procedure to expand Green's functions into continued fractions, after generating an appropriate basis set for a tridiagonal representation of a given operator.

Consider an operator H (we suppose here it is Hermitian) and a normalized seed state $|f_0\rangle$. We construct a *hierarchy of orthonormal states* according to the recursion scheme:

$$|f_0\rangle \qquad b_1|f_1\rangle = H|f_0\rangle - a_0|f_0\rangle$$
$$b_2|f_2\rangle = H|f_1\rangle - a_1|f_1\rangle - b_1|f_0\rangle$$
$$\cdots\cdots\cdots\cdots\cdots\cdots\cdots\cdots\cdots\cdots$$
$$b_{n+1}|f_{n+1}\rangle = H|f_n\rangle - a_n|f_n\rangle - b_n|f_{n-1}\rangle \qquad (3.21)$$

where

$$a_n = \langle f_n|H|f_n\rangle \qquad (3.22)$$

$$b_n = \langle f_{n-1}|H|f_n\rangle \qquad (3.23)$$

We have generated a chain of variables* with diagonal matrix elements a_n and nearest-neighbor interaction b_n. The explicit calculation of a_n and b_n can be performed with standard programs[20] for any operator expressed in a local basis. In Fig. 2 we give a schematic picture of the chain-model variables generated by the recursion method in a simple two-dimensional square lattice.

The sequence of states $\{|f_0\rangle, |f_1\rangle, \ldots\}$ defined by Eqs. (3.21)–(3.23) is formally equivalent to the sequence of vectors $\{|x_0\rangle, |x_1\rangle, \ldots\}$ defined by Eqs. (3.14)–(3.16) of the Lanczos method; the only (inessential) difference is

---

*The counterpart in the theory of orthogonal polynomials of the recursion relations (3.21) are the set of orthonormal polynomials $b_{n+1}P_{n+1}(E) = (E - a_n)P_n(E) - b_n P_{n-1}(E)$. Instead of Eq. (3.21), one can as well define the chain of variables $\{f_i\}$ via the relation $|f_{n+1}\rangle = H|f_n\rangle - a_n|f_n\rangle - b_n^2|f_{n-1}\rangle$, whose counterpart in the theory of orthogonal polynomials is the set of *monic* orthogonal polynomials $\Pi_{n+1}(E) = (E - a_n)\Pi_n(E) - b_n^2\Pi_{n-1}(E)$. See Chapter III for a thorough discussion.

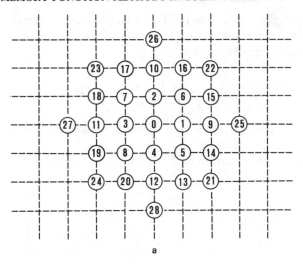

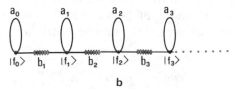

Figure 2.   Two-dimensional square lattice (a) in the recursion scheme representation (b). After defining $|\psi_0\rangle = |f_0\rangle = |\phi_0\rangle$; $\psi_1 = \Sigma_1^4 |\phi_i\rangle$; $|\psi_2\rangle = \Sigma_5^8 |\phi_i\rangle$, $|\psi_3\rangle = \Sigma_9^{12}|\phi_i\rangle$, $\psi_4 = \Sigma_{13}^{20}|\phi_i\rangle$, $\psi_5 = \Sigma_{21}^{24}|\phi_i\rangle$, $|\psi_6\rangle = \Sigma_{25}^{28}|\phi_i\rangle$; the first few states f_i are given by $|f_0\rangle = |\psi_0\rangle$, $|f_1\rangle = |\psi_1\rangle$, $|f_2\rangle = 2|\psi_2\rangle + |\psi_3\rangle$, $|f_3\rangle = 2|\psi_4\rangle + |\psi_6\rangle$, and the first few parameters b_i^2 are $b_1^2 = 4$, $b_2^2 = 5$, $b_3^2 = \frac{19}{5}, \ldots$.

that we have preferred to build in from the very beginning the normalization requirement, so that $\{|f_i\rangle\}$ constitute an orthonormal set, with nearest-neighbor interaction.

This chain model is mathematically equivalent to expressing the Hamiltonian H in a tridiagonal Jacobi form:

$$H = \begin{pmatrix} a_0 & b_1 & 0 & 0 & \cdot \\ b_1 & a_1 & b_2 & 0 & \cdot \\ 0 & b_2 & a_2 & b_3 & \cdot \\ 0 & 0 & b_3 & \cdot & \cdot \\ \cdot & \cdot & \cdot & \cdot & \cdot \end{pmatrix} \qquad (3.24)$$

As opposed to the Lanczos method, in which the diagonalization of (3.24) is performed, the recursion method focuses on the construction of the diagonal Green's-function matrix element

$$G_{00}(E) = \langle f_0 | \frac{1}{E-H} | f_0 \rangle \qquad (\text{Im } E > 0) \qquad (3.25)$$

This approach is much more convenient for handling operators with continua in the energy spectrum.

The matrix representation of $E - H$ in the set $\{|f_0\rangle, |f_1\rangle, \dots\}$ is

$$E - H = \begin{pmatrix} E - a_0 & -b_1 & 0 & \cdot \\ -b_1 & E - a_1 & -b_2 & \cdot \\ 0 & -b_2 & E - a_2 & \cdot \\ \cdot & \cdot & \cdot & \cdot \end{pmatrix} \qquad (3.26)$$

Let us denote by $D_0(E)$ the determinant of (3.26), by $D_1(E)$ the determinant of (3.26) when the first row and column are omitted, by $D_2(E)$ the determinant when the first and second rows and columns are omitted, etc. From the definition of the elements of the inverse matrix we have

$$\langle f_0 | \frac{1}{E-H} | f_0 \rangle = \frac{D_1(E)}{D_0(E)}$$

Using elementary properties of determinants,

$$D_0(E) = (E - a_0) D_1(E) - b_1^2 D_2(E)$$

we have

$$G_{00}(E) = \frac{1}{D_0(E)/D_1(E)} = \frac{1}{E - a_0 - b_1^2 D_2(E)/D_1(E)}$$

The procedure can be iterated to obtain the familiar continued-fraction expansion of the Green's function $G_{00}(E)$:

$$G_{00}(E) = \cfrac{1}{E - a_0 - \cfrac{b_1^2}{E - a_1 - \cfrac{b_2^2}{E - a_2 - \cdots}}} \qquad (3.27)$$

From expression (3.27), the physical quantities of interest can be calculated as presented in the discussion of Eq. (3.3).

In several problems of solid state physics we need not only diagonal but also off-diagonal matrix elements of the Green's function. The off-diagonal matrix elements of the resolvent can be conveniently expressed in the form[6]

$$G_{ab}(E) = \langle f_a | \frac{1}{E-H} | f_b \rangle$$

$$\equiv \frac{1}{4} \langle f_a + f_b | (E-H)^{-1} | f_a + f_b \rangle - \frac{1}{4} \langle f_a - f_b | (E-H)^{-1} | f_a - f_b \rangle$$

$$- \frac{i}{4} \langle f_a + if_b | (E-H)^{-1} | f_a + if_b \rangle + \frac{i}{4} \langle f_a - if_b | (E-H)^{-1} | f_a - if_b \rangle$$

$$(3.28)$$

where each of the four terms on the right-hand side are diagonal elements of $G(E)$ and can thus be evaluated with the recursion method, taking as seed state the linear combinations $|f_a + f_b\rangle$, $|f_a - f_b\rangle$, $|f_a + if_b\rangle$, $|f_a - if_b\rangle$, respectively.

The extension of the recursion method to non-Hermitian operators possessing real eigenvalues has been carried out by introducing an appropriate biorthogonal basis set[6,21] in close analogy with the unsymmetric Lanczos procedure.[16] Non-Hermitian operators with real eigenvalues are encountered, for instance, in the chemical pseudopotential theory.[9] Notice that the two-sided recursion method in formulation (3.18) is also valid for relaxation operators, as previously discussed.

Finally, we wish to note that the chain of variables defined by Eqs. (3.21) form a complete set for describing the time evolution of the initial state $|f_0\rangle$. In fact, the time evolution in a successive infinitesimal interval is determined from knowledge of $|f_0\rangle$ and $-iH|f_0\rangle$; in the next step we also need $(-iH)^2|f_0\rangle$, and so on. This is the origin of the deep relationship between the recursion method and the memory function formalism that we are going to illustrate.

D. Projective Techniques and the Memory Function Formalism

The problem of calculating the Green's function of a given operator can be considered from the elegant and general point of view of projective techniques and memory function formalism, as borrowed from statistical mechanics. In this formalism the calculation of the diagonal Green's-function matrix element takes on the simple pictorial aspect[22] of selecting an orbital of interest and considering the other degrees of freedom of the crystal as a highly non-Markovian reservoir affecting the dynamics of the starting state.

The advantages of this kind of formulation stand out not only in terms of elegance and beauty (the moment method, the Lanczos method, and the recursion method are relevant but particular cases of the memory function equations), but also in the possibility of providing insight into a number of problems, such as the asymptotic behavior of continued fraction parameters and their relationship with moments, the possible inclusion of nonlinear effects, the introduction of the concept of random forces, and so on.

The content of this section is, to a large extent, a matter of paraphrasing methods for the calculation of correlation functions using the projective techniques of Zwanzig[7] and Mori[8] and appropriate developments.[23] However, the specific path followed here has some significant differences, makes closer contact with the familiar Dyson equation, and considers in detail the problem of non-Hermitian operators and biorthogonal sets. We feel it is worthwhile to give a fully self-contained presentation, since the corresponding literature is spread over many articles in apparently different contexts.

Consider a state of interest $|f_0\rangle$ (normalized to 1) driven by an operator H, that we suppose to be Hermitian. Consider the standard dynamic equation

$$\frac{d|f_0(t)\rangle}{dt} = -iH|f_0(t)\rangle \tag{3.29}$$

and the correlation function

$$\Phi_0(t) = \frac{\langle f_0|f_0(t)\rangle}{\langle f_0|f_0\rangle} = \frac{\langle f_0|e^{-iHt}|f_0\rangle}{\langle f_0|f_0\rangle} \tag{3.30}$$

(when a state does not explicitly contain the time, it is implicitly assumed it refers to $t = 0$).

According to the methods of statistical mechanics, we define the projection operator P_0 and the complementary projector Q_0 as

$$P_0 = \frac{|f_0\rangle\langle f_0|}{\langle f_0|f_0\rangle} \quad \text{and} \quad Q_0 = 1 - P_0 \tag{3.31}$$

It is then possible to transform the equation of motion (3.29) for $|f_0(t)\rangle$ into an equivalent but more effective form involving the correlation function of the orbital of interest and the dynamics of its complementary counterpart. The procedure can be iterated on the complementary counterpart, and one arrives at a hierarchy of interdependent equations by the following steps.

From Eq. (3.29) we have

$$\frac{d|f_0(t)\rangle}{dt} = (-iH)e^{-iHt}|f_0\rangle = e^{-iHt}(-iH)|f_0\rangle$$

Inserting the operator $P_0 + Q_0 = 1$, we obtain

$$\frac{d|f_0(t)\rangle}{dt} = e^{-iHt}P_0(-iH)|f_0\rangle + e^{-iHt}(1-P_0)(-iH)|f_0\rangle \quad (3.32)$$

It is convenient to define the state $|f_1\rangle$, the operator H_1, the projected dynamic evolution $|f_1(t)\rangle$, and the correlation function $\Phi_1(t)$ as

$$|f_1\rangle = (1-P_0)H|f_0\rangle \qquad H_1 = (1-P_0)H$$

$$|f_1(t)\rangle = e^{-iH_1 t}|f_1\rangle \qquad P_0|f_1(t)\rangle \equiv 0 \qquad \Phi_1(t) = \frac{\langle f_1|f_1(t)\rangle}{\langle f_1|f_1\rangle} \quad (3.33)$$

We denote by a_0 and b_1^2 the matrix elements

$$a_0 = \langle f_0|H|f_0\rangle$$
$$b_1^2 = \langle f_1|f_1\rangle = \langle f_0|H|f_1\rangle \quad (3.34)$$

We can write Eq. (3.32) in the form

$$\frac{d|f_0(t)\rangle}{dt} = -ia_0|f_0(t)\rangle - ie^{-iHt}|f_1\rangle \quad (3.35)$$

We now exploit the integral identity:

$$e^{(1-P_0)(-iH)t} = e^{-iHt} - \int_0^t e^{-iH(t-\tau)}P_0(-iH)e^{(1-P_0)(-iH)\tau}\,d\tau \quad (3.36)$$

Equation (3.36) can be easily demonstrated by performing the Laplace transform of both members [i.e., multiplying by $\exp(-zt)$ and integrating from 0 to ∞ with $\mathrm{Re}\,z > 0$]:

$$\frac{1}{z-(1-P_0)(-iH)} = \frac{1}{z-(-iH)} - \frac{1}{z-(-iH)}$$

$$\cdot P_0(-iH)\frac{1}{z-(1-P_0)(-iH)} \quad (3.37)$$

Equation (3.37) is clearly an identity, as can be seen by multiplying it by

$[z-(-iH)]$ on the left and by $[z-(1-P_0)(-iH)]$ on the right; actually, Eq. (3.37) is nothing more than a Dyson equation for the operator $(1-P_0)(-iH) \equiv (-iH) - P_0(-iH)$.

Equation (3.35), upon inserting Eq. (3.36), becomes

$$\frac{d|f_0(t)\rangle}{dt} = -ia_0|f_0(t)\rangle - i|f_1(t)\rangle - i\int_0^t e^{-iH(t-\tau)}|f_0\rangle\langle f_0| - iH|f_1(\tau)\rangle\, d\tau$$

Since

$$\langle f_0|H|f_1(\tau)\rangle = \langle f_0|H(1-P_0)|f_1(\tau)\rangle$$
$$= \langle f_1|f_1(\tau)\rangle = b_1^2\Phi_1(\tau)$$

we have

$$\frac{d|f_0(t)\rangle}{dt} = -ia_0|f_0(t)\rangle - b_1^2\int_0^t |f_0(t-\tau)\rangle\Phi_1(\tau)\, d\tau - i|f_1(t)\rangle$$

$$(3.38)$$

The Langevin-type dynamic equation (3.38) is fully equivalent to the dynamic equation (3.29) but leads naturally to an iterative procedure.

It is convenient to define a hierarchy of orthogonal states $\{|f_0\rangle, |f_1\rangle, |f_2\rangle, \dots\}$, of projection operators $\{P_i\}$ and projected Hamiltonians $\{H_i\}$ as follows:

$$|f_0\rangle \qquad P_0 = \frac{|f_0\rangle\langle f_0|}{\langle f_0|f_0\rangle} \qquad (3.39a)$$

$$|f_1\rangle = (1-P_0)H|f_0\rangle$$

$$H_1 = (1-P_0)H \qquad P_1 = \frac{|f_1\rangle\langle f_1|}{\langle f_1|f_1\rangle} \qquad (3.39b)$$

$$|f_2\rangle = (1-P_0)(1-P_1)H|f_1\rangle$$

$$H_2 = (1-P_0)(1-P_1)H \qquad P_2 = \frac{|f_2\rangle\langle f_2|}{\langle f_2|f_2\rangle} \qquad (3.39c)$$

. . . .

$$|f_{n+1}\rangle = (1-P_0)(1-P_1)\cdots(1-P_n)H|f_n\rangle \qquad (3.39d)$$

$$H_{n+1} = (1-P_0)(1-P_1)\cdots(1-P_n)H \qquad P_{n+1} = \frac{|f_{n+1}\rangle\langle f_{n+1}|}{\langle f_{n+1}|f_{n+1}\rangle}$$

By successive application of the above-described procedures, we obtain the result that the equation of motion of the variables $|f_n(t)\rangle$ driven by the corresponding projected Hamiltonian H_n has the form

$$\frac{d|f_n(t)\rangle}{dt} = -ia_n|f_n(t)\rangle - b_{n+1}^2 \int_0^t |f_n(t-\tau)\rangle \Phi_{n+1}(\tau)\, d\tau - i|f_{n+1}(t)\rangle$$

(3.40)

where

$$a_n = \frac{\langle f_n|H|f_n\rangle}{\langle f_n|f_n\rangle}$$

(3.41)

$$b_{n+1}^2 = \frac{\langle f_{n+1}|f_{n+1}\rangle}{\langle f_n|f_n\rangle}$$

(3.42)

and

$$\Phi_{n+1}(t) = \frac{\langle f_{n+1}|f_{n+1}(t)\rangle}{\langle f_{n+1}|f_{n+1}\rangle} = \frac{\langle f_{n+1}|e^{-iH_{n+1}t}|f_{n+1}\rangle}{\langle f_{n+1}|f_{n+1}\rangle}$$

(3.43)

Multiplying the set of interdependent equations (3.40) by $\langle f_n|$, we see that the correlation functions $\{\Phi_0(t), \Phi_1(t), \Phi_2(t), \ldots\}$ satisfy the set of Volterra equations

$$\frac{d}{dt}\Phi_n(t) = -ia_n\Phi_n(t) - b_{n+1}^2 \int_0^t \Phi_n(t-\tau)\Phi_{n+1}(\tau)\, d\tau$$

(3.44)

Indicating by $\hat{\Phi}_n(z)$ the Laplace transform

$$\hat{\Phi}_n(z) = \int_0^\infty e^{-zt}\Phi_n(t)\, dt \qquad (\mathrm{Re}\, z > 0)$$

(3.45)

and performing the Laplace transform of Eq. (3.44), we obtain

$$z \cdot \hat{\Phi}_n(z) - 1 = -ia_n\hat{\Phi}_n(z) - b_{n+1}^2 \hat{\Phi}_n(z)\hat{\Phi}_{n+1}(z)$$

(3.46)

From Eqs. (3.46) we arrive thus at the continued fraction expansion

$$\hat{\Phi}_0(z) = \cfrac{1}{z + ia_0 + \cfrac{b_1^2}{z + ia_1 + \cfrac{b_2^2}{z + ia_2 + \cdots}}}$$

(3.47)

The Green's-function diagonal matrix element $G_{00}(E)$ on the orbital of interest and the correlation function $\hat{\Phi}_0(z)$ are related via

$$G_{00}(E) = \langle f_0 | \frac{1}{E-H} | f_0 \rangle \equiv (-i)\hat{\Phi}_0(-iE) \qquad (\text{Im } E > 0)$$

We then have

$$G_{00}(E) = \cfrac{1}{E - a_0 - \cfrac{b_1^2}{E - a_1 - \cfrac{b_2^2}{E - a_2 - \cdots}}} \qquad (3.48)$$

where the parameters a_n, b_n^2 are defined by Eqs. (3.41) and (3.42).

To verify the equivalence of the memory function approach to the recursion method or the Lanczos method, it is sufficient to note that the state $|f_{n+1}\rangle$ defined via Eq. (3.39d) coincides with the state

$$|f_{n+1}\rangle = (1 - P_n)(1 - P_{n-1})H|f_n\rangle \qquad (3.49)$$

in which the orthogonalization is confined to the *two immediate predecessors*. In fact, it can be verified that $|f_{n+1}\rangle$ as defined by Eq. (3.49) is automatically orthogonal to $|f_{n-2}\rangle$, $|f_{n-3}\rangle$,..., $|f_0\rangle$. Equation (3.49) can be recast in the form

$$|f_{n+1}\rangle = (1 - P_n - P_{n-1})H|f_n\rangle = H|f_n\rangle - a_n|f_n\rangle - b_n^2|f_{n-1}\rangle \quad (3.50)$$

which is perfectly equivalent to the Lanczos method [Eq. (3.14)] and to the recursion method [Eq. (3.21)] (except for the inessential difference in the orthonormalization already discussed; see footnote, p. 146).

To establish the relation between the memory function formalism and the moments, consider the Volterra integro-differential equations (3.44) for $\Phi_0(t)$:

$$\frac{d\Phi_0(t)}{dt} = -ia_0\Phi_0(t) - b_1^2\int_0^t \Phi_0(t-\tau)\Phi_1(\tau)\,d\tau \qquad (3.51)$$

Following an early idea in the literature,[24] it is convenient to expand both the correlation function $\Phi_0(t)$ and its memory function $\Phi_1(t)$ in a Taylor series using their respective moments:

$$\Phi_0(t) = \frac{\langle f_0 | e^{-iHt} | f_0 \rangle}{\langle f_0 | f_0 \rangle} = \sum_{n=0}^{\infty} \frac{\mu_n}{n!} (-it)^n \tag{3.52}$$

$$\Phi_1(t) = \frac{\langle f_1 | e^{-iH_1 t} | f_1 \rangle}{\langle f_1 | f_1 \rangle} = \sum_{n=0}^{\infty} \frac{\sigma_n}{n!} (-it)^n \tag{3.53}$$

Substituting the power moment expansion (3.53) and (3.52) into (3.51), one obtains[25]

$$a_0 = \mu_1 \qquad b_1^2 = \mu_2 - \mu_1^2$$

$$\sigma_n = \frac{\mu_{n+2} - \mu_1 \mu_{n+1}}{\mu_2 - \mu_1^2} - \sum_{m=0}^{n-1} \sigma_m \mu_{n-m} \qquad \text{if } b_1^2 \neq 0$$

$$\mu_n = \mu_1^n \qquad \text{if } b_1^2 = 0 \tag{3.54}$$

The procedure can be iterated, considering now the Volterra integro-differential equation (3.44) for $\Phi_1(t)$, then for $\Phi_2(t)$, and so on. The implications of Eq. (3.54) have been discussed in more detail in Chapter III, with comments on product-difference (PD) algorithms.

The memory function formalism leads to several advantages, both from a formal point of view and from a practical point of view. It makes transparent the relationship between the recursion method, the moment method, and the Lanczos method on the one hand and the projective methods of nonequilibrium statistical mechanics on the other. Also the ad hoc use of Padé approximants of type $[n/n+1]$, often adopted in the literature without true justification, now appears natural, since the approximants of the J-fraction (3.48) encountered in continued fraction expansions of autocorrelation functions are just of the type $[n/n+1]$. The mathematical apparatus of continued fractions can be profitably used to investigate properties of Green's functions and to embody in the formalism the physical information pertinent to specific models. Last but not least, the memory function formalism provides a new and simple PD algorithm to relate moments to continued fraction parameters.

Another remarkable advantage of the formalism is that the generalization to the non-Hermitian relaxation operator is straightforward. If H is not Hermitian, that is, if $H \neq H^\dagger$, we must go through the entire demonstration of this section; we can easily verify the usefulness of introducing the following biorthogonal basis set:

$$|f_0\rangle \qquad\qquad P_0 = \frac{|f_0\rangle\langle \tilde{f}_0|}{\langle \tilde{f}_0|f_0\rangle}$$

$$\langle \tilde{f}_0| \equiv \langle f_0|$$

$$|f_1\rangle = (1 - P_0)H|f_0\rangle \qquad P_1 = \frac{|f_1\rangle\langle \tilde{f}_1|}{\langle \tilde{f}_1|f_1\rangle}$$

$$\langle \tilde{f}_1| = \langle \tilde{f}_0|H(1 - P_0)$$

$$(3.55)$$

$$|f_2\rangle = (1 - P_0)(1 - P_1)H|f_1\rangle \qquad P_2 = \frac{|f_2\rangle\langle \tilde{f}_2|}{\langle \tilde{f}_2|f_2\rangle}$$

$$\langle \tilde{f}_2| = \langle \tilde{f}_1|H(1 - P_1)(1 - P_0)$$

· ·

All the relationships of the present section maintain their validity provided that the $\langle f_i|$ are systematically replaced by $\langle \tilde{f}_i|$. In particular, the parameters a_n and b_n^2 of the continued fraction expansion (3.48) are now given by [compare with Eqs. (3.41) and (3.42)]

$$a_n = \frac{\langle \tilde{f}_n|H|f_n\rangle}{\langle \tilde{f}_n|f_n\rangle} \qquad\qquad (3.56)$$

$$b_{n+1}^2 = \frac{\langle \tilde{f}_{n+1}|f_{n+1}\rangle}{\langle \tilde{f}_n|f_n\rangle} \qquad\qquad (3.57)$$

The parameters a_n and b_n^2 are not necessarily real, and b_n^2 can even be negative. The properties of continued fractions in relation to relaxation operators are described in Chapter III.

It is easily seen by inspection that the biorthogonal basis set definition (3.55) coincides with the definition (3.18) given in the discussion of the Lanczos method. We recall that the dynamics of operators (Liouville equations) or probabilities (Fokker-Planck equations) have a mathematical structure similar to Eq. (3.29) and can thus be treated with the same techniques (see, e.g., Chapter I) once an appropriate generalization of a scalar product is performed. For instance, this same formalism has been successfully adopted to model phonon thermal baths[26] and to include, in principle, anharmonicity effects in the interesting aspects of lattice dynamics and atom-solid collisions.[27]

Before concluding this section we think it instructive to give an example helpful to readers not acquainted with the use of memory function techniques.

Illustrative Example

A simple example will illustrate the memory function techniques of obtaining the Green's function of a non-Hermitian operator.

Consider a three-level model Hamiltonian with eigenstates $|\phi_1\rangle$, $|\phi_2\rangle$, $|\phi_3\rangle$ and eigenvalues E_1, E_2, E_3:

$$H = E_1|\phi_1\rangle\langle\phi_1| + E_2|\phi_2\rangle\langle\phi_2| + E_3|\phi_3\rangle\langle\phi_3| \qquad (3.58)$$

As a numerical example, we take

$$E_1 = 0 \qquad E_2 = 6 \qquad E_3 = -6i \qquad (3.59)$$

The fact that E_3 is a complex number implies that H is a non-Hermitian operator.

As a state of interest we take, for instance,

$$|f_0\rangle = \frac{1}{\sqrt{3}}(|\phi_1\rangle + |\phi_2\rangle + |\phi_3\rangle) \qquad (3.60)$$

Our aim is to evaluate explicitly the Green's-function matrix element

$$G_{00}(E) = \langle f_0| \frac{1}{E-H} |f_0\rangle \qquad (3.61)$$

We do this first directly, then by the moment method, and finally by using projective techniques and biorthogonal sets.

We begin with the *direct approach*, which is possible because of the simplicity of the model. We put (3.60) into (3.61) and use the values of E_i given in Eq. (3.59) to obtain

$$G_{00}(E) = \frac{1}{3}\left(\frac{1}{E} + \frac{1}{E-6} + \frac{1}{E+6i} \right) \qquad (3.62)$$

By straightforward manipulations on (3.62) we have

$$G_{00}(E) = \cfrac{1}{\cfrac{E^3 - 6(1-i)E^2 - 36iE}{E^2 - 4(1-i)E - 12i}}$$

Taking the quotient of polynomials and treating the rest in a similar manner, we obtain for $G_{00}(E)$ the expression

$$G_{00}(E) = \cfrac{1}{E - 2(1-i) - \cfrac{8i}{E - 7(1-i) - \cfrac{54i}{E + 3(1-i)}}} \qquad (3.63)$$

The parameters of the continued fraction (3.63) are

$$a_0 = 2(1-i) \qquad a_1 = 7(1-i) \qquad a_2 = -3(1-i)$$
$$b_1^2 = 8i \qquad b_2^2 = 54i \qquad b_n^2 = 0 \qquad (n \geq 3) \qquad (3.64)$$

We now apply the *moment method technique*. The moments of H with respect to the state of interest $|f_0\rangle$ are simply

$$\mu_n = \frac{\langle f_0|H^n|f_0\rangle}{\langle f_0|f_0\rangle} = \frac{1}{3}6^n[1 + (-i)^n] \qquad n = 1, 2, \ldots \qquad (3.65)$$

A number of moments (3.65) are reported for convenience in the first row of Table I. Using the memory function PD algorithm (see Chapter III and ref. 25), summarized by Eq. (3.54), we obtain the moments for the first few memory functions (also reported in Table I), and hence the continued fraction parameters (3.64) are recovered. Notice also that the last row of Table I is constituted by powers of the same number, as foreseen by Eq. (3.54) in the case of exact truncation.

Finally, we solve the same problem using *projective techniques*. We set up as initial states of the two-sided chain (3.55):

$$|f_0\rangle = \frac{1}{\sqrt{3}}(|\phi_1\rangle + |\phi_2\rangle + |\phi_3\rangle) \qquad \langle \tilde{f}_0| = \langle f_0|$$

TABLE I

Product-Difference Memory Function Procedure for Evaluating the Parameters
of the Continued Fraction Expansion[a] Starting from the Moments
$\mu_n = (1/3)6^n[1 + (-i)^n]$

0	1	2	3	4	5	6	7	8	9
1	$2(1-i)$	0	$72(1+i)$	864	$2592(1-i)$	0	$93,312(1+i)$	559,872	$3,359,232(1-i)$
1	$7(1-i)$	$-44i$	$-92(1+i)$	-208	$272(1-i)$	$-4672i$	$-15,424(1+i)$		
1	$-3(1-i)$	$-18i$	$54(1+i)$	-324	$972(1-i)$				
		$a_0 = 2(1-i)$	$a_1 = 7(1-i)$	$a_2 = -3(1-i)$		$b_1^2 = 8i$	$b_2^2 = 54i$	$b_3^2 = 0$	

[a] The appearance of exact powers of $-3(1-i)$ in the third row is related to the exact truncation of the continued fraction at the $n = 3$ step.

We have

$$a_0 = \langle \tilde{f}_0 | H | f_0 \rangle = \frac{1}{3}(6 - 6i) = 2(1 - i)$$

in agreement with Eq. (3.64). Then

$$|f_1\rangle = (1 - P_0)H|f_0\rangle = H|f_0\rangle - a_0|f_0\rangle$$

$$= \frac{2}{\sqrt{3}}\left[(-1+i)|\phi_1\rangle - (1+2i)|\phi_2\rangle + (2+i)|\phi_3\rangle\right]$$

and similarly

$$\langle \tilde{f}_1| = \langle \tilde{f}_0|H(1 - P_0) = \langle \tilde{f}_0|H - \langle \tilde{f}_0|a_0$$

$$= \frac{2}{\sqrt{3}}\left[\langle \phi_1|(-1+i) - \langle \phi_2|(1+2i) + \langle \phi_3|(2+i)\right]$$

We can now obtain

$$b_1^2 = \frac{\langle \tilde{f}_1 | f_1 \rangle}{\langle \tilde{f}_0 | f_0 \rangle} = 8i \qquad a_1 = \frac{\langle \tilde{f}_1 | H | f_1 \rangle}{\langle \tilde{f}_1 | f_1 \rangle} = 7(1 - i)$$

The next step functions are

$$|f_2\rangle = 6\sqrt{3}\,(1+i)\left[-(1+i)|\phi_1\rangle + |\phi_2\rangle + i|\phi_3\rangle\right]$$

$$\langle \tilde{f}_2| = 6\sqrt{3}\,(1+i)\left[-\langle \phi_1|(1+i) + \langle \phi_2| + \langle \phi_3|i\right]$$

and the remaining parameters of the continued fractions can be calculated and coincide, of course, with (3.64).

IV. ASYMPTOTIC BEHAVIOR OF CONTINUED FRACTION COEFFICIENTS

In periodic systems it is well known that the allowed energy levels are grouped into energy bands. Thus the Green's function corresponding to any reasonably general starting state (a reasonably general starting state is any state whose dynamic evolution is expected to spread to all band eigenfunctions) must have a *cut on the real axis corresponding to the allowed energy regions*. The band edges thus play a central role in determining the asymptotic behavior of continued fraction parameters in perfect crystals. Since the presence of allowed and forbidden energies is a consequence of translational symmetry, we can also say that the *crystal symmetry has a quite important effect in continued fraction tails*. Also, in a solid with impurities, producing a number of additional levels in the energy gap but no change in band edges, the asymptotic behavior of the Green's function is still determined by the perfect crystal only, in spite of the fact that the whole solid-plus-impurity system is aperiodic.

We consider first the simple situation of a solid with a single band extending between the edges E_1 and E_2. In this case the asymptotic values of a_n and b_n^2 are related to the bounds E_1, E_2 by the formulas

$$a = \lim_{n \to \infty} a_n = \tfrac{1}{2}(E_1 + E_2) \tag{4.1}$$

$$b^2 = \lim_{n \to \infty} b_n^2 = \left(\frac{E_2 - E_1}{4} \right)^2 \tag{4.2}$$

This simple and useful result can be obtained following, *mutatis mutandae*, an early argument of Van Vleck.[28] Consider, in fact, the Green's function

$$G_{00}(E) = \cfrac{1}{E - a_0 - \cfrac{b_1^2}{E - a_1 - \cfrac{b_2^2}{E - a_2 - \cdots}}} \tag{4.3}$$

and the tail of order n,

$$t_n(E) = \frac{b_n^2}{E - a_n -} \frac{b_{n+1}^2}{E - a_{n+1} -} \cdots$$

We have exactly

$$t_n(E) = \frac{b_n^2}{E - a_n - t_{n+1}(E)} \tag{4.4}$$

Assuming that the asymptotic limits of a_n and b_n^2 exist and are a and b^2, respectively, we see from (4.4) that the asymptotic tail $t(E)$ satisfies the self-consistent equation

$$t(E) = \frac{b^2}{E - a - t(E)} \tag{4.5}$$

This gives the square root terminator

$$t(E) = \tfrac{1}{2}\left[E - a - \sqrt{(E-a)^2 - 4b^2} \right] \qquad (4.6)$$

[The $-$ sign is chosen in such a way that $\text{Im } t(E) < 0$, $\text{Im } E \rightarrow 0+$ in order to preserve the Herglotz property of $G_{00}(E)$.] The density of states expressed by the standard equation

$$n(E) = -\frac{1}{\pi}\text{Im}\, G_{00}(E) \qquad \text{Im } E \rightarrow 0+$$

exhibits a continuum of states in the range $E_1 < E < E_2$, where $E_1 = a - 2b$, $E_2 = a + 2b$; Eqs. (4.1) and (4.2) are thus proved. We see that in the case of a single connected band, a is just the middle point of the band and $2b$ the band half-width.

These theoretical considerations, corroborated by experience based on numerical experiments, show that the asymptotic values a and b^2 are independent of the particular form of the connected spectrum and depend only on its bounds. Moreover, as observed by Gaspard and Cyrot-Lackmann,[4] van Hove singularities[2] in the spectrum produce *damped oscillations* in the continued fraction coefficients, with a frequency related to the position of the singularity and with a decay toward the asymptotic values, whose rate depends on the strength of the singularity. This oscillating behavior has been theoretically proved by Hodges,[29] who also derives expressions for the amplitude, phase, and decay of these oscillations. Beer and Pettifor[30] have given a workable procedure to guess the asymptotic values a and b^2 from the knowledge of the lowest-level parameters $\{a_0, a_1, \ldots, a_n\}$, $\{b_1^2, b_2^2, \ldots, b_n^2\}$. They use a square root terminator of the continued fraction combined with a suitable balance of the eigenvalues of the associated tridiagonal matrix.

The situation becomes much more complicated in the presence of two or more connected bands; the band gaps cause asymptotically *undamped oscillations*.[31-33] We refer to the papers of Magnus[32] and Turchi et al.[33] for a very detailed theoretical and numerical analysis of the asymptotic behavior of the continued fraction coefficients.

Here we comment briefly on a two-band model crystal with a single energy gap. A schematic picture for the density of states is shown in Fig. 3. Following the notations of Turchi et al.,[33] we denote by E_1, E_2, E_3, E_4 the energy bounds, by W the half-width of the total spectrum, and by G the half-width of the gap.

The presence of the gap produces undamped oscillations in the continued fraction parameters. In particular, a_n oscillates between $\tfrac{1}{2}(E_1 + E_4) \pm G$ and b_n between $\tfrac{1}{2}(W \pm G)$, while the multiplicity of the period of oscillations de-

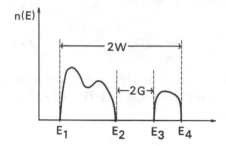

Figure 3. Schematic representation of the density of states of a two-band model crystal with a single energy gap. The notations used are the same of Turchi et al.[33]

pends only on the relative positions of the bounds E_1, E_2, E_3, E_4. Turchi et al.[33] also give algebraic conditions on the bounds in the case of integer (double, triple,...) periods.

The simple form of the terminator $t(E)$, expressed by Eq. (4.6) for a connected band, is lost in the presence of gaps within bands. In the literature,[32,33] analytic expressions for the tail behavior of the infinite continued fraction are provided for the case of a single or several gaps. For the single gap we quote the expression

$$
t_n(E) = \frac{E^2 - \dfrac{1}{2}\displaystyle\sum_{i=1}^{4} E_i E + \dfrac{1}{4}\displaystyle\sum_{i<j}^{4} E_i E_j - \dfrac{1}{8}\displaystyle\sum_{i=1}^{4} E_i^2 + 2b_n^2 - X^{1/2}(E)}{2\left(E - a_{n-1} - \dfrac{1}{2}\displaystyle\sum_{i=1}^{4} E_i\right)}
$$

(4.7)

where

$$
X(E) = (E - E_1)(E - E_2)(E - E_3)(E - E_4)
$$

Expression (4.7) allows one to approximate the tail of order n of the continued fraction (4.3) in terms of the last known parameters a_{n-1} and b_n^2 and in terms of the bounds E_1, E_2, E_3, E_4.

In spite of these useful investigations, further work must still be done to assess the accuracy in realistic situations such as those encountered in band structure calculations on real crystals. Notice, furthermore, that any kind of extrapolation of continued fraction parameters must be consistent with the theory of error bounds as provided by the first few exact available moments.

Finally, it is important to note that experience with different distribution functions shows that a large number of parameters a_n and b_n^2 must be known in order to be able to test their asymptotic behavior with reasonable accuracy. In this respect, stable algorithms and/or extended precision numerical procedures[34] are obligatory steps.

V. APPLICATIONS OF MEMORY FUNCTION METHODS IN PERIODIC AND APERIODIC SOLIDS

A. Introductory Remarks

In this section we give a few illustrative applications of memory function methods in solids. The subject, even when confined to electronic structure, is too wide for a thorough account, and by necessity a selection has to be made.

We remark that memory function methods have been extensively applied to perfect crystals, where they are alternative tools to the traditional band structure methods. In the presence of translational symmetry and long-range order, however, the use of the memory function techniques is not essential but rather a matter of convenience (or taste) in a number of situations. We thus focus on the study of the electronic structure in systems that are aperiodic because of impurities: our unorthodox way of looking at defects is to consider them as a source of a given frozen-in disorder in an otherwise perfect crystal lattice. The more general case of the presence of stochastic disorder will be discussed in Section VI.

The study of the electronic structure of impurities and defects in solids has a long tradition, both because of its own intrinsic theoretical interest and because of the technological importance in improving the performance of solid state devices. Lattice defects can be point defects (such as substitutional or interstital foreign atoms, vacancies, antisite defects in composite lattices), line defects (such as dislocations), planar defects (such as boundaries, adatom surfaces, stacking faults corresponding to misplaced planes of atoms), and so on.

The problems associated with defects are quite numerous, and a wealth of theoretical techniques have been devised for their solution.[35] We shall confine our attention to the Green's function formalism because of both the level of accuracy and sophistication of recent applications.[36,37] and the promise that memory function methods hold in this field.

B. Green's Function Formulation for the Impurity Problem

We write the Schrödinger equation for a crystal with an impurity in the form

$$\left[-\frac{\hbar^2}{2m}\nabla^2 + V_c + U \right]\psi = E\psi \qquad (5.1)$$

where V_c denotes the potential of the perfect lattice and U the additional potential due to the impurity. We write the Hamiltonian H as

$$H = H^0 + U \tag{5.2}$$

where H^0 is the Hamiltonian of the crystal without disturbance.

We assume that the crystal Hamiltonian H^0 is expressed in a local basis representation in the form of Eq. (2.4):

$$H^0 = \sum_{\mu m} E_{\mu m} |\phi_{\mu m}\rangle\langle\phi_{\mu m}| + \sum_{\mu m \neq \nu n} T_{\mu m, \nu n} |\phi_{\mu m}\rangle\langle\phi_{\nu n}| \tag{5.3}$$

The same representation is also used for U:

$$U = \sum_{\alpha m, \beta n} \gamma_{\alpha m, \beta n} |\phi_{\alpha m}\rangle\langle\phi_{\beta n}| \tag{5.4}$$

The range in coordinate space of the defect potential $U(\mathbf{r})$ determines the *impurity subspace* actually needed to describe the disturbance in Eq. (5.4).

It is convenient to consider some general formal properties of the Green's function for the operators H^0 and H. Corresponding to the operator H^0 one introduces the Green's function

$$G^0(E) = \frac{1}{E - H^0} \tag{5.5}$$

for energies E with an infinitesimal positive imaginary part. Similarly, for the operator H, one defines

$$G(E) = \frac{1}{E - H} \tag{5.6}$$

Combining Eqs. (5.2), (5.5), and (5.6), we obtain the Dyson equation

$$G(E) = G^0(E) + G^0(E) U G(E) \tag{5.7}$$

or equivalently

$$G(E) = \frac{1}{1 - G^0(E) U} G^0(E) \tag{5.8}$$

Equation (5.7) can be used to evaluate the matrix elements of G within the impurity subspace, and then, if needed, any other matrix element of G.

From Eq. (5.8) the condition for the existence of bound states (if any) within the band gaps is

$$\det\|1 - G^0(E)U\| = 0 \tag{5.9}$$

When the defect potential $U(\mathbf{r})$ has a *short-range* nature, the order of the determinantal equation (5.9) is manageable since the operator $1 - G^0(E)U$ needs to be represented only in the impurity subspace.

It is convenient to continue considering operator equations in general form. Following the literature,[35] we define the quantity

$$Q(E) = \det\|1 - G^0(E)U\| \tag{5.10}$$

and the phase shift associated with this scattering problem as

$$\delta(E) = -\arg Q(E) = -\arctan\frac{\operatorname{Im} Q(E)}{\operatorname{Re} Q(E)} \tag{5.11}$$

The defect-induced change in the density of states

$$\Delta(E) = D(E) - D^0(E) \tag{5.12}$$

is given, both in allowed and forbidden energy regions, by

$$\Delta(E) = \frac{1}{\pi}\frac{d\delta(E)}{dE} \tag{5.13}$$

[In Eq. (5.13) a factor 2 must be added if spin degeneracy is taken into account.]

Equation (5.13) can be arrived at by the following considerations. Recall that the spectral density $D(E)$ of a given operator H is connected to its Green's function $G(E)$ by the relation

$$D(E) = -\frac{1}{\pi}\operatorname{Im} \operatorname{Tr} G(E) \qquad (\operatorname{Im} E \to 0+) \tag{5.14}$$

We have

$$D(E) = -\frac{1}{\pi}\operatorname{Im} \operatorname{Tr}\left[-\frac{d}{dE}\ln G(E)\right]$$
$$= \frac{1}{\pi}\operatorname{Im}\frac{d}{dE}\ln \det G(E) \tag{5.15}$$

It is simple to verify that Eq. (5.15) becomes an identity if one uses the diagonal representation of $G(E)$.

Using (5.15) and a similar expression for $D^0(E)$, it follows that

$$\Delta(E) = D(E) - D^0(E) = \frac{1}{\pi} \operatorname{Im} \frac{d}{dE} \ln \frac{\det G(E)}{\det G^0(E)}$$

$$= -\frac{1}{\pi} \operatorname{Im} \frac{d}{dE} \ln Q(E) \qquad (5.16)$$

where in the last passage use has been made of Eqs. (5.8) and (5.10). Equation (5.13) follows upon remarking that if z is a complex number, then $\operatorname{Im} \ln z = \arg z \pmod{2\pi}$.

In a self-consistent calculation of defects, the initial guess of the potential U in the form (5.4) has to be checked or improved toward self-consistency; this requires a calculation of the density matrix of the imperfect crystal. This can be done as follows.

Consider the operator H of the crystal-plus-impurity system; let us denote by $|\psi_i\rangle$ and E_i its eigenstates and eigenfunctions. The (spinless) density matrix operator for the ground state is

$$\rho = 2 \sum_{\text{occupied}} |\psi_i\rangle\langle\psi_i|$$

where the sum is confined to states whose energy E_i is below the Fermi energy E_F and the factor 2 takes into account spin degeneracy. In a real-space representation, the spinless density matrix is

$$\rho(\mathbf{r},\mathbf{r}') = 2 \sum_{\text{occupied}} \psi_i(\mathbf{r})\psi_i^*(\mathbf{r}') \qquad (5.17)$$

We can formally express the eigenstates $|\psi_i\rangle$ as linear combinations of the localized orbitals $|\phi_{\mu m}\rangle$

$$|\psi_i\rangle = \sum_{\mu m} c_{i,\mu m}|\phi_{\mu m}\rangle \qquad (5.18)$$

Inserting (5.18) into (5.17), we have

$$\rho(\mathbf{r},\mathbf{r}') = 2 \sum_{E_i < E_F} \psi_i(\mathbf{r})\psi_i^*(\mathbf{r}')$$

$$= 2 \sum_{\mu m, \nu n} \sum_{E_i < E_F} c_{i,\mu m} c_{i,\nu n}^* \phi_{\mu m}(\mathbf{r})\phi_{\nu n}^*(\mathbf{r}')$$

$$= 2 \sum_{\mu m, \nu n} \phi_{\mu m}(\mathbf{r})\rho_{\mu m,\nu n}\phi_{\nu n}^*(\mathbf{r}') \qquad (5.19)$$

where the matrix elements ρ in the localized basis are defined as

$$\rho_{\mu m, \nu n} = \sum_{E_i < E_F} c_{i,\mu m} c^*_{i,\nu n} \tag{5.20}$$

In spite of the apparently awkward aspect of $\rho_{\mu m, \nu n}$, whose determination seems to require the eigenvectors of H, it is possible to calculate it via this elegant trick. Consider the Green's-function matrix elements

$$G_{\mu m, \nu n}(E) = \left\langle \phi_{\mu m} \left| \frac{1}{E-H} \right| \phi_{\nu n} \right\rangle = \sum_i \frac{1}{E-E_i} \langle \phi_{\mu m} | \psi_i \rangle \langle \psi_i | \phi_{\nu n} \rangle$$

$$= \sum_i \frac{1}{E-E_i} c_{i,\mu m} c^*_{i,\nu n} \tag{5.21}$$

For simplicity we suppose that in our case the localized orbitals $|\phi_{\mu n}\rangle$ and the coefficients $c_{i,\mu m}$ of (5.18) are real quantities. Comparing (5.21) and (5.20), we obtain at once

$$\rho_{\mu m, \nu n} = -\frac{1}{\pi} \int_{-\infty}^{E_F} \operatorname{Im} G_{\mu m, \nu n}(E)\, dE \tag{5.22}$$

which is a workable formula for the calculation of the density matrix of the actual crystal. From the change in the density matrix, a new one-electron perturbation operator can be constructed and represented in a localized basis. In principle, the procedure can be iterated to self-consistency. The size of the problem in the Green's-function method is dictated by the range of the perturbation potential, even for self-consistent calculations. Investigations along similar lines are very complicated and not routine but have produced interesting results for reasonably localized impurities and neutral vacancies.

C. The Recursion Method for the Localized and Extended Impurity Problem: Applications and Perspectives

The Green's-function formalism for impurities in its fully self-consistent formulation or in some simplified version has been used to treat short-range defect potentials. In this case the operator equations can be represented by a small basis set, restricted essentially to the impurity subspace. In addition to the matrix elements of U, one must calculate the matrix elements of $G^0(E)$. The latter are independent of the impurity disturbance and need only be calculated in the impurity subspace. Since the operator H^0 refers to the perfect crystal, it can be diagonalized with the standard methods of band theory. Denoting the resulting Bloch functions by $\psi_{n\mathbf{k}}$, we have

$$G^0(E) = \sum_{n\mathbf{k}} \frac{|\psi_{n\mathbf{k}}\rangle\langle\psi_{n\mathbf{k}}|}{E - E_{n\mathbf{k}}} \qquad (5.23)$$

From Eq. (5.23) or appropriate manipulations of it, the matrix elements of the Green's operator $G^0(E)$ between any pair of atomic-like basis orbitals can be obtained.

What we have described is the commonly adopted procedure for the evaluation of the Green's operator $G^0(E)$. We notice, however, that the matrix elements of G^0 can also be conveniently calculated using an alternative approach based on memory function methods. Finally, integrals such as those appearing in Eq. (5.22) can be evaluated by Gaussian quadrature and checked against error bounds [see Chapter III; of course, off-diagonal elements must be preliminarily expressed in terms of appropriate diagonal elements as in Eq. (3.28)].

A great simplification occurs when the disturbance of the impurity is confined to a single site and to a single orbital (or to a single appropriately symmetrized combination of orbitals). The Fredholm determinantal equation (5.9) is one-dimensional and becomes

$$1 - G_{00}^0(E)U_0 = 0 \qquad (5.24)$$

where U_0 is the impurity energy change of the orbital $|\phi_0\rangle$. Thus we obtain a Koster-Slater[38] type of equation:

$$\frac{1}{U_0} = G_{00}^0(E) \qquad (5.25)$$

The quantity U_0 is often assumed to be equal to the difference between the ionization potentials of the impurity atom and the perfect crystal atom. The above procedure, introduced by Dow and coworkers,[39] has had remarkable success in elucidating the overall trends of deep impurity levels or deep core excitons.

Equations of type (5.25) are also encountered in the modellistic description of vacancies. To simulate a vacancy, either one can increase by an arbitrary amount the energy of the orbitals on the vacancy site or, equivalently, one can remove the corresponding orbitals and take as the state of interest an appropriate combination of orbitals centered on nearest-neighbor positions. From Eq. (5.25), using the limit $U_0 \to \infty$, we see that the vacancy defect energy is determined by the zeros of $G_{00}^0(E)$.

As an example of the study of vacancies and self-interstitial impurities by the continued fraction expansion of Eq. (5.25), we mention the work of Kauffer et al.[40] These authors consider impurities in silicon and set up a model tight-binding Hamiltonian with s-p hybridization, which satisfactorily describes the valence and conduction bands of the perfect crystal. A cluster of 2545 atoms is generated, and vacancies (or self-interstitial impurities) are introduced at the center of the cluster. One then takes as a seed state an appropriate orbital or symmetrized combination of orbitals, and the recursion method is started. Though self-consistent potential modifications are neglected in this paper,[40] the model leads to qualitatively satisfactory results within a simple physical picture.

Summing up, we see that the traditional approach to impurity problems within the Green's-function formalism exploits the basic idea of splitting the problem into a perfect crystal described by the operator G^0 and a perturbation described by the operator U. The matrix elements of G^0 are then calculated, usually by direct diagonalization of H^0 or by means of the recursion method. Following this traditional line of attack, one does not fully exploit the power of the memory function methods. They appear at most as an auxiliary (but not really essential) tool used to calculate the matrix elements of G^0.

A less orthodox line of attack, as yet not explored to its full potential, applies from the beginning the recursion method to the solid-plus-impurity system. The direct use of memory function methods to the perturbed solid is no more difficult than for the perfect solid, with the advantage of overcoming the traditional separation of the actual Hamiltonian into a "perfect part" and a "perturbed part." In fact, such a separation, to make any practical sense, requires that the "perturbed part" be localized in real space, a restriction hardly met when treating impurities with a coulombic tail.

For extended impurities it is convenient to use the recursion method and to calculate directly the needed matrix elements of G, from which the projected density of states can be obtained and the effect of the impurity inferred.

It may seem that an extended impurity problem could require an impossibly large cluster size. This is not true, however, basically because of the physical implications of translational symmetry away from the impurity region. In fact, the asymptotic limits of the continued fraction parameters and the cut in the real energy axis are determined by the perfect crystal only; this allows a guideline for appropriate extrapolation of the recursion coefficients.

Only fragmentary and semiqualitative calculations have been published using this line of approach.[41,42] In the work of Bylander and Rehr,[41] the starting model is a crystal with one orbital per site described by a tight-binding Hamiltonian

$$H^0 = \sum_m E_0|\phi_m\rangle\langle\phi_m| + \sum_{m \neq n} T_{mn}|\phi_m\rangle\langle\phi_n|$$

The extended potential with a coulombic tail of the impurity centered at the origin is simulated by

$$U = U_0|\phi_0\rangle\langle\phi_0| - \sum_{m \neq 0} \frac{e^2}{\varepsilon\tau_m}|\phi_m\rangle\langle\phi_m|$$

where ε is the static dielectric constant of the crystal. Starting from the site impurity orbital, the recursion method is set up to evaluate

$$G_{00}(E) = \langle\phi_0|\frac{1}{E - H}|\phi_0\rangle$$

where $H = H^0 + U$. The impurity levels are inferred from the poles of $G_{00}(E)$ in the forbidden energy regions. The comparison with standard effective mass equations allows a better qualitative understanding of the comparative role of short-range and long-range contributions. This type of approach can be generalized in a self-consistent way as shown by Graft et al.[42], also in the case of superlattices. This approach should be of value in clarifying the role of many-valley interference effects[43] on the deep or shallow nature of excitons and of single or multiple ionized impurities.

Finally, we wish to briefly consider the case of a surface impurity. A physical quantity of interest is the energy gained in an adatom chemisorption process. We consider a simplified but workable model, which gives at least a guideline on how to face more complicated situations.[6]

Consider an absorbed atom described by a single orbital ϕ_A which interacts with only *one* orbital, say ϕ_1, of a given solid S. The Hamiltonian of the solid S plus a noninteracting adatom is

$$H^0 = \varepsilon_A|\phi_A\rangle\langle\phi_A| + \sum_{i,j=1}^N T_{ij}|\phi_i\rangle\langle\phi_j| \qquad (5.26)$$

The interaction U is of the form

$$U = \gamma(|\phi_A\rangle\langle\phi_1| + |\phi_1\rangle\langle\phi_A|) \qquad (5.27)$$

assuming γ real.

As is usual in the Green's function approach, we consider the operator

$$M = 1 - G^0 U \tag{5.28}$$

in a representation formed by the states $|\phi_A\rangle$ and $|\phi_1\rangle$, the only ones appearing in the impurity subspace according to Eq. (5.27). We have

$$
\begin{aligned}
M_{AA} &= 1 - \langle \phi_A | G^0 U | \phi_A \rangle = 1 \\
M_{11} &= 1 - \langle \phi_1 | G^0 U | \phi_1 \rangle = 1 \\
M_{A1} &= - \langle \phi_A | G^0 U | \phi_1 \rangle = - \gamma G^0_{AA}(E) = - \gamma \frac{1}{E - \varepsilon_A} \\
M_{1A} &= - \langle \phi_1 | G^0 U | \phi_A \rangle = - \gamma G^0_{11}(E)
\end{aligned}
\tag{5.29}
$$

all other off-diagonal matrix elements being zero.

From (5.28) and (5.29), we have

$$
\begin{aligned}
Q(E) &= \det \| 1 - G^0 U \| = 1 - \gamma^2 G^0_{AA}(E) G^0_{11}(E) \\
&= 1 - \frac{\gamma^2}{E - \varepsilon_A} G^0_{11}(E) = \frac{E - \varepsilon_A - \gamma^2 G^0_{11}(E)}{E - \varepsilon_A}
\end{aligned}
$$

Using Eq. (5.13) we can now calculate the defect-induced change in the density of states and hence the energy change in the adatom chemisorption process. The above results must be appropriately generalized if interactions of a more complicated nature are to be considered, but it is still possible to calculate physical quantities of interest, such as change of energy and ionicity.

VI. ELECTRONS IN LATTICES WITH STOCHASTIC DISORDER

A. Introductory Remarks and Diagonal Disorder

In recent years there has been an explosion of interest in the electron properties of disordered lattices. The more common line of approach to this kind of problem is to study the mean resolvent of the random medium,[44] and the memory function methods can be of remarkable help for this purpose. Otherwise one can investigate by the memory function methods (basically the recursion method) a number of judiciously selected configurations; this line of approach is particularly promising because it allows one to overcome some of the limitations inherent in the mean field theories. In this section we describe the use of continued fractions in connection with standard multiple scattering theories (typically ATA, average t-matrix approximation, and CPA, coherent potential approximation), as well as recent developments in which memory function methods are used directly to investigate an appropriate number of configurations for the ensemble-averaging process.

For the sake of simplicity, we consider a tight-binding Hamiltonian with one orbital per site, at regular lattice positions τ_n. We consider specifically the case of diagonal disorder, neglecting at this stage off-diagonal and environmental disorder. The simplest model Hamiltonian which embodies the above features is of the type

$$H = H^0 + W \tag{6.1a}$$

$$H^0 = \sum_{m \neq n} T_{mn} |\phi_m\rangle \langle \phi_n| \tag{6.1b}$$

$$W = \sum_m w_m |\phi_m\rangle \langle \phi_m| \tag{6.1c}$$

where we have separated the off-diagonal operator H^0, which has crystalline periodicity, from the diagonal operator W, which contains stochastic parameters. We suppose that the energies w_m are distributed randomly among the lattice sites τ_m according to a given distribution function $P(w)$. For example, in the case of a random substitutional binary alloy $A_x B_{1-x}$, the stochastic variable w can take either the value ε_A or ε_B in proportion to the concentration $c_A = x$ and $c_B = 1 - x$, respectively. In the case of the Anderson model,[45] w is a continuous variable with a rectangular distribution function within two given bounds. In the case of the Sumi-Toyozawa[46] model for electron (or exciton) energies modulated by a phonon field, a Gaussian distribution of the stochastic variable w is assumed.

It is evident that in a random system the exact eigenstates depend on the actual configuration and, besides being complicated, would be of little help even if known. It is thus convenient to consider some tractable approximation of the mean resolvent to obtain the density of states and spectral weights.

We start from the Dyson equation, corresponding to the split (6.1a):

$$G = G^0 + G^0 W G \tag{6.2}$$

Using standard iterative procedures, and provided the resulting series converges, we have

$$G = G^0 + G^0 W G^0 + G^0 W G^0 W G^0 + \cdots \tag{6.3}$$

In the local basis representation, Eq. (6.3) takes the form

$$G_{ll'} = G_{ll'}^0 + \sum_m G_{lm}^0 w_m G_{ml'}^0 + \sum_{mn} G_{lm}^0 w_m G_{mn}^0 w_n G_{nl'}^0 + \sum_{mnp} \cdots \qquad (6.4)$$

It is convenient to define a t matrix, which describes the *multiple scattering from the same site without excursion to the other sites*:

$$t_l = w_l + w_l G_{ll}^0 w_l + w_l G_{ll}^0 w_l G_{ll}^0 w_l + \cdots$$
$$= w_l + w_l G_{ll}^0 t_l \qquad (6.5)$$

From Eq. (6.5), we can also write

$$t_l(E) = \frac{w_l}{1 - w_l G_{00}^0(E)} \qquad (6.6)$$

where we have taken advantage of the translational invariance of the operator H^0 and hence of the fact that diagonal matrix elements G_{ll}^0 are all equal. The series (6.4) in terms of t_l can be recast in the form

$$G_{ll'} = G_{ll'}^0 + \sum_m G_{lm}^0 t_m G_{ml'}^0 + \sum_m \sum_{\substack{n \\ (n \neq m)}} G_{lm}^0 t_m G_{mn}^0 t_n G_{nl'}^0 + \sum_m \sum_{\substack{n \\ (n \neq m)}} \sum_{\substack{p \\ (p \neq n)}} \cdots$$

$$(6.7)$$

When w_l is a stochastic variable described by a probability distribution $P(w)$, we can consider the average t-matrix

$$\bar{t}_l(E) = \bar{t}(E) = \int \frac{w}{1 - w G_{00}^0(E)} P(w)\, dw \qquad (6.8)$$

We make the ensemble average of Eq. (6.7) and suppose that t_m are decoupled so that each t_m is replaced by $\bar{t}$. In reality, in Eq. (6.7) only *immediately successive* indices cannot repeat, and the first corrections are of third order in the t matrix; this is an advantage with respect to the ensemble average of Eq. (6.4) in the virtual crystal approximation (VCA), because first corrections would be of second order in the w matrix. In the ATA we replace each t_m by $\bar{t}$ in Eq. (6.7); the result corresponds to Eq. (6.4) with $w_m = \bar{t}/(1 + G_{00}^0 \bar{t})$. We obtain

$$G^{ATA}(E) = G^0(E - \Sigma) \qquad (6.9)$$

$$\Sigma = \Sigma(E) = \frac{\bar{t}(E)}{1 + G_{00}^0(E)\bar{t}(E)} \qquad (6.10)$$

The mean resolvent $G^{\text{ATA}}(E)$ of the random medium can be worked out explicitly if the Green's function of the perfect crystal and the distribution function for w are known. The matrix element $G_{00}^0(E)$, which plays a central role in the theoretical treatments of disorder, can be evaluated by standard expansion in continued fractions or by diagonalizing (using $\mathbf{k}$ methods) the periodic Hamiltonian H^0. It must be said, however, that the ATA is often a poor approximation, and it may even happen that it violates basic analytic requirements (Herglotz properties). It is thus convenient to describe briefly the next major development known as the coherent potential approximation; in the CPA the first correction to the ensemble average of (6.7) is of fourth order in the t matrix, one order more than ATA, which in turn is one order more than VCA.

B. Coherent Potential Approximation and the Recursion Method

The coherent potential approximation (CPA) was originally developed by Soven[47] for the electronic problem and by Taylor[48] for the phonon problem. The basic assumption of CPA is that a random medium with diagonal disorder can be described by an effective periodic Hamiltonian of the type

$$H^{\text{CPA}} = \sum_{m \neq n} T_{mn}|\phi_m\rangle\langle\phi_n| + \sum_m \Sigma(E)|\phi_m\rangle\langle\phi_m| \qquad (6.11)$$

where an energy-dependent coherent potential $\Sigma(E)$ has been added on every site. The effective potential $\Sigma(E)$ has to be determined in such a way that at every energy the ensemble average of the site t matrix vanishes.

To write down the above condition mathematically, consider the CPA crystal (which is a perfectly periodic crystal), a given site $\tau = 0$, and a given actual value w of the orbital energy at $\tau = 0$. Relative to the effective medium, we have

$$w = \Sigma + (w - \Sigma)$$

and the actual system consists of a perturbing potential $w - \Sigma$ on the CPA crystal at the $\tau = 0$ site. The t_0 matrix for an electron propagating according to $G^{\text{CPA}}(E) \equiv G^0(E - \Sigma)$ is

$$t_0(E) = \frac{w - \Sigma}{1 - (w - \Sigma)G_{00}^{\text{CPA}}(E)} \qquad (6.12)$$

The requirement that the ensemble average of Eq. (6.12) has to vanish leads to the implicit condition for $\Sigma(E)$:

$$\int \frac{w - \Sigma}{1 - (w - \Sigma)G_{00}^0(E - \Sigma)} P(w)\, dw \equiv 0 \qquad (6.13)$$

The knowledge of the diagonal matrix elements of the Green's function of the perfect crystal and the solution of the self-consistent CPA equation (6.13) are the basic ingredients for performing actual calculations.

For the case of a binary substitutionally disordered alloy composed of atoms A and B with concentration c_A and $c_B = 1 - c_A$, Eq. (6.13) becomes

$$\frac{\varepsilon_A - \Sigma}{1 - (\varepsilon_A - \Sigma)G_{00}^0(E - \Sigma)} c_A + \frac{\varepsilon_B - \Sigma}{1 - (\varepsilon_B - \Sigma)G_{00}^0(E - \Sigma)} c_B = 0$$

or, equivalently,

$$\bar{\varepsilon} - \Sigma = (\varepsilon_A - \Sigma)(\varepsilon_B - \Sigma)G_{00}^0(E - \Sigma) \qquad (6.14)$$

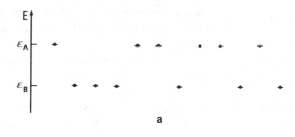

a

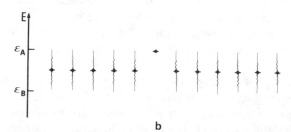

b

Figure 4. Schematic representation of the coherent potential approximation for a substitutionally disordered alloy. The vertical strip in (b) denotes the effective self-energy, which is a complex quantity, to be determined by the compatibility requirement between the local and average description.

where $\bar{\varepsilon} = c_A \varepsilon_A + c_B \varepsilon_B$, and ε_A and ε_B are the orbital energies of atoms A and B.

The above equations can be generalized to the case of disordered semiconducting alloys described by tight-binding models with several orbitals in the unit cell. In the literature the CPA tight-binding method has found wide applications[49] in semiconducting alloys and in the important class of "tunable gap" compounds such as $Hg_{1-x}Cd_xTe$.

Another interesting case, in which the self-consistent solution of Eq. (6.13) is reasonably simple, concerns a diagonal disorder described by a Gaussian distribution $P(w)$. A Gaussian distribution has been introduced by Sumi and Toyozawa[46] to simulate the fluctuation of the electron (or exciton) orbital energy due to a phonon field and explain the Urbach absorption tail of a number of semiconductors.

The technique of combining CPA concepts with a continued fraction analysis of the Green's function should also be invaluable in connection with the study of disorder with "augmented space" techniques.[50] Progress is expected when memory function methods are systematically used in the mentioned areas of investigation.

In what has been said until now, the role of the recursion method has been confined to a useful, but only auxiliary, tool in calculating quantities needed within the ATA and CPA formalism. However, the recursion method can be used in a more sophisticated manner as an alternative tool to the mean field theories. This is quite important in view of the fact that systematic improvements beyond the CPA are lacking.

In the case of binary alloys, for instance, the direct application of the recursion procedure is physically more transparent than CPA and superior in that detailed local configurations can be included.

The pattern to follow is described below. One begins by constructing a tight-binding model for the binary alloy with the extreme one-component situations accurately reproduced. Then a large cluster is generated, of the order of thousands of cells or so (with cyclic boundary conditions when necessary) with the atomic species at the lattice points chosen randomly according to the assigned concentration. The local density of states can be calculated for any site and any orbital; then the average over all sites can be performed.

The procedure of averaging over sites can be carried out much faster if one starts from the very beginning with a *state taken as the sum of localized orbitals multiplied by random plus or minus signs*[6] (or more generally multiplied by random phase factors). The Green's function corresponding to such a seed state is the sum of site diagonal elements plus off-diagonal elements; the latter tend to cancel out because of the random choice of phases. The cancellation is complete if the cluster is sufficiently large; if not, the degree of cancellation can be made satisfactory by a proper ensemble of a limited number of starting states (often 5 or 10 or so). A very interesting and recent study along the above-described lines[51] concerns the alloy $Pb_{1-x}Sr_xS$, whose gap varies continuously from 0.4 eV in PbS to 4.6 eV in SrS.

C. Localization and Spectra of Disordered Systems

The "mean field" approximate theories, ATA and CPA, are final simplifications of multiple scattering approaches to simulate a real disordered system by an equivalent effective system.

Anderson's[45] simple model to describe the electrons in a random potential shows that localization is a typical phenomenon whose nature can be understood only taking into account the degree of randomness of the system. Using a tight-binding Hamiltonian with constant hopping matrix elements V between adjacent sites and orbital energies uniformly distributed between $-W/2$ and $W/2$, Anderson studied the modifications of the electronic diffusion in the random crystal in terms of the stability of localized states with respect to the ratio W/V.

Since this original work, several approaches have been used to study the problem of localization; there are a number of papers and reviews on this subject.[52] Whereas it is recognized that a finite disorder at all sites of a one-dimensional system localizes the electron states, the problem is still open to a definite theoretical and numerical result for higher dimensions.

Infinite order perturbation theory has been suggested[45] to argue when localization takes place in a disordered crystal. The problems connected with the treatment of singularities in infinite summations and with convergence criteria, also exploiting renormalization techniques, have been critically reviewed by Licciardello and Economou.[53] Also, the work of Abou-Chacra et al.,[54] who used the idea of a Cayley-tree lattice to determine in a self-consistent way the position of the mobility edges, is in the spirit of infinite-order perturbation techniques. A critical review of the basic arguments used to identify localization and the relationship with spectra, absence of diffusion, and Green's function has been presented by Kimball,[55] and approximate analytic theories have also been attempted.[56]

The importance of the localization problem has also stimulated numerical studies and computer simulations.[57,58] By far the most useful and most often employed approach to perform numerical calculations on disordered systems is the recursion method. It shows the great advantage of a reduction of the multidimensional random lattice to a semiinfinite linear chain. Haydock[59] first applied the recursion method to the Anderson model for small disorder, exploiting perturbation theory on the linear chain[60] and providing a connection between the asymptotic values of the continued fraction coefficients and localization edges. Hodges et al.[61] argued that such asymptotic values should be related to Lifshitz bounds for the density of states. In fact, it is still an unresolved problem the critical study of the coefficients a_n and b_n, of their oscillating behavior, and of the rate of decrease of their amplitude and a connection with Anderson's localization theory.

An efficient terminator technique is certainly desirable in the application of recursion methods to the study of disordered systems. It has been shown recently that a self-consistently determined terminator[62] can be fruitfully applied to calculate the electronic states in the Anderson model and to evaluate the vibrational spectrum of lattices with isotopic disorder. The basic idea is to extend the procedures discussed in Section IV to ensemble averages. In this case a useful generalization of Eq. (4.5), satisfied by the terminator $t(E)$, is

$$\sum_c w_c \cfrac{1}{E - a_0^c - \cfrac{b_1^{2c}}{E - \cfrac{\ddots}{\cfrac{b_n^{2c}}{E - a_n^c - t(E)}}}} \equiv \sum_c w_c \cfrac{1}{E - a_0^c - \cfrac{b_1^{2c}}{E - \cfrac{\ddots}{\cfrac{b_{n+1}^{2c}}{E - a_{n+1}^c - t(E)}}}}$$

$$(6.15)$$

where c indicates different configurations and w_c are the corresponding probability weights. It is easily verified that in the case of a single configuration and existence of the limits a_n and b_n^2 for $n \to \infty$, Eq. (6.15) gives back the standard square root terminator (4.6).

It would be impossible to discuss in depth the many interesting contributions in this restless area of research, and what is mentioned here has only an indicative value. In this state of affairs, further progress is likely to need more and more accurate and heavy calculations; alternative or complementary procedures based on functional analysis theories[32,63] and orthogonal polynomial behavior could be of major help with the many questions still open to investigation.

References

1. See, for instance, W. Jones and N. H. March, *Theoretical Solid State Physics*, Vols. I and II, Wiley, New York, 1973.
2. See, for instance, F. Bassani and G. Pastori Parravicini, *Electronic States and Optical Transitions in Solids*, Pergamon Press, Oxford 1975; Nauka, Moscow, 1982 (in Russian, V. L. Bonch-Bruevich, ed).

3. F. Cyrot-Lackmann, *Adv. Phys.*, **16**, 393 (1967).

4. J. P. Gaspard and F. Cyrot-Lackmann, *J. Phys.*, **C6**, 3077 (1973).

5. R. Haydock, V. Heine, and M. J. Kelly, *J. Phys.*, **C5**, 2845 (1972); **C8**, 2591 (1975).

6. D. W. Bullett, R. Haydock, V. Heine, and M. J. Kelly, *Solid State Phys.*, **35**, (1980).

7. R. Zwanzig, in *Lectures in Theoretical Physics*, Vol. 3, W. E. Brittin, B. W. Downs, and J. Downs, eds., Interscience, New York, 1961, p. 106.

8. H. Mori, *Prog. Theor. Phys.*, **33**, 423 (1965); **34**, 399 (1965).

9. P. W. Anderson, *Phys. Rev.*, **181**, 25 (1969); J. D. Weeks, P. W. Anderson, and A.G.H. Davidson, *J. Chem. Phys.*, **58**, 1388 (1973). See also D. W. Bullett, *J. Phys.*, **C8**, 2695 (1975).

10. E. Doni, R. Girlanda, V. Grasso, A. Balzarotti, and M. Piacentini, *Nuovo Cimento*, **B51**, 154 (1979); E. Doni, R. Girlanda, and L. Resca, *J. Phys. Soc. Japan*, **49**, Suppl. A, 89 (1980).

11. F. Ducastelle and F. Cyrot-Lackmann, *J. Phys. Chem. Solids*, **31**, 1295 (1970); **32**, 285 (1971).

12. See, for instance, C. López, F. Yndeáin, and F. J. Yndeáin, *J. Phys.*, **C7**, 61 (1974); F. Yndeáin and F. J. Yndeáin, *ibid.*, **8**, 434 (1975); J. P. Gaspard in *Electronic Structure of Defects and of Disordered Structures*, Aussois Summer School, Les Ulis, Les Editions de Physique, 1981, p. 279. For an interesting application of the moment method to vibronic models see M. Cini, *Phys. Rev.*, **B29**, 547 (1984).

13. C. Domb, *Adv. Phys.*, **9**, 149, 245 (1960); *Adv. Chem. Phys.*, **15**, 229 (1969).

14. See, for instance, J. C. Wheeler, *J. Chem. Phys.*, **80**, 472 (1984) and references therein.

15. P. Lambin and J. P. Gaspard, *Phys. Rev.*, **B26**, 4356 (1982).

16. C. Lanczos, *J. Res. Nat. Bur. Stand.*, **45**, 255 (1950).

17. C. Lanczos, *J. Res. Nat. Bur. Stand.*, **49**, 33 (1952), *Applied Analysis*, Prentice-Hall, Englewood Cliffs, NJ, 1956.

18. See, for instance, J. H. Wilkinson, *The Algebraic Eigenvalue Problem*, Clarendon Press, Oxford, 1965; C. C. Paige, *J. Inst. Math. Appl.*, **10**, 373 (1972); **18**, 341 (1976).

19. See, for instance, K. M. Leung and D. L. Huber, *Phys. Rev.*, **B19**, 5483 (1979); **B20**, 1139 (1979); M. C. M. O'Brien, *J. Phys.*, **C16**, 85, 6345 (1983); H. Köppel, W. Domcke, and L. S. Cederbaum, *Adv. Chem. Phys.*, **57**, 59 (1984). For theoretical and experimental aspects of the physics of colour centers see for instance L. Martinelli, G. Pastori Parravicini, P. L. Soriani, Phys. Rev. B (1985) to appear, and G. Baldacchini, U. M. Grassano, M. Meucci, A. Scacco, F. Somma and M. Tonelli, J. Luminescence **31** and **32**, 154 (1984).

20. C. M. M. Nex, Cambridge Recursion Method Library, cited in *Solid State Phys.*, **35**, 78 (1980).

21. R. Haydock and M. J. Kelly, *J. Phys.*, **C8**, L290 (1975).

22. P. Grigolini, G. Grosso, and G. Pastori Parravicini, *Solid State Commun.*, **46**, 881 (1983).

23. P. Grigolini, *Nuovo Cimento*, **B63**, 174 (1981); *J. Stat. Phys.*, **27**, 283 (1982); M. W. Evans, G. J. Evans, W. T. Coffey, and P. Grigolini, *Molecular Dynamics*, Wiley, New York, 1982, and references quoted therein.

24. M. Bixon and R. Zwanzig, *J. Stat. Phys.*, **3**, 245 (1971). See also M. W. Evans, *Chem. Phys. Lett.*, **39**, 601 (1976).

25. P. Grigolini, G. Grosso, G. Pastori Parravicini, and M. Sparpaglione, *Phys. Rev.*, **B27**, 7342 (1983).

26. P. Grigolini and G. Pastori Parravicini, *Solid State Commun.*, **39**, 123 (1981); *Phys. Rev.*, **B25**, 5180 (1982).

27. S. A. Adelman, *Adv. Chem. Phys.*, **44**, 143 (1980); **53**, 61 (1983).

28. E. B. Van Vleck, *Trans. Amer. Math. Soc.*, **5**, 253 (1904).

29. C. H. Hodges, *J. Phys. Lett.*, **38**, L187 (1977). For other conjectures on the asymptotic behavior in periodic and aperiodic systems see A. Trias, M. Kiwi, and M. Weissmann, *Phys. Rev.*, **B28**, 1859 (1983). See also: G. Grosso, G. Pastori Parravicini and A. Testa, Phys. Rev. B (1985) to appear.

30. N. Beer and D. G. Pettifor, Newsletter **4**, 19 (1982) (Daresbury Laboratory); see also G. Allan, *J. Phys.*, **C17**, 3945 (1984); D. G. Pettifor and D. L. Weaire eds., *Recursion Method and its Applications*, Springer Series in Solid State Sciences, Vol. 58, Berlin 1985.

31. U. H. Gläser and P. Rennert, *J. Phys.*, **F11**, 2063 (1981).

32. A. Magnus, in *Padè Approximation and its Applications, Lecture Notes in Mathematics*, 765, L. Wuytack, ed., Springer, Berlin, 1979, p. 150; in *Padè Approximation and its Applications, Lecture Notes in Mathematics*, 1071 H. Werner, H. J. Bünger eds. Springer, Berlin 1984, p. 213.

33. P. Turchi, F. Ducastelle, and G. Tréglia, *J. Phys.*, **C15**, 2891 (1982).

34. See, for instance, R. P. Brent, *ACM Trans. Math. Software*, **4**, 1, 71 (1978). P. Giannozzi, G. Grosso and G. Pastori Parravicini, *Phys. Stat. Sol.*, (b) **128**, 643 (1985).

35. For reviews, see F. Bassani, G. Iadonisi, and B. Preziosi, *Rep. Prog. Phys.*, **37**, 1099 (1974); S. T. Pantelides, *Rev. Mod. Phys.*, **50**, 797 (1978); M. Lannoo and J. Bourgoin, *Point Defects in Semiconductors*, Vol. I, *Theoretical Aspects*, Vol. II, *Experimental Aspects*, Springer Verlag, Berlin, Vol. I, 1981; Vol. II, 1983; M. Jaros, *Deep Levels in Semiconductors*, Adam Hilger, Bristol, 1982.

36. J. Bernholc and S. T. Pantelides, *Phys. Rev.*, **B18**, 1780 (1978); J. Bernholc, N. O. Lipari, and S. T. Pantelides, *ibid.*, **B21**, 3545 (1980); J. Bernholc, N. O. Lipari, S. T. Pantelides, and M. Scheffer, *ibid.*, **B26**, 5706 (1982); D. P. DiVincenzo, J. Bernholc, and M. H. Brodsky, *ibid.*, **B28**, 3246 (1983).

37. G. A. Baraff and M. Schlüter, *Phys. Rev.*, **B19**, 4965 (1979); G. A. Baraff, E. O. Kane, and M. Schlüter, *ibid.*, **B21**, 3563 (1980); 5662 (1980); G. B. Bachelet, G. A. Baraff, and M. Schlüter, *ibid.*, **B24**, 915, 4736 (1981).

38. G. F. Koster and J. C. Slater, *Phys. Rev.*, **95**, 1167 (1954); **96**, 1208 (1954). See also J. Callaway, *J. Math. Phys.*, **5**, 783 (1964).

39. H. P. Hjalmarson, P. Vogl, D. J. Wolford, and J. D. Dow, *Phys. Rev. Lett.*, **44**, 810 (1980); H. P. Hjalmarson, H. Büttner, and J. D. Dow, *Phys. Rev.*, **B24**, 6010 (1981); S. Y. Ren, R. E. Allen, J. D. Dow, and I. Lefkowitz, *ibid.*, **B25**, 1205 (1982). See also J. Robertson, *ibid.*, **B28**, 3378, 4647 (1983).

40. E. Kauffer, P. Pêcheur, and M. Gerl, *J. Phys.*, **C9**, 2319 (1976); *Phys. Rev.*, **B15**, 4107 (1977); J. van der Rest and P. Pêcheur, *J. Phys. Chem. Solids*, **45**, 563 (1984). For native defects in III–V semiconductors, see for instance, P. J. Lin-Chung and T. L. Reinecke, *Phys. Rev.*, **B27**, 1101 (1983).

41. D. M. Bylander and J. J. Rehr, *J. Phys.*, **C13**, 4157 (1980).

42. R. Graft, G. Grosso, G. Pastori Parravicini, and L. Resca, *Solid State Commun.* **51**, 247 (1984). For application to superlattices see R. Graft, G. Pastori Parravicini and L. Resca, *Solid State Commun.*, **54**, 115 (1985).

43. L. Resca, *Phys. Rev.*, **B26**, 3238 (1982); **B29**, 866 (1984) and references therein.

44. See, for instance, J. M. Ziman, *Models of Disorder*, Cambridge University Press, 1979.

45. P. W. Anderson, *Phys. Rev.*, **109**, 1492 (1958); *Physica*, **B + C 117 + 118**, 30 (1983).

46. H. Sumi and Y. Toyozawa, *J. Phys. Soc. Japan*, **31**, 342 (1971); H. Sumi, *ibid.*, **32**, 616 (1972); Y. Toyozawa, *Physica*, **B + C 117 + 118**, 23, 290 (1983).

47. P. Soven, *Phys. Rev.*, **156**, 809 (1967); **178**, 1136 (1969).

48. D. W. Taylor, *Phys. Rev.*, **156**, 1017 (1967).

49. See, for instance, K. C. Hass, H. Ehrenreich, and B. Velický, *Phys. Rev.*, **B27**, 1088 (1983); **B29**, 3697 (1984); H. Ehrenreich and K. C. Hass, *J. Vac. Sci. Technol.*, **21**, 133 (1982); D. S. Montgomery, *J. Phys.*, **C16**, 2923 (1983), and references quoted therein.

50. See, for instance, A. Mookerjee, *J. Phys.*, **C6**, 1340 (1973); R. Mills and P. Ratana-vararaksa, *Phys. Rev.*, **B18**, 5291 (1978); T. Kaplan, P. L. Leath, L. J. Gray, and H. W. Diehl, *ibid.*, **B21**, 4230 (1980); A. Mookerjee and M. Yussouff, *J. Phys.*, **C17**, 1009 (1984).

51. L. C. Davis, *Phys. Rev.*, **B28**, 6961 (1983).

52. See, for instance, K. Ishii, *Suppl. Progr. Theor. Phys.*, **53**, 77 (1973); D. J. Thouless, *Phys. Rep.*, **13**, 93 (1974); D. J. Thouless, in *Ill-Condensed Matter*, *Les Houches Summer School Proceedings*, R. Balian, R. Maynard, and G. Toulouse, eds., North-Holland, Amsterdam, 1979, p. 1; R. J. Elliot, J. A. Krumhansl, and P. L. Leath, *Rev. Mod. Phys.*, **46**, 465 (1974); A. Nitzan, K. F. Freed, and M. H. Cohen, *Phys. Rev.*, **B15**, 4476 (1977); E. Abrahams, P. W. Anderson, D. C. Licciardello, and T. V. Ramakrishnan, *Phys. Rev. Lett.*, **42**, 673 (1979); P. Erdös and R. C. Herndon, *Adv. Phys.*, **31**, 65 (1982); M. Ya Azbel, *Phys. Rev.*, **B28**, 4106 (1983).

53. D. C. Licciardello and E. N. Economou, *Phys. Rev.*, **B11**, 3697 (1975).

54. R. Abou-Chacra, P. W. Anderson, and D. J. Thouless, *J. Phys.*, **C6**, 1734 (1973).

55. J. C. Kimball, *J. Phys.*, **C11**, 4347 (1978); **C13**, 5701 (1980).

56. W. Götze, *Solid State Commun.*, **27**, 1393 (1978); W. Götze, P. Prelovšek, and P. Wölfle, *ibid.*, **30**, 369 (1979); A. Gold, S. J. Allen, B. A. Wilson, and D. C. Tsui, *Phys. Rev.*, **B25**, 3519 (1982).

57. J. Stein and U. Krey, *Z. Phys.*, **B34**, 287 (1979); **B37**, 13 (1980), and references quoted therein.

58. D. Weaire and B. Kramer, *J. Non-Cryst. Solids*, **32**, 131 (1979).

59. R. Haydock, *Phil. Mag.*, **37**, 97 (1978).

60. R. Haydock, *J. Phys.*, **A10**, 461 (1977).

61. C. H. Hodges, D. Weaire, and N. Papadopoulos, *J. Phys.*, **C13**, 4311 (1980), and references therein.

62. J. J. Sinai, C. Wongtawatnugool, and S. Y. Wu, *Phys. Rev.*, **B26**, 1829 (1982); J. J. Sinai and S. Y. Wu, *ibid.*, **B28**, 4261 (1983).

63. A. Magnus in *Padé Approximation and its Applications, Lecture Notes in Mathematics 888*, M. G. de Bruin and H. von Rossum, eds., Springer, Berlin, 1981, p. 309.

V

MOLECULAR DYNAMICS: INTENSE EXTERNAL FIELDS

CONTENTS

I. HISTORICAL DEVELOPMENT

The first worker to consider the detailed effect of a strong external force field on the molecular dynamics of an isotropic molecular liquid seems to have been Benoit.[1] This treatment may be developed considerably and extended.[2-4] In this review, however, the technique of computer simulation is used in combination with "reduced" model theory (see Chapter I) in an attempt to probe more deeply into the fundamental physical characteristics of the molecular liquid state.

This chapter is dedicated to the South Wales Branch of the National Union of Mineworkers.

By employing a very strong external field, a gedankexperiment may be set up whereby the natural thermal motion of the molecules is put in competition with the aligning effect of the field. This method reveals some properties of the molecular liquid state which are otherwise hidden. In order to explain the observable effects of the applied fields, it is necessary to use equations of motion more generally valid than those of Benoit. These equations may be incorporated within the general structure of "reduced" model theory[5-7] (RMT) and illustrate the use of RMT in the context of liquid-state molecular dynamics. (Elsewhere in this volume RMT is applied to problems in other fields of physics where consideration of stochastic processes is necessary.) In this chapter modifications to the standard methods[8] are described which enable the detailed study of field-on molecular dynamics.

In general there is no limit to the types of fields that can be considered in the gedankexperiment, or to the field strength, or to the permutations and combinations. As the field strength is increased, orientational characteristics at field-on equilibrium must be describable through Langevin functions of various kinds, and the molecular dynamics by the use of RMT. One of the first breakthroughs in this context came in the mid-1970s with the development by Grigolini[9-12] of a theory to describe the competition between preparation and relaxation in the field of radiationless decay. The case of excitation with strong fields was also considered. The field has a direct effect on a variable of interest, such as the molecular angular velocity ω, and isolates (or decouples) the characteristics of this variable from others in the molecular ensemble, or thermal bath. This decoupling effect was predicted[9] theoretically by Grigolini in 1976 in the context of spin relaxation[13] and radiationless decay in molecules.[9] Numerical evidence for Grigolini's predictions was provided by Evans,[14-18] using molecular dynamics computer simulation, and by Bagchi and Oxtoby,[19-20] using computer simulation of stochastic forces in a different context.

Recently, another fundamental property of the liquid state has been discovered by comparing the orientational fall transient from field-on to field-off equilibrium with the relevant orientational autocorrelation function.[21,22] According to classical (linear) fluctuation-dissipation theory, the fall-transient and autocorrelation function (acf) are identical. Computer simulations show that the fall transient decays to equilibrium more quickly, and sometimes much more quickly, than the acf when the field initially applied to the molecular liquid becomes greater than the thermal kT in energetic terms. This speeding up of the fall transient is referred to in this review as the *deexcitation effect*. The RMT is capable of exploring the numerical results analytically using new concepts of nonlinearity in the molecular liquid state. One of the results of this gedankexperiment may be described in simple terms as follows. A Langevin equation of the type

$$\dot{\omega}(t) = -\gamma\omega + \mathbf{f}(t) \qquad (1)$$

is a description of the field-off molecular dynamics which always implies that the normalized fall transient and the relevant field-off acf must be identical experimental decay functions of time. In the simple equation (1.1), ω is the molecular angular velocity, γ a normalized friction coefficient, and $\mathbf{f}(t)$ a stochastic angular acceleration term. The new developments of nonlinear physics exemplified by and incorporated in RMT lead in certain long-time limits[22] to a fundamentally nonlinear structure such as

$$\dot{\omega}(t) = -\gamma\omega(t) + \gamma'\omega(t)\langle\omega^2(t)\rangle + \mathbf{f}(t) \qquad (2)$$

Note that, as discussed at length in Chapter VI, this equation has to be limited to providing information on $\langle\omega(t)\rangle$. The influence of the nonlinear term on the time evolution of $\langle\omega^2(t)\rangle$ is not compatible with the attainment of a canonical equilibrium. This type of equation lends itself to a qualitative description of deexcitation effects as observed by computer simulation. This is described further in Section IV.A with reference to the general analytical constraints imposed by RMT on *any theory* of the molecular liquid state.

This review is developed as follows. In Section II the fundamental idea of field-on computer simulation is described with reference to the molecular dynamics of an achiral molecule of C_{2v} symmetry. The field-on computer-simulated orientational characteristics are tested with the equilibrium theory of statistical mechanics as embodied in Langevin and Kielich functions[18] of applied field strength. The characteristics of the molecular dynamics at field-on equilibrium are quantified in terms of auto- and cross-correlation functions. Of particular interest in this context is the autocorrelation function of the molecular angular velocity ω. This acf develops oscillations under the influence of the field which decay *slowly* as a function of time. *The envelope of these oscillations is longer lived than the acf at field-off equilibrium and is also dependent on the field strength.*[17] These are essentially the characteristics of the Grigolini decoupling effect.[9] The theorems embodied in RMT allow Grigolini and his coworkers to describe these effects analytically[21] and, in principle, quantitatively. Theories such as those of van Kampen and Praestgaard,[23] which are purely phenomenological in nature and fall outside the RMT framework, do not describe the decoupling effect and should therefore be modified a little according to the numerical results. The deexcitation effect is introduced[14] with reference to achiral triatomics of C_{2v} symmetry. The three-dimensional molecular dynamics in this case is a formidable challenge to three-dimensional analytical theory because of the important role played by rotation-translation coupling.[24] The RMT can, however, follow the guidelines laid down by the numerical method qualitatively, by restricting the analytical treatment to two dimensions as described in Section III. In this limit a version of Eq. (2) is successful[22] in describing the numerical deexcitation effect at least qualitatively. This equation is a limiting form of the nonlinear two-dimensional itinerant oscillator,[25] an RMT-allowed mathematical model of the liquid state.

In Section IV the computer simulation is extended to describe the effects of excitation in chiral molecules and racemic mixtures of enantiomers. The modification of the dynamical properties brought about by mixing two enantiomers in equimolar proportion may be explained in terms of rotation-translation coupling.[24] The application of an external field in this context amplifies the difference between the field-on acf's and cross-correlation of enantiomer and racemic mixture and provides a method of studying experimentally the fundamental phenomenon of rotation-translation coupling in the molecular liquid state of matter.

The mathematical and computational techniques developed in this review may hopefully be transferred in the future to deal with the problems of technical and industrial importance that depend for their description on nonlinear stochastic differential equations. Of particular importance in this context is the nonlinear technique needed for the description of hysteresis in the Josephson tunneling junction.[26] The circuitry of the newer generation of computers and integrated circuits in general is reaching the atomic level in dimensionality. In this limit the field strengths involved in calculating the boundary characteristics of the circuitry are very large. Stochastic equations are necessary for a complete description of the phenomena involved. The methods used in this review may also be used to describe the diffusion of charge carriers in semiconductors, chemical reaction rates, superionic conductors, dipole-dipole interaction in dielectric relaxation, cycle slips in second-order phase-locked loops, and other phenomena which involve the interaction of fields and stochastic variables. Several of these topics are described in detail using RMT elsewhere in this volume. A generalization of the available methods of solution will have a variety of other applications to problems of practical importance in a range of disciplines. Add to this the power of computer simulation, and it is possible to begin to construct a sound and general theoretical basis for a number of useful phenomena.

To conclude this introductory section, the *simplest* type of equation that should be considered for a useful description is a stochastic differential equation such as

$$\frac{d^2x(t)}{dt^2} + \zeta\frac{dx(t)}{dt} + \frac{dV(x)}{dx} = \lambda(t) \tag{3}$$

This is an equation written, for example, for a Brownian particle of mass m moving along the x axis under the influence of a potential $V(x)$. Here $\lambda(t)$ is a white-noise driving force (a stochastic variable) coming from the Brownian movement of the surroundings, and $\zeta\dot{x}(t)$ is the systematic friction force. This equation can be solved exactly only in the known special cases $V = 0$ and $V = \gamma x^2$, where γ is a coefficient independent of time. Equation (3) is the Langevin equation equivalent to the Kramers equation[5,6]

$$\frac{\partial G}{\partial t} + \dot{x}\frac{\partial G}{\partial x} - \frac{1}{m}\frac{\partial V}{\partial x}\frac{\partial G}{\partial \dot{x}} = \frac{\zeta}{m}\frac{\partial}{\partial \dot{x}}\left(\dot{x}G + \frac{kT}{m}\frac{\partial G}{\partial \dot{x}}\right) \tag{4}$$

for the complete phase-space conditional probability density function $G(x, \dot{x}, t\mid x(0), \dot{x}(0), 0)$ under the boundary condition $G(x, \dot{x}, 0) = \delta(x - x_0)\delta(\dot{x} - \dot{x}_0)$. *The same equations*, written for two-dimensional rotational Brownian motion in N-potential wells of the cosine type have been used by Reid[26] as a good description of the absorption spectra of dipolar molecular liquids in the frequency range from static terahertz (far infrared).[5]

These equations *cannot*, however, describe the deexcitation effect without some development within the general RMT structure. The RMT therefore provides a more generally valid description of liquid-state molecular dynamics than Eqs. (3) or (4). It is probable that the same is true for Josephson junction hysteresis in the presence of fields and related phenomena such as those described in the other reviews of this special volume.

The powerful continued fraction procedure (CFP) described by Grosso and Pastori Parravicini in Chapter III may be used to solve Eqs. (3) and (4). An alternative approach has been provided by Ferrario et al.,[25] who have computed a variety of numerically derived orientation and velocity acf's for a simple cosine potential and the more complicated cosinal itinerant oscillator, another RMT-allowed structure. In Chapter VI, Ferrario et al. describe deexcitation effects from the two-dimensional disk-annulus itinerant oscillator also studied by Brot and coworkers.[27]

Coffey, Rybarsch, and Schroer[3] have considered the numerical solution of the Kramers equations for two-dimensional rotation, which is the exact rotational counterpart of Eq. (3) or Eq. (4). These equations are also soluble numerically by use of the CFP for any value of field strength embodied in the potential V. When the field is very intense—for example, in boundaries of integrated circuits—the rise transient after switching on the field becomes oscillatory, as demonstrated analytically by Coffey et al.[3] and numerically by Evans (see Section IV). At field-on equilibrium, the Grigolini decoupling effect appears from a slightly more complicated RMT structure[21] than the level embodied in Eq. (3) or (4). As we have mentioned, a greater degree of sophistication is also required for an analytical description of the deexcitation effect (Section III).

The properties of equations such as (3) and (4) which are not allowed by RMT are understood satisfactorily only in the relatively uninteresting linear case where, for example, rise and fall transients mirror each other as exponentials.[1] When this frontier is crossed, the applied field strength is such that it is able to compete effectively with the intermolecular forces in liquids. This competition provides us with information about the nature of a molecular liquid which is otherwise unobtainable experimentally. This is probably also the case for *internal* fields, such as described by Onsager for liquids,[28,29] for various kinds of internal fields in integrated computer circuits, activated polymers,[30] one-dimensional conductors, amorphous solids, and materials of interest to information technology. The chapters by Grosso and Pastori Parravicini in this volume describe with the CFP some important phenomena of the solid state of matter in a slightly different context.

II. COMPUTER SIMULATION

Computer simulation of the effect of intense externally applied force fields on molecular liquids has been initiated recently by Evans in a series of articles.[14-18] The purpose of the computer simulation is to produce a self-consistent numerical data for discussion by analytical theoreticians and experimentalists in this area. Without the nonlinear potential term V the equations of Section I produce results which are wholly at odds with, for example, spectral data in the far infrared.[5] This is understood now to be due to the neglect of inertial effects and to the assumption that the friction coefficient does not evolve with time, that is, there are no memory effects. In the language of RMT, the infinite Mori continued fraction has been truncated at the first approximant, using Langevin's original dramatic separation of time scales or ("adiabatic elimination"), into a viscous drag coefficient varying infinitely slowly with time and a white-noise term [i.e., one whose autocorrelation function is a delta function decaying infinitely quickly with time, Eq. (1)]. A great deal of effort has gone into rectifying this situation by introducing memory effects in an attempt to reproduce, at least qualitatively, the available far-infrared data as measured and catalogued, for example, by Reid.[31] Another hidden assumption in purely rotational or purely translational equations such as Eq. (1) is that the rotational motion of a molecule may be decoupled from the translation of its own center of mass. Computer simulation has been used[24] to show clearly, and for the first time, that there is a well-defined cross-correlation between molecular rotation and translation for all molecular symmetries. In chiral molecules, the difference in the spectral properties of enantiomers and their racemic mixture may be used in principle as an experimental method of detecting this cross-correlation. McConnell et al.[32] have shown how the simplest diffusional theory increases greatly in complexity when inertial effects are taken into consideration, even in the case $V = 0$, that is, for free diffusion governed by Eq. (1). Inertial effects alone *cannot*, however, explain the basic spectral features of dipolar liquids in the far infrared[5] without consideration of the memory effects incorporated in RMT in continued fraction form. Some simple models of the liquid state such as the Calderwood/Coffey itinerant oscillator[33] are RMT-allowed structures and therefore approximants of the appropriate RMT matrix continued fractions. Other models such as the Brownian motion in N-potential wells[23,26,27] are not allowed by RMT and are not generally descriptive of the molecular liquid state. The decoupling and deexcitation effects discovered recently by computer simulation provide these analytical structures with stringent tests. There is no complete analytical method yet available to describe even in rudimentary terms the detailed statistical cross-correlations[34] between rotation and translation in molecular liquids.

Itinerant oscillator theory is now available in a variety of forms, equivalent to various RMT continued fraction approximants.[6] The analytical theory is reviewed elsewhere in this volume. The original two-dimensional linear itinerant oscillator of Calderwood and Coffey[33] is able to qualitatively reproduce the available far-infrared spectra of molecular liquids[5] by simulating the librational (torsional oscillatory) motion of a caged and diffusing molecule. The equations of this version of the itinerant oscillator are soluble numerically using the RMT Pisa algorithm. The linear limit of the itinerant oscillator (essentially speaking, the limit $\sin\theta = \theta$, where θ is the amplitude of the libration) lends itself to analytical solution as described by Calderwood and Coffey.[33] For the case $\sin\theta \neq \theta$, the itinerant oscillator equations become insoluble analytically but have been solved numerically by Ferrario et al.[25] Evans[35] and Coffey[36] discovered independently a formal identity between the linear, two-dimensional itinerant oscillator and the three-variable Mori theory (a linear limit of RMT) in 1976–1977. This is known now to be an example of the structural constraints which RMT imposes on any mathematical model of the molecular liquid state.

These remarks apply to a molecular ensemble free of strong internal or external fields and associated potential terms V. With the imposition of this extra variable, the challenge to analytical theory becomes more difficult to meet[4] and there is a need for a clear and complementary method of approaching the problem numerically[22] using computer simulation.[14–18]

A. Basic Techniques

The vectorization of computer simulation algorithms for new generations of supercomputers is described by Heyes and Fincham in Vol. 63 of this series. Algorithm descriptions are available in the *Quarterly Review* of the

SERC CCP5 group, and actual listings are available from this group (c/o Daresbury Laboratory of SERC, Warrington, UK). The basic algorithms can be adapted simply for the problem at hand by incorporating an extra torque in the appropriate program loop. The torque may be applied to all or some of the molecules used in the simulation. The technique and first results are described in a series of five papers by Evans,[14-18] and this section and the following are based on these articles. These methods of computer simulation are used now for the description of problems of industrial importance, such as Josephson junction hysteresis,[26] nonlinear phenomena in advanced computer circuitry, or soliton diffusion in activated polymers.

The liquid phase of molecular matter is usually isotropic at equilibrium but becomes birefringent in response to an externally applied torque. The computer can be used to simulate (1) the development of this birefringence —the rise transient; (2) the properties of the liquid at equilibrium under the influence of an arbitrarily strong torque; and (3) the return to equilibrium when the torques are removed instantaneously—the fall transient. Evans initially considered[14] the general case of the asymmetric top (C_{2v} symmetry) diffusing in three-dimensional space and made no assumptions about the nature of the rotational and translational motion other than those inherent in the simulation technique itself. A sample of 108 such molecules was taken, each molecule's orientation described by three unit vectors, e_A, e_B, and e_C, parallel to its principal moment-of-inertia axes.

The total thermal energy of the 108 interacting molecules is stabilized by a temperature-rescaling routine written into the program. This is the case with and without an externally applied torque. A uniaxial external force F is applied to each molecule generating a torque $-e_B \times F$ via the arm e_B embedded in the molecule. The direction e_B is chosen for convenience and need not be along an axis of the principal moment-of-inertia frame. The torque therefore takes effect in the laboratory frame in a direction mutually orthogonal to e_B and F with a magnitude $|e_B||F|\sin\theta$, where θ is the angle between e_B and F for each of the 108 molecules. If F is the force due to an electric field, then a torque of this kind would be generated from a molecular dipole unit vector coincident with e_B. In general, F can be an electric, magnetic, electromagnetic, or other type of force field. The computer simulation method is then used to investigate the response of the molecular liquid sample to this field of force. This response can be measured in a number of ways:

1. *Through time autocorrelation and cross-correlation functions at equilibrium.* These can be computed for and among a variety of vectors associated with the molecular motion. They are equilibrium properties, and the fluctuation-dissipation theorem relates them to transient properties such as

$\Delta n(t)/n$, the transiently induced birefringence. By this means, *deterministic* equations of motion, without stochastic terms, can be used via computer simulation to produce spectral features. As we have seen, a stochastic equation such as Eq. (1) is based on assumptions which are supported neither by spectral analysis nor by computer simulation of free molecular diffusion. The field-on simulation allows us the direct use of more realistic functions for the description of intermolecular interaction than any diffusional equation which uses stochastically generated intermolecular force fields.

The time autocorrelation functions of primary interest to the various spectroscopic probes now available are the orientational acf's, for example,

$$\langle P_1(t)\rangle = \langle e_B(t)\cdot e_B(0)\rangle$$

$$\langle P_2(t)\rangle = \tfrac{1}{2}\left[3\langle e_B(t)\cdot e_B(0)\rangle^2 - 1\right]$$

These are nonexponential near the time origin, and these initial oscillations become more pronounced in the torque-on case[14,17] before the acf's settle into the familiar exponential decay. These oscillations are the signature of the far-infrared frequency region whose power absorption coefficient is essentially the Fourier transform of acf's such as $\langle e_B(t)\cdot e_B(0)\rangle$. This acf developes oscillations at equilibrium in the torque-on case, which means that the far-infrared spectrum of a molecular liquid subjected to a strong enough force field would sharpen and shift to higher frequencies. If the external torque becomes much larger than the internal (mean intermolecular) torque, the orientational dynamics will become coherent as in an ideal gas. However, the torque-on far-infrared frequency in the liquid is much higher than in the ideal gas at the same temperature. Because of the involvement of the molecular angular velocity ω in the kinematic equation[5] linking $\dot{e}_B$ to e_B $[\dot{e}_B = \omega \times e_B]$, the far-infrared spectrum is sensitive to the effect of intense external fields and the way in which these reveal some fundamental properties of the molecular liquid state of matter.

The angular velocity and angular momentum acf's themselves are important to any dynamical theory of molecular liquids but are very difficult to extract directly from spectral data. The only reliable method available seems to be spin-rotation nuclear magnetic relaxation. (An approximate method is via Fourier transformation of far-infrared spectra.) The simulated torque-on acf's in this case become considerably more oscillatory, and, *which is important*, the envelope of its decay becomes longer-lived as the field strength increases. This is dealt with analytically in Section III. In this case, computer simulation is particularly useful because it may be used to complement the analytical theory in its search for the forest among the trees. Results such as these for autocorrelation functions therefore supplement our analytical knowledge of liquid-phase anisotropy. They are not concerned primarily with reproducing any experimental data which may be available but with simulation of some functions more or less fundamental to the evaluation of the ideas embodied in Eq. (1) and its descendants. The computer enables us to use an external torque of arbitrary strength and origin with a minimum of mathematical elaboration. The most intense field strengths are probably electromagnetic in nature, and an obvious target for computer simulation would be laser-induced far-infrared absorption and dispersion anisotropy.

2. *Through monitoring rise and fall transients*. The simulation is intended to reproduce the conditions under which the linear equation (1) is no longer valid.

Another advantage of the simulation is its ability to make direct tests on the range of validity of basic thermodynamical theorems such as the fluctuation-dissipation theorem. In the second paper of the series by Evans,[15] he considers these points for the simplest type of torque mentioned above, $-\mathbf{e}_B \times \mathbf{F}$. Consider the return to equilibrium of a dynamical variable A after taking off at $t = 0$ the constant torque applied prior to this instant in time. If the torque is removed instantaneously, the first fluctuation-dissipation theorem implies that the normalized fall transient will decay with the same dependence as the autocorrelation function $\langle A(t)A(0)\rangle - A_{eq}^2 / \langle A^2(0)\rangle - A_{eq}^2$. Therefore,

$$\frac{\langle A(t)\rangle - A_{eq}}{\langle A(0)\rangle - A_{eq}} = \frac{\langle A(t)A(0)\rangle - A_{eq}^2}{\langle A^2(0)\rangle - A_{eq}^2}$$

where A_{eq} is the equilibrium average value as $t \to \infty$ after the arbitrary initial $t = 0$.

Theories by Coffey et al.,[3,4] Morita et al.,[2] and Grigolini et al.[22] are available that link the properties of rise and fall transients from stochastic equations of molecular diffusion. In particular, Coffey has shown[4-6] that the Debye diffusion equation produces, in general, rise and fall transients which are dependent on the external field strength and duration. The computer simulation shows clearly[15] that the fluctuation-dissipation theorem becomes inapplicable as the applied field strength increases. Fall transients such as $\langle e_{BZ}\rangle$ decay faster than the equilibrium correlation function $\langle \mathbf{e}_B(t)\cdot\mathbf{e}_B(0)\rangle$. The transient is a simple average over 108 molecules which is equivalent to making the usual single-molecule thermodynamic average in the limit $t \to \infty$ of the transient. Similarly, the transient average $[3\langle e_{BZ}(t)e_{BZ}(t)\rangle - 1]/[3\langle e_{BZ}(0)e_{BZ}(0)\rangle - 1]$ decays more quickly than the equilibrium-averaged autocorrelation function $\{3[\langle \mathbf{e}_B(t)\cdot\mathbf{e}_B(0)\rangle]^2 - 1\}/\{3[\langle \mathbf{e}_B(0)\cdot\mathbf{e}_B(0)\rangle]^2 - 1\}$.

This is the first problem, therefore, to be tackled by the analytical approach and is a fundamental hurdle to be overcome when dealing with data which is in any sense nonlinear. Indeed the RMT structure built up by Grigolini and coworkers implies rigorously and generally that the rules governing molecular diffusion are nonlinear in many interesting and different ways. If the Langevin equation becomes *nonlinear* in *the friction term*, the transient falls away more quickly than the equilibrium autocorrelation function.[6,22] This type of nonlinearity automatically takes care of the other features of the computer simulation, such as non-Markovian and non-Gaussian statistics.

The results of analytical theory are expressed essentially as averages of the type $\langle A^n \rangle$, where A is a dynamical vector, usually an orientation axis of the molecule in which there is embedded a unit vector such as $\mathbf{e}_B$.

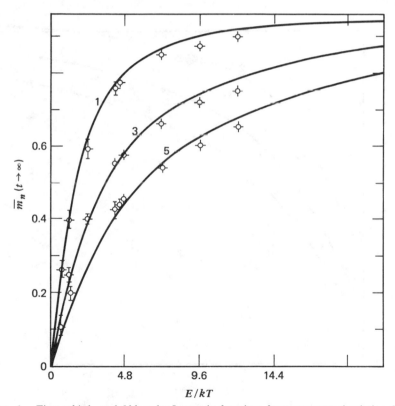

Figure 1.   First-, third-, and fifth-order Langevin functions from computer simulation. [Reproduced by permission from *J. Chem. Phys.*, **76**, 5482 (1982).]

This orientation is defined with respect to the static, laboratory frame of reference. Using stochastic concepts, these averages may be derived for positive integral n. The computer simulation allows the calculation of rise and fall transients such as $\langle e_{BZ}^n \rangle$, and this was carried through by Evans[15] for $n = 1$ to 6. The simulated rise and fall transients for $\langle e_{BZ} \rangle (n = 1)$ are not mirror images, the fall transients being the longer-lived and, in common with the equilibrium acf of $\mathbf{e}_B$, nonexponential. The transient curves are not as smooth as the acf's because in the former case $\langle \ \rangle$ means an average over the 108 molecules only, and in the latter a standard running time average. Therefore, the short-time oscillation in the rise transient may be different for different square-wave field amplitude and duration. The final levels attained by the simulated rise transients as $t \to \infty$ depend on the field strength and duration, as first predicted by Coffey. If, for example, $\langle e_{BZ} \rangle_{t \to \infty}$ is plotted against E/kT, where E is the perturbation energy (Fig. 1), we obtain the Langevin function $L(\mu F/kT)$, where μF is the effective energy due to the

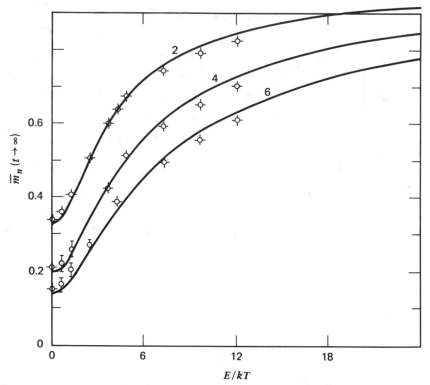

Figure 2.   Second-, fourth-, and sixth-order Langevin functions from computer simulation. [Reproduced by permission from *J. Chem. Phys.*, **76**, 5482 (1982).]

applied torque and kT is the mean thermal energy. Similarly, as n is increased from 1 to 6, similar higher-order Langevin functions are obtained from the simulation (Fig. 2). Their self-consistency was checked by Evans,[14] using a nonlinear least mean squares fitting routine to match the *complete set* of computer simulation data to form $L_n(a)$, where L_n denotes the Langevin function of order n:

$$L_1(a) = \frac{e^a + e^{-a}}{e^a - e^{-a}} - \frac{1}{a}$$

$$L_2(a) = 1 - \frac{2L_1(a)}{a}$$

$$L_3(a) = \frac{e^a + e^{-a}}{e^a - e^{-a}} - \frac{3}{a}L_2(a)$$

$$L_4(a) = 1 - \frac{4}{a}L_3(a)$$

$$L_5(a) = \frac{e^a + e^{-a}}{e^a - e^{-a}} - \frac{5}{a}L_4(a)$$

$$L_6(a) = 1 - \frac{6}{a}L_5(a)$$

Obviously, at a given field strength, the same value of a should and does come out of fitting all six computer-generated $\langle e_{BZ}^n \rangle$. This is a good check of the consistency of the computer simulation method, and it would be interesting to make further computations along these lines with different types and durations of external force fields.

1.   Polarizability and Hyperpolarizability

If an externally applied dielectric field is strong enough, each molecule is distorted in such a way that the effective arm (the total dipole moment) must be described as

$$\mathbf{m} = \boldsymbol{\mu} + \boldsymbol{\alpha} \cdot \mathbf{E}_0 + \tfrac{1}{2}\boldsymbol{\beta} : \mathbf{E}_0\mathbf{E}_0 + \tfrac{1}{6}\boldsymbol{\gamma} : \mathbf{E}_0\mathbf{E}_0\mathbf{E}_0 + \cdots \tag{5}$$

where $\boldsymbol{\mu}$ is the field-off (permanent) dipole moment, $\boldsymbol{\alpha}$ the polarizability, $\boldsymbol{\beta}$ the second, and so on. Here $\mathbf{E}_0$ is the applied electric field. The torque on each molecule is then $\mathbf{m} \times \mathbf{E}_0$. If $\mathbf{E}_0$ is not very large, then in energetic terms the $\boldsymbol{\mu}$ part of the right-hand side of (5) predominates. In condensed phases of matter, where the interaction of molecules is the important feature, $\mathbf{m}$ is supplemented by terms arising from the fields of other molecules, and the internal field $\mathbf{E}(t)$ at any molecule at a given t is different from $\mathbf{E}_0$. The field $\mathbf{E}(t)$ fluctuates with time, and therefore so does $\mathbf{m}$.

The fourth article of the series considers[18] the development of liquid anisotropy with a torque of the form $(\boldsymbol{\alpha} \cdot \mathbf{E}_0) \times \mathbf{E}_0$ and the field dependence of the relevant order parameters $\langle e_{AZ}^{2n} \rangle$ on E_Z^2. Here E_Z is the Z component of $\mathbf{E}_0$ in the laboratory frame. The angled brackets $\langle \; \rangle$ denote a simple arithmetic average over 108 molecules. The problem in molecular polarizability of non-pair-additive interaction is avoided because of its huge complexity and secondary importance in this context.

In the simplest case this kind of torque may be written in the form

$$- E_Z^2 (\alpha_2 - \alpha_1)\sin\theta \cos\theta \, \mathbf{u} \tag{6}$$

where θ is the angle between the field direction Z and $\mathbf{e}_A$ is a unit vector in the A axis of the principal moment of inertia frame. α_1 and α_2 are components of the molecular polarizability tensor along and perpendicular to $\mathbf{e}_A$. It is assumed for simplicity's sake that the two components of α perpendicular to $\mathbf{e}_A$ are equal. Here $\mathbf{u}$ is a unit vector defining the direction of the imposed torque. By definition

$$\cos\theta = \frac{\mathbf{e}_A \cdot \mathbf{E}_0}{|\mathbf{e}_A||\mathbf{E}_0|} = \frac{e_{AZ} E_Z}{|\mathbf{E}_0|} = e_{AZ}$$

$$\sin\theta \, \mathbf{u} = \frac{\mathbf{e}_A \times \mathbf{E}_0}{|\mathbf{e}_A||\mathbf{E}_0|} = \mathbf{i} e_{AY} - \mathbf{j} e_{AX}$$

where $\mathbf{i}$ and $\mathbf{j}$ are unit vectors. Therefore, the torque is

$$- E_Z^2 (\alpha_2 - \alpha_1)(\mathbf{i} e_{AZ} e_{AY} - \mathbf{j} e_{AZ} e_{AX}) \tag{7}$$

for each molecule at each instant t. Here $\mathbf{i}$ is a unit vector in the X direction. Clearly, the elements of Eq. (7) are defined in the static (laboratory) frame of reference.

Kielich[37] has calculated the "Langevin functions" generated by this type of torque, where $\mathbf{E}_0$ is an electric field. The torque produces the results

$$\langle e_{AZ}^{2n+1} \rangle = 0$$

for positive integral n, that is, its effect is detectable only through even-order averages over $\langle e_{AZ}^{2n} \rangle$. The computer simulation, to be valid, must therefore reproduce these features exactly. Defining $q = (\alpha_2 - \alpha_1)E_Z^2 2kT$, Kielich produces $(q = |q|)$

$$L_1(\pm q) = 0$$

$$L_2(\pm q) = +\frac{1}{2q} \pm \frac{1}{2q^{1/2}I(\pm q)}$$

$$L_3(\pm q) = 0$$

$$L_4(\pm q) = \frac{3}{4q^2} \pm \frac{2q+3}{4q^{3/2}I(\pm q)}$$

$$I(\pm q) = e^{\mp q}\int_0^{\sqrt{q}} e^{\pm x^2}\, dx$$

so that $L_1(\pm q)$ describes the q dependence of $\langle e_{AZ}\rangle$, $L_2(\pm q)$ that of (e_{AZ}^2), and so on. For small values of q,

$$L_2(\pm q) = \tfrac{1}{3} \pm \frac{4q}{45} + \frac{8q^2}{945} \pm \frac{16q^2}{14175} + \cdots$$

$$L_4(\pm q) = \tfrac{1}{5} + \frac{8q}{105} + \frac{16q^2}{1575} \mp \frac{32q^3}{51975} - \cdots$$

As $E_Z \to \infty$, $L_2(+q) \to 1$; and $L_2(-q) \to 0$, as do the L_4 functions. We shall call the L_2 and L_4 functions the second- and fourth-order Kielich functions, respectively.

The great advantage of the computer simulation method is that if it passes the test represented by these analytical results it may be used to investigate the molecular dynamics for all strengths of E_Z without any of the formidable difficulties associated with the stochastic theory, typified in the tours de force by Morita and coworkers.[2] Evans verifies in his article[18] that both second- and fourth-order Kielich functions *can* be generated self-consistently by computer simulation. The method used is analogous to the earlier work on Langevin functions.[14] He then illustrates the power of the computer simulation by investigating the effect of an arbitrarily strong second-order torque on the statistical correlation between a molecule's center-of-mass translational velocity **v** and its *own* angular velocity **ω** an interval t later in time. The computations are carried out for an asymmetric top diffusing in three dimensions. It is possible to quantify this correlation [which invalidates the Langevin equation (1)] only in a *moving* frame of reference. This may be illustrated with the following frame transformation. Let v_X, v_Y, and v_Z be the components of **v** in the laboratory frame (X, Y, Z). Define a moving (molecular) frame of reference with respect to unit vectors $\mathbf{e}_A$, $\mathbf{e}_B$, and $\mathbf{e}_C$ along the principal moment-of-inertia axes of the molecule under consideration. The components of **v** in the moving frame of the principal moments of inertia are then

$$v_A = v_X e_{AX} + v_Y e_{AY} + v_Z e_{AZ}$$
$$v_B = v_X e_{BX} + v_Y e_{BY} + v_Z e_{BZ}$$
$$v_C = v_X e_{CX} + v_Y e_{CY} + v_Z e_{CZ}$$

and similarly for $\boldsymbol{\omega}$. By applying symmetry rules of parity, time reversal, and so on, to our C_{2V}-symmetry molecule, we deduce that the cross-correlation function elements

$$\langle v_C(t)\omega_B(0)\rangle \quad \text{and} \quad \langle v_B(t)\omega_C(0)\rangle$$

exist for $t > 0$. All others vanish for all t. Both these functions exist at equilibrium in the molecular ensemble in the presence *and* absence of the external force field. In the latter case, both first-order correlation functions oscillate, the (C, B) function attaining[18] a maximum of $+0.22$ and a minimum of -0.04 with an uncertainty of ± 0.01 or thereabouts. It is clear, therefore, even *before* applying our field, that the classical theory of rotational Brownian motion, so long accepted as the basis of subjects such as dielectric relaxation, cannot be used to describe these rototranslational signatures. These respond to a field corresponding to $q = 17.5$ by becoming more oscillatory, the (CB) element this time having a positive peak of $+0.12$ and a negative peak of -0.13. If we take into consideration the fact that cooperativity in, for example, a mesophase is known to magnify single-molecule properties enormously (e.g., the Kerr constant[37] $\to \infty$) is is likely that rototranslation is the key to explaining many of the spectral features observable in the aligned (or unaligned) mesophase. First-order mixed autocorrelation functions of linear and angular velocity vanish for all t in the *laboratory* frame (X, Y, Z) for achiral molecular symmetries. For this reason, they have eluded detection until the recent suggestion to use moving frames by Ryckaert and coworkers.[39] The parity rule does *not* imply that the effects of rototranslation can be ignored. On the contrary, they are critically important and underpin every type of spectrum whose origins can be traced to molecular diffusion.[5,6] With the use of electric, and probably magnetic, fields, first-order linear, angular velocity acf's become observable in the laboratory frame. This has been confirmed recently by computer simulation (unpublished work).

B. Nonlinear Molecular Dynamics under Intense Force Fields

These dynamics are investigated in detail in the fifth part of the series.[17] This article was intended to pave the way numerically for future advances in the analytical theory of nonlinear molecular dynamics. The simplest type of

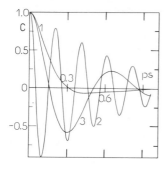

Figure 3. High-field components of the angular velocity acf. [Reproduced by permission from *J. Chem. Phys.*, **78**, 925 (1983).]

torque $(-\mathbf{e}_A \times \mathbf{F})$ is used to produce a variety of auto- and cross-correlation functions in the laboratory and moving frames of reference. In this section we describe briefly the features that emerge from the numerical analysis.

1. The velocity and angular velocity autocorrelation functions become more oscillatory as the externally applied torque increases in strength (Fig 3). Moving-frame component acf's such as $\langle v_A(t)v_A(0)\rangle/\langle v_A^2\rangle$ have different time dependencies, that is, $\langle v_A(t)v_A(0)\rangle/\langle v_A^2\rangle \neq \langle v_B(t)v_B(0)\rangle/\langle v_B^2\rangle \neq \langle v_C(t)v_C(0)\rangle/\langle v_C^2\rangle$, and so on, and the differences become more pronounced as the torque increases. Cross-correlation functions such as $\langle v_C(t)\omega_B(0)\rangle$ or $\langle \omega_C(t)v_B(0)\rangle$ become more oscillatory as the field strength increases to saturation level (about $34kT$).

2. The statistics governing the evolution of the center-of-mass velocity are not Gaussian. This is revealed by an examination of the kinetic energy acf $\langle v^2(t)v^2(0)\rangle/\langle v^4(0)\rangle$, which is invariant to frame transformation. The computer simulation produces accurately the $t \to \infty$ limit ($\frac{3}{5}$) of this acf even in the presence of a very strong field, energetically equivalent to $35.0kT$. In the interval between $t = 0$ and $t \to \infty$, the kinetic energy acf is different from the theoretical curve from Gaussian statistics. One of the interesting side effects of nonlinear RMT is its ability to reproduce this behavior analytically as described by Grigolini in the opening chapter of this volume.

3. The rotational velocity acf becomes rapidly oscillatory under a $35kT$ field. This means the *far-infrared absorption*[5] would be shifted to higher frequencies and sharpened considerably (Fig. 4).

4. The field-on angular velocity and angular momentum acf's in the laboratory frame are oscillatory, and the envelope of the decay of these oscillations becomes field-dependent as mentioned already.

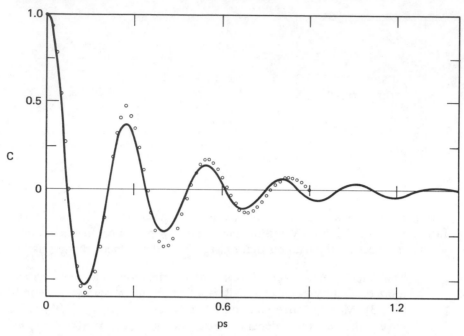

Figure 4. Rotational velocity (—) and angular velocity (○) a.c.f.'s under the influence of a strong external field. [Reproduced by permission from *J. Chem. Phys.*, **78**, 925 (1983).]

This "decoupling" effect[17] verifies Grigolini's prediction[9] of 1976 made in the context of vibrational relaxation[19,20] using RMT.

C. Transients Induced by Alternating Fields

Finally, we mention the results of article three[16] in the series, which considers an alternating external field $F = F_0 \cos \omega t$. The theory, for example, of the Kerr effect in an alternating electric field is confined to the case where the field is weak and has been reviewed by Kielich.[37] Two processes have been identified according to the value of ω.

1. If ω is much smaller than the characteristic frequency ω_0 of oscillations of the molecule, the field will have a reorienting effect, causing some polarization in the medium. If the oscillating, orienting field is made stronger, this polarization effect approaches saturation, and linear response theory is no longer applicable.
2. If ω is much higher than ω_0, no orientation of the electric dipoles is possible, and its only effect is electron distortion within each molecule.

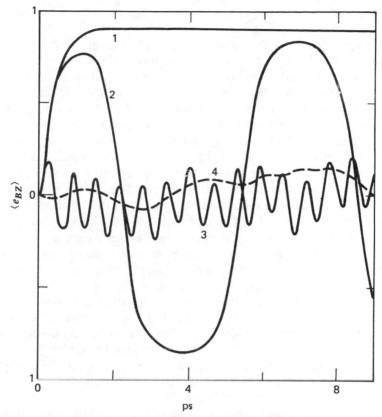

Figure 5.   Dispersion of the rise transients induced with an alternating field of increasing frequency (lowest frequency, curve 1). [Reproduced by permission from *J. Chem. Phys.*, **77**, 4632 (1982).]

The computer can be used to investigate in detail phenomena 1 and 2, which represent a dispersion of the rise transient for a given field strength as the frequency increases. Results were obtained[16] for $\omega = 0$, 10^{12}, 10^{13}, and 10^{14} radians per second, frequencies which take us through the dispersion region because the molecular dynamics simulation takes place on a picosecond time scale. In the static field ($\omega = 0$) we observe a conventional rise transient (Fig. 5). In an alternating field at frequencies below about 10^{11} rad/s, the plateau value of this transient is reached periodically as the external field reverses direction. However, as the alternating field frequency approaches the far-infrared/microwave region, the movement of the molecules in response to the strong external alternating field of force produces a transient which becomes rapidly oscillatory but which fails to reach the plateau level imposed by the static external field. In other words, the transient response is dispersed.

The form taken by each transient is the result of competition between the aligning field and the thermal opposition. The transient oscillations are not simple and are relatively long-lived in comparison with the rise time in the static field. The amplitude of the transient decreases (as mentioned already) as the field frequency increases, but even at $\omega = 10^{14}$ rad/s a long-lived response is measurable above the noise of the computer simulation.

The higher order transients $\langle e_{BZ}^n \rangle$ ($n = 2, 3, \dots$) are of interest in the theory of the Kerr effect, depolarized light scattering from birefringent liquids, spin-spin NMR, incoherent inelastic scattering, and so on. The mean value of $\langle e_{BZ}^2 \rangle$ from the simulation is shifted upwards from $\frac{1}{3}$ (the $t = 0$ value) by the alternating field. The mean value of $\langle e_{BZ} \rangle$, on the other hand, is very small, although the amplitude of the oscillations in this function is much larger. This suggests that only second-order transients (or even-order transients) produce birefringence in an achiral molecular liquid subjected to an alternating field (e.g., an electromagnetic field). This is the case experimentally in laser-induced birefringence when the molecules are achiral.

The external field used in this type of simulation breaks time-reversal symmetry, and a magnetic field would break parity-reversal symmetry. Lorentz forces on partial charges located on atoms of each molecule would make the latter precess along the magnetic field pseudo-vector. It is expected that a simulation of such a system would produce insight into new couplings via nonvanishing mixed autocorrelation functions such as the moving-frame functions described already (see Section IV).

A clue to the analytical understanding of these numerical results comes from the fact that the amplitude of the transients for high ω appears to be random. This could be accounted for by a theory based on amplitude randomness in an externally driven pendulum system.[4,6]

III. ANALYSIS OF THE NUMERICAL DATA WITH RMT

A. The Decoupling Effect

The nature of the excitation has a profound influence on the subsequent relaxation of molecular liquid systems, as the molecular dynamics simulations show. This influence can be exerted at field-on equilibrium and in decay transients (the deexcitation effect). Grigolini has shown that the effect of high-intensity excitation is to slow the time decay of the envelope of such oscillatory functions as the angular velocity autocorrelation function.[5] The effect of high-intensity pulses is the same as that of ultrafast (subpicosecond laser) pulses. The computer simulation by Abbot and Oxtoby[40] shows that in vibrational relaxation an increase in the time duration of the exciting pulses marks the oscillatory decay behavior of highly non-Markovian systems, that is, those in which memory effects are important. These results have been explained analytically by Grigolini, using RMT as described in this section in the related context of rototranslational relaxation of molecular liquids in the presence of very strong external fields. Some further evidence for the validity of Grigolini's RMT has been put forward by Bagchi and Oxtoby,[19,10] using Monte Carlo methods.

Both vibrational and rotovibrational relaxation can be described analytically as multiplicative stochastic processes. For these processes, RMT is equivalent to the stochastic Liouville equation of Kubo,[41] with the added feature that RMT takes into account the back-reaction from the molecule under consideration on the thermal bath. The stochastic Liouville equation has been used successfully to describe decoupling in the transient field-on condition[5] and the effect of preparation on decay.[19,20,42] When dealing with liquid-state molecular dynamics, RMT provides a rigorous justification for itinerant oscillator theory,[25,33] widely applied to experimental data by Evans and coworkers.[5,31] This implies analytically that decoupling effects should be exhibited in molecular liquids treated with strong fields. In the absence of experimental data, the computer runs described earlier[14-18] amount to an independent means of verifying Grigolini's predictions. In this context note that the simulation of Oxtoby and coworkers[19,20] are "semistochastic" and serve a similar purpose.

1. Decoupling Effects in the Quasi-Markov Limit

The simplest RMT structure needed to describe field-on computer simulations[17] is the essentially two-dimensional equation

$$\frac{dC(t)}{dt} = -\int_0^t \phi(t-\tau)C(\tau)\,d\tau + i\omega_1 C(t) \tag{8}$$

with frequency $\omega_1 = (\mu E/I)^{1/2}$, where μ and I denote the dipole and the moment of inertia, respectively, of a given molecule. This describes a torque applied externally, as in the simulation, to each molecule of the sample.

Equation (8) can be rewritten as

$$\frac{d\tilde{C}(t)}{dt} = -\int_0^t \exp[-i(t-\tau)\omega_1]\phi(t-\tau)\tilde{C}(\tau)\,d\tau \tag{9}$$

with the autocorrelation function written in the "interaction picture" as

$$\tilde{C}(t) = \exp[i\omega_1 t]C(t) \tag{10}$$

Assume, for the memory function, that

$$\phi(t) = \phi_0 \exp(-\gamma_a t) \qquad (11)$$

with γ_a large enough to approach the Markov limit for $C(t)$, that is, with the $\phi(t)$ a quasi-delta function in time. This implies in the original frame of reference that

$$\frac{dC(t)}{dt} = \left[i\Omega(\omega_1) - \Gamma(\omega_1) \right] C(t) \qquad (12)$$

where

$$\Omega(\omega_1) = \omega_1 + \frac{\omega_1 \phi_0}{\omega_1^2 + \gamma_a^2} \qquad (13)$$

$$\Gamma(\omega_1) = \frac{\gamma_a \phi_0}{\gamma_a^2 + \omega_1^2} \qquad (14)$$

For finite γ_a, the decay of the oscillation envelope becomes slower as the frequency ω_1 increases. This is the Grigolini decoupling effect in its simplest form and is qualitatively in agreement[21] with the computer simulations of Section II. *Note that there is no decoupling effect in Markovian systems*[23] (i.e., those where $\gamma_a \to \infty$).

When the system under consideration becomes significantly non-Markovian, we have to release the implicit assumption that the memory kernel is unaffected by external excitation. The three-dimensional computer simulation[14] of dichloromethane of Section II is described, for simplicity, with a liquid sample consisting of disks with moment of inertia I constrained to move on a plane surface and colliding with each other. Each disk has an electric dipole μ in one of its diameters. Using the Mori basis set provides the equation of motion:

$$\frac{d\theta}{dt} = iL_0\theta \qquad (15)$$

where θ is the orientation angle of the tagged molecule and L_0 is the

Liouvillian of the model system. This can be approximated by

$$\frac{df_0}{dt} = f_1$$

$$\frac{df_1}{dt} = -\Delta_0^2 f_0 + f_2$$

$$\frac{df_2}{dt} = -\Delta_1^2 f_1 + f_3 \tag{16}$$

$$\frac{df_3}{dt} = -\Delta_2^2 f_2 - \gamma f_3 + F(t)$$

where $f_0 \equiv \theta$, $f_1 \equiv \omega \equiv \dot{\theta}$. In the absence of an external field, $\Delta_0 = 0$. In the presence of the field we have to add to iL_0 the following interaction:

$$\mathscr{L}_1 \equiv iL_1 = -\left(\frac{E\mu}{I}\right)\sum_i \sin\theta_i \doteq -\omega_1^2 \sum_i \theta_i \tag{17}$$

θ is the angle between the ith dipole and the direction of the electric field. The linear approximation is justified when $\langle \omega^2 \rangle \ll \omega_1^2$.

The dynamics of an infinite chain of states is simulated by the friction γ and stochastic force $F(t)$ in the final equation of set (16). If these dynamics involve frequencies much higher than ω_1, that is, if they occur on a very much faster time scale, the expansion of L_1 can be limited to the four equations of set (16), assuming that γ and F are unaffected by the field. With these assumptions, Eq. (16) is equivalent to the model of the itinerant oscillator:[33]

$$\dot{\theta} = \omega$$

$$\dot{\omega} = -\Delta_1^2(\theta - \psi) - \omega_1^2\theta$$

$$\dot{\psi} = \nu \tag{18}$$

$$\dot{\nu} = \Delta_2^2(\theta - \psi) - \omega_2^2\psi - \gamma\nu + f(t)$$

When $\gamma \gg \Delta_2$ the equivalent of the microscopic time γ_a^{-1} is $\Gamma^{-1} \equiv \gamma/\Delta_2^2$. Decoupling effects are present when $\omega_1 \doteq \Gamma$. To obtain an approximate value of Γ we can use the experimental data as follows. First, we evaluate the value of decay of the oscillation envelopes of the angular velocity autocorrelation function as a function of ω_1. Equation (14) shows that this is, approximately, a Lorentzian, the linewidth of which provides the approximate expression for Γ. The agreement with the numerical decoupling effect is quantitatively good[21] when the ratio ω_2^2/ω_1^2 is assumed to be equal to 8.5. Simple Markovian models cannot account for decoupling effects.

B. Deexcitation Effects

These simulations test the basic validity of the fluctuation-dissipation theorem:

$$\Delta_A(t) = C_A(t) \tag{19}$$

where

$$\Delta_A(t) = \frac{\langle A(t) \rangle}{\langle A(0) \rangle} \tag{20}$$

$$C_A(t) = \frac{\langle A(t) A(0) \rangle_{eq}}{\langle A^2 \rangle_{eq}} \tag{21}$$

The autocorrelation function $C_A(t)$ is evaluated at equilibrium, whereas $\Delta_A(t)$ is a transient property requiring preliminary excitation of the variable of interest A. Evans investigated this problem by monitoring via computer simulation[15] the time behavior of a liquid sample after the instantaneous removal of a strong external field of force $\mathbf{E}$. He found that at the point $\mu E / kT = 12$, $\Delta_A(t)$ decays considerably faster than $C_A(t)$. Here μ is the dipole of the tagged molecule and μE is the energy associated with the field of force $\mathbf{E}$. In that case A is the component of the dipole along the Z axis.

1. Linear and Nonlinear Systems Far from Equilibrium

Under a wide range of conditions (described elsewhere in this volume), the fundamental equation of motion

$$\frac{dA}{dt} = iLA \tag{22}$$

can be recast in the general form[44,45]

$$\frac{dA}{dt} = \lambda A - \int_0^t \phi_A(t - \tau) A(\tau) \, d\tau + F_A(t) \tag{23}$$

Assuming that the stochastic force vanishes on average:

$$\langle F_A(t) \rangle = 0 \tag{24}$$

By a simple Laplace transform, from Eq. (23), we have

$$\Delta_A(t) = \mathscr{F}_A(t) \tag{25}$$

where

$$\mathscr{F}_A(t) = \mathscr{L}_a^{-1}[s - \lambda + \hat{\phi}_A(s)] \tag{26}$$

and $\hat{\phi}_A(s)$ denotes the Laplace transform $\mathscr{L}_a[\phi_A(t)]$. A further basic requirement for Eq. (19) to be valid is that $\phi_A(t)$ is a genuine equilibrium property. Zwanzig[44] has shown that in the nonlinear case the memory function $\phi_A(t)$ also depends on the variable of interest A. This means that if we excite that variable, we can destroy the basic condition that $\phi_A(t)$ must be a genuine equilibrium property. This prevents $\Delta_A(t)$ from being identifiable with the equilibrium property $C_A(t)$, as implied by the fluctuation-dissipation theorem.

The fundamental reason why excitation has an influence on the decay process is the breakdown of condition (24). This has profound consequences on relaxation when the time scale of the stochastic force $F_A(t)$ is not well separated from that of the variable of interest A, that is, when memory effects are important. The deexcitation effect is a direct consequence of the invalidation of Eq. (24) by intense external force fields and was first investigated by Grigolini and coworkers following the computer simulations described in Section II.

The preparation effect which results in the breakdown of Eq. (24) is relevant only when considering the intense non-Markovian case of Section II (even in the linear case). A review of these effects can be found in ref. 7.

When dealing with a nonlinear case, even when a large time scale separation between the system of interest and thermal bath is available, a new preparation effect takes place related to the fact that the memory kernel Φ_A depends on excitation.

The argument by Zwanzig links the violation of condition (24) with a well-defined nonlinearity in the stochastic equations describing the system under consideration (an ensemble of molecules). One way of introducing this type of nonlinearity is to use the nonlinear itinerant oscillator

$$
\begin{aligned}
\dot{\theta} &= \omega \\
\dot{\omega} &= -\kappa \sin[N(\theta - \psi + \xi)] - \omega_1^2 \sin\theta \\
\dot{\psi} &= \nu \\
\dot{\nu} &= \kappa \sin[N(\theta - \psi + \xi)] - \omega_2^2 \sin\psi - \Gamma_\nu \nu + f(t)
\end{aligned}
\tag{27}
$$

The derivation of these equations is described elsewhere in this volume. ξ is the average value of the angle between real and "virtual" dipole *when no external field is present*.

To simplify the calculation, Grigolini assumed[22] ν to be an infinitely fast variable, thereby providing

$$\dot{\theta} = \omega$$

$$\dot{\omega} = -\kappa \sin(N\Delta) - \omega_1^2 \sin\theta$$

$$\dot{\Delta} = -\frac{\kappa}{\Gamma_\nu}\sin(N\Delta) + \omega$$

$$\qquad\qquad (28)$$

$$-\frac{\omega_2^2}{\Gamma_\nu}\sin[N(\theta + \xi - \Delta)] + \frac{f(t)}{\Gamma_\nu}$$

$$\Delta = \theta - \psi + \xi$$

which is the nonlinear extension of the equivalent second-order Mori approximant discussed by Berne and Harp.[46] In the absence of an external field, a simplified picture approximately equivalent to Eq. (28) can be obtained, provided Δ is must higher in frequency than ω. This assumption allows the use of the adiabatic elimination procedure (AEP) developed in Chapter II within the context of RMT. It is then possible to show that the *long-time* behavior of Eq. (28) is roughly equivalent to that provided by the following Markovian, but nonlinear, equation:[47-49]

$$\dot{\omega} = -\gamma\omega + \gamma'\omega^3 + F(t) \qquad\qquad (29)$$

where

$$\gamma = \kappa N \frac{\langle \cos(N\Delta)\rangle}{\Gamma}$$

$$\gamma' = \frac{\kappa N^3}{6}\frac{\langle \cos(N\Delta)\rangle}{\Gamma^3}$$

$$\Gamma \equiv \frac{\kappa N}{\Gamma_\nu}$$

Note that this description is qualitatively correct, but flawed. Equation (29) does not satisfy the requirement of canonical equilibrium, and the correction is evaluated via linearization which underestimates the intensity of the correction itself. In Chapter VI a more accurate approach is shown. The arguments of this section are correct only from a qualitative point of view.

The details of how $F(t)$ is related to $f(t)$ are given elsewhere in this volume. Equation (29) may be used to describe the deexcitation effects described in Section II. A more accurate analytical description is provided by the numerical solution of the set of equations making up the nonlinear itinerant oscillator using continued fraction analysis.

2. Comparison with Computer Simulation

Assume that θ is a slow relaxation variable compared with ω. Using the adiabatic elimination procedure (AEP)[47,48] described in Chapter II, we obtain for the probability distribution of the variable θ, $\sigma(\theta; t)$ the following, generally valid, equation of motion:

$$\frac{\partial \sigma(\theta; t)}{\partial t} = \sum_{n=2}^{\infty} \int_0^t ds\, \phi_n(t, s) \frac{\partial^n \sigma(\theta; s)}{\partial \theta^n} \qquad (30)$$

This result is obtained by taking into account only the contribution $\mathscr{L}_1^2$ of the theory developed in Chapter II. If θ is almost Markovian, the higher order derivatives are negligible, leaving

$$\frac{\partial \sigma(\theta; t)}{\partial t} = D(t) \frac{\partial^2 \sigma(\theta; t)}{\partial \theta^2} \qquad (31)$$

where

$$D(t) \equiv \int_0^t ds \left[\langle \omega(t)\omega(s) \rangle - \langle \omega(t) \rangle \langle \omega(s) \rangle \right] \qquad (32)$$

Due to the overall symmetry constraints on a nonrotating molecular liquid sample, $\langle \omega(t) \rangle - \langle \omega(s) \rangle = 0$ even in the transient region. It is difficult to evaluate the nonstationary correlation function:

$$\phi(t, \tau) = \langle \omega(t)\omega(t - \tau) \rangle$$

$$= \int d\omega\, d\Delta \left(e^{\Gamma_0^+ \tau} \omega \right) \omega \sigma_B(\omega, \Delta; t) \qquad (33)$$

where Γ_0 is the Fokker-Planck operator of the (ω, Δ) system, and

$$\sigma_B(\omega, \Delta; t) = K e^{\Gamma_0 t} \int d\theta \exp\left(\langle \omega^2 \rangle_{eq}^{-1} \left[\frac{\kappa}{N} \cos(N\Delta) + \omega_1^2 \cos\theta \right.\right.$$

$$\left.\left. + \omega_2^2 \cos(\theta + \xi - \Delta) - \tfrac{1}{2}\omega^2 \right] \right) \qquad (34)$$

that is, the state of the (ω, Δ) system at a time t far from the sudden removal of the field, K is a suitable normalization factor, and $\langle \omega^2 \rangle_{eq}$ denotes the equilibrium value of $\langle \omega^2 \rangle$, which is dependent on whether or not the external field is present.

To evaluate this nonstationary correlation function, we shall assume that the variable ω is much slower than the "virtual" variable Δ. This allows us to envisage a simplified approach to the evaluation of the transient correlation function of Eq. (33). Soon after the sudden removal of the external field, Δ obtains its equilibrium distribution, thereby determining a change in the ω distribution from the equilibrium state. Neglecting the short-time contribution to $\phi(t, \tau)$, we can write

$$\phi(t, \tau) = K' \int d\omega \, d\Delta \left(e^{\Gamma_0^+ \tau} \omega \right) \omega \exp \left[-\frac{\omega^2}{2} \langle \omega^2(0) \rangle \right] \sigma_{eq}(\Delta) \qquad (35)$$

where K' is a suitable normalization factor, $\sigma_{eq}(\Delta) \propto \exp[k \cos(N\Delta)/N \langle \omega^2 \rangle_{eq}]$, and $\langle \omega^2(0) \rangle$ denotes the value of $\langle \omega^2(t) \rangle$ at the time t far from the sudden removal of the external field when this average value attains its largest deviation from $\langle \omega^2 \rangle_{eq}$.

It must therefore be understood that this is the new time origin. t is much shorter than the relaxation time of the variable ω. Throughout the analysis of this section, a parameter of basic importance is

$$R \equiv \langle \omega^2(0) \rangle - \langle \omega^2 \rangle_{eq} \qquad (36)$$

The absolute value of this parameter depends strongly on the intensity of the "virtual" dipole. By numerical means, we can show that for positive R', nonvanishing values of ξ are required. This can be understood on physical grounds through the fact that for $\xi = 0$, strong fields do not change the average value of the angle between real and "virtual" dipole, although they prevent large fluctuations about the mean value. The field-on equilibrium distribution is therefore associated with values of the corresponding potential energy smaller than those in the absence of the field. Assuming that $N = 1$, $\xi = \pi$, the effect of strong fields is to reduce the angle between real and "virtual" dipole from π to 0, thereby resulting in an effect which is the reverse of that described above. This makes it possible to obtain positive values for R. A more accurate discussion of this issue will be given in Chapter VI.

Focusing our attention on the slow reequilibrium process of ω, and using the results of AEP, we can derive from Eq. (33) the new equation

$$\phi(t, \tau) = \int d\omega \, \omega \left(e^{\Gamma_{ad}^+ \tau} \omega \right) \exp \left[-\frac{\omega^2}{\langle \omega^2(t) \rangle} \right] \qquad (37)$$

where Γ_{ad} is the effective operator associated with Eq. (37). To evaluate explicitly the time dependence of $\langle \omega^2(t) \rangle$, we can use the following mean field approximation to Eq. (37):

$$\dot{\omega} = -\gamma \omega + \gamma' \langle \omega^2(t) \rangle \omega + F(t) \qquad (38)$$

and solve this equation with the method of Suzuki. This results in

$$\left\langle \omega^2(t) \right\rangle = \frac{\left\langle \omega^2 \right\rangle_{eq} + R\left(\gamma + \left\langle \omega^2 \right\rangle_{eq}\gamma'\right)e^{-2\gamma t}/(\gamma - R\gamma')}{1 + R\gamma'e^{-2\gamma t}/(\gamma - R\gamma')} \tag{39}$$

In strongly nonlinear systems (large γ') and/or with significant excitation (large positive values of R), Eq. (39) locks the process of energy exchange between ω and its thermal bath. *This accelerates the decay of $\left\langle \cos\theta(t) \right\rangle$ according to Eq. (31).*

Equation (37) gives the result

$$\phi(t,\tau) = \left\langle \omega^2(t) \right\rangle \exp\left[-\left(\gamma - \gamma'\left\langle \omega^2(t) \right\rangle \right)\tau \right] \tag{40}$$

Equations (31) and (32) then result in

$$\left\langle \cos\theta(t) \right\rangle = \exp\left(-\int_0^t D(t')\,dt' \right)\left\langle \cos\theta(0) \right\rangle \tag{41}$$

for the fall transient of $\left\langle \cos\theta(t) \right\rangle$, where

$$D(t) = \int_0^t \phi(t,\tau)\,d\tau \tag{42}$$

and $\phi(t,\tau)$ is defined by Eqs. (40) and (41). Note that in the absence of excitation ($R = 0$), Eq. (42) reduces to the well-known Kubo result for the stochastic oscillator.[41] The analytical theory summarized briefly in these equations accounts qualitatively for the numerical deexcitation effect.[22] A quantitative description can be obtained under carefully controlled numerical conditions using the CFP to solve the nonlinear diffusion equations and from a more detailed discussion of the dependence of R on both ω_2 and ξ.

The RMT is a method for finding simplified models that satisfy the rigorous formal constraints imposed by the most general theories of relaxation. The structure of RMT implies that only some models of the liquid state can be accepted. Those that are invalidated by RMT cannot reproduce some basic features such as the deexcitation effect. Numerical methods such as computer simulation are now becoming important in uncovering some basic properties of the liquid state—properties that can be analyzed with RMT. An experiment could be made to detect the deexcitation effect using strong, ultrafast (subpicosecond) laser radiation (e.g., Nd-YAg laser with a streak camera detection system for second-order rise and fall transients).

The formal structure of RMT at short times implies that *dissipative processes cannot affect some variables of interest directly*. In this way it is possible to account for decoupling phenomena described earlier in this review. When the reduced model is given a nonlinear character, the long-time behavior of the corresponding part of interest of the system satisfies the constraints of the rigorous, generally valid, theory of relaxation. *The slowing down of averages associated with ω, and therefore the acceleration of the fall transient* $\langle \cos \theta(t) \rangle$, *are both natural outcomes of RMT*. The simplified theory of this section deals with the long-time limit of the more general theory, because in this limit the equations are tractable analytically. The CFP is capable of solving the equations without this constraint. An investigation of the role of the "virtual" dipole is necessary, because the computer simulation of the excitation effect implies that ξ does not vanish, that is, the angle between the real and virtual dipoles does not vanish on average. The theory of this section can be checked by using the CFP which, as shown in Chapter I, stems from the same theoretical background as that behind the RMT.

IV. FIELD EFFECTS IN CHIRAL MOLECULES: COMPUTER SIMULATION

Several interrelated problems in vapor and condensed phase molecular dynamics have been brought together recently using the computer simulation of optically active molecules. It is well known[50] that the average interaction potential between one R enantiomer and another R enantiomer is slightly different from that between R and S enantiomers. This is sometimes described in textbooks and review articles as "chiral discrimination." The effects[51] of chiral discrimination are sometimes very small, because the potential energy difference $\Delta V = V(R\text{-}R) - V(R\text{-}S)$ is only a few calories per mole. Nevertheless, it ought to be detectable spectroscopically with techniques already available such as infrared/radio-frequency double resonance and nonlinear, inverse lamb dip spectroscopy of dilute chiral gases. The effects of chiral discrimination in the Hamiltonian governing the molecular dynamics of a mixture of R and S camphor vapors will result in these types of ultra-high-resolution spectra being clearly different for RS and R (or S) camphor vapors. The reason for this is that the dynamical effect of intermolecular collisions will be different. In order to explain this effect it is not sufficient to rely on theories of collision-induced line broadening[5] that leave out of consideration the statistical correlation between molecular rotation and translation. (This type of theory is exemplified by the well-known m and J diffusion models.[5]) The statistical correlation between **v**, the center-of-mass translational molecular velocity and **J** the molecular angular momentum is identical for R and S camphor and exists at all vapor pressures. It vanishes in the theoretical limit of infinite dilution, because in this limit, and in this limit only, the intermolecular potential energy vanishes. The correlation may be quantified with a moving frame matrix such as $C_{tr}(t) = \langle \mathbf{v}(t)\mathbf{J}^T(0)\rangle_m$, where $\langle \ \rangle_m$ denotes running time averaging with both $\mathbf{v}$ and $\mathbf{J}$ defined in this moving frame, for convenience that of the principal molecular moments of inertia. R and S enantiomers are physically different mirror images, and the cross-correlation matrix C_{tr} contains elements that are asymmetric in their time dependence.[24] The overall symmetry of C_{tr} for the R enantiomer is different from that of its S counterpart. The molecular dynamics for the R and S enantiomers are therefore different when looked at from the correct (and unique) viewpoint embodied in C_{tr}. If the R and S enantiomers are mixed in equal proportion, to give the racemic mixture, the asymmetric elements in C_{tr} cancel and vanish for all t, leaving other elements intact.[24] This is, essentially speaking, the effect of chiral discrimination in a dynamical context. The cross-correlation matrix C_{tr} behaves as it does because the R and S stereoisomers are mirror images, whose dynamical properties *in the laboratory frame* are all precisely identical to an observer who does not have the use of polarized radiation. A theory of molecular dynamics which does not consider C_{tr} is left with an impossible task—that of describing R and S identically and at the same time explaining why the laboratory-frame dynamical properties of mixtures of R and S are different because of chiral discrimination. It follows that these observable differences cannot be explained without considering C_{tr}. Therefore, these differences may be used to measure C_{tr} and cross-correlation functions like it. This must, and can only, be done by taking chiral discrimination fully into account in the Hamiltonian. Therefore the interpretation of spectral differences between RS and R or S is in principle a method of measuring those energy differences known as chiral discrimination.

Boiling point differences between RS and R or S liquids are known, and there are well-known density and refractive index differences at constant temperature. These are usually small, but sometimes there are differences of tens of degrees in the melting points of racemic mixture and enantiomer. Examples are the lactic acids, canadines, and 2-chlorobutanes, a relatively simple molecular structure where there is a 9-K difference[52] between the melting points of enantiomers and racemic mixture. In the solid state it is well known that these differences can be amplified greatly.[53,54] The spectral differences between RS and R or S in the molecular crystalline state are pronounced. In the infrared there are, for example, pronounced displacements in the frequency of fundamentals and lattice modes.[50,51] In the supercooled and glassy condition, various spectral properties of enantiomer and racemic mixtures are known which prove conclusively that there are large observable differences. Camphor provides a good example, with its well-characterized "pseudo-eutectic" between rotator and λ phases at 1 bar below the normal melting points. Rossitter[54] has unearthed a spectacular difference in the Cole-Cole plot of RS and R or S camphor in the λ phase. The existence of Cole-Cole arcs for both enantiomers and racemic mixture implies the existence of a molecular dynamical process of some kind. Evans[53,55] has recently corroborated these findings to a limited extent in the far infrared in the room-temperature rotator phase, providing another spectral angle on the molecular dynamics. The far infrared results of Evans and radio-frequency data of Rossitter imply that the absorption of the R and S enantiomers is identical to unpolarized radiation but that the absorption of the racemic mixture (RS camphor) is distinctly different at essentially the same temperature. These results cannot be explained without recognizing the role of C_{tr} and taking chiral discrimination into account. In this context it is interesting that elements of C_{tr}, when properly normalized, increase in amplitude in the supercooled condition below the normal melting point for both chiral and achiral molecules.

Therefore, by looking at supercooled and glassy chiral molecules we maximize our chances of seeing laboratory-frame differences between an enantiomer and its racemic mixture. These differences cannot be explained in the theory of molecular diffusion without accounting for C_{tr}. The only theoretical method available for doing this in detail is computer simulation, and in this section we use this method to investigate the effect of chiral discrimination in the laboratory and moving frames of reference for some correlation functions of sec-butyl chloride supercooled in the amorphous condition to 50 K. This molecule is chosen for investigation because it is the simplest chiral system in which the difference between the melting points of enantiomers and racemic mixture is known to enough accuracy to verify that there is a large effect on the melting point of chiral discrimination.

Specifically in this section we investigate laboratory and moving-frame differences between the molecular dynamics of the sec-butyl chloride enantiomers and racemic mixture, both at equilibrium in the supercooled condition and in their response to an external electric field.

The details of the pair potential used in the simulations are given in Table I. This consists of an all-$trans$ model of the sec-butyl chloride molecule with six moieties. The intermolecular pair potential is then built up with 36 site-site terms per molecular pair. Each site-site term is composed of two parts: Lennard-Jones and charge-charge. In this way, chiral discrimination is built in to the potential in a natural way. The phase-space average R-R (or S-S) potential is different from the equivalent in R-S interactions. The algorithm transforms this into dynamical time-correlation functions.

The effect of chiral discrimination at equilibrium is illustrated in Fig. 6 for the angular velocity acf of both enantiomers and racemic mixture at

TABLE I

Site-Site Potential for *trans-sec*-Butyl Chloride: Lennard-Jones and Partial Charge Terms

Moiety	$x(e_3)$,[a] Å	$y(e_1)$,[a] Å	(R) $z(e_2)$,[a] Å	(S) $z(e_2)$,[a] Å	ϵ/k, K	σ, Å	q, $|e|$
CH_3—	−2.14	−0.29	0.48	−0.48	158.6	4.0	0.04
CH_2—	−0.87	0.32	1.09	−1.09	158.6	4.0	0.028
Cl	0.40	0.00	−1.27	1.27	127.9	3.6	−0.177
C	0.40	−0.29	0.48	−0.48	35.8	3.4	0.001
H	0.40	−1.39	0.69	−0.69	10.0	2.8	0.068
CH_3—	1.67	0.32	1.09	−1.09	158.6	4.0	0.04

[a]Coordinates relative to center-of-mass, principal molecular moment-of-inertia frame.

50 K, in the supercooled liquid condition of the 2-chlorobutanes. There is little difference among the three cases, and this is buried in the statistical noise generated by the computer run. In the case of the center-of-mass velocity acf's in the same, equilibrium, condition (Fig. 7a) the difference between enantiomers and racemic mixture does, however, fall outside the noise, and this is an indication that the mixture of right and left stereoisomers has dynamical properties different from those of each component.

The effect of the strong external field is to accentuate this difference— using the method first developed for achiral molecules described earlier in this review—that of aligning the molecules in the "molecular dynamics" cube with an externally applied torque.[14-18] This may be used to simulate the effect of an electric field on an assembly of dipolar molecules using second-order

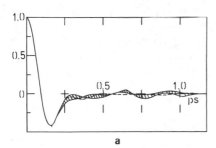
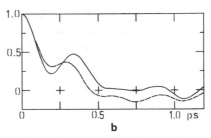

Figure 6. (*a*) 2-Chlorobutane at 50 K, 6×6 site–site potential, angular velocity autocorrelation functions. Crosshatching indicates computer noise difference between R and S enantiomers. (——) Racemic mixture. (*b*) As for (*a*), under the influence of a strong field **E**, producing a torque $-e_3 \times E$ in each molecule of the molecular dynamics sample. (1) (——) Racemic mixture; (2) (---) R enantiomer. Ordinate: Normalized correlation function; abscissa: time, ps.

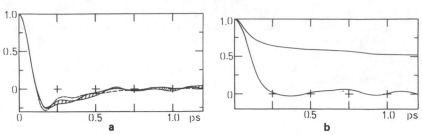

Figure 7.   (a) As for fig. 6(a): center-of-mass linear velocity acf's (b) As for Fig. 6(b): racemic
mixture. Ordinate: Normalized correlation function; abscissa: time, ps.

averages, or, a train of Gwatt electromagnetic pulses from a mode-locked
laser. These methods have been described earlier, and here we concentrate
only on the results of applying the torque to the e_3 vector of the 2-chloro-
butane molecules—enantiomers and racemic mixture.

The application of a powerful enough torque in this way reveals much
about the hidden theoretical structure of the molecular liquid state, lending
support to the reduced model theory of Grigolini.[5-7] For the supercooled
2-chlorobutanes we use an applied torque powerful enough to result in
saturation of the component in the laboratory from $\langle e_{3z} \rangle = 1.0$. This is
achieved quite easily because of the low temperature (50 K). Figure 6b shows
that the effect on the angular velocity is pronounced, this function taking on
a complex oscillatory form quite different from its counterpart at equilibrium (Fig. 6a). The decoupling effect of Grigolini is, however, small in this
case because of the low temperature and high packing density of the supercooled environment. The envelope of the acf under the influence of the field
is not much longer lived than that at equilibrium, although the structure of
the acf is changed. The chiral discrimination effect seems to be enhanced by
the external torque, that is, there is a greater difference between the acf's of
enantiomer and racemic mixture. This should be checked, however, with
more powerful computers and longer runs, to improve the statistics. Figure

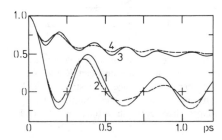

Figure 8.   As for Fig. 6(b), angular
momentum acf's. (1) Racemic mixture; (2) R
enantiomer; (3) second moment,[5] racemic mixture; (4) second moment,[5] R enantiomer.
Ordinate: Normalized correlation function; abscissa: time, ps.

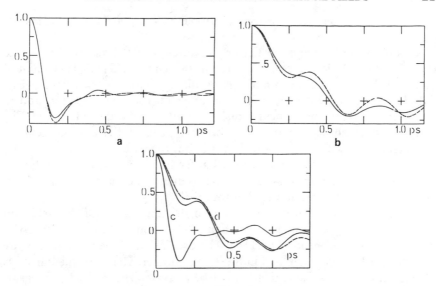

Figure 9.   (*a*) As for Fig. 6(*a*), rotational velocity acf $\langle \dot{\mathbf{e}}_2(t)\cdot\dot{\mathbf{e}}_2(0)\rangle/\langle \dot{\mathbf{e}}_2^2\rangle$. (——) R enantiomer, (---) racemic mixture. (*b*) As for Fig. 9*a*, aligned with an external field. (——) Enantiomer: (---) racemic mixture. Ordinate: Normalized correlation function; abscissa: time, ps. (*c*) R enantiomer $\langle \dot{\mathbf{e}}_1(t)\cdot\dot{\mathbf{e}}(0)\rangle/\langle \dot{\mathbf{e}}_1^2\rangle$ isotropic sample. (*d*) Aligned sample. (---) R enantiomer; (——) racemic mixture.

7*b* shows that the parallel effect on the center-of-mass velocity acf is much smaller, making the acf less oscillatory. It should be noted that the kinetic energy acf $\langle v^2(t)v^2(0)\rangle/\langle v^4\rangle$ approaches a limit of 0.5 as $t \rightarrow \infty$. For an isotropic three-dimensional system the levels dictated[38] by Gaussian statistics is 0.6, that for a two-dimensional sample 0.5. This shows that the saturation by a field along the z axis has reduced the dimensionality of the molecular dynamics, one dimension being constrained by the field.

Some features of the Grigolini decoupling effect are shown more clearly in the angular momentum acf's (Fig. 8) for the aligned samples. These acf's are much different from their counterparts of Fig. 6, being more oscillatory and longer lived. The envelope of the oscillations measures the extent of the decoupling from the molecular thermal bath, as defined by Grigolini and coworkers. Again the effects of chiral discrimination seem to be visible clearly in the laboratory frame as a difference in period and amplitude of the acf oscillations for enantiomer and racemic mixture. The rotational velocity acf's exemplified in Fig. 9 are observable experimentally by far-infrared spectroscopy and also show the Grigolini decoupling effect in both enantiomer and racemic mixture (Figs. 9*a* and *b*). In this case also the effect of chiral discrimination is clear, as in the rotational velocity acf's of $\dot{e}_1$ (Figs. 9c and d).

Chiral discrimination is most clearly manifested in the moving frame, and the sec-butyl chlorides are no exception in this context. The (3,1) element of $\langle \mathbf{v}(t)\mathbf{J}^T(0)\rangle_m$ for the S enantiomer is oscillatory with a positive peak (Fig. 10). The same element for the R enantiomer mirrors this behavior, and in the racemic mixture the element vanishes for all t. In the aligned sample (Fig. 10b) the normalized amplitude of the first peak of the cross-correlation function is reduced by a factor of about 5, but thereafter there are signs that the function is considerably more oscillatory in nature. Again the behavior of the R enantiomer mirrors that of the S enantiomer, and the same function in the racemic mixture vanishes for all t.

The transient interval of time between the application of the field and saturation (Fig. 11a) lasts for less than 1.0 ps, and in this period the rise transient oscillates deeply (Fig. 11b). The oscillation of the racemic mixture is significantly deeper than that in the R enantiomer. The experimental study of transients such as these, then, might be a convenient method of measuring the dynamical effect of chiral discrimination in the liquid state. Deep transient oscillations such as these have been foreseen theoretically by Coffey and coworkers using the theory of Brownian motion.[3] The equivalent fall transients (Fig. 11b) are much longer lived than the rise transients and are not oscillatory. They decay more quickly than the equilibrium acf's. The effect of chiral discrimination in Fig. 11b is evident. Note that the system

$$\dot{\theta} = \omega$$

$$\dot{\omega} = -\gamma\omega - \frac{\mu E \sin\theta}{I} + f(t)$$

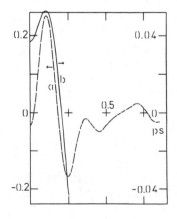

Figure 10. (a) (3,1) element of the cross-correlation function for the R enantiomer, $\langle \mathbf{v}(t)\mathbf{J}^T(0)\rangle$, $(\langle v_3(t)J_1(0)\rangle/(\langle v_3^2\rangle^{1/2}\langle J_1^2\rangle^{1/2}))$. (b) The same under the aligning field. Ordinate: Normalized cross-correlation function; abscissa, time, ps.

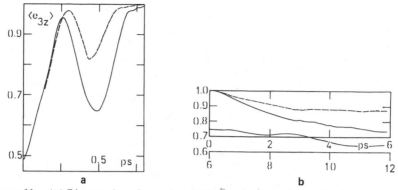

Figure 11. (a) Rise transients for sec-butyl chloride, $\langle e_{3z} \rangle$. (——) racemic mixture; (---) R enantiomer. (b) Fall transients: Note the chiral discrimination working through into the laboratory frame as a difference in the final levels attained by each transient. Ordinate: $\langle e_{3z} \rangle$; abscissa: time, ps.

as E becomes very large, results in the oscillating behavior of $\cos \theta$ as a trivial consequence of the fact that when $\omega_1 = (\mu E/I)^{1/2} \gg \gamma$ this is an inertial regime ($\cos \theta$ becomes non-Markovian as a consequence of the coupling $\mu E \sin \theta$). Finally, in Fig. 12 we illustrate the presence of rotation-translation coupling through the propeller effect. As the name implies, this is the translation of molecules induced by an applied torque. Baranova et al.[56] have suggested that this phenomenon could be used to separate R and S enantiomers physically from a racemic mixture by applying an alternating, circularly polarized, field. The simulations of Fig. 12 seem to imply that there is a very small net translational drift in the sample. The simple center-of-mass velocity components and $\langle v_x \rangle$, $\langle v_y \rangle$, and $\langle v_z \rangle$ gradually drift away from their initial zero values. The behavior of $\langle v_z \rangle$ in this respect for the racemic mixture and R enantiomer is different (Fig. 12a)—the components drift off in opposite directions. The equivalent drift in $\langle v_x \rangle$ is faster for the R enantiomer (Fig. 12b). The drift in $\langle v_y \rangle$ (Fig. 12c) is similar for the R enantiomer and the racemic mixture. Note that these results appear in the laboratory frame, so that a strong enough uniaxial electric field such as the one used in this paper seems to cause a *very small* amount of bulk movement of both the R enantiomer and racemic mixture, *but movement in different directions*. This is because the torque components applied externally in the simulation cause translation in different directions for the R and S enantiomers. The net translation in the racemic mixture is the resultant of these two vectors. Similar results are observable by computer simulation with circularly polarized fields. In conclusion, the application of an external torque seems to emphasize the role of rotation/translation and chiral discrimination in the

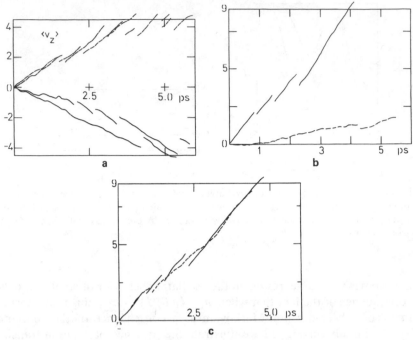

Figure 12.   Velocity drift with applied field. (a) $\langle v_z \rangle$ (in program units; two runs); (---) R enantiomer; (——) racemic mixture. (b) $\langle v_x \rangle$; (——) R enantiomer; (---) racemic mixture. (c) $\langle v_y \rangle$; (——) R enantiomer; (---) racemic mixture. Abscissa time, ps.

condensed state dynamics of optically active molecules. *The results reported in this section are preliminary in nature* and should be supplemented by computations using better statistics for the calculation of equilibrium correlation functions and rise and fall transients.

A.   Rise and Fall Transients in Fluorochloroacetonitrile

The real enantiomers of the fluorochloroacetonitrile molecule have been synthesized recently and are available for spectral investigation of their molecular dynamics. The first results on the simulation of liquid fluorochloroacetonitrile at 133 K, 1 bar, are reported in this section in the form of rise and fall transients for the enantiomers and racemic mixture. The site-site parameters for this simulation are given in Table II. The site symmetry in fluorochloroacetonitrile is such that the net dipole moment reverses in direction from R to S enantiomer. In the R and S enantiomers the transients are mirror images, and transient averages of the type[14-18] $\langle e_{2z}^{2n+1} \rangle$ vanish for all n in the racemic mixture. The effect of the field is therefore to produce a large *laboratory-frame* difference between the R and S enantiomers and mixtures.

TABLE II

Site-Site Potential for Fluorochloroacetonitrile: Lennard-Jones and Partial Charge Terms

Atom	$x(e_3)$,[a] Å	$y(e_1)$,[a] Å	(R) $z(e_2)$,[a] Å	(S) $z(e_2)$, Å	ϵ/k, K	σ, Å	$q/\lvert e\rvert$
N	0.75	-2.28	0.35	-0.35	47.8	3.0	-0.16
C_1	0.27	-1.29	-0.03	0.03	35.8	3.4	-0.02
C_2	-0.34	-0.05	-0.51	0.51	35.8	3.4	0.03
H	-0.34	-0.38	-1.59	1.59	10.0	2.8	0.51
Cl	0.61	1.33	0.07	-0.07	127.9	3.6	-0.16
F	-1.64	0.06	0.02	-0.02	54.9	2.7	-0.20

[a]Atomic coordinates w.r.t center of mass, principal molecular moment-of-inertia frame.

Therefore, the external electric field can be used to identify the differences between the molecular dynamics of enantiomers and their mixtures. The rise transients from the simulation show a strong field dependence, becoming much more rapid as the field strength increases.[2] The R and S enantiomers mirror each other in their field dependence for a averages of the form $\langle e_{2z}^{2n+1}\rangle$ and are identical for even averages of the type $\langle e_{2z}^{2n}\rangle$. *The fall transients decay much more rapidly than the equivalent autocorrelation functions at equilibrium.* The effect is pronounced for saturating field strengths (i.e., $\langle e_{2z}\rangle \doteq \pm 1.0$). As the external field strength decreases, the fall transients approach the equilibrium autocorrelation function. The deexcitation effect is clearly, therefore, a function of the external field first applied to the sample and then removed, that is, the fall transient behavior from the computer simulation depends on the history of the field application.

1. Rototranslational Correlations in the Laboratory Frame of Reference

To end this section and the review, we mention briefly the first results from the simulation on *laboratory-frame cross-correlation* of the type $\langle \mathbf{v}(t)\mathbf{J}^T(0)\rangle$. Here $\mathbf{v}$ is the molecular center-of-mass linear velocity and $\mathbf{J}$ is the molecular angular momentum in the usual laboratory frame of reference. For chiral molecules the center-of-mass linear velocity $\mathbf{v}$ seems to be correlated *directly* in the laboratory frame with the molecule's own angular momentum $\mathbf{J}$ at different points t in the time evolution of the molecular ensemble. This is true in both the presence and absence of an external electric field. *These results illustrate the first direct observation of elements of* $\langle \mathbf{v}(t)\mathbf{J}^T(0)\rangle$ *in the laboratory frame of reference.* The racemic modification of physical and molecular dynamical properties depends, therefore, on the theorem $\langle \mathbf{v}(t)\mathbf{J}^T(0)\rangle \neq 0$ in both static and moving frames of reference. An external electric field enhances considerably the magnitude of the cross-correlations.

222   M. W. EVANS

Acknowledgments

The SERC and University of Wales are thanked for generous financial support.

References

1. H. Benoit, *Ann. Phys.*, **6**, 561 (1951).
2. A. Morita and H. Watanabe, *J. Chem. Phys.*, **70**, 4708 (1979).
3. W. T. Coffey, C. Rybarsch, and W. Schroer, *Chem. Phys. Lett.*, **99**, 31 (1983).
4. W. T. Coffey and B. V. Paranjape, *Proc. Roy. Irish Acad.*, **78A**, 17 (1978).
5. M. W. Evans, G. J. Evans, W. T. Coffey and P. Grigolini, *Molecular Dynamics*, Wiley-Interscience, New York, 1982, Chapters 9 and 10.
6. W. T. Coffey, M. W. Evans, and P. Grigolini, *Molecular Diffusion and Spectra*, Wiley-Interscience, New York, 1984, Chapters 7–9.
7. P. Grigolini, *Nuovo Cimento*, **63B**, 174 (1981).
8. D. Fincham and D. Heyes, (Vol. 63 of this series).
9. P. Grigolini, *Mol. Phys.*, **31**, 1717 (1976).
10. P. Grigolini and A. Lami, *Chem. Phys.*, **30**, 61 (1978).
11. P. Grigolini, *Chem. Phys. Lett.*, **47**, 483 (1977).
12. P. Grigolini, *Chem. Phys.*, **38**, 389 (1979).
13. See ref. 5, Chapter 9.
14. M. W. Evans, *J. Chem. Phys.*, **76**, 5473 (1982).
15. M. W. Evans, *J. Chem. Phys.*, **76**, 5480 (1982).
16. M. W. Evans, *J. Chem. Phys.*, **77**, 4632 (1982).
17. M. W. Evans, *J. Chem. Phys.*, **78**, 925 (1983).
18. M. W. Evans, *J. Chem. Phys.*, **78**, 5403 (1983).
19. B. Bagchi and D. W. Oxtoby, *J. Phys. Chem.*, **86**, 2197 (1982).
20. B. Bagchi and D. W. Oxtoby, *J. Chem. Phys.*, **77**, 1391 (1982).
21. M. W. Evans, P. Grigolini, and F. Marchesoni, *Chem. Phys. Lett.*, **95**, 544 (1983).
22. M. W. Evans, P. Grigolini, and F. Marchesoni, *Chem. Phys. Lett.*, **95**, 548 (1983).
23. E. Praestgaard and N. G. van Kampen, *Mol. Phys.*, **43**, 33 (1981).
24. M. W. Evans, *Phys. Rev. Lett.*, **50**, 371 (1983).
25. M. Ferrario, M. W. Evans, and W. T. Coffey, *Adv. Mol. Rel. Int. Proc.*, **23**, 143 (1982).
26. C. J. Reid, *Mol. Phys.*, **49**, 331 (1983).
27. C. Brot, G. Bossis, and P. Hesse-Bezot, *Mol. Phys.*, **40**, 1053 (1980).
28. L. Onsager, *J. Amer. Chem. Soc.*, **58**, 1486 (1936).
29. See ref. 6, appendix to Chapter 5.
30. M. Andretta et al., (Vol. 63 of this series).
31. C. J. Reid, Ph.D. thesis, University of Wales, presented in brief as Chapter 4 of ref. 5.
32. J. T. Lewis, J. R. McConnell, and B. K. P. Scaife, *Proc. Roy. Irish Acad.*, **76A**, 43 (1976).
33. J. H. Calderwood and W. T. Coffey, *Proc. Roy. Soc.*, *A*, **356**, 269 (1977).
34. M. W. Evans, *Phys. Rev. A*, **30(4)**, 2062 (1984).
35. M. W. Evans, *Chem. Phys. Lett.*, **50**, 371 (1983).

36. W. T. Coffey, *Mol. Phys.*, **38**, 437 (1979).

37. S. Kielich, in *Dielectric and Related Molecular Processes*, Vol. 1, M. Davies, ed., The Chemical Society, London, 1972.

38. B. J. Berne and R. Pecora, *Dynamic Light Scattering with Reference to Physics, Chemistry and Biology*, Wiley-Interscience, New York, 1976.

39. J. P. Ryckaert, A. Bellemans, and G. Ciccotti, *Mol. Phys.*, **44**, 979 (1981).

40. R. J Abbott and D. W. Oxtoby, *J. Chem. Phys.*, **72**, 3972 (1980).

41. R. Kubo, *Adv. Chem. Phys.*, **15**, 101 (1969).

42. P. Grigolini, *J. Chem. Phys.*, **74**, 1517 (1981).

43. M. Ferrario and P. Grigolini, *J. Chem. Phys.*, **74**, 235 (1981).

44. R. Zwanzig, *Lecture Notes in Physics*, **132**, 198 (1980).

45. P. Grigolini, *J. Stat. Phys.*, **27**, 283 (1982).

46. B. J. Berne and G. D. Harp, *Adv. Chem. Phys.*, **17**, 143 (1968).

47. U. Balucani, R. Vallauri, V. Tognetti, P. Grigolini, and P. Marin, *Z. Phys.*, **B49**, 181 (1982).

48. S. Faetti, P. Grigolini, and F. Marchesoni, *Z. Phys.*, **B47**, 353 (1982).

49. A. Balucani, V. Tognetti, R. Vallauri, P. Grigolini, and M. P. Lombardo, *Phys. Lett.*, **86A**, 426 (1981).

50. S. F. Mason, *Molecular Optical Activity and Chiral Discriminations*, C.U.P., Cambridge, (1982).

51. J. Jacques, A. Collet, and S. H. Wilen, *Enantiomers, Racemates and Resolutions*, Wiley, New York, 1981.

52. J. Timmermans and F. Martin, *J. Chim. Phys.*, **25**, 411 (1928).

53. G. J. Evans, *J. Chem. Soc., Faraday Trans II*, in press.

54. V. Rossitter, *J. Phys.* **D5**, 1969 (1972).

55. M. W. Evans, *J. Chem. Soc., Faraday Trans II*, **79**, 719 (1983).

56. N. B. Baranova and B. Y. Zeldovich, *Chem. Phys. Lett.*, **57**, 435 (1978).

VI

NONLINEAR EFFECTS IN MOLECULAR DYNAMICS OF THE LIQUID STATE

M. FERRARIO, P. GRIGOLINI, A. TANI, R. VALLAURI, and
B. ZAMBON

CONTENTS

I. INTRODUCTION

The main aim of this paper is to discuss the role played by nonlinearity in molecular dynamics in the liquid state. Earlier theoretical investigations on this subject (reported by M. W. Evans in Chapter V) are complicated by the fact that they have to interpret computer simulations of three-dimensional systems. As nonlinearity by itself involves technical and conceptual difficulties, it would be especially advantageous to focus on many-particle models which are capable of exhibiting the characteristic dynamical properties of the liquid state without complications due to the multidimensional character of the system.

Such systems do exist. However, so far they have received only very limited attention, preventing us from making a complete and exhaustive investigation of whether or not the reduced model theory (RMT) can account reliably for their main properties.

We shall limit ourselves to arguing that the results of the investigations so far indicate substantial agreement between theory and experiment, at least at a qualitative level. The various sections of this paper should be regarded as parts of a single demonstration supporting this point of view.

First (Sections II–V), we shall tackle the problem of translation. The simplest way of doing this is to study one-dimensional chains of particles. Bishop et al.[1] have shown via computer simulation that one-dimensional Lennard-Jones systems exhibit the same dynamic properties as real three-dimensional liquids. This makes our investigations less academic than they seem at a purely intuitive level, as physical intuition would refuse to take as a "liquid" sample a chain of particles which cannot bypass each other.

Section II is devoted to reviewing the basic ideas of the RMT, having in mind a linear chain of particles coupled with each other via linear interactions. Section III is devoted to illustrating the results of computer experiments that we did to supplement those of the interesting paper of Bishop et al.[1] with additional information as to whether or not the non-Gaussian features of the velocity variable are the same as those of the real three-dimensional fluids. We shall show that these are qualitatively similar, although in the one-dimensional case the non-Gaussian character is more intense and much more persistent.

Section IV is devoted to modeling the mesoscopic level of a nonlinear one-dimensional system. This term means a regime intermediate between the short-time one where the particles collide with each other via their Lennard-Jones interactions, the details of this collisional regime being beyond our capability of description, and the "macroscopic" level of description guaranteed by computer simulations when exploring the time decay of the velocity correlation functions. In Section V we study the long-time behavior of the nonlinear version of the itinerant oscillator. The particular form chosen for the "virtual" potential is proven to be responsible for whether the decay after excitation is faster or slower than the corresponding equilibrium correlation function.

In the second place, we shall study rotational dynamics. Rotational processes are of fundamental importance for dielectric relaxation. To shed light on some controversial issues in dielectric relaxation, Brot and co-workers[2] did a computer simulation of a system of disks interacting via both Lennard-Jones potentials and electric dipole-dipole couplings. This is precisely the kind of simplified microscopic system we need to firmly establish a nonlinear theory for rotation.

Section VI is devoted to describing the details of this "experiment," which will be widely used to monitor equilibrium and nonequilibrium properties. These will prove to be in excellent agreement with the predictions of the nonlinear itinerant oscillator, thereby providing a convincing account for the effects discussed by Evans in Chapter V (which are recovered in this two-dimensional case). Section VII is devoted to a critical discussion of the RMT in the light of the "experimental results" reported here.

II. LINEAR ONE-DIMENSIONAL SYSTEMS AND THE RMT

Let us consider the simple one-dimensional system, a scheme of which is given in Fig. 1. This system is described by the following Newton equations:

$$\dot{R}_i = v_i - v_{i-1}$$

$$m_i \dot{v}_i = k(R_{i+1} - R_i) \qquad m_i = \begin{cases} m & \text{for } i \neq b \\ M & \text{for } i = b \end{cases} \tag{2.1}$$

where R_i is the displacement of the particle i from particle $i-1$, k is the

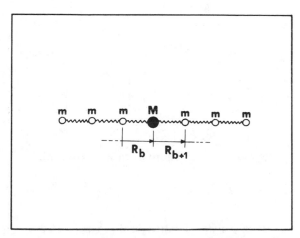

Figure 1. A schematic picture of the one-dimensional system under study in the first part of this chapter. The particle of interest corresponds to an index $i = b$ and has mass M, while the particles of the thermal bath have mass m. We use as space coordinates the displacement of particle i from particle $i-1$. The particles of the system interact via a nearest-neighbor interaction. We shall consider both the case where this interaction is linear (Section II) and where it is of the Lennard-Jones type (Section IV).

force constant, and m is the mass of all the particles in the chain except the particle of interest, which has mass equal to M. This linear chain is reminiscent of that studied in Chapter II (see Section II of that chapter). Here, however, we have changed our notation for the displacements in order to avoid confusion with the parameters of the Mori expansion that will be used in this section. Also, the "Brownian" particle, labeled with b, is at a generic position of an infinite chain of particles rather than being the extreme particle of a semiinfinite one. The corresponding Liouville equation then reads

$$\frac{\partial}{\partial t}\rho(v_i, R_i, t) = L\rho(v_i, R_i, t) \tag{2.2}$$

the dynamical operator being expressed by

$$L = -\sum_{i=-\infty}^{+\infty}\left\{(v_i - v_{i-1})\frac{\partial}{\partial R_i} - \frac{kR_i}{m_i}\left(\frac{\partial}{\partial v_i} - \frac{\partial}{\partial v_{i-1}}\right)\right\}$$

$$m_i = \begin{cases} M & \text{for } i = b \\ m & \text{for } i \neq b \end{cases} \tag{2.3}$$

It is straightforward to check that the equilibrium solution to Eq. (2.2) is given by the following canonical expression:

$$\rho_{eq}(v_i, R_i) \propto \exp\left[-\frac{v_b^2}{2\langle v_b^2\rangle_{eq}}\right]$$

$$\times \prod_{\substack{i=-\infty \\ i \neq b}}^{+\infty}\exp\left[-\frac{v_i^2}{2\langle v_i^2\rangle_{eq}}\right]\prod_{i=-\infty}^{+\infty}\exp\left[-\frac{(R_i - \overline{R})^2}{2\langle(R_i - \overline{R})^2\rangle_{eq}}\right]$$

$$\tag{2.4}$$

satisfying the celebrated energy equipartition relationship

$$k\langle(R_i - \overline{R})^2\rangle_{eq} = m\langle v_j^2\rangle_{eq} = M\langle v_b^2\rangle_{eq} \qquad j \neq b \tag{2.5}$$

where $\overline{R}$ is the average distance between the particles in the chain.

We are interested in evaluating the velocity autocorrelation function

$$\Phi_b(t) = \frac{\langle v_b(0)v_b(t)\rangle_{eq}}{\langle v_b^2(0)\rangle_{eq}} \tag{2.6}$$

where $\langle \dots \rangle_{eq}$ means averaging on the equilibrium distribution of Eq. (2.4). We can undertake this evaluation by applying the Mori-like theory of Chapter I (which in the present case coincides with the standard Mori theory). We then have

$$f_0 \equiv v_b \tag{2.7}$$

To generate the vector f_n from f_{n-1}, we have to take the part of the derivative of f_{n-1} orthogonal to the vector space spanned by the vectors f_{n-1}, $f_{n-2}, \dots, f_0$, the scalar product being defined by

$$\langle A|B \rangle \equiv \langle A|B \rangle_{eq} \equiv \int A(v_i; R_i) B(v_i; R_i) \rho_{eq}(v_i; R_i) \prod_{i=-\infty}^{\infty} dv_i \, dR_i \tag{2.8}$$

When applying this definition we get

$$f_{2n} = \frac{k}{M} \left(\frac{k}{m} \right)^{n-1} (v_{b+n} + v_{b-n}) \qquad n \neq 0 \tag{2.9}$$

$$f_{2n+1} = \frac{k}{M} \left(\frac{k}{m} \right)^{n} (R_{b+n+1} - R_{b-n}) \tag{2.10}$$

The expansion parameters of the Mori continued fraction are therefore given by

$$\Delta_1^2 = \frac{2k}{M}$$

$$\Delta_2^2 = \Delta_3^2 = \cdots = \Delta_i^2 = \cdots = \frac{k}{m} \tag{2.11}$$

This result implies that the energy equipartition relationship of Eq. (2.5) applies as well as the general definitions of Chapter I. Note that for $M \gg m$ the variable v_b turns out to be coupled weakly to the thermal bath. This condition generates that time-scale separation which is indispensable for recovering an exponential time decay. To recover the standard Brownian motion we have therefore to assume that the "Brownian" particle be given a macroscopic size. In the linear case, when $M = m$ we have no chance of recovering the properties of the standard Brownian motion. In the next two sections we shall show that microscopic nonlinearity, on the contrary, may allow that the Markov characters of the standard Brownian motion be recovered with increasing temperature.

As a matter of fact, from Eq. (2.11) the Laplace transform of the correlation function $\Phi_b(t)$, $\hat{\Phi}_b(z)$, turns out to be

$$\hat{\Phi}_b(z) = \frac{1}{z + 2\hat{\varphi}(z)} \tag{2.12}$$

where

$$\hat{\varphi}(z) = \frac{k}{\left.\frac{M}{z}\right|} + \frac{k}{\left.\frac{m}{z}\right|} + \frac{k}{\left.\frac{m}{z}\right|} + \frac{k}{\left.\frac{m}{z}\right|} + \cdots \tag{2.13}$$

When $M = m$ (see also Grosso and Pastori Parravicini, Chapter IV),

$$\hat{\varphi}(z) = \frac{k/m}{z + \hat{\varphi}(z)} \tag{2.14}$$

This leads, after easy algebra, to

$$\hat{\varphi}(z) = -\frac{z}{2} + \frac{1}{2}\left(z^2 + 4\frac{k}{m}\right)^{1/2} \tag{2.15}$$

which results in

$$\hat{\varphi}(z) = \frac{1}{\left(z^2 + 4k/m\right)^{1/2}} \tag{2.16}$$

It is well known[3] that this is the Laplace transform of

$$\Phi_b(t) = J_0(\omega_0 t) \tag{2.17}$$

where

$$\omega_0 = 2\left(\frac{k}{m}\right)^{1/2} \tag{2.18}$$

This is a special case of the more general result obtained by Cukier et al.,[4] which reads

$$\Phi_b(t) = J_0(\omega_0 t) + \left[\frac{2(1-\mu)}{1-2\mu}\right] \sum_{l=1}^{\infty} (1-2\mu)^l J_{2l}(\omega_0 t) \tag{2.19}$$

where

$$\mu = \frac{m}{M} \tag{2.20}$$

and $J_{2l}(\omega_0 t)$ denotes a Bessel function of the $(2l)$th order (J_0 is the corresponding zeroth-order Bessel function). As remarked above, for $\mu \to 0$ this tends precisely to the exponential result[4]

$$\Phi_b(t) = \exp\{-\mu\omega_0 t\} \tag{2.21}$$

which is the case of best convergence of the Mori chain.

Let us now relate the discussion of the linear version of the RMT to the case $\mu = 1$, which is expected to be the case of worst convergence.

When using the Mori basis set to expand the Liouvillian operator of Eq. (2.2) we get (see Chapter I and ref. 5)

$$\frac{d}{dt}\mathbf{A} = \begin{vmatrix} 0 & 1 & 0 & 0 & 0 & \vdots \\ -2\Delta^2 & 0 & 1 & 0 & 0 & \vdots \\ 0 & -\Delta^2 & 0 & 1 & 0 & \vdots \\ 0 & 0 & -\Delta^2 & 0 & 1 & \vdots \\ \cdots & \cdots & \cdots & \cdots & \cdots & \cdots \end{vmatrix} \mathbf{A} \tag{2.22}$$

where

$$\mathbf{A} = \begin{vmatrix} f_0 \\ f_1 \\ \vdots \\ f_i \\ \vdots \end{vmatrix} \qquad \Delta^2 = \frac{k}{m} \tag{2.23}$$

and v_b has to be identified with f_0 of Eq. (2.7). As is well known,[5] the third-order truncation is equivalent to the model of the itinerant oscillator. This example of a linear reduced model is described by

$$\begin{aligned} \dot{x} &= v \\ \dot{v} &= -2\Delta^2(x-y) \\ \dot{y} &= w \\ \dot{w} &= 2\Delta^2(x-y) - \gamma_2 w + f_2(t) \end{aligned} \tag{2.24}$$

where $f_2(t)$ is assumed to be a white Gaussian noise defined by

$$\langle f_2(0)f_2(t)\rangle = 2D\delta(t) \qquad D = \gamma_2\langle v^2\rangle_{eq} \tag{2.25}$$

In Fig. 2 we compare the itinerant oscillator with the exact autocorrelation function $J_0(\omega_0 t)$.

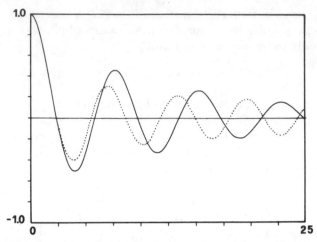

Figure 2.   The time behavior of the correlation function $\Phi_b(t)$, Eq. (2.6), with $\mu = 1$, $\omega_0 = 1$, $\Delta = 0.5$. The dotted curve denotes the result provided by the linear itinerant oscillator (three states of the Mori chain). The solid curve denotes the exact result provided by the function $J_0(\omega_0 t)$, Eq. (2.17). The expansion coefficients of the itinerant oscillator are rigorously derived from the chain of Eq. (2.22). The friction γ_2 is determined so as to get the exact value of the diffusion coefficient. This is defined by $D = \int_0^\infty \langle v(0)v(t) \rangle \, dt$ and is equal to $1/2\Delta$ for the exact solution $J_0(\omega_0 t)$. In the case of the itinerant oscillator described by Eq. (2.24), it becomes $D = \gamma_2 / 2\Delta^2$. This leads to $\gamma_2 = \Delta$.

When aiming at a more satisfactory simulation of the exact results, we need to include more steps of the Mori continued fraction. This can be interpreted as a reduced model with more than one virtual particle. We have found that an accurate simulation of the exact result (i.e., a curve completely overlapping that of Fig. 2) in the linear case requires as many as 10 virtual particles.

This suggests that the time behavior of the velocity correlation functions in the liquid state can exhibit only modest agreement with the results of computer molecular dynamics if based on the linear itinerant oscillator. Better agreement would be found when dealing with the case of almost exponential correlation functions. This regime, however, for $\mu = 1$, cannot be explored by using microscopic models such as that of Fig. 1. In fact, in the case of $\mu = 1$ the properties of the system do not change with varying temperature. The more realistic case of 1-d Lennard-Jones chains, on the contrary, exhibits a transition to an almost exponential regime with increasing temperature. This depends on the intimately nonlinear nature of the Lennard-Jones systems, which should therefore be taken into account. The standard Mori approach, however, is fraught with conceptual difficulty when applied to a nonlinear microscopic system.

The linear version of the RMT belongs to the same family of modeling methods as Adelman's approach to simulating the medium where a chemical reaction takes place with a few degrees of freedom.[6]

Computer simulations (see Section III) reveal, however, that in the liquid state the variable velocity is neither Markovian nor Gaussian, thereby making it necessary to discard modeling approaches that are linear in nature.

An important feature of this paper is the development of an entirely new, *nonlinear*, approach to liquid-state modeling. In a sense this is an extension of the preceding linear modeling and reduces to it in the appropriate limits.

Before facing this problem, we study to what extent the nonlinear nature of the microscopic model affects the decay of the velocity autocorrelation functions. This will be the subject of Section III. Section III will also provide a detailed account of the computer simulation technique on which we shall base our arguments concerning translational dynamics.

III. COMPUTER SIMULATION OF THE ONE-DIMENSIONAL LENNARD-JONES SYSTEM

A. The System

Our system consists of $N = 500$ particles bounded to move over a linear chain of length L on which periodic boundary conditions are imposed. The particles interact, only with their first neighbor, through a Lennard-Jones (L-J) potential given by

$$\varphi(r) = 4\left\{ \left(\frac{1}{r}\right)^{12} - \left(\frac{1}{r}\right)^{6} \right\} \qquad \text{for } r \leq r_c$$

$$\varphi(r) = \varphi(r_c) \qquad \text{for } r > r_c; \ r_c = 3.5 \qquad (3.1)$$

where r is the molecular separation. This potential is expressed in the usual reduced units for an L-J system. In these units the length is measured in units of σ and the time in units of $\sigma(m/\varepsilon)^{1/2}$, where ε and σ are the usual L-J parameters and m is the atomic mass.

The equations of motion for our system are given by

$$\ddot{r}_i = f_i - f_{i-1} = F_i \qquad (3.2)$$

where f_i is the total force acting on particle i derived from Eq. (3.1) of the L-J potential given above, and r_i is the distance of the ith particle in the chain from the $(i-1)$th. The boundary conditions are expressed through the fact that r_1 is the distance of particle 1 from particle N.

Using relative distances of the particles instead of absolute coordinates is a convenient choice, especially for long chains. In this case the increment of the position is added to a much smaller quantity than the absolute position of the particle, thereby reducing the round-off errors.

With regard to the thermodynamic properties of the system, we measure the density as the inverse of the average distance between the particles and the temperature as their average kinetic energy.

The initial configuration of the system is chosen such that all the particles are equally spaced and the velocities are extracted from a Gaussian distribution relative to the temperature of interest.

B. The Algorithm

The algorithm chosen to integrate the Newton equations of motion is a slight variation of the popular Verlet algorithm.[9] If $r_{i,n+1}$, $r_{i,n}$, and $r_{i,n-1}$ indicate the relative position of the particle i at the time steps $n+1$, n, and $n-1$, respectively, then

$$r_{i,n+1} = 2r_{i,n} - r_{i,n-1} + F_{i,n}h^2 + \left(F_{i,n+1} + F_{i,n-1} - 2F_{i,n} \right)\frac{h^2}{12} + O(h^6)$$

$$(3.3)$$

which is equivalent to taking into account also the second derivative of the total force acting on the particles. Equation (3.3) can be translated into a predictor-corrector-like algorithm as follows. The predicted positions at time $n+1$ are evaluated by using the Verlet algorithm, which permits us to compute $F_{i,n+1}$ and the positions $r_{i,n+1}$ with formula (3.2). The procedure should be repeated until the new value of $r_{i,n+1}$ does not differ appreciably from the previous one. We observe, however, that one iteration is sufficient to achieve the desired accuracy.

The velocity at time step $n+1$ is obtained by using Simpson's integration rule:

$$v_{i,n+1} = v_{i,n-1} + \left(F_{i,n+1} + 4F_{i,n} + F_{i,n-1} \right)\frac{h}{3} + O(h^5) \qquad (3.4)$$

Although it is not really necessary to have the velocity computed with such a high order of accuracy, this helps to give a precise monitoring of the total energy of the system step by step.

We wish to point out here that the use of a higher order algorithm turns out to be essential for an accurate integration of the motion in contrast to the two-dimensional case where at the same average kinetic energy and "number of collisions" the Verlet algorithm is sufficiently correct. In our opinion, this is due to the fact that in the one-dimensional case the collisions are always frontal, whereas in the two-dimensional case this kind of collision is an exception. Therefore, in the two-dimensional case, a centrifugal potential must be added to the L-J potential, making the particles approach the inversion point at a smaller velocity than in the corresponding one-dimensional case, which reduces the computational error involved. A comparison performed in our system shows that in the case of density 0.5 and temperature 0.67 the energy was conserved within 0.2×10^{-3} by using Verlet's algorithm, whereas the accuracy in the energy obtained with the improved method was 1.25×10^{-5}. We also noted that in order to obtain the same accuracy with Verlet's algorithm, the time step must be decreased by a factor of 5.

C.   Check of the Equilibrium Properties

The thermodynamic equilibrium of the system was controlled by examining the molecular velocity distribution (MVD) and the first-neighbor distribution. In the case where only a first-neighbor interaction is present, the latter, as will be shown in Section IV, has an analytical expression given by

$$d(r) = \frac{1}{n}\exp\{-\beta[\varphi(r)+p\cdot r]\}$$
$$\beta = \langle v^2 \rangle_{eq}^{-1}$$

(3.5)

where n is the normalization factor and p is a constant external force which

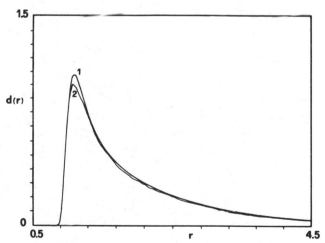

Figure 3.   First-neighbor distribution function $d(r)$ [(1) theoretical; (2) "experimental"], corresponding to a simulation run of 10,000 steps ($h = 5 \times 10^{-3}$, $\langle v^2 \rangle_{eq} = 2.25$, $\rho = 0.5$, $p = 1.95$). The pressure is determined by using the state equation corresponding to the density of the system.

is to simulate pressure. In Fig. 3 we show the distribution $d(r)$, as computed from the simulation, together with the theoretical curve corresponding to the temperature and density of our system.

It is worth noting here that this distribution allows large fluctuations of the density, as is also revealed from a view of the trajectories followed by the particles. These are shown in Fig. 4 for a run of 8000 steps. In addition to irregular patterns, we note that some of the trajectories are strongly correlated: there is evidence that strong collisions occur for which two clusters of particles move far apart, giving rise to high density fluctuations.

The molecular velocity distribution was monitored by averaging over the total number of time steps and all the particles. It was found to reproduce the Maxwell distribution relative to the average temperature of the system (the value of $\langle v^2 \rangle^2 / \langle v^4 \rangle - \frac{1}{3}$ was about 0.01). We wish to point out, however, that the thermalization in the one-dimensional case offered some difficulties, in contrast with the two-dimensional case, where it requires usually only a few hundred time steps. This occurred principally at low densities where the collisions between the particles are mainly binary collisions. In this case, with our starting conditions, the distribution was found to be

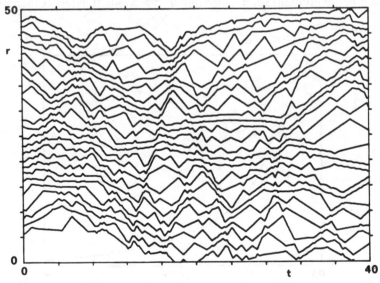

Figure 4.   Trajectories followed for 24 particles of the chain for a time of 40 reduced units (8000 time steps with $h = 0.005$). The abscissa represents the time and the ordinate the position of the particles in the chain. The density and temperature for this run were, respectively, $\rho = 0.5$ and $\langle v^2 \rangle_{\text{eq}} = 2.25$.

different from a Maxwellian even after very long runs, for example, 20,000 steps. In this case we adopted a thermalization procedure by starting from a higher density configuration, which can be brought more easily to equilibrium, and successively changed its length adiabatically up to the desired length, in a manner which is reminiscent of the temperature-scaling procedure widely used in molecular dynamics. Using this method, the equilibrium configuration was achieved even at very low densities.

D.   Standard Observations in Molecular Dynamics

In Fig. 5 we show the function $g(r)$, which represents the local density of particles around a given particle normalized in such a way that $\lim_{r \to \infty} g(r) = 1$. We note here that in the case of nearest-neighbor interaction this function could be calculated starting from $d(r)$ because of the independence of relative positions of adjacent particles.

The other function generally reported in molecular dynamics (MD) is the normalized velocity autocorrelation function (NVAF). We show it in Fig. 6 for $\rho = 0.5$ and $T = 2.25$. We computed it by averaging over all the particles and equally spaced time origins of 10–30 time steps, depending on the trade-off between accuracy and computer time, according to the formula

$$\psi(t) = \frac{\displaystyle\sum_{m=1, m_s}^{M} \sum_{i=1}^{N} v_i(t+m)v_i(m)}{\displaystyle\sum_{m=1, m_s}^{M} \sum_{i=1}^{N} v_i^2(m)} \qquad (3.6)$$

where m_s is the spacing between the time origins. We evaluated also the non-Gaussian properties

$$\eta(t) = \frac{\left\langle v^3 v(t) \right\rangle_{eq}}{\left\langle v^4 \right\rangle_{eq}} - \frac{\left\langle vv(t) \right\rangle_{eq}}{\left\langle v^2 \right\rangle_{eq}} \qquad (3.7)$$

$$\varepsilon(t) = \frac{\left\langle v^2 v^2(t) \right\rangle_{eq}}{\left\langle v^4 \right\rangle_{eq}} - \frac{2}{3}\left[\left(\frac{\left\langle vv(t) \right\rangle_{eq}}{\left\langle v^2 \right\rangle_{eq}} \right)^2 + 1 \right] \qquad (3.8)$$

The special relevance of Eqs. (3.7) and (3.8) to the problem of excitation will be discussed in Section V. As to the latter, it is evident that the Gaussian approximation underestimates $\langle v^2 v^2(t)\rangle/\langle v^4 \rangle$. Similar behavior for the three-dimensional case has been interpreted in terms of a reduced model whose virtual potential is softer than a harmonic one. To gain computa-

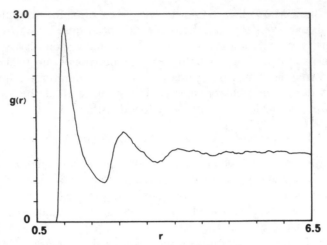

Figure 5. The function $g(r)$ calculated on a simulation of 15,000 steps with $h = 0.005$, $\rho = 0.5$, $\langle v^2 \rangle_{eq} = 0.67$.

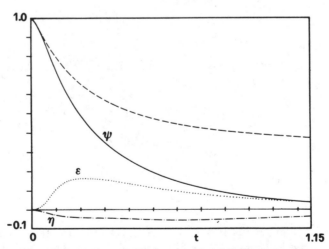

Figure 6. The velocity autocorrelation function (—), kinetic energy autocorrelation function (---), $\varepsilon(t)$ ($\cdots$), and $\eta(t)$ (—·—) computed from a run of 8000 steps ($h = 0.005$) with a spacing between time origins (m_s) of 10 steps. The density was $\rho = 0.5$ and the temperature $\langle v^2 \rangle_{eq} = 2.25$.

238

tional advantages provided by systems periodic in space, such a virtual potential was assumed there to be sinusoidal.[7,8] While this assumption is certainly sound in the rotational case (see Sections VI and VII), it does not make sense in the translational case, where it appears unrealistic that the "virtual" particle can depart without limits from the real one. Furthermore, this investigation was fraught with the usual inconvenience stemming from the comparison between a theoretical one-dimensional model and a three-dimensional real system. In the next section we will come back to a more rigorous foundation of this model, and we will point out the theoretical difficulties that still haunt this subject.

E.   Results of an Excitation Experiment

In order to investigate the microscopic basis of an RMT for one-dimensional systems, we examined the behavior of excited particles in interaction with a thermal bath. By excited particles we mean a set of particles whose velocity is well above the thermal one, the thermal bath being constituted by the system itself. We examined the correlation function for two sets of particles. Namely, by indicating

$$\frac{\langle vv(t) \rangle}{\langle v^2 \rangle}\bigg|_{\rho} = \frac{\sum\limits_{i,\,m} v_i(m) v_i(t+m)}{\sum\limits_{i,\,m} v_i^2(m)} \tag{3.9}$$

which is the correlation function of particles whose velocity at the time origin is statistically distributed according to a distribution function ρ, we measured this function for ρ_1 and ρ_2 given by

$$\rho_1 = \begin{cases} \exp\left\{-\dfrac{v^2}{2\langle v^2 \rangle_{eq}}\right\} & \text{for } |v| \geq v_{exc} \\[2mm] 0 & \text{for } |v| < v_{exc} \end{cases} \tag{3.10}$$

$$\rho_2 = \exp\left\{-\frac{v^2}{2\langle v^2 \rangle_{exc}}\right\} \quad \text{with } \langle v^2 \rangle_{exc} \gg \langle v^2 \rangle_{eq} \tag{3.11}$$

We note that in both cases the decay is faster than the equilibrium NVAF. We also observe that in the case of ρ_1 there is a monotonic tendency for the NVAF to decay faster as the value of v_{exc} is increased. The NVAF corresponding to ρ_1 is shown in Fig. 7 and that to ρ_2 in Fig. 8. It is worth remarking that the accelerated decay monitored by these two "experiments"

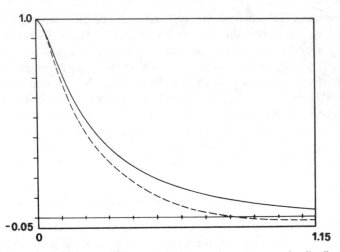

Figure 7. Decay of an excited autocorrelation function relative to the distribution ρ_1 in Section III (---) and the corresponding equilibrium autocorrelation function (——). $\rho = 0.5$, $\langle v^2 \rangle_{eq} = 2.25$, $v_{exc}/\langle v^2 \rangle_{eq}^{1/2} = 3.0$.

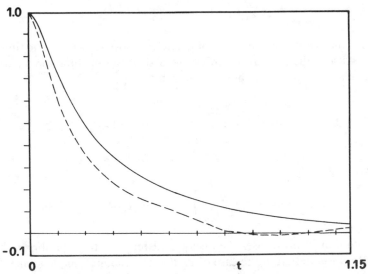

Figure 8. Autocorrelation function relative to the distribution ρ_2 of Section III with $r \equiv \langle v^2 \rangle_{exc}/\langle v^2 \rangle_{eq} - 1 = 3.0$ (---) and its corresponding equilibrium autocorrelation function (——).

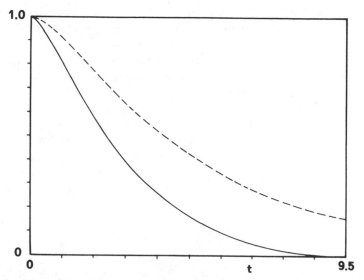

Figure 9.   Decay of the autocorrelation function of R_0 (solid line) and $R_a - \langle R_a \rangle_{eq}$ (dashed line).

(and especially that concerning ρ_2) is in complete agreement with the fact that $\eta(t)$ is negative [see Eq. (5.66b)].

Another piece of information that we wanted to extract from our experiments was connected with the dynamic behavior of spatial variables. If we consider three successive particles in the chain and we denote by R_0 the distance of the middle one from the center of mass of the other two and by R_a the distance between these two, we can compute the normalized autocorrelation function of these two variables. They are shown in Fig. 9; as can be immediately observed, they decay to zero on a time scale which is much greater than that of the velocity variable. Also, the center of mass decays faster than R_a. In the next section we shall argue that this suggests that the "virtual" potential characterizing the itinerant oscillator model has to be assumed to be fluctuating around a mean shape, which, moreover, will be shown to be nonlinear and softer than its harmonic approximation.

IV.   THE EFFECTIVE NONLINEAR POTENTIAL

The chain of equations corresponding to the experimental result of Section III, expressed in reduced units, reads

$$\dot{v}_i = \varphi'(R_{i+1}) - \varphi'(R_i) \qquad \dot{R}_i = v_i - v_{i-1} \qquad (4.1)$$

where R_i is the displacement of particle i from particle $i-1$ and $f(R_i) = -\varphi'(R_i)$ is the L-J force exerted by particle $i-1$ over particle i. As the preceding experiment concerns the case $M = m$, any given particle of the chain can be chosen to represent the Brownian particle. We assume, therefore, that the bth particle of the sample is the Brownian particle. With respect to our chain, we must consider that two external constant forces p are applied to each end of the chain in order to simulate the pressure forces. This is equivalent to setting the periodic boundary conditions in our computer experiment. In the case of an L-J potential, this is needed to prevent the chain from breaking apart under a fluctuation large enough to give one portion of the chain the translational energy sufficient to leave the other part. Following these considerations, the Liouvillian of our system becomes

$$L = -\sum_{i=1}^{N} \left\{ (v_i - v_{i-1}) \frac{\partial}{\partial R_i} + (f(R_i) - p)\left(\frac{\partial}{\partial v_i} - \frac{\partial}{\partial v_{i-1}} \right) \right\} \quad (4.2)$$

It is easy to check that this Liouvillian allows the equilibrium distribution

$$\rho_{eq} = \prod_{i=1}^{N} \exp\left\{ -\beta \frac{v_i^2}{2} \right\} \exp\left\{ -\beta [\varphi(R_i) + pR_i] \right\}$$

$$\beta = \langle v^2 \rangle_{eq}^{-1} \quad (4.3)$$

which permits us to determine the equation of state of a linear chain for any given interparticle potential. In accordance with the choice of the bth particle as the Brownian particle, we can split the Liouvillian as follows:

$$L = L_b + L_{bath}$$

$$L_b = -[f(R_b) - f(R_{b+1})] \frac{\partial}{\partial v_b} - v_b \left(\frac{\partial}{\partial R_b} - \frac{\partial}{\partial R_{b+1}} \right)$$

$$L_{bath} = -v_{b+1} \frac{\partial}{\partial R_{b+1}} + v_{b-1} \frac{\partial}{\partial R_b} + [f(R_b) - p] \frac{\partial}{\partial v_{b-1}}$$

$$-[f(R_{b+1}) - p] \frac{\partial}{\partial v_{b+1}} - \sum_{\substack{i=1 \\ i \neq b \\ i \neq b+1}}^{N} \left\{ (v_i - v_{i-1}) \frac{\partial}{\partial R_i} \right. \quad (4.4)$$

$$\left. + [f(R_i) - p]\left(\frac{\partial}{\partial v_i} - \frac{\partial}{\partial v_{i-1}} \right) \right\}$$

It is convenient to associate with each particle i two coordinates R_{0i} and

R_{ai} defined as

$$R_{0i} = \frac{R_i - R_{i+1}}{2} \qquad R_{ai} = \frac{R_i + R_{i+1}}{2} \tag{4.5}$$

which are nothing but the distance between the two particles neighboring the particle i and the distance of particle i from their center of mass. Expressed in terms of these new variables, and dropping the index b, L_b becomes

$$L = -\left\{ f(R_a + R_0) - f(R_a - R_0) \right\} \frac{\partial}{\partial v} - v \frac{\partial}{\partial R_0} \tag{4.6}$$

We note from this Liouvillian that the motion of the variable v could be determined whenever the motion of R_a and R_0 were exactly known. This is, of course, not possible at this level of description, but nonetheless we can make some reasonable assumptions about this motion. We shall discuss here the simplest assumptions and their consequences on the behavior of the variable of interest v.

We note from Eq. (4.6) that while v and R_0 are mutually influenced, R_a is not driven by v directly but only through the thermal bath. Furthermore we note that, the probability distribution of R_a depends from R_0 according to

$$P_{R_0}(R_a) = \frac{\exp\left\{ -\beta\left[\varphi(R_a + R_0) + \varphi(R_a - R_0) \right] \right\} \exp\left\{ -2\beta p R_a \right\}}{\int_R \exp\left\{ -\beta\left[\varphi(R_a + R_0) + \varphi(R_a - R_0) \right] \right\} \exp\left\{ -2\beta p R_a \right\} dR_a} \tag{4.7}$$

where R denotes the integration region where the weighting function does not vanish. If we further assume that the variable R_a is much faster than R_0, we can perform an average over R_a and introduce an effective potential whose derivative is given by

$$-\varphi'_{\text{eff}}(R_0) = \int_R \left\{ f(R_a + R_0) - f(R_a - R_0) \right\} P_{R_0}(R_a) \, dR_a \tag{4.8}$$

It easy to check that this effective potential can be expressed in the more compact form by

$$\varphi_{\text{eff}}(R_0) = -\frac{1}{\beta} \ln\left[\frac{N(R_0)}{N(0)} \right] \tag{4.9}$$

where

$$N(R_0) = \int_R \exp\{-\beta[\varphi(R_a + R_0) + \varphi(R_a - R_0)]\}\exp\{-2\beta pR_a\}\, dR_a$$

(4.10)

We note here that in the case of a linear microscopic interaction, $P_{R_0}(R_a)$ does not depend on R_0, and this potential simply reduces to a linear one. The effective potential, in this case, turns out to be $\varphi_{\text{eff}} = kR_0^2$, which is exactly the virtual potential of the well-known linear itinerant oscillator (see Section II). In the more general case, the virtual potential is given by Eq. (4.9), and the Liouvillian reads

$$L_{\text{eff}} = \Omega(a_- A_+ - a_+ A_-) - \Gamma A_+ A_-$$

(4.11)

where

$$\Omega = \sqrt{\frac{\langle v^2 \rangle_{\text{eq}}}{\langle R_0^2 \rangle_{\text{eq}}}}$$

(4.12)

and the creation and annihilation operators are given by

$$a_- = \frac{v}{\langle v^2 \rangle_{\text{eq}}^{1/2}} + \langle v^2 \rangle_{\text{eq}}^{1/2} \frac{\partial}{\partial v}$$

$$a_+ = -\langle v^2 \rangle_{\text{eq}}^{1/2} \frac{\partial}{\partial v}$$

$$A_- = \langle R_0^2 \rangle_{\text{eq}}^{1/2} \frac{\partial}{\partial R_0} - \frac{\langle R_0^2 \rangle_{\text{eq}}^{1/2}}{\langle v^2 \rangle_{\text{eq}}} f_{\text{eff}}(R_0)$$

$$A_+ = -\langle R_0^2 \rangle_{\text{eq}}^{1/2} \frac{\partial}{\partial R_0}$$

(4.13)

This is the most general form of the Liouvillian when one aims to describe the process by only two variables and their canonical distribution has to be recovered. The first term of L_{eff} is the Liouvillian given by Eq. (4.6) after performing the average, and the second term is added to satisfy the equilibrium distributions of R_0 and v. The representation of L_{eff} in terms of the creation and annihilation operators will turn out to be useful in Section V, where some properties of the Liouvillian (4.11) will be discussed. Note that this represents a generalization of the standard approximation of the Mori theory in the case that only two variables are relevant, in other words when only two time scales are involved, the first one

$$T = \frac{1}{\Omega} \qquad (4.14)$$

characterizing the coupling strength of the variable of interest with its thermal bath, and the second one

$$\Gamma = \frac{1}{\tau_c} \qquad (4.15)$$

characterizing the time scale of the thermal bath.

Figure 10 clearly shows that the mean potential evaluated with Eq. (4.9) is softer than the harmonic approximation having the same second derivative at the origin $R_0 = 0$.

This result confirms the point of view of Balucani et al.[7,8] on the need of using a virtual potential softer than the harmonic one. On the other hand, this potential restricts the motion of the particle in a region around $R_b = 0$, which contrasts with the idea of a sinusoidal potential used in ref. 7. This

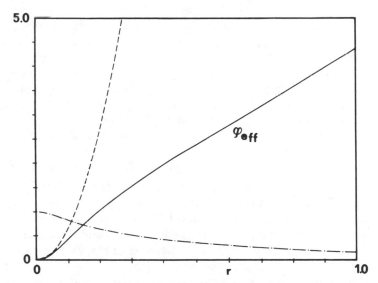

Figure 10.   Effective potential of Eq. (4.9) (——), its harmonic approximation (---), and the equilibrium distribution of R_0 (– · –) corresponding to $\rho = 0.5$, $\langle v^2 \rangle_{eq} = 2.25$, $p = 1.95$.

difficulty can be bypassed simply by shifting our attention from the translational to the rotational case that will be studied in Section VI.

In Section V we shall show that the system described by Eq. (4.11) coincides with a nonlinear extension of the popular model of the intinerant oscillator. Note that we are exploring a mesoscopic regime, implying averaging processes which significantly reduce the nonlinear character of the real microscopic interaction (consider, for instance, how nonlinear the L-J potential is).

As a final remark we wish to note that the result of the one-dimensional computer simulation shown in Fig. 9 suggests that the nonlinear version of the itinerant oscillator model should be improved by giving a fluctuating character to the virtual potential, since R_0 does not turn out to be a very fast variable. Therefore, the potential of Eq. (4.9) should be interpreted as an average potential around which an effective potential would fluctuate.

V.   THE NONLINEAR ITINERANT OSCILLATOR

This section plays a central role in this paper. The major issues under discussion here are:

1. It will be shown that the effective dynamical operator of the preceding section, obtained by averaging on R_a, v and R_0 being regarded the slowest variables of the system, is nothing but a nonlinear version of the celebrated itinerant oscillator (ref. 10; see also Chapter V).
2. The long-time properties of the nonlinear itinerant oscillator are studied.
3. Predictions are made about the results of ideal excitation experiments, preparing the system of tagged particles in an unstable initial distribution. The qualitative behavior of the system after this preparation phase is traced back to the form of the virtual potential, that is, the interaction between real and virtual particles.

Note that although this discussion will be limited to the translational case, the major results can be applied immediately to rotational dynamics. Transforming translation into angular velocities and axis coordinates into angular coordinates leaves unchanged the form of the equations involved. The only caveat which has to be used concerns the boundary conditions that have to be imposed so as to respect the periodic character of the rotational problems (rotation by 2π leaves these systems unchanged).

In the linear case the itinerant oscillator model is well known, and its rigorous foundation has been illustrated in Section II. Its nonlinear version reads

$$\dot{v} = \frac{-\varphi'(x-y)}{M}$$

$$\dot{w} = \frac{\varphi'(x-y)}{m'} - \gamma w + f(t) \tag{5.1}$$

where, as usual, $f(t)$ is regarded as being a Gaussian white noise defined by

$$\langle f(0)f(t)\rangle = 2D\delta(t) \tag{5.2}$$

and we denote by φ the virtual potential of this model. The nonlinear nature of this oscillator depends on the fact that

$$\varphi(R) = \frac{\alpha_1}{2}R^2 + \frac{\alpha_2}{4}R^4 + \frac{\alpha_3}{6}R^6 + \cdots \tag{5.3}$$

where

$$R = y - x \tag{5.4}$$

and in general

$$\varphi(R) \neq \frac{\alpha_1}{2}R^2 \tag{5.5}$$

To get complete equivalence with the effective Liouvillian of Section IV, we make the assumption that w is so fast as to replace Eq. (4.1) with

$$\dot{v} = -\frac{1}{M}\psi(R) \tag{5.6}$$

$$\dot{R} = v - \frac{1}{\gamma m'}\psi(R) + \frac{f(t)}{\gamma} \tag{5.7}$$

where

$$\psi(R) = \varphi'(R) \tag{5.8}$$

By adopting a description in terms of the creation and annihilation operators (see Section IV and Chapter 2)

$$a_- = \frac{v}{\langle v^2\rangle_{eq}^{1/2}} + \langle v^2\rangle_{eq}^{1/2}\frac{\partial}{\partial v} \tag{5.8a}$$

$$a_+ = -\langle v^2\rangle_{eq}^{1/2}\frac{\partial}{\partial v} \tag{5.8b}$$

the Fokker-Planck equation corresponding to the system of Eqs. (5.6) and (5.7) reads

$$\frac{\partial}{\partial t}\rho(v,R,t)=L\rho(v,R,t) \tag{5.9}$$

where

$$L=\Omega(a_-A_+-a_+A_-)-\Gamma A_+A_- \tag{5.10}$$

and

$$\Omega=\sqrt{\frac{\langle v^2\rangle_{eq}}{\langle R^2\rangle_{eq}}} \tag{5.11}$$

$$\Gamma=\frac{\langle v^2\rangle_{eq}}{\langle R^2\rangle_{eq}}\frac{M}{m'}\frac{1}{\gamma} \tag{5.12}$$

$$A_-=\langle R^2\rangle_{eq}^{1/2}\frac{\partial}{\partial R}+\frac{\langle R^2\rangle_{eq}^{1/2}}{M\langle v^2\rangle_{eq}}\psi(R) \tag{5.13}$$

$$A_+=-\langle R^2\rangle_{eq}^{1/2}\frac{\partial}{\partial R} \tag{5.14}$$

The dynamical operator L of Eq. (5.10) is therefore proved to be identical to the reduced one of Section IV provided that $\psi(R)$ be identified with the $\varphi'_{eff}(R)$ as defined via Eq. (4.8). Note that R, of course, has to be identified with R_0 of Section IV. We require also that M be identified with the mass of the tagged particle m. We see that our system is completely defined by the parameters $\langle v^2\rangle_{eq}$, m, Ω, Γ, and the effective potential of Eq. (4.8). $\langle R^2\rangle_{eq}$ can be defined via Eq. (5.11) as being $\langle R^2\rangle_{eq}=\langle v^2\rangle_{eq}/\Omega^2$. A further relevant parameter will be defined by Eq. (5.32a).

Note that whereas a_+ and a_- satisfy the commutation rule

$$[a_-,a_+]=1 \tag{5.15}$$

the A_+ and A_- do not unless $\psi''(R)$ is independent of R. Let us therefore split ψ into a harmonic and an anharmonic part defined by

$$\psi_0=\alpha_1 R \tag{5.16a}$$

and

$$\psi_1=\sum_{i=2}^{\infty}\alpha_i R^{2i-1} \tag{5.16b}$$

This allows us to define the following creation and annihilation operators:

$$b_- \equiv A_- - \frac{1}{\alpha_1} \frac{1}{\langle R^2 \rangle_{eq}^{1/2}} \psi_1(R) \tag{5.17a}$$

$$b_+ \equiv A_+ = -\langle R^2 \rangle_{eq}^{1/2} \frac{\partial}{\partial R} \tag{5.17b}$$

b_- can also be written

$$b_- = \langle R^2 \rangle_{eq}^{1/2} \frac{\partial}{\partial R} + \frac{R}{\langle R^2 \rangle_{eq}^{1/2}} \tag{5.18}$$

To get Eq. (5.18) from Eq. (5.17) we used

$$\alpha_1 \langle R^2 \rangle_{eq} = M \langle v^2 \rangle_{eq} \tag{5.19}$$

which implies that $\langle R^2 \rangle_{eq}$ has to be intended as the equilibrium value which would be attained if only the linear interaction were present. As a consequence, from (5.11) and (5.12) we have

$$\Omega = \sqrt{\frac{\alpha_1}{M}} \tag{5.20a}$$

$$\Gamma = \frac{\alpha_1}{\gamma m'} \tag{5.20b}$$

The operators b_+ and b_- satisfy the standard commutation rule

$$[b_-, b_+] = 1 \tag{5.21}$$

For the AEP of Chapter II to be applied, we need to split $\mathscr{L}$ of Eq. (5.10) into an unperturbed and a perturbation part as follows:

$$\mathscr{L}_0 = -\Gamma A_+ A_- \tag{5.22}$$

$$\mathscr{L}_1 = \Omega(a_- A_+ - a_+ A_-) \tag{5.23}$$

$\mathscr{L}_0$ and $\mathscr{L}_1$, in turn, have to be divided into an unperturbed and a perturbation part as follows:

$$\mathscr{L}_0 = \mathscr{L}_{harm} + \mathscr{L}_{anharm} \tag{5.24}$$

$$\mathscr{L}_{harm} = -\Gamma b_+ b_- \tag{5.25a}$$

$$\mathscr{L}_{anharm} = -\frac{\Gamma b_+ \psi_1(R)}{\alpha_1 \langle R^2 \rangle_{eq}^{1/2}} \tag{5.25b}$$

By applying the AEP of Chapter II up to $\mathscr{L}_1^4$, we arrive at

$$\frac{\partial}{\partial t}\sigma(v,t) = \left\{ \alpha a_+ a_- + \beta a_+ a_-^3 + \eta a_+ a_- a_+ a_- + \delta a_+^2 a_-^2 + \varepsilon a_+^3 a_- \right\}\sigma(v,t)$$

$$(5.26)$$

where

$$\alpha = -\Omega^2 \big[A_-(s_0)A_+(0) \big]^{(0)} \tag{5.27a}$$

$$\beta = -\Omega^4 \big[A_-(s_0)A_+(s_1)QA_+(s_2)A_+(0) \big]^{(0)} \tag{5.27b}$$

$$\eta = \Omega^4 \Big\{ \big[A_-(s_0)A_+(s_1)QA_-(s_2)A_+(0) \big]^{(0)}$$

$$+ \big[A_-(s_0)A_+(0) \big]^{(1)} \cdot \big[A_-(s_0)A_+(0) \big]^{(0)} \Big\} \tag{5.27c}$$

$$\delta = \Omega^4 \big[A_-(s_0)A_-(s_1)QA_+(s_2)A_+(0) \big]^{(0)} \tag{5.27d}$$

$$\varepsilon = -\Omega^4 \big[A_-(s_0)A_-(s_1)QA_-(s_2)A_+(0) \big]^{(0)} \tag{5.27e}$$

where

$$\big[B(s_0)C(s_1)\cdots D(s_n)F(0) \big]^{(m)}$$

$$\equiv \frac{d^m}{dz^m}\int_0^\infty ds_0\, e^{-zs_0}\int_0^{s_0} ds_1 \cdots \int_0^{s_{n-1}} ds_n\, B(s_0)C(s_1)\cdots D(s_n)F(0)$$

$$(5.28)$$

and

$$B(t) = \exp\{-\mathscr{L}_0 s\}\, B \exp\{\mathscr{L}_0 s\} \tag{5.29}$$

By using the commutation rules (5.15) we can rewrite Eq. (5.26) as follows:

$$\frac{\partial}{\partial t}\sigma(v,t) = \left\{ (\alpha+\eta)a_+ a_- + \beta a_+ a_-^3 + (\eta+\delta)a_+^2 a_-^2 + \varepsilon a_+^3 a_- \right\}\sigma(v,t)$$

$$(5.30)$$

When evaluating the average value of $\langle v(t)\rangle$, only the first two terms of Eq.

(5.30) contribute.

$$\frac{\partial}{\partial t}\langle v(t)\rangle = -\langle v^2\rangle_{eq}^{1/2}\int dv\, v\,\frac{d}{dv}\left\{(\alpha+\eta)\left[\frac{v}{\langle v^2\rangle_{eq}^{1/2}}+\langle v^2\rangle_{eq}^{1/2}\frac{d}{dv}\right]\right.$$

$$\left.+\beta\left[\frac{v}{\langle v^2\rangle_{eq}^{1/2}}+\langle v^2\rangle_{eq}^{1/2}\frac{d}{dv}\right]^3\right\}\sigma(v,t)$$

$$=\langle v^2\rangle_{eq}^{1/2}\int dv\left\{(\alpha+\eta)\left[\frac{v}{\langle v^2\rangle_{eq}^{1/2}}+\langle v^2\rangle_{eq}^{1/2}\frac{d}{dv}\right]\right.$$

$$\left.+\beta\left[\frac{v}{\langle v^2\rangle_{eq}^{1/2}}+\langle v^2\rangle_{eq}^{1/2}\frac{d}{dv}\right]\left[\frac{v}{\langle v^2\rangle_{eq}^{1/2}}+\langle v^2\rangle_{eq}^{1/2}\frac{d}{dv}\right]^2\right\}\sigma(v,t)$$

$$=\langle v^2\rangle_{eq}^{1/2}\int dv\left\{(\alpha+\eta)\frac{v}{\langle v^2\rangle_{eq}^{1/2}}+\beta\frac{v}{\langle v^2\rangle_{eq}^{1/2}}\right.$$

$$\left.\cdot\left[\frac{v}{\langle v^2\rangle_{eq}^{1/2}}+\langle v^2\rangle_{eq}^{1/2}\frac{d}{dv}\right]\cdot\left[\frac{v}{\langle v^2\rangle_{eq}^{1/2}}+\langle v^2\rangle_{eq}^{1/2}\frac{d}{dv}\right]\right\}\sigma(v,t)$$

$$=(\alpha+\eta)\langle v(t)\rangle+\beta\int dv\, v\langle v^2\rangle_{eq}^{1/2}\frac{d}{dv}$$

$$\cdot\left[\frac{v}{\langle v^2\rangle_{eq}^{1/2}}+\langle v^2\rangle_{eq}^{1/2}\frac{d}{dv}\right]\sigma(v,t)$$

$$+\beta\int v\frac{v}{\langle v^2\rangle_{eq}^{1/2}}\left[\frac{v}{\langle v^2\rangle_{eq}^{1/2}}+\langle v^2\rangle_{eq}^{1/2}\frac{d}{dv}\right]\sigma(v,t)$$

$$=(\alpha+\eta)\langle v(t)\rangle-\beta\int dv\langle v^2\rangle_{eq}^{1/2}\left[\frac{v}{\langle v^2\rangle_{eq}^{1/2}}+\langle v^2\rangle_{eq}^{1/2}\frac{d}{dv}\right]\sigma(v,t)$$

$$+\beta\int dv\frac{v^3}{\langle v^2\rangle_{eq}}\sigma(v,t)-2\beta\int dv\, v\sigma(v,t)$$

$$=(\alpha+\eta)\langle v(t)\rangle-3\beta\langle v(t)\rangle+\beta\frac{\langle v^3(t)\rangle}{\langle v^2\rangle_{eq}}\tag{5.31}$$

This is a result of special interest on which depends the long-time behavior of the system under study. It is also evident that the sign of β is decisive to determine what effects are produced by making $\langle v(0) \rangle \neq 0$ and $\langle v^3(0) \rangle \neq 0$.

Before going on with our calculation, let us remark that the three significant independent parameters of our system are the frequency Ω of Eq. (5.20a), the damping Γ of Eq. (5.20b), and the parameter W,

$$W = \frac{\alpha_2}{\alpha_1} \langle R^2 \rangle_{\mathrm{eq}} \tag{5.32a}$$

This parameter is a quantitative measure of the nonlinearity of our system. Note, indeed, that when only the first contribution to nonlinearity is taken into account,

$$\langle \psi' \rangle_{\mathrm{eq}} = \alpha_1 + 2\alpha_2 \langle R^2 \rangle_{\mathrm{eq}} \tag{5.32b}$$

W is therefore defined as the ratio of the mean value of the second term of ψ' to the first term, it being understood that this mean value is evaluated at the equilibrium distribution in the purely linear case. A further parameter of interest is

$$g = \frac{\Omega}{\Gamma} \tag{5.32c}$$

which measures the memory intensity.

In terms of the parameter W we have

$$A_- = b_- + \Delta A_-$$

$$\Delta A_- = \frac{1}{\alpha_1} \frac{1}{\langle R^2 \rangle_{\mathrm{eq}}^{1/2}} \alpha_2 \langle R^2 \rangle_{\mathrm{eq}}^{3/2} (b_+ + b_-)^3 \tag{5.33}$$

$$= W(b_+ + b_-)^3$$

when only the first nonlinear contribution is present. The operator $\mathscr{L}_0$ reads

$$\mathscr{L}_0 = \mathscr{L}_{\mathrm{harm}} + \mathscr{L}_{\mathrm{anharm}} \tag{5.34a}$$

$$\mathscr{L}_{\mathrm{harm}} = -\Gamma b_+ b_- \tag{5.34b}$$

$$\mathscr{L}_{\mathrm{anharm}} = -\Gamma W b_+ (b_+ + b_-)^3 \tag{5.34c}$$

We shall assume $\mathscr{L}_{\mathrm{anharm}}$ to be small compared to $\mathscr{L}_{\mathrm{harm}}$. This allows us to write the operator of time evolution as follows:

$$e^{\mathcal{L}_0 t} \simeq \exp[\mathcal{L}_{harm}t]\exp[\mathcal{L}_{anharm}t]\exp\left[-\tfrac{1}{2}[\mathcal{L}_{harm}, \mathcal{L}_{anharm}]t^2\right]$$

$$\simeq e^{\mathcal{L}_{harm}t}[1+\Phi(t)] \tag{5.35}$$

where

$$\Phi(t) = \mathcal{L}_{anharm}t - \tfrac{1}{2}[\mathcal{L}_{harm}, \mathcal{L}_{anharm}]t^2 \tag{5.36}$$

It is evident now that β plays a major role in defining the long-time behavior of our system. We believe that the reader will find it illuminating to follow the detailed calculations needed to get the value of β.

First of all, from Eqs. (5.27b) and (5.20a) we have

$$\beta = -\left(\frac{\alpha_1}{M}\right)^2\left[A_-(s_0)A_+(s_1)QA_+(s_2)A_+(0)\right] \tag{5.37a}$$

Note that b_+ and A_+ coincide. β can be written

$$\beta = -\left(\frac{\alpha_1}{M}\right)^2\left[\Delta A_-^{(3-)}(s_0)b_+(s_1)Qb_+(s_2)b_+(0)\right]$$

$$-\left(\frac{\alpha_1}{M}\right)^2\left[b_-e^{\Gamma_{harm}(s_0-s_1)}\Phi^{(2-)}(s_0-s_1)b_+e^{\Gamma_{harm}s_1}\right.$$

$$\left.\cdot Qe^{-\Gamma_{harm}s_2}b_+e^{\Gamma_{harm}s_2}b_+\right]$$

$$-\left(\frac{\alpha_1}{M}\right)^2\left[b_-e^{\Gamma_{harm}(s_0-s_1)}b_+e^{\Gamma_{harm}(s_1-s_2)}\Phi^{(2-)}(s_1-s_2)b_+e^{\Gamma_{harm}s_2}b_+\right] \tag{5.37b}$$

The indices $(3-),(2-)$ on ΔA_- and Φ indicate that only the terms involving three and two steps down have to be determined, that is,

$$\Delta A^{(3-)} = Wb_-^3 \tag{5.38}$$

$$\Phi^{(2-)} = -\Gamma Wb_+b_-^3 t - \tfrac{1}{2}\Gamma^2 W\{b_+b_-b_+b_-^3 - b_+b_-^3b_+b_-\}t^2 \tag{5.39}$$

The last contribution on the right-hand side of Eq. (5.37b) vanishes, since it happens that at a certain level b_-^3 is applied to $|2\rangle$.

After a tedious calculation, we get

$$\beta = -10W\frac{\Omega^4}{\Gamma^3} = -10W\Omega g^3 \tag{5.40a}$$

By following the same approach, we also get

$$\alpha = -\frac{\Omega^2}{\Gamma} = \Omega g \qquad (5.40b)$$

We do not evaluate η, but we are in a position to say that this is a contribution proportional to $W\Omega g^3$, which can therefore be disregarded.

We are now in a position to discuss qualitatively the effect on the relaxation of v of the excitation of this variable. Let us consider the case when the equilibrium distribution is, by using some preparation method, shifted by a small value $u \ll \langle v^2 \rangle_{eq}^{1/2}$. This means

$$\sigma(v,0) \propto \exp\left[-\frac{(v-u)^2}{2\langle v^2 \rangle_{eq}} \right] \qquad (5.41a)$$

Note that

$$\langle v(0) \rangle = u \qquad (5.41b)$$

$$\langle v^3(0) \rangle = 3\langle v^2 \rangle_{eq} u \qquad (5.41c)$$

From Eq. (5.31) we then get

$$\frac{\partial}{\partial t}\langle v(t) \rangle \bigg|_{t=0} = (\alpha + \eta)u - 3\beta u + 3\beta \frac{\langle v^2 \rangle_{eq}}{\langle v^2 \rangle_{eq}} u = (\alpha + \eta)u \quad (5.42)$$

This means that the effective damping at $t = 0$ is $\alpha + \eta$. Moreover, in general, when the starting point condition

$$\sigma(v,0) \propto \exp\left\{ -\frac{(v-u)^2}{2\langle v^2 \rangle_{exc}} \right\} \qquad (5.43)$$

is used, we get the following effective damping:

$$-\Gamma_{eff} = (\alpha + \eta) - 3\beta\left(\frac{\langle v^2 \rangle_{exc}}{\langle v^2 \rangle_{eq}} - 1 \right) \qquad (5.44)$$

Neglecting η and using both Eqs. (5.40a) and (4.50b), we get

$$\Gamma_{\text{eff}} = \frac{\Omega^2}{\Gamma} + 30W\frac{\Omega^4}{\Gamma^3}\left(\frac{\langle v^2\rangle_{\text{exc}}}{\langle v^2\rangle_{\text{eq}}} - 1\right)$$

$$= g\Omega\left(1 + 30Wg^2r\right) \tag{5.45}$$

where

$$r \equiv \frac{\langle v^2\rangle_{\text{exc}}}{\langle v^2\rangle_{\text{eq}}} - 1 \tag{5.46}$$

The remarks above suggest the following approximate solution:

$$\frac{d}{dt}\langle v(t)\rangle = -\Gamma(t)\langle v(t)\rangle \tag{5.47}$$

where

$$\Gamma(t) = \gamma - \gamma'\langle v^2(t)\rangle \tag{5.48a}$$

$$\gamma \equiv \frac{\Omega^2}{\Gamma} + 3\beta = \Omega g\left(1 - 30Wg^2\right) \tag{5.48b}$$

$$\gamma' \equiv \frac{3\beta}{\langle v^2\rangle_{\text{eq}}} = -\frac{30Wg^3\Omega}{\langle v^2\rangle_{\text{eq}}} \tag{5.48c}$$

This is reminiscent of Suzuki's mean field approximation[11] except for the factor 3 [Suzuki would imply $\beta\langle v^3(t)\rangle = \langle v^2(t)\rangle\langle v(t)\rangle$, whereas we assumed $\langle v^3(t)\rangle = 3\langle v^2(t)\rangle\langle v(t)\rangle$], which has the effect of rendering the contribution of nonlinearity more significant. (Note that in Chapter V, Evans deals with an earlier version of this theory where the factor of 3 is omitted.)

To obtain an equation of motion of $\langle v^2(t)\rangle$ which is consistent with the assumption of Eq. (5.47), we can write

$$\frac{d}{dt}\langle v^2(t)\rangle = -2\Gamma(t)\langle v^2(t)\rangle + 2\varepsilon \tag{5.49}$$

where ε has to be given a suitable value so as to satisfy

$$\lim_{t \to \infty} \langle v^2(t)\rangle = \langle v^2\rangle_{\text{eq}} \tag{5.50}$$

as we know that this condition is satisfied by the contracted Fokker-Planck equation. We obtain, therefore,

$$\varepsilon = \left(\gamma - \gamma'\langle v^2 \rangle_{eq}\right)\langle v^2 \rangle_{eq} \tag{5.51}$$

Note that Eq. (5.49) can be written as

$$\frac{d}{dt}\langle v^2(t) \rangle = 2\gamma'\left(\langle v^2(t) \rangle - \alpha_1\right)\left(\langle v^2(t) \rangle - \alpha_2\right) \tag{5.52}$$

where

$$\alpha_1 = \langle v^2 \rangle_{eq} \tag{5.53}$$

$$\alpha_2 = \frac{\gamma}{\gamma'} - \langle v^2 \rangle_{eq} \tag{5.54}$$

From this equation we derive

$$\frac{d}{dt}\ln\frac{\langle v^2(t) \rangle - \alpha_1}{\langle v^2(t) \rangle - \alpha_2} = 2\gamma'(\alpha_1 - \alpha_2) = -2\gamma \tag{5.55}$$

which means

$$e^{-2\gamma t} = \frac{1}{C}\frac{\langle v^2(t) \rangle - \alpha_1}{\langle v^2(t) \rangle - \alpha_2} \tag{5.56}$$

and therefore

$$\langle v^2(t) \rangle = \frac{\alpha_1 - \alpha_2 C e^{-2\gamma t}}{1 - C e^{-2\gamma t}} \tag{5.57}$$

If the constant C is given a proper value so as to satisfy the initial condition $\langle v^2(0) \rangle = \langle v^2 \rangle_{exc}$, we obtain

$$\langle v^2(t) \rangle = \frac{\langle v^2 \rangle_{eq} + \left(\gamma/\gamma' - \langle v^2 \rangle_{eq}\right)\dfrac{\langle v^2 \rangle_{exc} - \langle v^2 \rangle_{eq}}{\gamma/\gamma' - \langle v^2 \rangle_{exc} - \langle v^2 \rangle_{eq}}e^{-2\gamma t}}{1 + \dfrac{\langle v^2 \rangle_{exc} - \langle v^2 \rangle_{eq}}{\gamma/\gamma' - \langle v^2 \rangle_{eq} - \langle v^2 \rangle_{exc}}e^{-2\gamma t}} \tag{5.58}$$

Let us now come back to Eq. (5.47). We obtain the final result

$$\langle v(t) \rangle = \langle v(0) \rangle \exp\left[-\gamma t + \gamma' \int_0^t \langle v^2(t') \rangle \, dt' \right] \qquad (5.59)$$

Due to the way we followed to arrive at Eq. (5.59), the effective damping at $t = 0$ [i.e., virtually that of Eq. (5.44)] is the result of a sort of coarse-grained measurement on the short time region of the corresponding equilibrium correlation function. In accordance with the linear response theory,[12] $\langle v(t) \rangle$ as given by Eq. (5.59) should tend to coincide with the corresponding equilibrium correlation function as $\langle v^2 \rangle_{exc} \rightarrow \langle v^2 \rangle_{eq}$.

The major result of this section is that it sheds further light on the microscopic mechanism behind the linear response theory.[12] Note that if $\beta = 0$, Eq. (5.31) becomes

$$\frac{d}{dt} \langle v(t) \rangle = -\gamma \langle v(t) \rangle \qquad (5.60)$$

which is the same result as that provided by the standard Langevin equation. In this case even very intense excitation processes would result in decay behavior indistinguishable from the equilibrium correlation function. In other words, the range of validity of the linear response theory[12] would be unlimited. The breakdown of the linear response theory depends on the fact that $\beta \neq 0$, which means that W does not vanish. However, at a fixed value of W, β becomes extremely small for $g \rightarrow 0$ [see, e.g., (5.40a)]. This is somewhat reminiscent of the well-known arguments of Van Kampen[13], according to whom the validity of the linear response theory relies on the same mechanism of randomization as that responsible for rendering the system memoryless. Figure 11 shows that the CFP calculations completely support this point of view, as the equilibrium velocity correlation function $\langle v(0)v(t) \rangle_{eq}$ at $W = 0$ is almost indistinguishable from that concerning the extremely nonlinear case $W = 1$. A strong excitation ($r = 3$), however, makes the case at $W = 1$ show its true nonlinear nature. This figure concerns the case of an itinerant oscillator with an effective potential harder than the linear one. In consequence, in line with our arguments above, the decay of $\langle v(t) \rangle$ after excitation is faster than the equilibrium correlation function.* However, we have found that the non-Gaussian property $\varepsilon(t)$ of this oscillator is greatly different from the experimental one (see Fig. 6). This supports our belief that the proper reduced model should be characterized by a potential softer than the linear one.

---

*The decay of $\langle v(t) \rangle$ closely parallels that of the excited correlation function [see Eqs. (5.66) and (5.66b)].

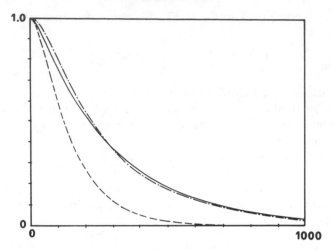

Figure 11. Theoretical calculations via the CFP for a nonlinear itinerant oscillator with effective potential harder than the harmonic one. The equilibrium autocorrelation function (——) and the excited autocorrelation function corresponding to an excited distribution ρ_2 (---) are shown in the case of $W = 1$, $\Omega = 0.01$, $\Gamma = 0.03$, and $r = 3$. For comparison with the linear case, we have also plotted the equilibrium autocorrelation function at $W = 0$ (— · —).

We observe at this point that Eq. (5.59) supplemented by Eq. (5.58) expresses the most recent analytical result obtained to account for the effects of nonlinear excitation.[14] Note, however, that the perturbation approach behind this equation means that it is unable to account for large deviations from linear response theory.[12] In other words, both the intrinsically nonlinear statistics of the system under study and the intensity of the external excitation have to be assumed to be quite small. The rotational counterpart of Eq. (5.58) (v is replaced by the angular velocity ω) has been checked[14] by comparison with the results obtained by applying the continued fraction procedure (CPF) (see Chapters III and IV). It has been shown that the deviation of the linear response theory from the CFP is intermediate between that predicted by Eq. (5.59) and that based on Suzuki's mean field approximation (Chapter V). (In agreement with the CFP, however, both predict that the decay of $\langle \omega(t) \rangle$ becomes slower with increases in the excitation parameter $r \equiv \langle \omega^2 \rangle_{exc} / \langle \omega^2 \rangle_{eq}$ $- 1$.)

The problem of the breakdown of linear response theory could be approached in a somewhat different way. Let us consider the case where $\langle v^2 \rangle_{exc}$ of Eq. (5.43) is very close to $\langle v^2 \rangle_{eq}$. This means that

$$\langle v^2 \rangle_{exc} = \langle v^2 \rangle_{eq} + \Delta \qquad (5.61a)$$

where

$$|\Delta| \ll \langle v^2 \rangle_{eq} \tag{5.61b}$$

Let us assume again

$$u^2 \ll \langle v^2 \rangle_{eq} \tag{5.62}$$

Equations (5.61) and (5.62) lead to the following perturbation expansion:

$$\sigma(v,0) = \sum_{r=0}^{\infty} C_r v^r \sigma_{eq}(v) \tag{5.63}$$

where

$$\sigma_{eq}(v) \propto \exp\left[-\frac{v^2}{2\langle v^2 \rangle_{eq}} \right] \tag{5.64}$$

With the assumption that the variable v is purely Markovian, Eq. (5.63) leads to

$$\langle v(t) \rangle = \sum_{r=0}^{\infty} C_{2r+1} \langle v^{2r+1}(0)v(t) \rangle \tag{5.65}$$

Explicitly evaluating the first two nonvanishing terms, we have, by paying some attention to the normalization factors

$$\langle v(t) \rangle = \frac{u}{\left(1 + \Delta/\langle v^2 \rangle_{eq}\right)^{3/2}} \left[\left(1 + \frac{3}{2}\frac{\Delta}{\langle v^2 \rangle_{eq}}\right) \frac{\langle v(0)v(t) \rangle_{eq}}{\langle v^2 \rangle_{eq}} \right.$$
$$\left. + \frac{3}{2}\frac{\Delta}{\langle v^2 \rangle_{eq}}\eta(t) \right] \tag{5.66}$$

where

$$\eta(t) = \frac{\langle v^3(0)v(t) \rangle_{eq}}{\langle v^4 \rangle_{eq}} - \frac{\langle v(0)v(t) \rangle_{eq}}{\langle v^2 \rangle_{eq}} \tag{5.66a}$$

By using the same method as that above, it is shown that the excited correlation function $\langle v(0)v(t) \rangle / \langle v^2 \rangle|_{\rho_2}$ reads in terms of equilibrium properties as follows:

$$\frac{\langle v(0)v(t)\rangle}{\langle v^2\rangle}\bigg|_{\rho_2} = \frac{\langle v(0)v(t)\rangle_{\rm eq}}{\langle v^2\rangle_{\rm eq}} + \frac{3}{2}\frac{\Delta}{\langle v^2\rangle_{\rm eq}}\eta(t)\left(1 + \frac{\Delta}{\langle v^2\rangle_{\rm eq}}\right)^{-3/2}$$

(5.66b)

It is possible to show in general that the excited correlation function coincides with the decay of $\langle v(t)\rangle$ in the limit of small u. By assuming that the system is also Gaussian, we have

$$\langle v^{2r+1}(0)v(t)\rangle_{\rm eq} = (2r+1)(2r-1)!!\langle v^2\rangle^r\langle v(0)v(t)\rangle_{\rm eq} \quad (5.67)$$

This means that in the linear Markovian case, excitation of whatever intensity has no effect on the decay of the system, which is proved to exhibit the same decay as the equilibrium correlation function. This is in line with the theoretical remarks made above.

In the nonlinear case, the function $\eta(t)$ does not vanish (at times intermediate between $t = 0$ and $t = \infty$). The itinerant oscillator with an effective potential harder than the linear one is shown to result in $\eta(t) < 0$ in accordance with the results of CFP calculations which show its decay after excitation to be faster than the corresponding equilibrium correlation function (see Fig. 11).

In spite of the fact that the decay after excitation of the hard-potential itinerant oscillator is similar to the experimental computer simulation result of Figs. 7 and 8, we do not believe that it is the reduced model equivalent to the one-dimensional many-particle model under study. As remarked above, indeed, the $\varepsilon(t)$ function is not correctly reproduced by this reduced model. The choice of a virtual potential softer than the linear one seems also to be in line with the point of view of Balucani et al.[7,8] They used an itinerant oscillator with a sinusoidal potential, which is the simplest one (to be studied via the use of CFP) to deal with the soft-potential itinerant oscillator. Note that the choice

$$\varphi(R) \equiv \alpha\frac{R^2}{2} - \beta\frac{R^4}{4}$$

(5.68)

would involve an irreversible escape from the potential well. The rate of this would become larger and larger with increasing values of the nonlinearity parameter W.* Unfortunately, the cosine model does not stop the two par-

---

*Figure 10 shows that our system is significantly nonlinear, thereby making the results largely dependent on the actual form of the "virtual" potential.

ticles from drifting infinitely apart either [even if this is made much more difficult than in the case of the potential of Eq. (5.68)].

It should be a matter for future investigation to establish whether or not the effective potential arrived at in the preceding section can fit the main requirements coming from the experimental results of Section III.

A great deal of the difficulty affecting the translational case can be by-passed simply by considering the rotational one. The cosine potential in such a case naturally fits the requirements that the system be left unchanged by rotating through 2π the real body around the virtual one [see Eq. (5.69)]. The rotational counterpart of Eq. (5.1) reads

$$\dot{w} = -\frac{\varphi'(\Delta)}{I} \qquad \dot{\nu} = \frac{\varphi'(\Delta)}{I'} - \gamma\nu + f(t) \qquad (5.69)$$

where

$$\Delta = \theta - \psi \qquad (5.70)$$

θ and ψ being the rotational coordinates of the real and virtual body, respectively. This is a disk, simulating the real molecule, rotating around an orthogonal axis and bounded via the potential $\varphi(\Delta)$ to an external annulus which rotates around the same axis. The annulus, which is subjected in turn to a standard fluctuation-dissipation process, simulates the thermal bath of the disk.

The remarks above show that it is quite natural in the present case to assume

$$\varphi(\Delta) \propto \sin \Delta \qquad (5.71)$$

thereby rendering straightforward the use of CFP. We are naturally led in this way to explore carefully via computer simulation the nonlinear properties of the two-dimensional system investigated by Brot and coworkers,[2] of which the reduced model of Eq. (5.69) seems to be a true simulation.

VI. ROTATIONAL DYNAMICS: COMPUTER SIMULATION OF EQUILIBRIUM AND NONEQUILIBRIUM PROPERTIES

In the first part of this section we will illustrate the computer techniques used to repeat in this two-dimensional case the experiment of monitoring the polarization after sudden removal of an electric field (i.e., the two-dimensional counterpart of a key experiment illustrated by Evans in Chapter V). In addition, we shall monitor several equilibrium properties which contribute to the physical mechanism behind the interesting effect of decay acceleration discovered by Evans.[15] The appealing physical interpretation reported in Chapter V relies on the model of the nonlinear itinerant oscillator [see Eq. (5.69)]. The basic steps of this interpretation are:

1. Equilibrium in the presence of an external field implies a potential energy of the variable much higher than at the field-off equilibrium.
2. After sudden removal of this field, the potential energy is rapidly converted into rotational kinetic energy.
3. The nonlinear itinerant oscillator implies a slower decay of angular velocity as the intensity of the rotational kinetic energy is increased, thereby speeding up the decay of the dipole vector.

In Section VI.B we shall check directly whether or not step 3 is supported by direct experiment of excitation of the angular velocity.

A. Computational Details

Molecular dynamics simulations have been carried out on an isolated sample of 313 Stockmayer molecules. Their interparticle potential $u_{ij}(r)$ is composed of two parts:

$$u_{ij}(r) = u_{ij}^*(r) + u_{ij}^{LR}(r) \qquad (6.1)$$

where $u_{ij}^{LR}(r)$, the most long-range part, is the bidimensional dipolar potential

$$u_{ij}^{LR}(r) = \frac{(\mu_i \cdot \mu_j)}{r^2} - 2(\mu_i \cdot \mathbf{r})(\mu_j \cdot \mathbf{r}) \qquad (6.2)$$

(see Bossis et al.[2a]) and $u_{ij}^*(r)$ a two-dimensional exp-4 (Lennard-Jones-like) potential defined as follows:

$$u_{ij}^*(r) = A \exp(-br) - \alpha \varepsilon_{ij} \left(\frac{\sigma}{r}\right)^4 \qquad (6.3)$$

The constants of Eq. (6.3) have been fixed so that $u_{ij}^*(\sigma) = 0$ and the minimum value $-\varepsilon_{ij}$ is reached at $r_{\min} = 2^{\frac{1}{4}}\sigma$.

The above short-range potential is the single most important difference between this system and those simulated by Bossis et al. in their work on the calculation of the dielectric constant via molecular dynamics simulation. The form (6.3) has been adopted to be consistent with the bidimensional electrostatics employed for the dipolar interaction.

A soft repulsive potential barrier, "the wall," surrounds the sample and keeps the molecules inside a circular region of radius $R = 13.2\sigma$. Its form is analogous to that of Eq. (6.3), corrected and truncated so that potential and force due to the wall are purely repulsive.

The values of the parameters that characterize the runs are in reduced units, $T = 1.79$ and $\mu = 1.35$. As a consequence, the value of the dipolar coupling parameter $\gamma = \mu^2 \rho / k_B T$ is smaller than that used by Bossis et al. (0.58 here, vs. 1 and 2) and should lead to a lower dielectric constant ε more representative of nonassociated polar liquids.

In the study of structural properties of fluids of particles interacting through orientation-dependent potentials, it is convenient to decompose the full pair-correlation function $h(1,2) = h(r_{12}, \Omega_1, \Omega_2)$:

$$h(1,2) = h_s(r_{12}) + h_D(r_{12}) D(1,2) + h_\Delta(r_{12}) \Delta(1,2) \qquad (6.4)$$

where, in two dimensions, the projections $h_D(r_{12})$ and $h_\Delta(r_{12})$ are defined by

$$h_D(r_{12}) = \frac{2}{\Omega^2} \int h(1,2) D(1,2) \, d\Omega_1 \, d\Omega_2 \qquad (6.5a)$$

$$h_\Delta(r_{12}) = \frac{2}{\Omega^2} \int h(1,2) \Delta(1,2) \, dr_1 \, d\Omega_2 \qquad (6.5b)$$

where

$$\Delta(1,2) = \mathbf{u}_1 \cdot \mathbf{u}_2 \qquad (6.6a)$$

and

$$D(1,2) = \frac{2(\mathbf{u}_1 \cdot \mathbf{r}_{12})(\mathbf{u}_2 \cdot \mathbf{r}_{12})}{r_{12}^2} - \mathbf{u}_1 \cdot \mathbf{u}_2 \qquad (6.6b)$$

with $\Omega = 2\pi$.

To avoid any structural perturbation due to density oscillations close to the border, already observed by Bossis et al.,[2a] the functions defined above have been calculated considering as reference molecules only those particles within 8σ from the center and for pair separation $r_{12} \le 11\sigma - r_{10}$, where r_{10} is the distance of particle 1 from the center.

B. Results

The results obtained for $g(r_{12}) = h(r_{12}) + 1$, $h_D(r_{12})$, and $h_\Delta(r_{12})$ are shown in Figs. 12, 13, and 14, respectively. When differences due to the applied static electric field are visible, functions corresponding to field intensity E_0 ranging from 0.5 to 2 (equivalent to 0.675–$2.7 k_B T$) are plotted.

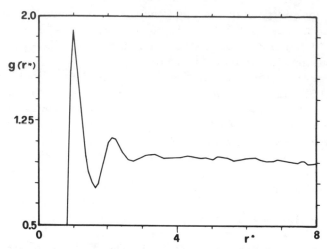

Figure 12. Radial distribution function $g(r_{12}) = h(r_{12}) + 1$ for a field intensity $E_0 = 1$. The $g(r_{12})$ corresponding to $E_0 = 0$ and $E_0 = 2$ are virtually coincident.

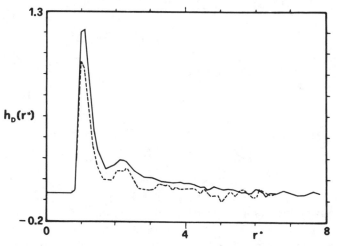

Figure 13. $h_D(r_{12})$ vs. r/σ at various field intensities; (——) $E_0 = 0$; (---) $E_0 = 2$.

As can be seen from Fig. 12, the positional structure of the fluid is substantially unaffected by the applied field. Only the $g(r_{12})$ relevant to the case $E_0 = 1$ shows a slightly reduced structure, namely, a lower first and second maximum. The difference between this $g(r_{12})$ and the others, however, is rather small and might well depend on the higher average temperature of the corresponding run, 1.79 here vs. 1.48 in ref. 2a.

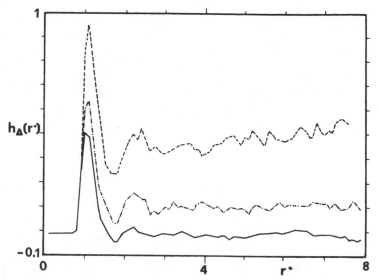

Figure 14. Vector correlation function $h_\Delta(r_{12})$ vs. r/σ. (——) $E_0 = 0$; (---) $E_0 = 2$; (—·—) $E_0 = 1$.

Moreover, we observe a large overall reduction of the intensity of the various peaks of all $g(r)$'s compared to those calculated in ref. 2a. In this case, too, there certainly is a temperature effect, but this is very likely a second-order effect. The reduced structure is primarily a consequence of the different short-range potentials employed here and in ref. 2a. In fact, a test run performed under the same ρ, T, and μ conditions as in that work, and with no applied field, led to a $g(r)$ with a maximum slightly larger than 2, while Bossis et al. would obtain a first peak intensity of ~ 3.5 for $\gamma = 0.58$ (see Fig. 8 of Bossis[2a]). The same test run also showed that the orientation structure of the fluid as revealed by $h_D(r)$ and $h_\Delta(r)$ is practically unaffected by the form of the short-range potential. We have found that the number, location, and intensity of the peaks in $h_D(r)$ and $h_\Delta(r)$ are very close to those reported by Bossis.[2a] So the differences one can observe in the functions of Figs. 13 and 14 at zero applied field with respect to the relevant functions of ref. 2a, that is, the remarkably lower first peak and lack of clearly defined peaks beyond $r \cong 2.5$, are to be traced back to the lower polarity of our sample, $\gamma = 0.58$ vs. 1 and 2. As for the sensitivity of $h_D(r)$ and $h_\Delta(r)$ to the intensity of the applied field, the latter function displays much larger variations. In $h_D(r)$, one can note that the effect of the applied field is restricted to a reduction of the intensity of the first peak, the difference at larger distances being blurred by the noise of the function. In $h_\Delta(r)$, conversely, a rise of the long-range part of the curve with the intensity of the applied field is clearly visible. This reflects the progressive alignment of the individual dipoles imposed by the external field, that in the limit of very high intensity would yield an asymptotic value of 1. On the other hand, the persistence in $h_\Delta(r)$ at $E_0 = 2$ of a first minimum and a second maximum located at the same r value of the corresponding part of $g(r)$ confirms that the local positional structure determined by the short-range part of the potential is not affected by the applied field. However, the tagged dipole "sees" a larger number of neighboring dipoles aligned in its direction than in the field-off case. In the present case, the circular symmetry of the particles eliminates any competition between short-range forces and external field. Actually, the disks can adjust themselves to the field without those short-range structural rearrangements that might be required, for example, in a fluid composed of dipolar rods.

We believe that these results confirm that the external field has a deep influence on the relative orientation between the tagged dipole and its neighbors. We guess that a sudden removal of the external field will produce a fast local rearrangement (much faster than the microscopic depolarization) that will in turn excite the angular velocity ω. In accordance with our remarks in Section V, we are then led naturally to study the non-Gaussian property of the angular velocity.

Among the various autocorrelation functions (ACF) we have computed, attention is devoted here to the individual dipole ACF, $C_\mu(t) = \langle \mu\mu(t) \rangle_{eq} / \langle \mu^2 \rangle_{eq}$, the angular velocity ACF, $C_\omega(t) = \langle \omega\omega(t) \rangle_{eq} / \langle \omega^2 \rangle_{eq}$, and especially the non-Gaussian properties

$$\varepsilon(t) = \frac{\langle \omega^2\omega^2(t) \rangle_{eq}}{\langle \omega^4 \rangle_{eq}} - \frac{2}{3}\left[\left(\frac{\langle \omega\omega(t) \rangle_{eq}}{\langle \omega^2 \rangle_{eq}} \right)^2 + 1 \right] \qquad (6.7)$$

$$\eta(t) = \frac{\langle \omega^3\omega(t) \rangle_{eq}}{\langle \omega^4 \rangle_{eq}} - \frac{\langle \omega\omega(t) \rangle_{eq}}{\langle \omega^2 \rangle_{eq}} \qquad (6.8)$$

The non-Gaussian properties $\varepsilon(t)$ and $\eta(t)$ are reported in Figs. 15 and 18 together with the angular velocity and (Fig. 15) rotational kinetic energy ACF. It can be seen that no cage effect or negative portion is present in $C_\omega(t)$ due to the lower polarity, that is, lower μ and ρ and higher T, of the present sample compared to that of ref. 2b.

Moreover, the Gaussian prediction for $C_{\omega^2}(t)$ is accurate in the very short and long time limits but fails at intermediate times. Thus, the non-Gaussian behavior observed here is quite similar, qualitatively, to that reported by

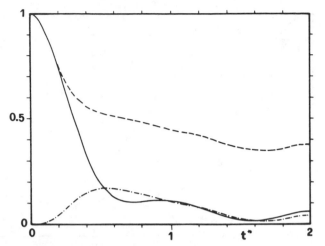

Figure 15. (——) Angular velocity autocorrelation function (acf), (---) rotational kinetic energy acf, and (– ·–) its non-Gaussian component [see Eq. (6.7)].

Balucani et al.[7,8] for the translational kinetic energy ACF in a model of argon at different ρ and T conditions. The same behavior is exhibited by the one-dimensional translation case of Section III. The function $\varepsilon(t)$ (6.7) rises quickly and decays slowly after the maximum that is located here at $t^* = 0.55$. Note that the same behavior is exhibited by $\eta(t)$. This will produce excitation effects of the opposite kind to those of the translational case, which is characterized by a negative $\eta(t)$ (see Section III).

The individual dipole ACF is compared to $C_\mu = \langle \mathbf{M} \cdot \mathbf{M}(t) \rangle_{eq} / \langle \mathbf{M}^2 \rangle_{eq}$ and $C_m(t) = \langle \mathbf{m} \cdot \mathbf{m}(t) \rangle_{eq} / \langle \mathbf{m}^2 \rangle_{eq}$ in Fig. 16. Here $\mathbf{M}$ is the mean dipole induced by the reference dipole μ_i in the whole disk of radius R, and $\mathbf{m}$ is the corresponding dipole induced in a smaller region s centered on μ_i. The latter quantity clearly depends on the radius of s, r_s, so that it is convenient to extrapolate it to $r_s/R = 0$. However, the $C_m(t)$ reported in Fig. 16, which actually relates to $r_s/R = 0.3$, according to ref. 2b, can already be considered the $r_s/R = 0$ limit. The results plotted in Fig. 16 confirm that Onsager's model assumption that $C_\mu(t)$ and $C_m(t)$ are identical is not correct, even at lower polarities than those of ref. 2b. Furthermore, we can note that the trend manifested in ref. 2b by $C_\mu(t)$ and $C_m(t)$ with γ, that is, faster decaying $C_M(t)$ and slower $C_m(t)$ with increasing γ, is not observed here. Our $C_M(t)$ at $\gamma = 0.58$, for example, is as fast as the one reported in ref. 2b for $\gamma = 1$, while $C_m(t)$ is slightly slower than the corresponding function. This might be a consequence of the different short-range potential employed in our simulation that leads to a looser positional ordering compared to that of ref. 2a.

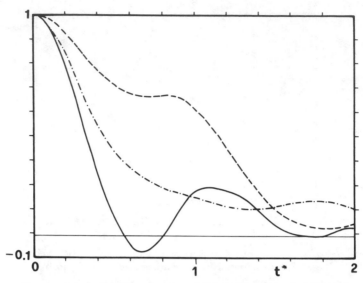

Figure 16. (—) Autocorrelation function of the electric moment of the total sample C_μ, (---) acf of the electric moment m of an inner smaller sphere $C_m(t)$ for $r/R = 0.3$, and (– · –) individual dipole acf.

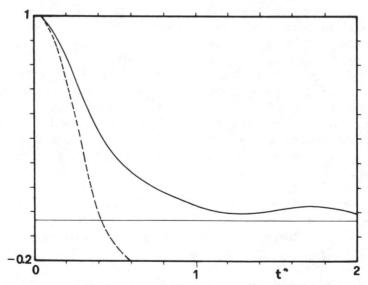

Figure 17. Comparison of (—) the equilibrium individual dipole acf with (---) the time decay of $\theta(t)$, the average angle between the individual dipoles and the external field, after a sudden removal of the field $E_0 = 0.5$.

Finally, no rototranslational couplings of the kind reported in ref. 16 (a three-dimensional system) have been observed, either in a laboratory fixed frame or in a moving one, as expected because of the circular symmetry of the disks.

The behavior of $\langle \cos\theta(t)\rangle$, the average angle between the individual dipoles and the external field E_0, after the sudden removal of the field has been followed and compared to the equilibrium ACF of the individual dipole, $C_\mu(t)$. The results are shown in Fig. 17 and relate to average values computed on several independent trajectories. As one can see, the nonequilibrium function decays faster than the equilibrium dipole ACF. This is a major result of this paper. It confirms that the accelerated decay of dipole orientation detected by Evans in a three-dimensional case[15] is a general physical feature. We believe that the root cause of this can be traced to the non-Gaussian property $\eta(t)$. In Section VI.C we shall make a direct check of this interpretation by directly monitoring the angular velocity $\omega(t)$ when evolving in an excited condition.

C. Excited Initial Distribution and Corresponding Angular Velocity Correlation Function

The two-dimensional system described above is the simplest source of information on reorientation dynamics. We are interested in the behavior of the system when it is excited out of the equilibrium state. The 313 disk molecules are our thermal bath described by a temperature T and by the properties shown in the equilibrium MD runs. The variable of interest to be excited is the angular velocity ω, and the dynamical behavior of ω itself is the relevant signal.

From the equilibrium MD runs we saw that

1. The angular velocity equilibrium distribution $\sigma_{eq}(\omega)$ is a Gaussian:

$$\sigma_{eq}(\omega) = \exp\left\{-\frac{\omega^2}{2\langle\omega^2\rangle_{eq}}\right\}$$

2. The dynamical relaxation of ω is not Gaussian [remember the non-Gaussian properties $\eta(t)$ and $\varepsilon(t)$].

The investigation of the behavior of the excited system requires a parallel equilibrium simulation describing our "thermal bath." A MD run at $T=1.9$, $R=15$, $\mu=1.63$ has been carried out for 2000 steps (after equilibration). Every 100 steps a complete configuration $\{z_i, v_i, \omega_i, u_i\}$ of the system has been stored as the starting point for an excited run. A full set of relevant autocorrelation functions has been extracted from the run, and they are used for comparison with the corresponding excited quantities.

A number of nonequilibrium runs have been carried out, each starting from a different initial phase configuration out of that mentioned above. In each run, of length 100 MD steps, a few particles were randomly chosen out of the 313. The choice was made (1) excluding particles too close to the MD disk edge by imposing a maximum $R_{max} = 13\sigma$ on the particle position, and (2) avoiding the choice of particles too close to each other by imposing a minimum distance of 3σ between them.

Such requirements always limited the number of extracted particles to a maximum of seven for each run (often no more than four particles have been extracted). Although one would have preferred to excite one particle at a time, practical considerations about computer timing have forced us to the more efficient technique mentioned above. However, it must be stressed that conditions 1) and 2) assure that the noise introduced by using more than a single excitation is negligible. This has been checked numerically.

Averaging over single excitations is required to clear the output signal of intrinsic thermal fluctuations. These can be described by the "thermal bath" mean kinetic energy $k_B T$, and if we call N the total number of single excitations performed, we expect for the angular velocity ω a noise of the order of $[k_B T / (N-1)]^{1/2}$ as compared to the nonequilibrium excited average $\langle \omega^2 \rangle_{exc}$.

D. The Excitation

We want to extract the relevant behavior of particles described by a time $t = 0$ distribution different from the equilibrium one $\sigma_{eq}(\omega)$. This is made by:

1. Extracting a few particles at random.
2. Producing an excitation of their angular velocity, that is, far-from-equilibrium fluctuation.
3. Following the motion of those particles as driven by the thermal bath formed by the "remainder" of the particles. As an important condition this thermal bath must not be taken out of equilibrium.

We have described step 1 already. Step 2 is accomplished by using at time $t = 0$ a perturbation of the disk particle angular velocity of Dirac delta form in time. As we wanted a $\sigma_{exc}(\omega, t = 0)$ not too far from equilibrium Gaussian characteristics (see Section V), we gave $\sigma_{exc}(\omega)$ a Gaussian shape with a width larger than the equilibrium one, that is,

$$\sigma_{exc}(\omega) \propto \exp\left(\frac{-\omega^2}{(1+\alpha)\langle \omega^2 \rangle_{eq}} \right) \tag{6.9}$$

This is immediately accomplished with the perturbation

$$[\omega_i(0^+) - \omega_i(0^-)] \equiv \Delta\omega_i(t) = \alpha\omega_i(0) \tag{6.10}$$

where i is the extracted particle index.

The use of Verlet[9] and Singer's algorithm makes it necessary to use extra care in integrating the equation of motion. Ciccotti et al.[17] have shown how to do it in the case of Verlet's algorithm.[9] As to the rotational equation of motion, we followed a similar procedure using the quantities

$$\Delta u_x(t) = -\Delta\omega_i(t)u_y(0)$$
$$\Delta u_y(t) = \Delta\omega_i(t)u_x(0) \tag{6.11}$$

which are equivalent to Eq. (6.10).

We recall here that a $\delta(t)$ perturbation is included at time $t = 0$ in Verlet's algorithm by following the formula

$$x(0 + \Delta t) = 2x(0) - x(-\Delta t) + [\dot{x}(0^+) - \dot{x}(0^-)]\,\Delta t + \ddot{x}(0)\,\Delta t^2 + O(\Delta t^4) \tag{6.12}$$

when $\dot{x}(0^+) = \dot{x}(0^-)$. The same holds for the third-order time derivative $\dddot{x}$.

The condition in step 3 about the thermal bath can be checked immediately. Our perturbation, in fact, invokes an energetic contribution of $n\alpha k_B T/2$, where n is the number of extracted particles in each excitation run ($n < 7$). If we compare this to the total kinetic energy, we have that the condition

$$\frac{n\alpha k_B T}{2} \ll \frac{3}{2} N_{\text{part}} k_B T \tag{6.13}$$

is well satisfied for small α. We have used $\alpha = 1$ and $\alpha = 2$. For each of these two values, about $N = 50$ single particle excitations have been produced, their motion followed for 100 steps and then averaged to give the results shown in Fig. 18.

This experimental result has to be compared with the prediction of the model of Eq. (5.69). This model can be studied by using the CFP.[14] The results of ref. 14 are illustrated by Figs. 19 and 20. This figures show that the qualitative agreement between the theory based on the nonlinear itinerant oscillator of Eq. (5.69) and the experiment of Fig. 18 is indeed satisfactory.

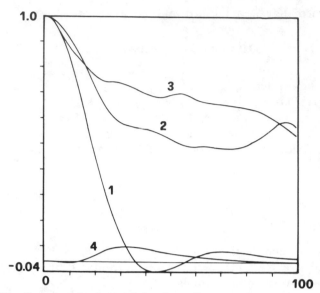

Figure 18. Computer simulation of the effects of excitation on the decay of the angular velocity ω. Curve 1 denotes the equilibrium correlation function $\langle \omega\omega(t)\rangle_{\text{eq}}/\langle \omega^2\rangle_{\text{eq}}$. Curves 2 and 3 denote the correlation function $\langle \omega\omega(t)\rangle_{\text{eq}}/\langle \omega^2\rangle_{\text{eq}}$ evaluated on a Gaussian distribution with $\sigma = 4\langle \omega^2\rangle_{\text{eq}}$ and $\sigma = 9\langle \omega^2\rangle_{\text{eq}}$, respectively. Curve 4 denotes the non-Gaussian function $\eta(t)$ (evaluated at equilibrium).

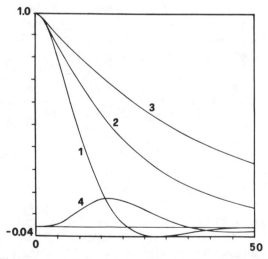

Figure 19. Theoretical calculations via the CFP applied to Eq. (5.69) (v being assumed as infinitely fast) of the effects of excitation on the decay of the angular velocity ω. Curves 1 to 4 have the same meaning as in Fig. 18, and curves 2 and 3 correspond to the same excitation parameter as those of Fig. 18. The memory and nonlinearity parameters which best reproduce the experimental results are $g = 1.45$ and $W = 0.0680$.

272

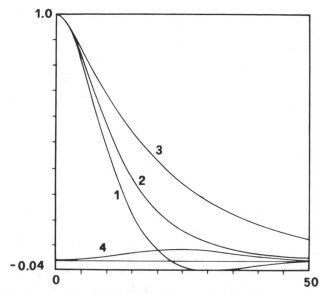

Figure 20.   Theoretical calculations via CFP applied to Eq. (5.69) (v being assumed to be infinitely fast) of the effects of excitation on the decay of the angular velocity ω. The curves 1 and 4 have the same meaning as in Figure 18. $g = 0.62$, $W = 0.0421$.

On the one hand, the fact that the decay of the correlation function of the angular velocity ω becomes slower with increasing width of the Gaussian distribution of the tagged molecules is in line with the fact that $\eta(t) \geq 0$ (see Fig. 15). The model of the nonlinear itinerant oscillator of Eq. (5.69) satisfies this non-Gaussian property (see Figs. 19 and 20). This can also be arrived at by applying to the rotational case the theoretical investigation of Section V.

It is now clear that the analysis of Chapter V on a three-dimensional computer experiment is basically correct. An external electric field significantly affects the average relative orientation of two neighboring dipoles (as also revealed by Fig. 14). After sudden removal of this field, the system undergoes a reequilibration process which allows the field-off conditions to be recovered. Throughout this reequilibration process, the angular velocity appears to be slower than at equilibrium. This is confirmed by the accelerated decay of electric polarization shown in Fig. 17. From the results obtained in this section we understand that for the angular velocity to become slower, this variable must absorb energy from its thermal bath. The angle ξ defined in Chapter V was introduced to simulate precisely the excitation process of this key variable. As argued in Chapter V, this excitation should take place in the early phase of the depolarization process.

VII. CONCLUDING REMARKS

The breakdown of linear response theory[4] is due to non-Gaussian properties. These can be revealed unambiguously via computer simulation. These non-Gaussian properties, in turn, depend on the fact that the macroscopic time scale is not infinitely longer than the microscopic one. In a sense, non-Gaussian and non-Markovian properties are closely related to each other. A strong non-Markovian character could alone produce large deviations from the linear response theory, but merely as an effect of preparation.[18] In this chapter we have focused our attention on the breakdown of linear response theory due to the nonlinear statistics in the system under study. Therefore we have particularly studied decay processes as close as possible to the exponential regime or those with weak oscillation properties. These quasi-Markovian systems are especially suitable for a purely analytical study. A purely analytical study, however, can be carried out only in those regions where the deviations from linear response theory are extremely weak. Analytical expressions only indicate a trend of the system, which is then examined via the CFP and computer experiments in the region where these effects are more marked, and therefore analytical theory is invalidated.

General agreement is obtained between theory and computer experiments. Translational and rotational dynamics, as monitored via computer experiments, are characterized by negative and positive η functions, respectively. Theory then predicts that excitation produces decays faster and slower than the corresponding equilibrium correlation functions. This is completely supported by the excitation experiments illustrated in this chapter.

These nonlinear statistics stem from a mesoscopic level, where the original microscopic nonlinear characteristics appear to be greatly reduced in effect. A rigorous approach to the mesoscopic level from the microscopic one is still an open problem. The study of the translational case indicates that the well-known model of the itinerant oscillator should be supplemented by additional properties such as the fluctuating character of the virtual potential itself, as well as the nonlinear character of this potential. This establishes contact in a very interesting way with Chapter X (Faetti et al.) and Chapter VII (Bertolini et al.). Indeed, in Chapter VII it will be shown that a fluctuating character in the virtual potential is required to simulate the effect of hydrogen-bond dynamics in associated liquids.

The results of computer experiments indicate that the virtual potential must be softer than the linear approximation. However, a simple analytical expression for this, which is also sound from a physical point of view, can be obtained only in the rotational case, where complete agreement between theory and experiment can be achieved. Comparably good agreement in the translational case requires further investigation.

References

1. M. Bishop, M. Derosa, and J. Lalli, *J. Stat. Phys.*, **25**, 229 (1981).

2. (a) G. Bossis, *Mol. Phys.*, **38**, 2023 (1979). (b) G. Bossis, B. Quantrec, and C. Brot, *Mol. Phys.*, **89**, 1233 (1980). (c) C. Brot, G. Bossis, and C. Hesse-Bezot, *Mol. Phys.*, **40**, 1053 (1980).

3. M. Abramowitz and I. A. Stegun, *Handbook of Mathematical Functions*, Dover, New York, 1964.

4. R. I. Cukier, . E. Shuler, and J. D. Weeks, *J. Stat. Phys.*, **5**, 99 (1972).

5. M. Ferrario and P. Grigolini, *J. Math. Phys.*, **20**, 2567 (1979).

6. S. A. Adelman, *Adv. Chem. Phys.*, **53**, 61 (1983).

7. U. Balucani, V. Tognetti, R. Vallauri, P. Grigolini, and P. Marin, *Z. Phys.*, **49b**, 181 (1982).

8. U. Balucani, V. Tognetti, R. Vallauri, P. Grigolini, and M. P. Lombardo, *Phys. Lett.*, **86A**(8), 426 (1981).

9. L. Verlet, *Phys. Rev.*, **159**, 98 (1968).

10. M. W. Evans, G. J. Evans, W. T. Coffey, and P. Grigolini, *Molecular Dynamics*, Wiley-Interscience, New York, 1982.

11. M. Suzuki, *Adv. Chem. Phys.*, **46**, 426 (1981).

12. R. Kubo, *J. Phys. Soc. Japan*, **12**, 570 (1957).

13. N. G. Van Kampen, *Phys. Norvegica*, **5**, 279 (1979); for a discussion of Van Kampen's arguments, see also G. Jaccucci, *Physica*, **118A**, 157 (1983) and references therein.

14. M. Leoncini, Ph.D. thesis, 1984; M. Ferrario, P. Grigolini, M. Leoncini, L. Pardi, and A. Tani, *Mol. Phys.*, **53**, 1251 (1984).

15. M. W. Evans, *J. Chem. Phys.*, **76**, 5480 (1982).

16. J. P. Ryckaert, A. Bellemans, and G. Ciccotti, *Mol. Phys.*, **44**, 979 (1981).

17. G. Ciccotti, G. Jaccucci, and I. R. McDonald, *J. Stat. Phys.*, **21**, 1 (1977).

18. P. Grigolini, *J. Chem. Phys.*, **74**, 1517 (1981).

VII

DYNAMICAL PROPERTIES OF HYDROGEN-BONDED LIQUIDS

D. BERTOLINI, M. CASSETTARI, M. FERRARIO,
P. GRIGOLINI, and G. SALVETTI

CONTENTS

I. INTRODUCTION

Chapters V and VI have demonstrated the state of the art when the reduced model theory (RMT) is applied to normal (unassociated) liquids. The remaining difficulties in this field might have discouraged us from facing a still more difficult subject in this chapter: the dynamics of water and other hydrogen-bonded liquids. On the contrary, we shall show that the RMT can provide a good theoretical background for the description of the macroscopic effect of hydrogen-bond dynamics.

A somewhat surprising result of Chapter VI is that the foundation of a nonlinear version of the itinerant oscillator is unavoidably accompanied by a fluctuating character of the "virtual" potential itself. In this chapter we

277

shall give an appealing physical meaning to the results derived in Chapter VI from the joint use of analytical theory and computer simulation. We shall show indeed that this fluctuating character of the virtual potential naturally stems from the H-bond dynamics.

To make clear this intriguing feature we shall detail all the physical aspects of our theory with special emphasis on water, even if the model can be used to describe other H-bonded liquids.

Both in its stable and metastable (supercooled) conditions, water has[1,2] properties which turn out to be anomalous even when compared with other H-bonded liquids (alcohols, for instance). These characteristic features can be traced back to the indigenous aqueous properties of connectivity and to cooperative phenomena[3] of the H-bond network. As to the long-time diffusional properties, the remarkable deviation from Arrhenius behavior in the supercooled region can be related to some structural contributions coming from the hydrogen bond and its dynamics within the network. Agreement on these points is fairly general. However, one can find in the literature only a few attempts at developing an exhaustive theory for rotational diffusion (dielectric relaxation) and translational diffusion (self-diffusion). In spite of the wealth of information available on the network statistics (provided mainly by computer simulation), only Angell[4] and, more recently, Stanley and Teixeira[5] have treated the long-time diffusional properties. There is available, moreover, an interesting attempt by Chu and Sposito[6]; these authors explored the short-time regime and developed a "relaxing cage" model, enabling them to reproduce the translational velocity autocorrelation function[7] through a suitable weighting of the vibrational modes of ice Ih. In a subsequent paper[8], Sposito explores the long-time behavior using a hydrodynamic approach. However, the small corrections to the classical relations introduced to describe water behavior do not involve H-bond dynamics.

Our aim will be to obtain a coherent description of dielectric relaxation and self-diffusion based on H-bond dynamics in order to explain the experimental data at temperatures where the hydrogen bond plays a fundamental role. An important aspect of our treatment will be the possibility of relating the long-time to the short-time dynamics.

Following Eisemberg and Kauzmann,[9] we consider for simplicity's sake at least three types of "dynamic structures" for water:

1. The *instantaneous structure* (*structure I*) can be thought of as a snapshot of the network with an "exposure time" which is short compared with the molecular vibration times, that is, $\tau_I = 10^{-14}-10^{-16}$ s (this is precisely the order of magnitude of the time steps used in computer simulation "experiments").

2. The *mean vibrational structure* (*structure V*) is that obtained with an "exposure time" of the same order of magnitude as the vibrational correlation time τ_V ($\tau_V \simeq 10^{-13}$ s is the time scale of infrared and Raman spectroscopy[10,11]). Recently, time-averaged molecular dynamics experiments[12,13] have allowed a statistical study of this type of structure.

3. The *mean diffusional structure* (*structure D*) is that obtained with an "exposure time" of the order of the orientational correlation time (or of the translational counterpart) of the molecule. This structure has also been explored by computer simulation experiments,[7,14-17] but the agreement between the results and experiment is not always satisfactory.

For ice Ih, at the melting temperature, $\tau_V = 10^{-13}$ s and $\tau_D = 10^{-5}$ s. In contrast, in water, owing to the disorder of the liquid phase the two times are $\tau_V = 2$–3×10^{-13} s and $\tau_D = 1.8 \times 10^{-11}$ s, respectively. The mean lifetime of the hydrogen bond τ_{HB} assumes values[18] intermediate between the characteristic time of structure V and structure D ($\tau_{HB} = 9 \times 10^{-13}$ s). We aim to describe the dynamic properties of water in structure D, by means of a theory that can account for diffusional phenomena in the different molecular environments; to do this we shall have to apply suitable time-average procedures to processes the time scale of which is shorter than τ_{HB}.

We shall characterize a molecular environment by considering the number of intact H bonds with which the tagged molecule is engaged at time t; if the maximum number of bonds is four, we will need to introduce a stochastic variable η to explore the environments with $0, 1, 2, 3, 4$ bonds, respectively. We shall also write a master equation for η in which three parameters are present:

p_B = the probability that a randomly considered hydrogen bond is intact

τ_f = the mean time during which the hydrogen bond remains broken

("death" time)

τ_{HB} = the mean "lifetime" of the hydrogen bond

We shall show that two of these parameters are enough to describe the H-bond dynamics (τ_{HB} and p_B, for example). We shall be able to calculate these two parameters from experimental data (density and depolarized Rayleigh scattering at various temperatures) and to compare the predictions of our theory with the observed diffusional properties of water. Moreover, it will become clear that the assumption of a discrete variable η essentially means that the short-time details (over a time scale of the order of τ_V) of the rotational and translational velocity correlation functions are not important in determining the long-time dynamics.

There are models for water in the literature which aim at building up an effective Hamiltonian[19] from a description of structure V. These models must give the same results for long-time dynamics as those that will emerge straightforwardly from our statistical model, based on H-bond dynamics. This must be expected in the temperature range at which the structural features of the H-bond network are dominant. These literature models are limited to describing the short-time dynamics and some thermodynamic properties of water.

The RMT, being a compromise between a priori theory and a purely phenomenological approach, is suitable for introducing H-bond dynamics into a diffusional framework. In the RMT the tagged molecule is thought of as interacting with a virtual (i.e., effective) rototranslating entity, thereby simulating the detailed interaction of the molecule with its complex molecular environment. By writing the Fokker-Planck equations corresponding to the stochastic equations of motion of the molecule and the virtual body, and by applying AEP (discussed in detail in Chapter II), a technique that allows us to obtain straightforwardly the time-scale separation described already, one obtains two diffusion equations—rotational and translational. These equations depend on the characteristic parameters of H-bond dynamics and on the diffusion coefficients in the five molecular environments defined above.

II. HYDROGEN-BOND DEFINITION AND STATISTICS

In the last 15 years, various potentials for water have been used in computer simulation experiments; a recent review on this topic is available in ref. 20, and a comparative analysis in ref. 21. We shall report some experimental results from which two possible definitions of the hydrogen bond must be inferred.

1. *Hydrogen-Bond Energetic Definition*: The hydrogen bond is intact if the energy V is less than a threshold value V_{HB}; otherwise the bond is broken. The probability $p(V)$ that the interaction energy between two water molecules is V, as obtained from computer simulation, is shown at various temperatures in Fig. 1. The hydrogen bond is related to the presence of a maximum in the attractive part of the interaction.

2. *Hydrogen-Bond Geometric Definition*: The hydrogen bond is intact if the distance R between two molecules is shorter than a value R_{max} and at the same time the angles in Fig. 2a are smaller than some

threshold values δ_0 and θ_0; otherwise the bond is broken. In Fig. 2b we have drawn the curve $X(R)$, that is the "quasi-component distribution function" obtained in ref. 22 at 10°C using a Rahman/ Stillinger type 2 potential.[7]

A statistical description of the H-bond network can be based on both the H-bond definitions[22,23]; different results are expected if the threshold parameters change (for example, the mean number n_{HB} of bonds per molecule is sensible to the threshold values). Mezei and Beveridge[22] have compared these two definitions of the hydrogen bond and the various potentials used for water. In every case, independently of the potential used, one obtains from computer simulation typical curves for statistical quantities such as those in Figs. 1 and 2, in which significant maxima, connected to hydrogen bonds, are present. These maxima are broadened by the spatial and orientational disorder of the liquid phase. If the same quantities are calculated by averaging over times of the order of τ_V, a narrowing of the maxima occurs and an increase of n_{HB} results, compared with the value of the corre-

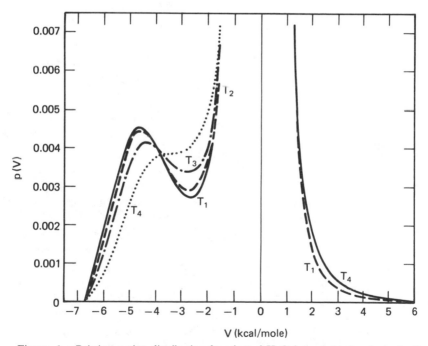

Figure 1. Pair-interaction distribution function of H_2O from molecular dynamics "experiments" with $\rho = 1$ g/cm^3 at $T_1 = -3°C$, $T_2 = 10°C$, $T_3 = 41°C$, $T_4 = 118°C$. (From Stillinger and Rahman 1974.[7])

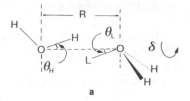

a

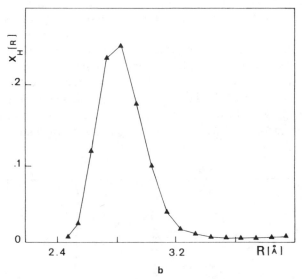

b

Figure 2. (a) Geometrical parameters used in the hydrogen bond definition. (b) Quasi-component distribution functions for the parameter R; ST2 water at 10°C. (From Mezei and Beveridge 1981[22].)

sponding quantity for structure I using the same threshold parameters for the H-bond definition.[12,13]

Definitions 1 and 2 are on-off definitions which lead us to associate with the hydrogen bond some characteristic parameters, such as mean energy of the bond, mean bond-breaking energy, probability p_B that a bond is randomly intact, and mean lifetime. The definition of these parameters is well grounded physically when the maxima in Figs. 1 and 2 belong to narrow curves, that is, when it is possible to define a mean value representative of the majority of the H-bond population. The statistical description of the H-bond network will be greatly simplified (independently of the potential, one assumes) if a sufficiently "strong" definition of the bond is used ($V_{HB} =$

-3–4 kcal/mol, or the corresponding geometric definition). With this assumption for the hydrogen bond, it turns out that a water molecule is able to form in the liquid at most four bonds and the fractions of molecules with i intact hydrogen bonds f_i ($i = 0, 1, \ldots, 4$) follow a binomial distribution[5,22,24]:

$$f_i = \binom{4}{i} p_B^i (1 - p_B)^{4-i} \tag{2.1}$$

where $p_B = n_{HB}/4$.

In this picture, liquid water maintains a "memory" of the tetrahedral structure of ice Ih and, moreover, in the liquid phase, the four hydrogen bonds (the "natural bonds") are virtually equivalent.

Using a "weak" definition of the bond, a greater number of hydrogen bonds than the "natural" is obtained, and the quantity n_{HB} tends to the number of nearest neighbors of a molecule. This is only slightly dependent on temperature.[25] Statistical data have been reported recently[21] for the H-bond network obtained at 25°C using various different potentials and a "weak" definition of the bond ($V_{HB} = -2.25$ kcal/mol). The f_i values obtained from computer simulation data with different potentials are listed in Table I.

If one considers, among the six bonds suggested by computer experiment, four bonds to be equivalent, with probability p_B, and the other two with probabilities p_B' and p_B'', respectively, the mole fractions f_i can be calculated

TABLE I

The Molar Fraction f_i of Water Molecules with i Intact Bonds at 25°C[a]

	(1)	(2)	(3)	(4)	(5)
f_0	0	0.0003	0	0	0.0006
f_1	0.007	0.0067	0.008	0.01	0.012
f_2	0.068	0.0666	0.088	0.096	0.0939
f_3	0.294	0.2976	0.337	0.341	0.3322
f_4	0.521	0.5232	0.468	0.462	0.4689
f_5	0.104	0.1012	0.095	0.086	0.0886
f_6	0.007	0.0044	0.004	0.004	0.0038

[a](1) From a computer simulation experiment using the ST2 potential (see ref. 21). (2) Calculated from Eq. (2.2) with $p_B = 0.867$, $p_B' = 0.13$, $p_B'' = 0.06$. (3) From a computer simulation experiment using TIPS4 potential (see ref. 21). (4) From a computer simulation experiment using SPC potential (see ref. 21). (5) Calculated from Eq. (2.2) with $p_B = 0.837$, $p_B' = 0.13$, $p_B'' = 0.06$.

with the following equations:

$$f_0 = (1 - p_B)^4 d$$
$$f_1 = 4p_B(1 - p_B)^3 d + (1 - p_B)^4 m$$
$$f_2 = 6p_B^2(1 - p_B)^2 d + 4p_B(1 - p_B)^3 m + (1 - p_B)^4 s$$
$$f_3 = 4p_B^3(1 - p_B)d + 4p_B(1 - p_B)^3 s + 6p_B^2(1 - p_B)^2 m \qquad (2.2)$$
$$f_4 = p_B^4 d + 4p_B^3(1 - p_B)m + 6p_B^2(1 - p_B)^2 s$$
$$f_5 = p_B^4 m + 4p_B^3(1 - p_B)s$$
$$f_6 = p_B^4 s$$

where

$$d \equiv (1 - p_B')(1 - p_B'')$$
$$m \equiv p_B'(1 - p_B'') + p_B''(1 - p_B')$$
$$s \equiv p_B' p_B''$$

These equations reduce to the binomial distribution of order 6 if $p_B = p_B' = p_B''$ and to the binomial distribution of order 4 if $p_B' = p_B'' = 0$.

In columns 2 and 5 of Table I are listed the f_i values calculated by Eqs. (2.2) using the sets of probability values p_B, p_B', and p_B'' in captions. The resulting agreement with "experiment" (i.e., computer simulation) is very satisfactory.

The two hydrogen bonds with probability p_B' and p_B'' are bonds with an attractive energy lower than that of the four equivalent bonds with probability p_B. In our picture of the dynamics we shall not consider the two low-probability bonds as important; this is equivalent to assuming a sufficiently "strong" definition of the hydrogen bond.

With these conditions we can go on to build up a model of the H-bond dynamics.

III.   HYDROGEN-BOND DYNAMICS

We shall introduce a stochastic variable η to simulate the random change with time of the number of hydrogen bonds of the tagged molecule. Five possible "states" for the molecule are obtained as the number of hydrogen bonds changes from 0 to 4. The corresponding state $\psi(t)$ of the liquid can be written

$$\psi(t) = \sum_{i=0}^{4} p_i(t)|\eta_i\rangle \qquad (3.1)$$

The time evolution of the probability $p_i(t)$ is given by the following master equation:

$$\frac{d}{dt}p_0(t) = -k_{01}p_0 + k_{10}p_1$$

$$\frac{d}{dt}p_1(t) = k_{01}p_0 - (k_{10} + k_{12})p_1 + k_{21}p_2$$

$$\frac{d}{dt}p_2(t) = k_{12}p_1 - (k_{21} + k_{23})p_2 + k_{32}p_3$$

$$\frac{d}{dt}p_3(t) = k_{23}p_2 - (k_{32} + k_{34})p_3 + k_{43}p_4$$

$$\frac{d}{dt}p_4(t) = k_{34}p_3 - k_{43}p_4$$

$$(3.2)$$

or, in compact notation,

$$\frac{d}{dt}p_i(t) = \sum_{j=0}^{4}(K_\eta)_{ij}p_j(t) = \left[K_\eta p(t)\right]_i \qquad (3.2')$$

where k_{ij} is the "transition rate" from the i to the j molecular state. In writing Eq. (3.2) we have assumed that simultaneous breaking or forming of two bonds are statistically negligible events.

Using the parameters τ_{HB} and τ_f, it is possible to define p_B through the following relations[26]:

$$p_B = \frac{\tau_{HB}}{\tau_{HB} + \tau_f} \qquad \text{or} \qquad \frac{p_B}{1 - p_B} = \frac{\tau_{HB}}{\tau_f} \qquad (3.3)$$

and to introduce the "transition rates" k_{ij} by the following simple considerations:

1. After a time equal to τ_{HB} (τ_f), one hydrogen bond of the molecule must have been broken (or formed) in the five molecular states. For example, for a molecule in state $|\eta_4\rangle$ there are four ways to get the state $|\eta_3\rangle$, each corresponding to breaking one of its hydrogen bonds. As there are ten different possibilities of breaking a hydrogen bond (4 in $4 \to 3$, 3 in $3 \to 2$, 2 in $2 \to 1$, 1 in $1 \to 0$), we get for the velocity k_{43}

$$k_{43} = \frac{4}{10}\frac{1}{\tau_{HB}} \qquad (3.4)$$

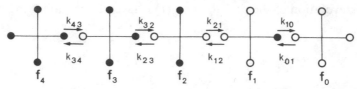

Figure 3.   A sketch of H-bond dynamics of H_2O. f_i is the mole fraction with i intact hydrogen bonds. (●) Intact hydrogen bond; (○) broken hydrogen bond. $k_{ii-1} = i/10\tau_{HB}$ ($i = 1,\ldots,4$); $k_{ii+1} = (4-i)/10\tau_f$ ($i = 0,\ldots,3$).

2.   The breaking and forming of a bond are statistically independent events.

For clarity's sake the k_{ij} are drawn, with their values as functions of τ_{HB}^{-1} and τ_f^{-1}, in Fig. 3 together with a sketch of the five molecular states involved in our picture for water. If the k_{ij} so defined are introduced into Eqs. (3.2), their equilibrium solution results in the binomial distribution obtained by computer simulation experiments for the fractions f_i. In the next section we shall present a model for the diffusional dynamics of water utilizing the master equations (3.2).

IV.   THE RMT AT WORK: LONG-TIME BEHAVIOR TAKING HYDROGEN-BOND DYNAMICS INTO ACCOUNT

As discussed in detail in Chapter VI, the RMT is especially suitable for organizing information provided both by computer and real experiment into a coherent, unified whole. The main steps of this procedure are as follows:

1.   The set of dynamical variables of interest is enlarged via inclusion of a few additional variables (usually termed auxiliary or virtual). This serves the twofold purpose of providing a simplified picture of the real thermal bath and recovering a distinct time-scale separation between relevant and irrelevant parts. In other words, the system of interest plus the set of virtual variables behaves like a mesoscopic system—a system with a time scale intermediate between the microscopic and the macroscopic. The first well-known example at the mesoscopic level is the Brownian particle of Einstein theory.[27]

2.   Projection techniques are applied to this mesoscopic system so as to serve the important purpose of exploring the long-time behavior with simple analytical expansions, which are, furthermore, closely related to the short-time behavior.

In this section we shall show how to apply these main ideas to water, availing ourselves of the prime results arrived at in the foregoing sections.

In Section III the H-bond dynamics are related to a stochastic variable η with five possible values $\eta_0, \eta_1, \ldots, \eta_4$, corresponding to the number of hydrogen bonds established by the tagged molecule with its neighbors. The RMT, on the other hand, suggests that we should simulate the environment of the tagged molecule through the intermediacy of a rototranslating virtual body, which is in turn driven by standard fluctuation-dissipation processes. The potential V governing the interaction between the real and virtual bodies will be assumed to depend on relative (orientation and translation) coordinates.

We are thus led to assume that V and the fluctuation-dissipation process driving the virtual body depends on the variable η so as to simulate the effects of the H-bond dynamics. For example, a strong interaction potential V accompanied by friction and stochastic torques (forces) of weak intensity simulates the solidlike properties of the environment when the tagged molecule is characterized by four hydrogen bonds. In this case a reasonable approximation is to simulate such an environment by ice Ih. In the opposite limit, a weak potential V with strong friction and stochastic torques (forces) simulates the diffusional properties of the unbounded molecule (liquid water at high temperature or in very dilute nonpolar solution).

The random motion of the stochastic variable η will also explore the states with intermediate values of potential and stochastic torques (forces). The macroscopic properties of the system will depend on suitable averages on all these states.

This leads to the following system of equations:

$$\dot{\boldsymbol{\mu}} = \boldsymbol{\omega} \times \boldsymbol{\mu}$$

$$\dot{\omega}_i = -\frac{\partial V/\partial\theta}{I_i^{(R)}} + r_i^{(R)}\omega_j\omega_k$$

$$\dot{\nu}_i = -\frac{\partial V/\partial\psi_i}{I_i^{(V)}} + r_i^{(V)}\nu_j\nu_k - \gamma_i^{(\nu)}(\eta(t),t)\nu_i + f_i^{(\nu)}(\eta(t),t)$$

$$\dot{\boldsymbol{\Delta}}_\Omega = \mathbf{k}(\nu_i,\omega_i,\boldsymbol{\Delta}_\Omega)$$

$$\dot{v}_i = -\frac{\partial V/\partial x_i}{m_R}$$

$$\dot{w}_i = -\frac{\partial V/\partial y_i}{m_V} - \gamma_i^{(w)}(\eta(t),t)w_i + f_i^{(w)}(\eta(t),t)$$

$$\dot{\boldsymbol{\Delta}}_T = \mathbf{w} - \mathbf{v} \tag{4.1}$$

The tagged molecule with mass m_R and moments of inertia $I_i^{(R)}$ interacts via the potential V with a virtual body with mass m_V and moments of inertia $I_i^{(V)}$. The moments of inertia are related to the corresponding principal axis systems. Δ_Ω and $\Delta_T \equiv \mathbf{y} - \mathbf{x}$ denote orientation and position of the virtual body relative to a real-body fixed frame. $\mathbf{x}$ and $\mathbf{y}$ are the coordinates of the centers of gravity of real and virtual body, respectively. The corresponding velocities are defined by $\mathbf{v} = \dot{\mathbf{x}}$, $\mathbf{w} = \dot{\mathbf{y}}$. The orientation of the principal axis system of the real body is referred to the laboratory frame via the Euler angles $\mathbf{\Omega} = (\alpha, \beta, \gamma)$, whereas the orientation of the virtual body is related to this principal axis system via $\Delta_\Omega = (\alpha', \beta', \gamma')$.

The best way of expressing the equation of motion of the angular velocity ω is to refer it to the principal axis system of the real body. This makes the nonlinear term $r_i^{(R)} \omega_j \omega_k$ (the Euler term) appear in the second equation of the set of (4.1).[28] $r_i^{(R)}$ is defined by $r_i^{(R)} \equiv (I_j^{(R)} - I_k^{(R)})/I_i^{(R)}$, (i, j, k) being a cyclic permutation of the indices $1, 2, 3$. The translational motion is described easily in the laboratory frame. The influence of the virtual body on the time evolution of ω is exerted via the derivative $\partial V/\partial \theta_i = \lim_{\Delta \theta_i \to 0} \Delta V/\Delta \theta_i$, where θ_i is an angle of rotation around the ith principal axis. In a similar way we can define v, $r_i^{(V)} v_j v_k$ via the infinitesimal rotation angle $\Delta \psi_i$ around the ith axis of the virtual body. The third and sixth equations of (4.1) are just standard Langevin equations, where $\gamma_i^{(\nu)}$ and $\gamma_i^{(w)}$ are the rotational and translational friction coefficients of the virtual body, and $f_i^{(\nu)}$, $f_i^{(w)}$ are the corresponding stochastic torques and forces. $f_i^{(\nu)}$ and $f_i^{(w)}$ are supposed to be Gaussian white noises defined by

$$\left\langle f_i^{(\nu)}(0) f_j^{(\nu)}(t) \right\rangle = 2 D_i^{(\nu)}(\eta) \cdot \delta_{ij} \cdot \delta(t) \qquad D_i^{(\nu)}(\eta) = \left\langle v_i^2 \right\rangle \gamma_i^{(\nu)}$$

$$\left\langle f_i^{(w)}(0) f_j^{(w)}(t) \right\rangle = 2 D_i^{(w)}(\eta) \cdot \delta_{ij} \cdot \delta(t) \qquad D_i^{(w)}(\eta) = \left\langle w_i^2 \right\rangle \gamma_i^{(w)}$$

The fourth equation of set (4.1) expresses the time derivative of Δ_Ω (the relative orientation between real and virtual body) in terms of $\mathbf{v}$, ω, and Δ_Ω themselves. For simplicity, the explicit analytical expression is omitted. The corresponding (very simple) translational term is given by the last equation of the set (4.1).

Although Eq. (4.1) formally depends on η, no relevant assumption has been made so far about the statistical properties needed to simulate correctly the environment of the tagged molecule.

In principle, since the potential V is not linear, Eqs. (4.1) can be used to simulate the non-Gaussian non-Markovian behavior of the variables ω and $\mathbf{v}$ as well as the rototranslational phenomena.[29] Note that this non-Gaussian, non-Markovian behavior depends on the presence of the virtual body and the nonlinear nature of the potential V in spite of the Markovian-Gaussian character of the fluctuation-dissipation process governing the stochastic torques $f_i^{(\nu)}$ and the stochastic forces $f_i^{(w)}$. Note also that the multiplicative nature (see Chapters II and X) of the stochastic variable η produces significant non-Gaussian effects on the variables of interest. This aspect will be discussed in Section VIII.

We shall not devote a great deal of attention to this problem (left for future investigation), focusing significant attention only on long-time behavior. This makes it legitimate to construct a rigorous "microscopic" theory for η with the remarks of the preceding section.

The Fokker-Planck equation to be associated with Eqs. (4.1) belongs to the family of stochastic Liouville equations.[30] Its explicit expression is

$$\frac{\partial}{\partial t}\rho(\boldsymbol{\Omega},\mathbf{x},\boldsymbol{\Delta_\Omega},\boldsymbol{\Delta}_T,\boldsymbol{\omega},\boldsymbol{\nu},\mathbf{v},\mathbf{w},\eta;t)=\Gamma\rho=(\Gamma_0+\Gamma_1)\rho(\ldots,\eta;t)$$

where Γ is the Fokker-Planck operator split into an unperturbed part Γ_0 and a perturbed part Γ_1. Their explicit expression is

$$\Gamma_0 \equiv \Phi_i^{(\mathrm{R})}\frac{\partial}{\partial\omega_i}-r_i^{(\mathrm{R})}\frac{\partial}{\partial\omega_i}\omega_j\omega_k+\Phi_i^{(\mathrm{V})}\frac{\partial}{\partial\nu_i}$$

$$-r_i^{(\mathrm{V})}\frac{\partial}{\partial\nu_i}\nu_j\nu_k+\gamma_i^{(\nu)}\frac{\partial}{\partial\nu_i}\nu_i+D_i^{(\nu)}\frac{\partial^2}{\partial\nu_i^2}+\psi_i^{(\mathrm{R})}\frac{\partial}{\partial\nu_i}+\psi_i^{(\mathrm{V})}\frac{\partial}{\partial w_i}$$

$$+\gamma_i^{(w)}\frac{\partial}{\partial w_i}w_i+D_i^{(w)}\frac{\partial^2}{\partial w_i^2}-\frac{\partial}{\partial\boldsymbol{\Delta_\Omega}}\mathbf{k}+(v_i-w_i)\frac{\partial}{\partial(\boldsymbol{\Delta}_T)_i}+K_\eta$$

$$\Gamma_1=-i\omega_iJ_i^{(\mathrm{R})}-v_i\frac{\partial}{\partial x_i}\qquad\qquad (4.2)$$

where

$$\Phi_i^{(\mathrm{R})}=\frac{\partial V/\partial\theta_i}{I_i^{(\mathrm{R})}}\qquad \Phi_i^{(\mathrm{V})}=\frac{\partial V/\partial\psi_i}{I_i^{(\mathrm{V})}}$$

$$\psi_i^{(\mathrm{R})}=\frac{\partial V/\partial x_i}{m_{\mathrm{R}}}\qquad \psi_i^{(\mathrm{V})}=\frac{\partial V/\partial y_i}{m_{\mathrm{V}}}$$

$iJ_i^{(\mathrm{R})}=\partial/\partial\theta_i$ defines the generator of the rotation $\Delta\theta_i$, and K_η is the operator defined by Eq. (3.2'). This operator drives the motion of the stochastic variable η and appears in Eqs. (4.2) according to the prescriptions of the stochastic Liouville equation.[30]

The slow variables of the system under study are $\mathbf{x}$, $\boldsymbol{\Omega}$, and η, thereby suggesting as the most suitable projection operator the following:

$$P\rho(\boldsymbol{\Omega},\mathbf{x},\boldsymbol{\Delta_\Omega},\boldsymbol{\Delta}_T,\boldsymbol{\omega},\boldsymbol{\nu},\mathbf{v},\mathbf{w},\eta;t)=\rho_{\mathrm{eq}}^{(B)}(\boldsymbol{\omega},\boldsymbol{\nu},\mathbf{v},\mathbf{w},\boldsymbol{\Delta}_T,\boldsymbol{\Delta_\Omega};\eta)\cdot\sigma(\boldsymbol{\Omega},\mathbf{x},\eta;t)$$

with

$$\sigma = \int d\omega \, d\mathbf{v} \, d\mathbf{v} \, d\mathbf{w} d\mathbf{\Delta}_T \, d\mathbf{\Delta}_\Omega \rho(\ldots, \eta; \, t)$$

where $\rho_{eq}^{(B)}$ denotes the equilibrium distribution of the system corresponding to $\Gamma_0 - K_\eta$ when the stochastic variable η assumes the value corresponding to a certain environment.

By applying the AEP of Chapter II to order Γ_1^2, we obtain for σ the following equation of motion:

$$\frac{\partial}{\partial t}\sigma = \left\{ -J_i \left(\int_0^\infty \langle \omega_i(0)\omega_i(t); \, \eta \rangle \, dt \right) J_i + \frac{\partial}{\partial x_i} \left(\int_0^\infty \langle v_i(0)v_i(t); \, \eta \rangle \, dt \right) \frac{\partial}{\partial x_i} \right.$$
$$\left. - iJ_i \left(\int_0^\infty \langle \omega_i(0)v_i(t) \rangle \, dt \right) \frac{\partial}{\partial x_i} + K_\eta \right\} \sigma \qquad (4.3)$$

The only "microscopic" feature still remembered by the system when described by Eq. (4.3) is that of the H-bond dynamics as simulated by the variable η. As mentioned in the introduction, the integrations in time appearing on the right-hand side of Eq. (4.3) are made legitimate by the fact that the correlation functions $\langle \omega_i(0)\omega_i(t); \, \eta \rangle$, $\langle v_i(0)v_i(t); \, \eta \rangle$ and $\langle \omega_i(0)v_i(t); \, \eta \rangle$ are supposed to be much faster than both the dielectric relaxation and self-diffusion processes. Henceforth we shall neglect the third term on the right-hand side of Eq. (4.3) concerning rototranslational phenomena. This assumption allows us to obtain two independent equations for rotation and translation, respectively.

As to rotational dynamics (and thereby dielectric relaxation), this implies that an equation of motion for the correlation function

$$\Phi_\mu^{(i)} = \langle \mathbf{\mu}(0) \cdot \mathbf{\mu}(t); \, \eta_i \rangle$$

can be derived. This is just the molecular dipole correlation function in an environment as defined when the variable η assumes the value η_i. In other words, to define this correlation function, we have to make an average over the molecules of the sample which by chance are found in this environment.

Let us come back to the problem of rotational diffusion. To derive the equation of motion of $\Phi_\mu^{(i)}(t)$, we have to remember first of all that the equations corresponding to the rotational part are written in the reference frame of the molecular principal axis. By making a transformation from this reference system to that of the laboratory, we have

$$\mu_k = \sum_l D_{kl}^{(1)}(\mathbf{\Omega}) T_l$$

where μ_k and T_l are the spherical components of the dipole moment in the laboratory and molecular frame of reference, respectively. The $D_{kl}^{(1)}$ are the first-rank Wigner matrices.[31] The rotational relaxation process is described practically only by the first term on the right-hand side of Eq. (4.3).

Let us denote by $\hat{\Phi}_\mu^{(i)}(z)$ the Laplace transform of $\Phi_\mu^{(i)}(t)$, where $z = i\omega$. Its explicit form (see Chapters I, III, and IV) is given by

$$\hat{\Phi}_\mu^{(i)}(z) = \cfrac{1}{z + \lambda_0(\eta_i) + \cfrac{\Delta_1^2(\eta_i)}{z + \lambda_1(\eta_i) + \Delta_2^2(\eta_i) \atop \ddots}}$$

The explicit expressions for the parameters λ_i and Δ_i^2 become more and more complicated as i increases. (Note, however, that the CFP of this volume should make it possible in principle to make such a calculation; see also Chapter III). For simplicity we limit ourselves to providing the explicit expressions of λ_0, which becomes

$$\lambda_0(\eta_i) = \frac{\displaystyle\sum_{jj'} \sum_{l,k=-1}^{1} T_l T_k \left(\int_0^\infty \langle \omega_j(0)\omega_{j'}(t); \eta_i \rangle\, dt \right) \cdot \int d\Omega\, D_{0l}^{(1)} J_j J_{j'} D_{0k}^{(1)}}{\displaystyle\sum_{jj'} \sum_{l,k=-1}^{1} T_l T_k \int d\Omega\, D_{0l}^{(1)} J_j J_{j'} D_{0k}^{(1)}}$$

If $\Delta_1^2/(\lambda_1 + \Delta_2^2) \ll 1$, $\lambda_0(\eta_i)$ is virtually the inverse of the time decay of the correlation function describing the state η_i, $D_R(\eta_i)$. A more general expression for this parameter is

$$D_R(\eta_i) = \left\{ \lim_{z \to 0} \hat{\Phi}_\mu^{(i)}(z) \right\}^{-1} = \lim_{z \to 0} \left\{ z + \lambda_0(\eta_i) + \cfrac{\Delta_1^2(\eta_i)}{z + \lambda_1(\eta_i) + \Delta_2^2(\eta_i) \atop \ddots} \right\}$$

When taking into account also the influence of the operator K_η driving the dynamics of the hydrogen bond and thereby changing the molecular environment with respect to time, one obtains the following equation for rotational diffusion:

$$\frac{d}{dt}\Phi_\mu^{(i)}(t) = -D_R(\eta_i)\Phi_\mu^{(i)}(t) + \sum_{j=0}^{4} (K_\eta)_{ij}\Phi_\mu^{(j)}(t) \qquad i = 0,1,\ldots,4$$

$$(4.4)$$

where $(K_\eta)_{ij}$ is defined by Eq. (3.2').

A description of translation is much simpler. By separating the translational from the rotational part in Eq. (4.3), we get

$$\frac{\partial}{\partial t}\sigma_T(\mathbf{x}, \eta;\, t) = \left\{ D_T(\eta)\nabla^2 + K_\eta \right\} \sigma_T(\mathbf{x}, \eta;\, t) \tag{4.5}$$

where

$$\sigma_T(\mathbf{x}, \eta;\, t) = \int d\mathbf{\Omega}\, \sigma(\mathbf{x}, \mathbf{\Omega}, \eta;\, t)$$

and

$$D_T(\eta) = \frac{1}{3}\int_0^\infty \langle \mathbf{v}(0)\cdot\mathbf{v}(t);\, \eta\rangle\, dt$$

Equations (4.4) and (4.5) are reminiscent of each other in the sense that the diffusional behavior of water in the long-time region results from different, randomly explored, environments, each one being characterized by well-defined diffusion coefficients. Therefore, the corresponding macroscopic correlation function turns out to be an average over several molecular environments, the permanence time in each environment and the transition rate from one to another being determined by a well-defined statistical process.

We shall show later that these simple equations provide a coherent explanation for water diffusion from the supercooled region to the boiling point, thereby showing that at least in this thermodynamic region the hydrogen bond plays an important role.

V. PARAMETERS OF HYDROGEN-BOND DYNAMICS

To compare the results of Eqs. (4.4) and (4.5) with experimental data, we must derive the parameters necessary to describe the H-bond dynamics (p_B and τ_{HB}). As mentioned in Section III, the number of unbroken (intact) bonds per molecule, n_{HB} (and therefore $p_B = n_{HB}/4$), obtained in the computer simulation experiments depend on the way the bond itself is defined and on the kind of potential used.

The soundest energetic definition seems to be that of assuming $V_{HB} = -3.5$ kcal/mol; this value corresponds to the temperature-invariant point exhibited by $p(V)$ (see Fig. 1).[7] However, the dependence of n_{HB} on temperature turns out to be more or less marked according to whether the definition is stronger or weaker. As a consequence p_B cannot be derived straightforwardly from the computer simulation alone, even if this turns out to be useful as a checking tool. Furthermore, in our opinion, a theory for dynamics should be comprehensive and coherent, able to account for the thermodynamic and structural properties of water.

In the last few years, several models have been developed which more or less allow the anomalies of the static behavior of liquid water to be explained qualitatively. These can be divided into the following groups:

1. Models that focus on the statistical consequences and correlations deriving from peculiar properties such as connectivity and the directional nature of the hydrogen bond.[5,32]
2. Models that aim to single out some cooperative mechanism driving the processes through which a network of bonds is built up.[33]
3. Models that provide a description of the hydrogen bond in the liquid state and rely on a mental picture of the liquid structure, thought of as being a continuum deformation of ice Ih (continuum network model) and obtained through special configurational averages of structure V.[13,19,34]

Models of type 1 are in line with a description such as ours relying on a discrete stochastic variable.

The third group of models succeeds in accounting for some thermodynamic properties in the stable region of water ($0°C < T < 100°C$). This could be used as a starting point for describing the short-time region by properly defining a continuous stochastic variable that could be made to depend suitably on the stochastic parameters of Eq. (4.1).

Whereas models of type 3, at least in their original forms, seem more suitable to describe "normal" water,[35] models 1 and 2 have been used specifically to support the attempts made to describe the anomalies of water in the low-temperature and supercooled regions. These anomalies can be traced to the existence of peculiar microregions in the liquid sample [patches f_4 (ref. 5) or bicyclic octamers[33]] which exhibit special features when compared with the remainder of the liquid sample. These microregions can be thought of as dynamic structures that are created or destroyed continuously with the characteristic times of the H-bond dynamics. In both the model groups such microregions become much wider as temperature decreases, especially when the supercooled region is attained. This should provoke a parallel increase of p_B. In spite of the connective and cooperative phenomena, a simple statistics can describe the network. This is again due to the fact that, from a statistical point of view, the four hydrogen bonds which can be established by the tagged molecule with its environment are indistinguishable from each other. Eventually, peculiar features, intramolecular in character, will be proven to be more relevant at higher temperatures where the dynamics is faster and the hydrogen bond less important. We have then to understand how, with changing temperature and pressure, the behavior of the parameters p_B and τ_{HB}, which are necessary for describing the hydrogen bond dynamics, can be obtained.

A. Calculation of the Probability p_B

The model which allows a quantitative evaluation of p_B is that of Stanley and Teixeira,[5] the most important aspects of which are the following:

1. Both in the stable and supercooled regions, water is found to be above the percolation threshold p_c ($p_c = 0.387$).
2. The effect of connectivity, resulting from the directional character of the hydrogen bond, is to compel the molecules of species f_4 to gather in microregions with different properties from the remainder of the liquid sample.
3. These regions provoke anomalous fluctuations of density and entropy, from which, at least at a qualitative level, an explanation can be derived for the increase of the isothermal compressibility and specific heat C_p in the metastable region.

If this model is accepted as reliable, the microregions of molecules with four intact hydrogen bonds can be thought of as having a density ρ_4 virtually equal to that of the tetragonal structure of ice Ih at the same temperature.

The remainder of the liquid sample will have an average density ρ_{gel} which can be derived by extrapolating to low temperatures the behavior of the density of water below the percolation threshold. In this region the connectivity phenomena concerning the fraction $f_4 = p_B^4$ are probably negligible. The average density can be written as follows:

$$\bar{\rho} = \rho_4 f_4 + \rho_{gel}(1 - f_4) \tag{5.1}$$

from which we derive

$$p_B = \frac{(\rho_{gel} - \bar{\rho})^{1/4}}{(\rho_{gel} - \rho_4)^{1/4}} \tag{5.2}$$

Equations such as (5.1) are also found in the two-states theories of water.[36,37] These theories aim at explaining all the properties of water via the peculiar features of an open (icelike) and closed species of water (the remainder of the liquid sample). According to these theoretical approaches, the thermodynamic parameters (density, enthalpy, dependence on temperature and pressure of the probability of belonging to one species, etc.) characteristic of the two species must be defined via a compromise. In contrast to what happens in the case of density, the definition of these parameters turns out to be unsatisfactory. Geometrical arguments show that it is reasonable to give the patches f_4 the same density as ice, but this assumption cannot clearly be extended to the other thermodynamic properties such as isothermal compressibility, molar entropy, and so on. Furthermore, Eq. (5.1) neglects the boundary phenomena (i.e., the phenomena of interaction between microregions and the remainder of the liquid sample). This approximation, which proves to be quite good for density,[38] does not apply at all, for example, to the molar enthalpy.[39] In consequence, Eq. (5.2) must be thought of as a reasonable way of calculating $p_B(T)$ when only the geometrical properties of the patches f_4 are taken into account.

When dealing with the problem of evaluating $p_B(T)$ via Eq. (5.2), the greatest difficulty lies in the evaluation of $\rho_{gel}(T)$. A physically well-grounded way of calculating this quantity is to fit the liquid density throughout the whole coexistence curve from temperature close to the critical point ($\sim 374°C$) to about $170°C$. It is reasonable to assume that in this region liquid water is found below the percolation threshold, that is, the fraction f_4 is quite small. By extrapolating the density behavior from this temperature range to temperatures lower than $170°C$ and to the supercooled region, one obtains the contribution of ρ_{gel} to the density of water at these temperatures.

The corresponding fitting procedure is like that normally used in the critical region, with an additional correction term of the second order in $t = (T_c - T)/T_c$:

$$\rho_L = \rho_{gel} = \rho_c \left[1 + B_0 t^\beta + B_1 t + B_2 t^2 \right]$$

with

$$\rho_c = 0.320 \text{ g/cm}^3 \qquad B_0 = 2.63458 \qquad B_1 = 0.474217$$

$$B_2 = -0.6424516 \qquad T_c = 374.15°C \qquad \beta = 0.37$$

When these parameters are used, the disagreement with the experimental data is within 0.1% throughout the range 372°C to 170°C, that is, it is of the same order of magnitude as the disagreement among the results of different experiments.[40,41]

As shown in Fig. 4, a systematic deviation of ρ_{gel} from $\bar{\rho}$ occurs in the region below 150°C.

Recently, Leblond and Hareng[42] have obtained almost the same results for ρ_{gel} (ρ_N in ref. 42) by extrapolating the expansivity of superheated water.

The ice density to be used is obtained by fitting the ice density from $-100°C$ to $0°C$ (ref. 9, p. 104) and extrapolating this fitting up to temperatures greater than $0°C$.

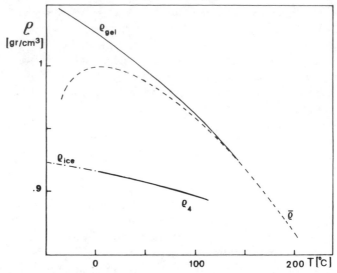

Figure 4. $\bar{\rho}$, density of water measured at different temperatures; ρ_4, mean density of the 4-bonded water fraction obtained by extrapolating the ρ_{ice} vs. T curve; ρ_{gel}, mean density of water obtained by extrapolating to low temperatures $\bar{\rho}(T)$ of the liquid. $\bar{\rho}(T)$ between the critical point and 170°C is the density of the liquid on the coexistence curve.

We use the following relationship:

$$\rho_4 = \rho_{ice}$$

$$= 0.91671 - 1.67378 \times 10^{-4}T - 3.58125 \times 10^{-7}T^2 + 2.0833 \times 10^{-10}T^3$$

with T expressed in °C.

The values of $\bar{\rho}$ assumed are from ref. 43 for water in the stable region and from ref. 44 for the supercooled region.

Using the values obtained for ρ_{gel}, ρ_4, and $\bar{\rho}$, we can calculate $p_B(T)$ from Eq. (5.2). The resulting $p_B(T)$ values are shown in Fig. 5 together with the values calculated using the computer simulation data (n_{HB}) of Yamamoto et al.[24] These authors used in their experiment an attractive cutoff potential ($R = 4.4$ Å) and an energetic definition with $V_{HB} = -3.5$ kcal/mol. The $p_B(T)$ values obtained by Angell[4] are also drawn in Fig. 5.

Comparing the values of $p_B(T)$ obtained from Eq. (5.2) with the p_B ($p_B = n_{HB}/4$) values resulting from various simulation experiments in structure I, it follows that the agreement is satisfactory if an energetic definition with a threshold $V_{HB} \cong -3.5$ kcal/mol is used, as in Yamamoto et al.[24] This value of V_{HB} is virtually equal to the value corresponding to the temperature-

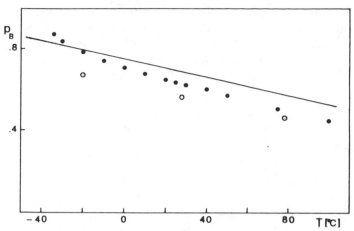

Figure 5.   The probability p_B vs. T for water; (●) from Eq. (5.2); (○) from ref. 17; (—) from ref. 4.

invariant region shown in Fig. 1; it represents a "strong" definition of the hydrogen bond.

If we calculate $p_B(T)$ using infrared and Raman spectroscopic data,[10,11] $p_B(T)$ values greater than those in Fig. 1 are obtained ($p_B = 0.9$ against the value 0.7 at 0°C); that is, these techniques give data comparable to those obtained using a "weak" definition of the bond. A possible explanation is

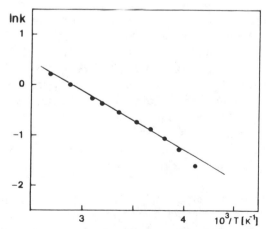

Figure 6. $\ln k = \ln(1 - p_B)/p_B$ vs. $10^3/T$ for water; ● calculated using p_B values in Table II, (—) Arrhenius fitting curve.

that the OH free fraction these techniques measure is lower than the real one, as very likely there are free OH's in the structured part which are "screened" by the environment.

Using the usual relation for the rate processes, one obtains the activation energy E_{HB} for H-bond formation with the equation

$$k(T) = \frac{1 - p_B(T)}{p_B(T)} = A e^{-E_{HB}/RT}$$

Figure 6 illustrates $k(T)$, which shows an Arrhenius trend in the range $-20°C$ to $+100°C$ with $A = 28$, a value which is much greater than $A = 1$ as predicted by two-state theory. The value of E_{HB} ($E_{HB} = 2.4$ kcal/mol) agrees with the one calculated with data from different experimental techniques.[10]

B. The Hydrogen Bond Lifetime τ_{HB}

The hydrogen bond lifetime τ_{HB}, a fundamental parameter in our picture for water, has also been considered by other authors, some of whom[18,45,46] suggested the use of depolarized Rayleigh scattering experiments to measure this quantity. The observed frequency spectrum consists of two distinct Lorentzians, the linewidth of the broader one being interpreted by these authors as the inverse of τ_{HB}.

This has been explained through the fact that the scattering amplitude is proportional to the optical anisotropy correlation function, a quantity which the breaking of a hydrogen bond modulates in time, a process that affects the optical anisotropy of water. We see, therefore, that the linewidth is the inverse of the average lifetime of the hydrogen bond.

Although three experimental investigations provide results slightly different from each other,[18,45,46] an Arrhenius-like behavior for τ_{HB} [$\tau_{HB} = A' \exp(E_s/RT)$ with $E_s = 2.5–3$ kcal/mol] is always observed. Figure 7 reports, vs. $10^3/T$, the Arrhenius fitting of the experimental τ_{HB} values obtained in ref. 18. Recently some authors[47,48] succeeded, using different potentials in molecular dynamics experiments, in evaluating the dependence on temperature of τ_{HB}. Although the τ_{HB} values so obtained are quite different from those in Fig. 7, the $\tau_{HB}(T)$ behavior is virtually the same as that of Fig. 7.

In Fig. 7 we have also shown $\tau_f(T)$, that is, the temperature dependence of the time during which the hydrogen bond remains broken. $\tau_f(T)$ has been calculated with Eq. (3.3), and it is nearly temperature-independent in the range considered ($E_{HB} \simeq E_s$).

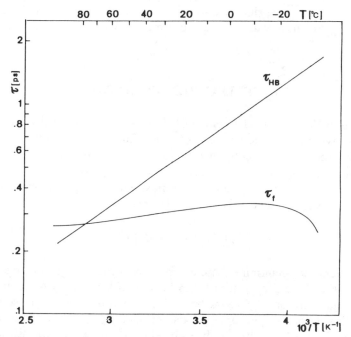

Figure 7. The mean lifetime of the hydrogen bond of water τ_{HB} vs. $10^3/T$ from ref. 18. The mean time during which the bond is broken τ_f vs. $10^3/T$ calculated from Eq. (3.3).

The different behavior of τ_{HB} and τ_f with varying temperature depends on the different meanings of these two times; the breaking of a hydrogen bond depends on thermal phenomena alone, while the formation process depends on both the same thermal phenomena and the number of broken hydrogen bonds available. These are phenomena of opposite types which in the stable region of water tend to cancel one another.

In the first approximation these two parameters would not depend on the environment. However, this contradicts physical intuition, which suggests, for example in the structured regions, a higher value of τ_{HB}. This is not the case, because discriminating among the different environments ultimately results in a deviation from the binomial distribution, in contradiction with computer simulation data. If a dependence on the environment is present it must be negligible so that a single value of τ_{HB}, characterizing all the environments, can be used. In contrast, the peculiar characteristics of the environment (geometric features of the network and spatial bond distribution) will affect the diffusional coefficients markedly. In consequence, five diffusional coefficients will be introduced for the five environments.

In the following we shall use for p_B the values resulting from Eq. (5.2) and for τ_{HB} the mean value of the experimental data reported in ref. 18 (see Fig. 7).

VI. DIELECTRIC RELAXATION

In Section IV we derived a system of differential equations describing the rotational dynamics of the dipole correlation functions of the tagged molecule in different environments. In Eq. (4.4) the quantities depending on the environment are $\Phi_\mu^{(i)}(t)$ and the diffusion coefficients $D_R(\eta_i)$.

The solution of Eq. (4.4) can be obtained with a diagonalization procedure which gives for the eigenvalues λ_i the equation

$$\det \left\| \left[D_R(\eta_i) - \lambda + K_\eta \right] I \right\| = 0 \tag{6.1}$$

In consequence, to determine these eigenvalues, which are the inverse of the relaxation times, we simply need to find a way of evaluating the coefficients of rotational diffusion $D_R(\eta_i)$ in the various environments and to solve Eq. (6.1). As an illustrative example, let us consider $D_R(\eta_4)$. The tetra-bonded molecules (f_4) tend to clump with each other, so we are led to assume that the diffusion coefficient in this environment is given by a mean value with the same order of magnitude as $1/\tau_D^{ice} = 10^{-7} \; \text{ps}^{-1}$ at 0°C. As to $D_R(\eta_0)$ (the rotational diffusion coefficient in the environment without intact hydrogen bonds), this coefficient must be of the same order of magnitude as the rotational diffusion time found in very dilute aqueous solutions in nonpolar liquids[49] (benzene, carbon tetrachloride, cyclohexane, etc), that is, $D_R^{sol}(\eta_0) = 25 \; \text{ps}^{-1}$ at 25°C. On the other hand, in the range from -30°C to 100°C the coefficients k_{ij}, which define the H-bond dynamics, turn out to be $0.1 < k_{ij} < 1.5 \; \text{ps}^{-1}$. As a consequence, $D_R(\eta_4) \ll k_{ij}$, whereas $D_R(\eta_0) \gg k_{ij}$. These results suggest a first-approximation calculation corresponding to the conditions

$$D_R(\eta_0), D_R(\eta_1) \gg k_{ij} \quad \text{and} \quad D_R(\eta_4), D_R(\eta_3), D_R(\eta_2) \ll k_{ij} \tag{6.2}$$

On intuitive grounds we can imagine the molecules with 2, 3, 4 intact hydrogen bonds as being dynamically hindered (note that this is the most structured part in the network), thereby characterized by a low diffusion coefficient, whereas molecules with 0 or 1 bonds and placed either outside or at the boundary of the most structured regions have larger diffusion coefficients.

This point of view is similar to both that of Magat and Reinisch[50] and that of Anderson.[51]

Equation (4.4) [obtained by applying the AEP to the Fokker-Planck equations of Eq. (4.1)] is a five-state generalization of Eq. 3 of ref. 51. Moreover, in Eq. (4.4) the transition probabilities from one state to another are expressed explicitly in terms of the parameters of the H-bond dynamics.

When the H-bond network is structured enough to bring water above the percolation threshold, the largest relaxation time can be regarded as being, at a first level of approximation, the time of residence in the structured region ("immobile" fraction in refs. 5 and 52). This relaxation time corresponds to the characteristic time of the principal dielectric relaxation band. The theoretical results obtained with the conditions (6.2) are illustrated in Fig. 8. Figure 8 also shows the experimental results of refs. 52 and 53. The full line corresponds to the lowest eigenvalue λ_0 in the set of five obtained from Eq. (6.1).

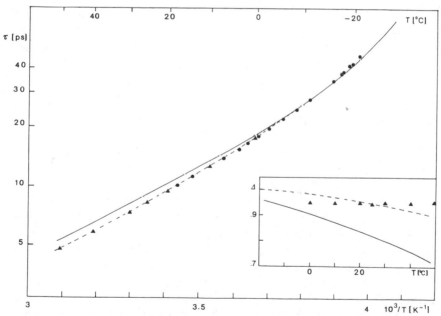

Figure 8. Dielectric relaxation time of water vs. $10^3/T$; (●) from ref. 52; (▲) from ref. 53; (——) $(\lambda_0)^{-1}$ as calculated from Eq. (6.1) using the conditions (6.2); (---) $(\lambda_0)^{-1}$ as calculated from Eq. (6.1) using the following values for the rotational diffusion parameters: $D_R(\eta_0) = 432\exp(-850/T)$ ps^{-1}; $D_R(\eta_1) = 154\exp(-1000/T)$ ps^{-1}; $D_R(\eta_2) = 45.2\exp(-1814/T)$ ps^{-1}; $D_R(\eta_3) = D_R(\eta_4) = 0$. Insert: The relative amplitude of the first dielectric relaxation band of water Δ_0 vs. T; (▲) $(\varepsilon_0 - \varepsilon_\infty)/(\varepsilon_0 - n^2)$ calculated from the experimental values in ref. 53; (——) Δ_0 amplitude calculated from Eq. (6.3); (---) Δ_0 amplitude calculated from Eq. (6.4).

As reasonably expected, the agreement between λ_0 and $1/\tau_D$ is especially good in the low-temperature region. Furthermore, the other eigenvalues are proven to correspond to frequencies much higher than the characteristic frequency of the first band. Therefore, this theoretical result reproduces the experimental data which, at microwave frequencies, can be fitted by a single dielectric relaxation time.

For a more accurate calculation, a knowledge of $D_R(\eta_i)$ is necessary, whereas with assumptions (6.2) the result is independent of the particular values of these coefficients.

A reasonable evaluation of $D_R(\eta_i)$ can be given on the following considerations and hypotheses:

1. $D_R(\eta_i)$ vs. T shows an Arrhenius-like behavior with an activation energy E_i increasing with i ($i = 1,\ldots,4$).
2. The condition $D_R(\eta_4)$, $D_R(\eta_3) \ll k_{ij}$ is valid.
3. The rotational diffusion process related to $D_R(\eta_1)$ is responsible for the observed deviations from ideal Debye behavior of water in the terahertz region.[53,54]
4. The activation energies of $D_R(\eta_0)$ and $D_R(\eta_1)$ are assumed equal to the corresponding activation energies of $D_T(\eta_0)$ and $D_T(\eta_1)$. (The physical basis of this assumption is the observed temperature independence of the quantity $\bar{D} \cdot \tau_D$ for a nonassociated liquid, $\bar{D}$ being the self-diffusion coefficient.)

The activation energies E_0 and E_1 used in this second calculation will be obtained in the following section.

To evaluate $D_R(\eta_2)$ we have not found an independent physical argument, so we obtain this coefficient from Eq. (6.1), having assigned to λ_0 the value $1/\tau_D$ and to the other $D_R(\eta_i)$ the values obtained using the arguments already given. The $D_R(\eta_2)$ so calculated increases with T as expected and also exhibits an Arrhenius trend with an activation energy $E = 3.5$ kcal/mol, about the same value observed for the dielectric relaxation of methyl alcohol. This is a particularly encouraging result, as in the H-bond structure of this alcohol, linear chains—that is environments like the $|\eta_2\rangle$ in water—are predominant. The values of $D_R(\eta_i)$ are listed in Table II.

To calculate the amplitudes we have used a procedure like that in ref. 51, assuming as the initial state $[\Phi_\mu^{(i)}(0) = 1]$ the state of the system having all the dipoles aligned and distributed in the various environments according to the binomial distribution. The total correlation function is therefore

$$\Phi(t) = \sum_{i=0}^{4} f_i \Phi_\mu^{(i)}(t) = \sum_{i=0}^{4} \sum_{j=0}^{4} F_{ij} e^{-\lambda_j t} = \sum_{j=0}^{4} \Delta_j e^{-\lambda_j t} \qquad (6.3)$$

TABLE II

Calculated and Measured Values of Some Dielectric Parameters of Liquid H_2O
at Various Temperatures[a]

T (°C)	(1) τ_{HB}, ps	(2) P_B	(3) $\tau = 1/\lambda_0$, ps	(4) τ_D, ps	(5) $D_R(\eta_0)$, ps^{-1}	(6) $D_R(\eta_1)$, ps^{-1}	(7) $D_R(\eta_2)$, ps^{-1}	(8) Δ_0	(9) $1/\lambda_1$, ps	(10) Δ_1
−30	1.504	0.835	77.2	83.3$^+$	13.1	2.52	0.026	1	0.275	0
−20	1.206	0.783	41.4	42.7	15.0	2.97	0.032	1	0.25	0
−10	0.983	0.741	26.5	26.3	17.1	3.45	0.042	1	0.224	0
0	0.814	0.706	18.5	17.7	19.2	3.96	0.06	0.98	0.20	0.02
10	0.683	0.675	13.7	12.7	21.5	4.51	0.078	0.973	0.178	0.027
20	0.579	0.645	10.4	9.36	23.8	5.1	0.105	0.956	0.16	0.04
25	0.536	0.631	9.16	8.27	25	5.4	0.112	0.948	0.152	0.045
30	0.497	0.618	8.14	7.3	26.2	5.7	0.127	0.938	0.144	0.051
40	0.431	0.591	6.5	5.81	28.6	6.32	0.143	0.918	0.131	0.063
50	0.377	0.566	5.3	4.75	31.1	6.98	0.157	0.897	0.12	0.077
75	0.278	0.505	3.36	2.91$^+$	37.6	8.71	0.247	0.829	0.097	0.12
100	0.214	0.446	2.27	1.93$^+$	44.3	10.6	0.35	0.746	0.08	0.165

[a](1) The H-bond mean lifetime, τ_{HB}, from ref. 18. (2) The probability that an hydrogen bond is randomly intact, P_B, from Eq. (5.2). (3) The relaxation time calculated from Eq. (6.1) with conditions (6.2). (4) Measured relaxation time τ_D, from refs. 52 and 53; ($+$) τ_D, values foreseen by our model using the parameters in columns 5–7, which list the rotational diffusion parameters utilized for a more accurate calculation of dielectric behavior. (8) The principal relaxation band amplitude Δ_0. (9) and (10) The characteristic time $1/\lambda_1$ and the amplitude Δ_1 of the only other significant band, as calculated from Eqs. (6.1) and (6.4), respectively.

where λ_j and Δ_j are the eigenvalues and related amplitudes, respectively.

The dielectric spectrum of water predicted by our model refined in this way exhibits some interesting features, in good agreement with those experimentally observed:

1. The amplitude of the first band is $\Delta_0 \cong f_4 + f_3 + f_2$.
2. There appears one second band, which becomes more and more important as temperature increases and which has the same temperature dependence as f_1. Its value always remains much lower than Δ_0.

These findings support the qualitative picture for dielectric relaxation in which the molecule is seen to spend a time equal to τ_D in the structured part of the liquid and, after having escaped, to relax essentially in the state $|\eta_1\rangle$.

However, in some features the calculated spectrum does not agree with experiment. Indeed, as the experimental first band amplitude is practically temperature-independent, being $(\varepsilon_0 - \varepsilon_\infty)/(\varepsilon_0 - n^2) = 0.95$ between 0°C and 50°C,[53] the calculated Δ_0 varies from 0.916 to 0.75 in the same temperature interval. The approximations made in developing the model, particularly the assumption that the dipole moment is the same in all the states $|\eta_i\rangle$, are responsible for these discrepancies. We expect, indeed, that μ in $|\eta_4\rangle$ is larger than in $|\eta_0\rangle$,[55,56] and it turns out that this assumption affects the amplitude Δ_i, not the eigenvalues λ_i. However, much information can be gained on the dynamic behavior of water, which, moreover, is not affected by the difficulties met in the case of a polar nonassociated liquid, where to link the microscopic to the macroscopic dielectric relaxation time the problem of reaction field and that of the dipolar long-range interactions must be accounted for.[57] In water, the dielectric relaxation behavior (principal band) is governed essentially by the H-bond network, where the molecule interacts only with its first neighbors through the hydrogen bond.

To account for the "true" dipole moment in the states $|\eta_i\rangle$, one must consider not only the first neighbors of the water molecule but also the small contribution of the higher-order neighbors (see ref. 9, p. 102), that is, one must face the same difficulties as those met when calculating the dipole static correlation function [the Kirkwood $g_k(\eta_i)$ factor].

After solving this problem we should have reduced our model to a five-states model for water, each state with its $g_k(\eta_i)$ and $\mu(\eta_i)$ factor, that is, to a model like those considered before by other authors.[55] In this case we can write the total correlation function $\Phi(t)$ as

$$\Phi(t) = \frac{\displaystyle\sum_{i=0}^{4} f_i g_k(\eta_i)\mu^2(\eta_i)\Phi_\mu^{(i)}(t)}{\displaystyle\sum_{i=0}^{4} f_i g_k(\eta_i)\mu^2(\eta_i)} \tag{6.4}$$

If we use the $\mu(\eta_i)$ and $g_k(\eta_i)$ values at $T = 25°C$ given in ref. 58, and if, for simplicity's sake, we assume that these are temperature-independent, we can calculate the amplitudes of the bands. The results are shown in the insert of Fig. 8.

This improvement in evaluating the "true" dipole moment affects the calculation in the correct sense; that is, the disagreement between the calculated band and the experimental one decreases. However, for a completely satisfactory model, one really allowing prediction of the observed dielectric spectrum of water, much work is still necessary. In this direction we mention two recent papers by Stillinger[56] and Carnie and Patey.[59] In Table II we have listed all the data obtained from our model.

VII. TRANSLATIONAL DIFFUSION

Starting from Eq. (4.5) and projecting onto the states $|\eta_i\rangle$, one obtains $\sigma(\mathbf{x}, \eta_i, t) = G_{si}(\mathbf{x}, t)$, that is the projected Van Hove function. This function gives the probability that a molecule in the state $|\eta_i\rangle$ is at x at time t if it was at the origin $(x = 0)$ at $t = 0$.

Equation (4.5) changes into the equation system

$$\frac{\partial}{\partial t} G_{si}(x, t) = D_T(\eta_i)\nabla^2 G_{si}(x, t) + \sum_{j=0}^{4} (K_\eta)_{ij} G_{sj}(x, t) \qquad (7.1)$$

where

$$D_T(\eta_i) = \frac{1}{3} \int_0^\infty \langle v(0) \cdot v(t); \eta_i \rangle \, dt \qquad i = 0, 1, \dots, 4$$

By transformation to k space, one obtains the following equation system for the intermediate scattering functions $F_{si}(k, t)$:

$$\frac{d}{dt} F_{si} = -D_T(\eta_i)k^2 F_{si}(k, t) + \sum_{j=0}^{4} (K_\eta)_{ij} F_{sj}(k, t) \qquad i = 0, 1, \dots, 4$$

$$(7.2)$$

We have thus a system similar to that solved already for rotational diffusion; the only difference is the presence of the factor k^2 in the terms containing the diffusion coefficient $D_T(\eta_i)$. From Eqs. (7.2), with the initial condition $F_{si}(k, 0) = 1$ (this is equivalent to putting $G_{si}(x, 0) = f_i \delta(x)$ and the normalization condition $\Sigma_{i=0}^{4} \int G_{si}(x, t) \, dx = 1$ in Eqs. (7.1), one obtains $F_{si}(k, t)$, and the result for the total intermediate scattering function $F_s(k, t)$ is

$$F_s(k, t) = \sum_{i=0}^{4} f_i F_{si}(k, t) \qquad (7.3)$$

If one returns to ω space by Fourier transforming, one obtains the self-dynamical form factor $S_s(k, \omega)$, the real part of which is

$$\mathrm{Re}\, S_s(k, \omega) = \sum_{i=0}^{4} \frac{A_i(k^2)\lambda_i^T(k^2)}{\omega^2 + \lambda_i^{T2}(k^2)} \qquad (7.4)$$

The diffusion coefficient $\overline{D}$ [defined as $\overline{D} = \lim_{k^2,\,\omega^2 \to 0} S_s(k, \omega)$] is given by the equation

$$\overline{D} = \sum_{i=0}^{4} D_T(\eta_i) f_i \qquad (7.5)$$

that is, the mean, weighted on the equilibrium distribution, of the diffusion coefficients in the various states $|\eta_i\rangle$. To evaluate the coefficients $D_T(\eta_i)$, we make the following assumptions:

1. The temperature behavior of $D_T(\eta_i)$ is of the Arrhenius type, and the activation energies E_i increase with i [$D_T(\eta_i) = A_i e^{-E_i/RT}$].
2. $D_T(\eta_4)$ and $D_T(\eta_3)$ have the value of the self-diffusion in ice Ih ($\cong 10^{-11}$ cm^2/s), that is, they are negligible.
3. $D_T(\eta_2)$ has the activation energy of self-diffusion in MeOH ($E_2 = 2.7$ kcal/mol).[60]
4. $D_T(\eta_0)$ and $D_T(\eta_1)$ are responsible for the "normal" behavior of water.[61]

The diffusion coefficient $\overline{D}$ of a nonassociated liquid along the coexis-

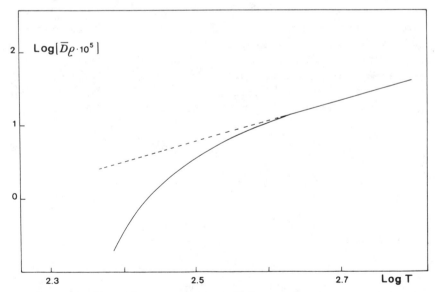

Figure 9. $\log(\overline{D}\rho \times 10^5)$ vs. $\log T$ for H$_2$O. $\overline{D}$ and ρ are in cgs units and T in kelvins. (---) The self-diffusion behavior of "normal" water [see Eq. (7.6)]; (——) the fitting curve of the experimental results (see ref. 61).

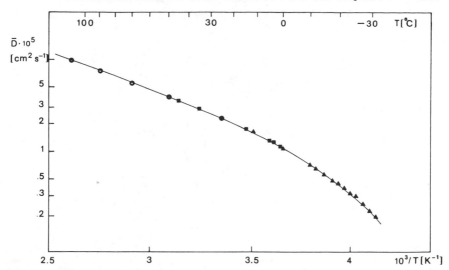

Figure 10.   Self-diffusion coefficient of water $\bar{D}$ vs. $10^3/T$; (◯) experimental values from ref. 64; (■) from ref. 63; (▲) from ref. 62; (——) calculated from Eq. (7.5) and the following values of the translational diffusion parameters: $D_T(\eta_0) = 130 \exp(-850/T) \times 10^{-5}$ cm^2/s; $D_T(\eta_1) = 174 \exp(-1000/T) \times 10^{-5}$ cm^2/s; $D_T(\eta_2) = 394.5 \exp(-1350/T) \times 10^{-5}$ cm^2/s; $D_T(\eta_3) = D_T(\eta_4) = 0$.

tence curve (for example, H_2S) is given by the relation

$$D_N\rho = AT^{2.8} \cong Be^{-E/RT} \tag{7.6}$$

with $E \cong 1.8$ kcal/mol. Liquid water, along its coexistence curve, begins to deviate from the "normal" behavior described by Eq. (7.6) at about 150°C (see Fig. 9).

For $T > 150$°C, water is essentially in the states $|\eta_0\rangle$ and $|\eta_1\rangle$; as the statistical weights of the two states are about equal, it follows that

$$D_T(\eta_0) + D_T(\eta_1) = 2D_N$$

Taking into account assumption 1 and the relations (7.6), if one utilizes the experimental data for $\bar{D}$ of water at two temperatures (for example, 0°C and 25°C), from Eq. (7.5) it is possible to calculate the parameters A_0, A_1, A_2.

In Fig. 10 we have drawn $\bar{D}(T)$ calculated from Eq. (7.5), assuming the following values for $D_T(\eta_i)$:

$$D_T(\eta_4) = D_T(\eta_3) = 0 \qquad D_T(\eta_0) = 130e^{-850/T}$$
$$D_T(\eta_1) = 174e^{-1000/T} \qquad D_T(\eta_2) = 394.5\, e^{-1350/T}(10^{-5}\ \text{cm}^2\ \text{s}^{-1})$$

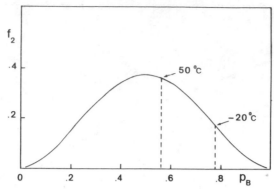

Figure 11.   The mole fraction of water f_2 vs. p_B.

The measured $\overline{D}(T)$ values are also shown.[62-64] The agreement between the calculated and experimental curves is satisfactory throughout the whole range of temperature investigated.

To explain, at least qualitatively, the pressure dependence of the self-diffusion of water, the temperature dependence of $D_T(\eta_2)\cdot f_2$ must be considered, as its contribution is always relevant in Eq. (7.5) ($D_T(\eta_2)\cdot f_2 = 0.7\overline{D}$ at $-20°C$ and $D_T(\eta_2)f_2 = 0.5\overline{D}$ at $50°C$). The diffusion coefficient of water increases rapidly with pressure (up to 1 kbar) in the supercooled region,[65] whereas at temperatures higher than $25°C$ it is virtually pressure-independent,[64] and for $T > 200°C$ it decreases as pressure increases. To rationalize this behavior we remark that an increase in pressure affects p_B in the same way as increasing temperature, that is, p_B decreases as these quantities increase. Moreover, the binomial distribution describing the structural equilibrium always remains valid.[24,66]

In Fig. 11 we have shown f_2 vs. p_B and its values at $-20°C$ and $+50°C$. If pressure increases, at $-20°C$ we obtain an increasing rate for f_2 much higher than that at $+50°C$, and f_2 dominates in the $D_T(\eta_2) f_2$ behavior and consequently in the $\overline{D}$ behavior. From this regime to that of high temperatures, where f_2 decreases with p_B and $D_T(\eta_2) f_2$ becomes largely irrelevant to determining $\overline{D}$ (which is expected to decrease with increasing pressure as in "normal" liquids), a temperature region must exist in which $\overline{D}$ remains practically constant with varying pressure, as experimentally observed.

In our picture the anomalous behavior of water is essentially due to the appearance in the liquid at $T < 150°C$ of the structured part (especially the states $|\eta_4\rangle$ and $|\eta_3\rangle$), the contribution of which results in a temperature dependence of $\overline{D}$ higher than that observed in a "normal" liquid (see Fig. 9).

Another singular aspect of the water behavior, at this point of our description, is that the product $\bar{D}\tau_D$ proves to be surprisingly constant with varying temperature,[52] as observed in "normal" liquids. The explanation of this comes directly from our model in which $\bar{D}$ and $\lambda_0 = 1/\tau_D$, obtained by solving two equations exhibiting the same formal structure:

$$\det \left\| \left[D_R(\eta_i) - \lambda + K_\eta \right] I \right\| = 0 \qquad \det \left\| \left[D_T(\eta_i) k^2 - \lambda + K_\eta \right] I \right\| = 0$$

must have the same temperature dependence.

In conclusion, even if we have not considered rototranslational coupling, the H-bond dynamics links together the rotation and diffusion coefficients as experimentally observed.

VIII.   NON-GAUSSIAN PROPERTIES AS AN EFFECT OF HYDROGEN-BOND DYNAMICS

In this section we shall show that the joint use of RMT and AEP succeed in explaining the strong deviation from Fick's law observed in quasi-elastic neutron scattering experiments, that is, the appearance of super-Burnett coefficients.[67]

In the case of simple liquids these deviations can be accounted for by using techniques based on kinetic and mode-mode coupling theories[67,68] and are related to the non-Gaussian properties of the fluid, as observed in computer simulation experiments (see Chapter VI).

We shall show how these deviations in water are essentially related to the "multiplicative" nature of the stochastic forces in the model used to picture the H-bond dynamics (Chapters II and X). Starting from Eq. (4.5), if the characteristic times of the H-bond dynamics are not infinitely small, super-Burnett coefficients appear in the diffusion equation. To do this, rather than performing calculation on the basis of the variable η defined by Eq. (3.2) we shall use as basis the operator K_η defined by Eq. (3.2′):

$$K_\eta | p_n \rangle = E_n | p_n \rangle$$
$$\langle \tilde{p}_n | K_\eta = \langle \tilde{p}_n | E_n$$

The eigenvector corresponding to the eigenvalue 0 is the equilibrium state $| P_0 \rangle$ (corresponding to the left state $\langle \tilde{P}_0 |$).

This leads us to define $\mathscr{L}$ as

$$\mathscr{L} = \mathscr{L}_a + \mathscr{L}_b + \mathscr{L}_1$$

where

$$\mathscr{L}_a = |p_0\rangle\langle\tilde{p}_0|D_T(\eta)|p_0\rangle\langle\tilde{p}_0|$$

$$\mathscr{L}_b = \sum_{n>0} |p_n\rangle\langle\tilde{p}_n|D_T(\eta)|p_n\rangle\langle\tilde{p}_n|$$

$$\mathscr{L}_1 = \sum_{n>0} \left\{ |p_0\rangle\langle\tilde{p}_0|D_T(\eta)|p_n\rangle\langle\tilde{p}_n| + |p_n\rangle\langle\tilde{p}_n|D_T(\eta)|p_0\rangle\tilde{p}_0| \right\}$$

Using the projection operator $P = |p_0\rangle\langle\tilde{p}_0|$ and the AEP technique, one obtains

$$\frac{\partial}{\partial t}\sigma_T = \left\{ \langle\tilde{p}_0|D(\eta)|p_0\rangle\nabla^2 + D^{(2)}\nabla^2\nabla^2 \right\}\sigma_T \qquad (8.1)$$

where

$$D^{(2)} = \sum_{n\neq 0} \langle\tilde{p}_0|D(\eta)|p_n\rangle\frac{1}{E_n}\langle\tilde{p}_n|D(\eta)|p_0\rangle$$

The second term in Eq. (8.1) represents a negative correction to Fick's law $(D^{(2)} > 0)$.

$D^{(2)}$ can be also written as

$$D^{(2)} = \sum_{ij} A_{ij}[\overline{D} - D_T(\eta_i)][\overline{D} - D_T(\eta_j)]$$

where

$$\overline{D} = \sum_{i=0}^{4} D_T(\eta_i)f_i$$

and its sign is the right one to account for the observed deviation from Fick's law.

The same result can also be obtained starting from Eq. (7.1) and using a diagonalization procedure at each temperature, when k^2 assumes different values. This allows us to calculate $A_i(k^2)$ and $\lambda_i^T(k^2)$ by Eqs. (7.3) and (7.4). Five Lorentzian curves $S_s(k,\omega)$ result; for $k^2 \to 0$ they reduce to $S_s = \overline{D}k^2/[\omega^2 + (\overline{D}k^2)]$, that is, Fick's law.

To compare the calculated values of the linewidth of the principal Lorentzian $\lambda_0^T(k^2)$ with those obtained by quasi-elastic neutron scattering, we have carried out the calculation using the same k^2 values as in these experiments. The calculated $\lambda_0^T(k^2)$ values and the experimental linewidth $\Delta E(k^2)$ of the corrected quasi-elastic neutron-scattering line[69,71] are shown in Fig. 12. The calculated deviation from Fick's law appears to be higher than

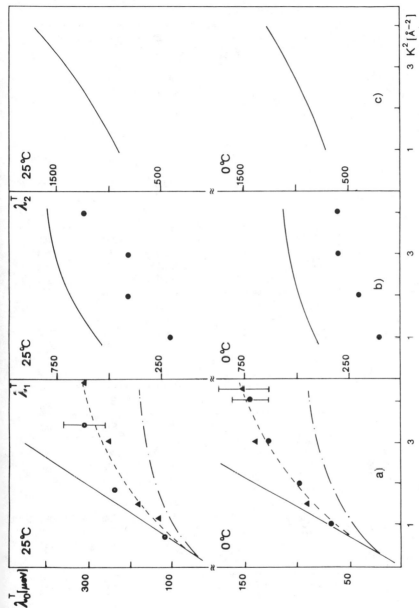

Figure 12. The half-widths λ_i^T of the Lorentzian lines contributing to the real part of the self- dynamical form factor $S_s(k, \omega)$ of water at 0°C and 25°C as a function of k^2 (k is the wave vector. (a) $\lambda_0 T$ vs. k^2: (○) Experimental values from ref. 74 (at 20°C); (●) from ref. 70; (▲) from ref. 71; (——) $\overline{D}k^2$ vs. k^2 (Fick's law); (---) fitting curve of the experimental points; (-·-) half-width of the first Lorentzian line as calculated from Eq. (7.4); (b) (●) Experimental half-width of the second Lorentzian line from ref. 71; (——) the same as calculated from Eq. (7.4); (c) (——) The half-width of the third Lorentzian line as calculated from Eq. (7.4).

TABLE III
Amplitudes A_i/λ_i^T and Half-Width λ_i^T of the Lorentzian Lines Contributing to the Real Part of the Self-Dynamical Form Factor $S_s(k, \omega)$ for Water at Various Temperatures for Different k^2 Values [Eq. (7.2)][a]

$K^2, \mathring{A}^{-2}$	A_0/λ_0^T, ps/rad	A_1/λ_1^T, ps/rad	A_2/λ_2^T, ps/rad	λ_0^T, MeV	λ_1^T, MeV	λ_2^T, MeV
			$-30°C$			
1	62.3	0.011	0.002	10.5	331	650
2	37.84	0.021	0.007	17	369	747
3	29.5	0.027	0.012	21.6	393	846
4	25.32	0.029	0.016	24.9	411	946
			$-20°C$			
1	32.67	0.029	0.005	19.8	341	641
2	20.43	0.047	0.017	30.7	392	760
3	16.24	0.052	0.028	37.7	423	882
4	14.1	0.051	0.036	42.5	443	1006
			$0°C$			
1	13.5	0.071	0.01	46.1	413	718
2	8.55	0.09	0.04	67.6	495.5	884
3	6.87	0.081	0.06	80	540	1064
4	6.01	0.072	0.071	87.5	565	1252
			$25°C$			
1	6.15	0.112	0.013	95.3	546	885
2	3.83	0.113	0.063	133.6	683	1116
3	3.04	0.084	0.093	153	746	1390
4	2.65	0.068	0.098	165	778	1674
			$50°C$			
1	3.27	0.143	0.011	167	700	1082
2	1.94	0.12	0.08	227	912.5	1379
3	1.509	0.073	0.113	255	996	1765
4	1.30	0.054	0.113	271	1034	2178

[a]Subscript 0 refers to the principal Lorentzian line and subscripts 1 and 2 to the only other two significant Lorentzian lines.

the measured one, even if the deviation is predicted in the right direction. However, a quantitative comparison is not meaningful, owing to the scatter in the experimental data.[72]

Recently, Chen et al.[71] have shown that the disagreement among the values of $\Delta E(k^2)$ given by different authors very likely can be explained by the presence of a second Lorentzian line in the observed spectrum, and thus the result depends on the instrumental linewidth used to analyze the spectrum. The contribution of the second Lorentzian line increases with temperature.

Chen et al.'s observations are well supported by the results of our model. Besides the principal Lorentzian line, two of the others are vanishingly small, the remaining two being wider than the principal Lorentzian line. Their amplitudes increase with T, as does the second Lorentzian line observed by Chen et al. The parameters of the three nonvanishingly calculated Lorentzian lines are listed in Table III, and the linewidths λ_i^T vs. k^2 of the two wider lines are shown in Fig. 12 together with the experimental results of ref. 69. The agreement so obtained suggests the opportunity of analyzing the experimental spectrum from quasi-electric neutron scattering using three Lorentzians.

The results listed in Table III can be discussed in the context of the "random jump" model.[73,74] This model foresees a linear trend for $\bar{D}k^2/\lambda_0^T(k^2)$ $=1+l_0^2 k^2$ where l_0 is the "jump" characteristic length, related to the probability density $P(l) = le^{-l/l_0}/l_0^2$. The predictions of this model agree with our results only at $T = 0°C$. For temperatures higher than $0°C$ the curve of $\bar{D}\cdot k^2/\lambda_0^T(k^2)$ vs. k^2 increases its slope with increasing k^2. In Fig. 13 we show l_0^2 vs. T; at the higher temperatures l_0 is obtained from the angular coefficient of the straight line tangent to the curve at low k^2 values. The resi-

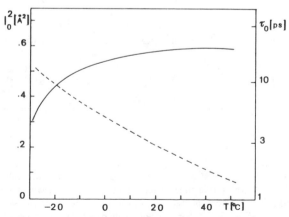

Figure 13. (——) the square of the "jump" characteristic length l_0^2 vs. temperature calculated from our model. (---) The calculated residence time $\tau_0 = l_0^2/\bar{D}$ vs. temperature.

dence time τ_0 is also drawn in Fig. 13; $\tau_0 = l_0^2/\overline{D}$ decreases, as expected, with increasing temperature.

IX. THE DYNAMICS OF OTHER HYDROGEN-BONDED LIQUIDS

The model we have reported, being based on H-bond dynamics, can be utilized on principle to describe the dynamic behavior of all H-bonded liquids. We shall present some examples which outline the effectiveness of this picture.

A. D$_2$O

The procedure described in Sections IV and V to calculate τ_D and $\overline{D}$ of water can be utilized without difficulty to calculate the same quantities for D$_2$O. The p_B calculation procedure is the same as that in Section V.A.

$\tau^{\text{D}_2\text{O}}_{\text{HB}}$, the mean lifetime of the bond in D$_2$O, has been measured in a depolarized Rayleigh scattering experiment by Danninger.[46] He obtained $\tau^{\text{D}_2\text{O}}_{\text{HB}}$ about 10% greater than $\tau^{\text{H}_2\text{O}}_{\text{HB}}$ at all the investigated temperatures.

If in addition to the assumptions equivalent to those discussed for H$_2$O, we assume for simplicity that the D_R and D_T values for D$_2$O are equal to those utilized for H$_2$O, then the parameters characterizing the D$_2$O diffusion dynamics can be calculated. In Table IV we list the calculated ratios $\lambda_0^{\text{D}_2\text{O}}/\lambda_0^{\text{H}_2\text{O}}$ and $\overline{D}_{\text{D}_2\text{O}}/\overline{D}_{\text{H}_2\text{O}}$ with the corresponding experimental ratios.[63,75]

The calculated ratio of the two translational diffusion coefficients is always slightly lower than the experimental one. The systematic deviation can be rationalized considering the approximation made by assuming the same

TABLE IV

The Ratio of the Characteristic Parameters of D$_2$O (1) and H$_2$O (2) at Various Temperatures[a]

T, °C	(1) $\tau_{\text{HB}}^{(1)}/\tau_{\text{HB}}^{(2)}$	(2) $p_B^{(1)}/p_B^{(2)}$	(3) $\lambda_0^{(2)}/\lambda_0^{(1)}$	(4) $\tau_D^{(1)}/\tau_D^{(2)}$	(5) $\overline{D}^{(2)}/\overline{D}^{(1)}$ (Calc.)	(6) $\overline{D}^{(2)}/\overline{D}^{(1)}$ (Exp.)
5	1.1	1.056	1.31	1.30*	1.25	1.29
25	1.1	1.06	1.26	1.275†	1.19	1.23
50	1.1	1.07	1.23	1.22	1.17	1.19

[a](1) H-bond lifetimes τ_{HB} from ref. 46. (2) Probability p_B calculated following the procedure in Section V.A. (3) Relaxation times calculated from Eq. (6.1) with the same conditions (6.2) for H$_2$O and D$_2$O. (4) Experimental values of dielectric relaxation times from ref. 75; (*) at 10°C; (†) extrapolated. (5) Self-diffusion coefficients calculated from Eq. (7.5) assuming for D$_2$O the same $D_T(\eta_i)$ used for H$_2$O. (6) Self-diffusion experimental data from ref. 63.

$D_T(\eta_i)$ values in the two liquids. In fact, $D_T(\eta_i)$ of D_2O, owing to its larger mass, is underestimated.

The substantial differences of τ_D and $\overline{D}$ in the two liquids are traced back, in our picture, to differences between the probabilities p_B; they are due to the presence of a structured part in H_2O smaller than that in D_2O.

B. Alcohols

We suggest that a picture of this type can also include the dynamics of the alcohols if the features of the hydrogen bonding in these liquids are taken into account. A Monte Carlo statistical simulation of liquid methanol and ethanol[76,77] gives the following results:

1. Liquid methanol and ethanol are found to consist primarily of long, winding H-bond chains with roughly linear hydrogen bonds.
2. The two hydrogen bonds in the chains can be considered equivalent with a probability p_B that a hydrogen bond is maintained. Molecules with three hydrogen bonds are also present. The number of these suggests that the third hydrogen bond cannot be considered equivalent to the other two; a probability p_B' for this bond must be introduced. By the way, it is possible that the presence of the third hydrogen bond is due to a weak definition of the bond in these simulation experiments.
3. A "waterlike" energetic definition of the hydrogen bond $V_{HB} = -2.25$ kcal/mol gives the f_i values listed in Table V for methanol and ethanol at 25°C.

TABLE V
The Fraction of Molecules with i Intact Bonds in MeOH
and EtOH at 25°C[a]

	MeOH		EtOH	
	(1)	(2)	(3)	(4)
f_0	0.025	0.0243	0.04	0.0397
f_1	0.242	0.231	0.29	0.2902
f_2	0.561	0.578	0.56	0.5606
f_3	0.165	0.168	0.11	0.1095
f_4	0	0	0	0

[a](1) From computer simulation experiment.[76] (2) Calculated from Eq. (9.1) with $p_B = 0.82$ and $p_B' = 0.25$. (3) From computer simulation experiment.[77] (4) Calculated from Eq. (9.1) with $p_B = 0.78$ and $p_B' = 0.18$.

Using the relations:

$$f_0 = (1 - p_B)^2 (1 - p'_B)$$

$$f_1 = 2 p_B (1 - p_B)(1 - p'_B) + p'_B (1 - p_B)^2$$

$$f_2 = p_B^2 (1 - p'_B) + 2 p_B (1 - p_B) p'_B$$

$$f_3 = p_B^2 p'_B \qquad (9.1)$$

one obtains for methanol at 25°C, $p_B = 0.82$ and $p'_B = 0.25$, and for ethanol at the same temperature, $p_B = 0.78$ and $p'_B = 0.18$.

The probability p'_B of the third hydrogen bond decreases from methanol to ethanol, and it is always lower than p_B; this is very likely ascribable to steric effects (increasing with the molecular weight of the alcohol) which hinder formation of the third hydrogen bond. So in an alcohol with a long alkyl chain we can assume $p'_B = 0$, and for a statistical description of the network the binomial distribution

$$f_0 = (1 - p_B)^2 \qquad f_1 = 2 p_B (1 - p_B) \qquad f_2 = p_B^2$$

Following a procedure like that for H_2O we shall define the H-bond dynamic parameters of the alcohols as sketched in Fig. 14.

To describe the dielectric relaxation, a system of equations corresponding to Eq. (4.4) is utilized:

$$\frac{d}{dt} \Phi_\mu^{(i)} = - D_R(\eta_i) \Phi_\mu^{(i)} + \sum_{j=0}^{2} (K_\eta)_{ij} \Phi_\mu^{(j)} \qquad i = 0, 1, 2 \qquad (9.2)$$

By using the experimental τ_D (refs. 79 and 80) and $D_R(\eta_0)$ (refs. 81 and 82) of n-pentanol and n-decanol and assuming for a qualitative calculation $p_B = 0.75$–0.8 at room temperature, we obtain the results $\tau_{HB} \simeq 60$ ps for pentanol and $\tau_{HB} \simeq 150$ ps for decanol, whereas for methanol and ethanol

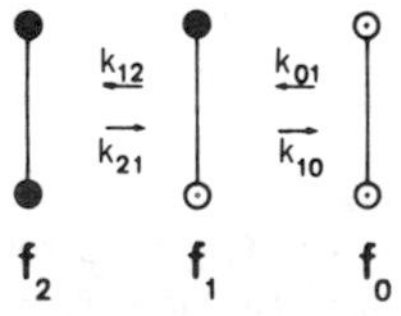

Figure 14.   A sketch of the H-bond dynamics for primary alcohols with long alkylic chains. f_i, mole fraction with i intact hydrogen bonds; ($\bullet$) intact hydrogen bond; ($\odot$) broken hydrogen bond. $k_{ii-1} = i/3\tau_{HB}$ ($i = 1, 2$); $k_{ii+1} = (2-i)/3\tau_f$ ($i = 0, 1$).

one obtains 2 ps and 7 ps, respectively. The fundamental result is that τ_{HB} increases with alkyl chain length.

X. CONCLUDING REMARKS

A major result of this work is that we quantitatively assessed the relevant role played by H-bond dynamics to determine the long-time diffusional properties. Indeed, both the rotational and translational diffusion processes are driven by a secondary time-dependent process: the H-bond dynamics simulated by the stochastic variable η.

The diffusion equations involved in our theoretical analysis, which are derived via the AEP from the fluctuating reduced model of Section IV, can be regarded as a five-state version of the Anderson two-state model[51] supplemented by a quantitative description of the secondary process. We would especially stress that, in accordance with the point of view of other authors,[50] the characteristic time of the principal dielectric relaxation band roughly coincides with the residence time in the structured part of the liquid.[83]

The picture of the translation closely parallels the rotational picture. In this case the counterpart of the Anderson model is the random jump model (see Section VIII). An important theoretical prediction of the reduced model of Section IV is the deviation from Fick's law, which is firmly supported by experiment. Note that this deviation precisely depends on the fluctuating nature of the reduced model. On the other hand, the theoretical analysis of Chapter VI shows that this multiplicative fluctuation must be traced back to the nonlinear nature of the microscopic interaction, therby rendering plausible the appearance of non-Gaussian microscopic properties.

Furthermore, in the picture of this chapter also, the appearance of more than one Lorentzian curve in the spectrum of the quasi-elastic neutron scattering (a result which has been recently confirmed by Teixeira et al.[84]) is related to the H-bond dynamics, whereas Teixeira et al. ascribe this effect to the orientational part of the intermediate scattering function. We believe that both physical causes can contribute to this effect.

The general strategy of attack to the molecular dynamics outlined by this and earlier chapters appears especially promising to shed further light into this stimulating field of research. In the special case of H-bonded liquids, a natural development of the ideas outlined here implies the replacement of the discrete variable η with a continuous variable, which in turn involves the replacement of the master equation method with a suitable Fokker-Planck equation. Moreover, this improvement of the theory is fundamental to exploring the short-time dynamics when the details of the correlation functions on the time scale of structure V of water must be accounted for.

References

1. C. A. Angell, in *Water, A Comprehensive Treatise*, Vol. 7, F. Franks, ed., Plenum, New York, 1982.

2. E. W. Lang and H. D. Ludemann, *Angew. Chem.*, **21**, 315 (1982)

3. C. A. Angell, *Ann. Rev. Phys. Chem.*, **34**, 593 (1983).

4. C. A. Angell, *J. Phys. Chem.*, **75**, 3698 (1971).

5. H. E. Stanley and J. Teixeira, *J. Chem. Phys.*, **73**, 3404 (1980).

6. S. Y. Chu and G. Sposito, *J. Chem. Phys.*, **69**, 2539 (1978).

7. F. H. Stillinger and A. Rahman, *J. Chem. Phys.*, **60**, 1545 (1974).

8. G. Sposito, *J. Chem. Phys.*, **74**, 6943 (1981).

9. D. Eisemberg and W. Kauzmann, *The Structure and Properties of Water*, Clarendon Press, Oxford, 1969, p. 219.

10. G. E. Walrafen, in *Water, A Comprehensive Treatise*, Vol. 1, F. Franks, ed., Plenum, New York, 1972.

11. W. A. P. Luck, in *Water, A Comprehensive Treatise*, Vol. 2, F. Franks, ed., Plenum, New York, 1973.

12. F. Hirata and P. J. Rossky, *J. Chem. Phys.*, **74**, 6867 (1981).

13. A. C. Belch, S. A. Rice, and M. G. Sceats, *J. Chem. Phys.*, **77**, 455 (1981).

14. F. H. Stillinger and A. Rahman, *J. Chem. Phys.*, **57**, 1281 (1972).

15. D. C. Rapaport and H. A. Scheraga, *Chem. Phys. Lett.*, **78**, 491 (1981).

16. R. W. Impey, P. A. Madden, and I. R. McDonald, *Mol. Phys.*, **46**, 513 (1982).

17. H. Tanaka, K. Nakanishi, and N. Watanabe, *J. Chem. Phys.*, **78**, 2626 (1983).

18. O. Conde and J. Teixeira, *J. Phys. (Paris)*, **44**, 525 (1983).

19. S. A. Rice, A. C. Belch, and M. G. Sceats, *Chem. Phys. Lett.*, **84**, 245 (1981); M. G. Sceats and S. A. Rice, in *Water, A Comprehensive Treatise*, Vol. 7, F. Franks, ed., Plenum, New York, 1982.

20. D. L. Beveridge, M. Mezei, P. K. Mehrotra, F. T. Marchese, G. Ravi-Shanker, T. Vasu, and S. Swaminatham, *Adv. Chem. Ser.* **204**, 297 (1983).

21. W. L. Jorgensen, J. Chandrasekhar, J. D. Madura, R. W. Impey, and M. L. Klein, *J. Chem. Phys.*, **79**, 926 (1983).

22. M. Mezei and D. L. Beveridge, *J. Chem. Phys.*, **74**, 622 (1981).

23. A. Geiger, F. H. Stillinger, and A. Rahman, *J. Chem. Phys.*, **70**, 4184 (1979).

24. Y. Kataoka, H. Hamada, S. Nose, and T. Yamamoto, *J. Chem. Phys.*, **77**, 5699 (1982).

25. A. H. Narten, M. D. Danford, and H. A. Levy, *Discuss, Farady Soc.*, **43**, 97 (1967).

26. H. G. Hertz, in *Water, A Comprehensive Treatise*, Vol. 3, F. Franks, ed., Plenum, New York, 1973.

27. S. Chandrasekhar, in *Selected Topics on Noise and Stochastic Processes*, N. Wax, ed., Dover, New York, 1959, p. 3.

28. P. H. Hubbard, *Phys. Rev.*, **A6**, 2421 (1972).

29. M. W. Evans, *Phys. Rev. Lett.*, **50**, 371 (1983).

30. R. Kubo, *Adv. Chem. Phys.*, **15**, 101 (1969).

31. M. E. Rose, *Elementary Theory of Angular Momentum*, Wiley, New York, 1957.

32. A. Geiger and H. E. Stanley, *Phys. Rev. Lett.*, **49**, 1749, 1895 (1982).

33. F. H. Stillinger, *Science*, **209**, 4455 (1980); in *Water in Polymers*, S. P. Rowland, ed., (Am. Chem. Soc., Washington, DC, 1980) p. 11–22.

34. A. C. Belch and S. A. Rice, *J. Chem. Phys.*, **78**, 4817 (1983) and references therein.

35. M. G. Sceats and S. A. Rice, *J. Chem. Phys.*, **72**, 3248 (1980).

36. C. M. Davies and J. Jarzynski, *Adv. Mol. Relax. Proc.*, **1**, 155 (1968).

37. W. Kauzmann, *L'Eau et les Systemes Biologiques, Colloque Internationaux de CNRS*, **246**, 63 (1976).

38. E. Trinh and A. Apfel, *J. Chem. Phys.*, **69**, 4245 (1978).

39. G. D'Arrigo, *Nuovo Cimento*, **61B**, 123 (1981).

40. R. C. Weast, ed., *Handbook of Chemistry and Physics*, 51st ed., Chemical Rubber Publishing Co., Cincinnatti, 1970–1971, Steam Table, page E12.

41. J. R. Heiks, M. K. Burnett, L. V. Jones, and E. Orban, *J. Phys. Chem.*, **58**, 488 (1954).

42. J. Leblond and M. Hareng, *J. Phys. (Paris)*, **45**, 373 (1984).

43. G. S. Kell, *J. Chem. Eng. Data*, **20**, 97 (1975).

44. B. V. Zhelenyi, *Russ. J. Chem. Phys.*, **42**, 950 (1968); **43**, 1311 (1969).

45. C. J. Montrose, J. A. Bucaro, J. Marshall-Coakley, and T. A. Litovitz, *J. Chem. Phys.*, **60**, 5025 (1974).

46. W. Danninger and G. Zundel, *J. Chem. Phys.*, **74**, 2769 (1981).

47. D. C. Rapaport, *Mol. Phys.*, **50**, 1151 (1983).

48. A. Geiger, P. Mausbach, J. Schnitker, R. L. Blumberg and H. E. Stanley, *J. Phys. (Paris)*, **C7**, 31 (1984).

49. C. J. Reid and M. W. Evans, in *Molecular Interactions*, Vol. 3, H. Ratajczak and W. J. Orville-Thomas, eds., Wiley, New York, 1982.

50. M. Magat and L. Reinisch, *L'Eau et les Systems Biologiques, Colloque Internationaux de CNRS*, **246**, 79 (1976).

51. J. E. Anderson, *J. Chem. Phys.*, **47**, 4879 (1967).

52. D. Bertolini, M. Cassettari, and G. Salvetti, *J. Chem. Phys.*, **76**, 3285 (1982).

53. U. Kaatze and V. Uhlendorf, *Z. Phys. Chem. NF*, **126**, 151 (1981).

54. M. N. Afsar and J. B. Hasted, *Infrared Phys.*, **18**, 835 (1978).

55. J. B. Hasted, in *Water, A Comprehensive Treatise*, Vol. 1, F. Franks, ed., Plenum, New York, 1972.

56. F. H. Stillinger, in *Liquid State of Matter: Fluids Simple and Complex*, E. W. Montroll and J. L. Lebowitz, eds., North-Holland, Amsterdam, 1982.

57. C. Brot, G. Bossis, and C. Hesse-Bezot, *Mol. Phys.*, **40**, 1053 (1980).

58. G. H. Haggis, J. B. Hasted, and T. J. Buchanan, *J. Chem. Phys.*, **20**, 1452 (1952).

59. S. L. Carnie and G. N. Patey, *Mol. Phys.*, **47**, 1129 (1982).

60. H. J. V. Tyrrel, *Diffusion and Heat Flow in Liquids*, Butterworth, London, 1961, p. 155.

61. F. Dupré, D. Piaggesi, and F. P. Ricci, *Phys. Lett.*, **80A**, 178 (1980).

62. K. T. Gillen, D. C. Douglass, and M. J. R. Hoch, *J. Chem. Phys.*, **57**, 5117 (1972).

63. R. Mills, *J. Phys. Chem.*, **77**, 685 (1973).

64. K. Krynicki, C. D. Green, and D. W. Sawyer, *Faraday Disc. Chem. Soc.*, **66**, 198 (1978).

65. C. A. Angell, E. D. Finch, L. A. Woolf, and P. Bach, *J. Chem. Phys.*, **61**, 4973 (1975).

66. F. H. Stillinger and A. Rahman, *J. Chem. Phys.*, **61**, 4973 (1975).

320 D. BERTOLINI ET AL.

67. I. De Schepper, "Generalized Hydrodynamics for the Diffusion Process," thesis, University of Nymegen, 1974.
68. J. P. Boon and S. Yip, *Molecular Hydrodynamics*, McGraw-Hill, New York, 1980.
69. K. E. Larson and U. Dahlborg, *Physica*, **30**, 1561 (1964).
70. G. J. Safford, P. S. Leung, A. W. Naumann, and P. C. Schaffer, *J. Chem. Phys.*, **50**, 4444 (1969).
71. S. H. Chen, J. Teixeira, and R. Nicklow, *Phys. Rev.*, **A26**, 3477 (1982).
72. P. Von Blanckenhagen, *Ber. Bunsenges. Phys. Chem.*, **76**, 891 (1972) and references therein.
73. P. A. Egelstaff, *An Introduction to the Liquid State*, Academic Press, London, 1967.
74. T. Springer, *Quasielastic Neutron Scattering for the Investigation of Diffusive Motions in Solids and Liquids*, Springer, Berlin, 1972.
75. C. H. Collie, J. B. Hasted, and D. M. Ritson, *Proc. Phys. Soc. (London)*, **60**, 145 (1948).
76. W. L. Jorgensen, *J. Am. Chem. Soc.*, **102**, 543 (1980).
77. W. L. Jorgensen, *J. Am. Chem. Soc.*, **103**, 345 (1981).
78. D. Bertolini, M. Cassettari, and G. Salvetti, *J. Chem. Phys.*, **78**, 365 (1983).
79. J. Peyrelasse, C. Boned, and J. P. Lepetit, *J. Phys. E: Sci. Instrum.*, **14**, 1002 (1981).
80. C. Boned and J. Peyrelasse, *J. Phys. E: Sci. Instrum.*, **15**, 534 (1982).
81. G. P. Johari and C. P. Smyth, *J. Am. Chem. Soc.*, **91**, 6215 (1969).
82. L. Glasser, J. Crossley, and C. P. Smyth, *J. Chem. Phys.*, **57**, 3977 (1972).
83. D. Bertolini, M. Cassettari, M. Ferrario, G. Salvetti, and P. Grigolini, *Chem. Phys. Lett.*, **98**, 548 (1983).
84. J. Teixeira, M. C. Bellissent-Funel, S. H. Chen, and A. J. Dianoux, *J. Phys. (Paris)*, **C7**, 65 (1984).

VIII

SLOW MOTION EPR SPECTRA
IN TERMS OF A GENERALIZED
LANGEVIN EQUATION

M. GIORDANO, P. GRIGOLINI, D. LEPORINI,
and P. MARIN

CONTENTS

I. INTRODUCTION

As we have emphasized in Chapters I, III, and IV, computational techniques based on the information derived from spectral moments have quite a long history. In the field of magnetic resonance,[1-27] these can be traced back to the pioneering work of Van Vleck.[1] In a more general form his basic idea can be expressed as follows. The normalized moment of a magnetic resonance absorption curve may be defined by the relation

$$\langle \omega^{2n} \rangle = \int_{-\infty}^{+\infty} g(\omega - \omega_0)(\omega - \omega_0)^{2n} \, d\omega \tag{1.1}$$

321

where $g(\omega - \omega_0)$ is the lineshape function with the line centered at frequency ω_0. The general expression for the $(2n)$th moment originally suggested by Waller[28] and proved by mathematical induction is

$$\left\langle \omega^{2n} \right\rangle = (i\hbar)^{-2n} \frac{\mathrm{Tr}\left\{ \left[\mathscr{H}, \left[\mathscr{H} \cdots \left[\mathscr{H}, S_x \right] \cdots \right] \right]^2 \right\}}{\mathrm{Tr}\left\{ S_x^2 \right\}} \qquad (1.2)$$

where $\mathscr{H}$ is the Hamiltonian of the spin system of physical interest, S_x is the x component of the total spin operator of the system. The merit of van Vleck's work was that of defining $\mathscr{H}$ as a suitably truncated Hamiltonian so as to disregard the contributions which come from satellite lines and introduce errors in the computation of the moments of the main lines.

Equation (1.2), in turn, can be replaced by an equivalent expression explicitly involving summation on the infinite spins surrounding a tagged spin. This expression involves only the parameters defining the dipolar interaction and the geometry of the crystal lattice under examination.

This expression is especially suitable for computational purposes. Nevertheless, its explicit form rapidly becomes cumbersome as n increases. To the best of our knowledge, the highest order moment so far expressed explicitly in terms of lattice geometry is in ref. 19. Knak Jensen and Hansen[19] evaluated the analytical expressions for the sixth and eighth moments of the magnetic resonance lines of a dipolar-coupled rigid lattice. In the greatly simplified one-dimensional case,[24b] however, one can extend their calculation up to the 20th moment.

The central idea of van Vleck has also been applied to the field of nuclear quadrupole resonances,[3,12] still in the solid state. We would also like to mention the well-known work of Gordon,[5-10] who studied molecular spectra in terms of moments. His investigation provides examples of application of the moment theory in the gas phase.[5]

Generally speaking, when the ideal scheme of a rigid regular lattice cannot apply, the rigorous evaluation of spectral moments becomes very complicated. Electronic paramagnetic resonance (EPR) spectroscopy, which is the subject of this chapter, provides an illustration of that difficulty: the spin degrees of freedom are coupled with molecular orientation, the dynamics of which depend on the interaction between the tagged molecule (the one containing the electronic spin) and the solvent molecules. This implies virtually that the problem of molecular dynamics in the liquid state must be solved analytically before further progress can be made. This is because the moments of the spin correlation function cannot be determined until those of the orientation and angular velocity correlation function have been analytically determined and evaluated. As pointed out in Chapter VI, this is not a practicable approach, because a reliable description of the time evolution of correlation functions is still a subject of investigation, and great efforts are currently being made to shed light on this problem. A vast simplification can be attained when the theory of the stochastic Liouville equation (SLE)[29] is used. This method is based on regarding the molecular orientation Ω as a set of stochastic variables driven by a standard[29] (or generalized[30]) Langevin equation. This means replacing the dynamical operator (Chapter I) that rigorously describes the dynamics of the thermal bath with the Fokker-Planck operator associated with the standard (or generalized) Langevin equation driving Ω. Note that Ω is a special case of the family of auxiliary (or virtual) variables b_v of Chapter I. [In the present case, the set of relevant variables a (Chapter I) is represented by the spin variables.] The RMT of Chapter I is a general approach, which when applied to EPR spectroscopy turns out to be equivalent to the SLE.[31]

The spectral moments can also be expressed in terms of the time derivatives evaluated at $t = 0$ of

$$\Phi(t) \equiv \frac{\langle S_x(0) S_x(t) \rangle}{\langle S_x S_x \rangle} \qquad (1.3)$$

where $\langle \cdots \rangle$ denotes an equilibrium average and

$$S_x(t) \equiv e^{\Gamma t} S_x \qquad (1.4)$$

In the case of NMR spectroscopy in the solid state, Γ is the rigorous Liouvillian $iL \equiv i\mathcal{H}^x$, where

$$\mathcal{H}^x A \equiv [\mathcal{H}, A] \qquad (1.5)$$

In the case of EPR spectroscopy in the liquid state, when the SLE theory is applied this is intended to be an effective Liouvillian operator (see also the next sections). Let us define the parameters s_n as follows:

$$s_n \equiv \left[\frac{d^n}{dt^n} \Phi(t) \right]_{t=0} \qquad (1.6)$$

These parameters can also be written as

$$s_n = \frac{\langle S_x \Gamma^n S_x \rangle}{\langle S_x^2 \rangle} \qquad (1.7)$$

From linear response theory[32] we derive

$$g(\omega) = \frac{1}{2\pi} \int_{-\infty}^{\infty} dt\, e^{-i\omega t} \Phi(t) \qquad (1.8)$$

The inverse Fourier transformation provides

$$\Phi(t) = \int_{-\infty}^{\infty} d\omega \, e^{i\omega t} g(\omega) \tag{1.9}$$

Evaluating the nth order time derivative at $t = 0$ of both terms of Eq. (1.9), we get

$$s_n = (i)^n \int_{-\infty}^{+\infty} d\omega \, \omega^n g(\omega) \tag{1.10}$$

When Γ is a rigorous Liouvillian, time inversion symmetry implies that the odd s_n parameters vanish. When n is even, the right-hand side of Eq. (1.10) can then be identified with the moments defined by Eq. (1.1). Note that in this case Eqs. (1.5), (1.7), and (1.10) allow us to recover Eq. (1.2). When Γ is an effective operator of the type provided by RMT, even the odd moments s_n can be different from zero.

For both NMR spectroscopy in the solid state and EPR spectroscopy in the liquid state it is possible to determine a fairly large number of parameters s_n. After solving this first problem, one has then to determine the absorption lineshapes in terms of this information. The main aim of this chapter is to show the algorithm of Chapter III at work. However, before launching into this illustration, we would like to mention the alternative approaches currently being used in the field of magnetic resonance.

1.   *The Mori Theory and Related Approaches.*   Lado, Memory, and Parker[15] have investigated the problem of NMR spectroscopy with a spin system driven by a rigorous Liouvillian. After applying the standard Mori procedure[33] to build up the best expansion basis set for that Liouvillian, they evaluated expansion parameters of the resulting continued fraction with the algebra of orthogonal polynomials.[34] In a sense, the same approach has been followed by Giordano et al.,[35] who studied EPR resonance phenomena by using a Mori-like approach, valid also in the case of effective Liouvillians. To determine the corresponding expansion parameters, they followed Dupuis,[36] who in turn based his algorithm on the same theoretical background as Lado et al., that is, the algebra of orthogonal polynomials[34] and the mathematical theory of approximants.[37] For a general discussion along the same lines, see also Lonke.[14]

2.   *The Lanczos Method.*   The use of classical moment theory involves large determinants, thereby implying delicate problems of numerical stability. This instability arises from very severe cancellations in the determinants, which in turn come from the properties of the moments themselves. Whitehead and Watt[38] bypassed this problem by using the Lanczos al-

gorithm.[39] Roughly speaking, the Lanczos method shares the basic idea of the Mori approach of expanding the dynamical operator in a suitable basis set so as to result in a tridiagonal form. The Mori approach is more attractive because of the attention devoted to deriving via this technique a generalized Langevin equation (Chapters I, III, and IV). The Lanczos method has recently been applied by Moro and Freed[40] to evaluating the EPR spectra.

As discussed in Chapter IV, the joint use of a method directly evaluating the continued fraction expansion parameters from a chain of "states" starting from the variable of interest (Lanczos method) and that of deriving them from the s_n parameters turns out to be suggested naturally by the generalized Langevin equation method. We would like to mention, for example, Lee and coworkers,[40c] who also derived the former method by the Mori generalized Langevin equation.

3. *The Padé Approximants.* This method has been applied to the field of electron spin resonance (ESR) spectra by Dammers et al.,[41] who succeeded in dealing with the problem of saturation.

The work of other authors cannot be clearly classified as belonging to one of these three main families. A quantum-statistical theory of longitudinal magnetic relaxation based on a continued fraction expansion has been given by Sauermann,[42] who also pointed out the equivalence between his continued fraction approach (based on the Mori scalar product) and the $[N|N]$ Padé approximants.

The use of a basis set reminiscent of that provided by Mori theory can be found in the work of Baram.[43]

We would also like to mention the informational approach,[44] which determines the most probable absorption lineshapes given a limited amount of information, that is, a finite number of spectral moments. The method is based on minimizing the informational entropy.[44]

We will not deal here with the subject of EPR spectroscopy of the solid state. In this field of investigation a kind of delta-like approach such as that recently proposed to deal with molecular dynamics in the liquid state[45] has developed naturally. According to the European Molecular Liquid Group (EMLG), the symbol Δ symbolizes the cooperative efforts of computer simulation, experiment, and theory. Knak Jensen and Hansen,[26] for instance, carried out a computer simulation of the dynamics of N identical spins placed in a rigid simple cubic lattice subject to an external magnetic field $\mathbf{B}_0$. A further example of numerical study is the paper of Sur and Lowe.[24a] Free-induction decay measurements,[22,23] on the other hand, represent the experimental corner of this ideal triangle, the theoretical corner of which is, of course, expressed by the theoretical papers mentioned above.

II. GENERAL THEORY

According to linear response theory,[32] the EPR spectrum can be expressed as (Re{...} means the real part of {...})

$$I(\omega) = \mathrm{Re}\mathbb{L}\Phi(t) = \mathrm{Re}\mathbb{L}\frac{\langle S_- S_+(t)\rangle}{\langle S_- S_+\rangle} \qquad (2.1)$$

where the problem of determining the lineshape is reduced to determining the Laplace transform $\mathbb{L}\Phi(t)$ of the equilibrium correlation function

$$\Phi(t) = \frac{\langle S_- S_+(t)\rangle}{\langle S_- S_+\rangle} \qquad (2.2)$$

where $S_\pm = S_x \pm iS_y$ are the ± 1 spherical components of the spin angular momentum S. The time evolution of the variables $S_\pm$ is described by the following equation of motion:

$$\frac{dS_\pm}{dt} = iLS_\pm \qquad (2.3)$$

where L is the rigorous quantum-mechanical Liouvillian superoperator defined as

$$iL \equiv i\mathscr{H}^\times = -i[\mathscr{H}, \ldots] \qquad (2.4)$$

$\mathscr{H}$, in turn, is given by

$$\mathscr{H} = \mathscr{H}_s + \mathscr{H}_I + \mathscr{H}_B$$

$\mathscr{H}_s$ is the intramolecular purely electronic and nuclear spin part of the total Hamiltonian, that is, the part which does not depend on the molecular orientation. $\mathscr{H}_B$ denotes the Hamiltonian of the irrelevant degrees of freedom. This chapter is developed as follows. First we consider the orientation $\Omega = (\alpha, \beta, \gamma)$ of the tagged molecule, where α, β, γ are the Euler angles defining the rotation from the molecular to the laboratory frame. Then we include in the set of irrelevant variables the angular velocity ω of the tagged molecule. The motion of ω is driven by the interaction with the solvent molecules. This interaction extends the "irrelevant" set of variables so as to involve the degrees of freedom of the solvent molecules. The solute is assumed to be so dilute that we can safely disregard the interaction between the tagged molecule and the other solute molecules. $\mathscr{H}_I$ denotes the interaction between $\mathscr{H}_s$ and $\mathscr{H}_B$. In this scheme we disregard the magnetic intermolecular interaction, so that the main contribution to the relaxation of our spin system comes from the interaction with the orientation degree of freedom.

Regarding the pair (Ω, ω) as a classical stochastic variable, we can proceed in the usual way[46] by defining the Fokker-Planck equation associated with the corresponding Langevin equation (see Chapter II). The friction and diffusion parts of the equation simulate the remaining "irrelevant" degrees of freedom. In Section III.E we shall illustrate how to proceed in that case. In general, this calculation involves too many variables to make a handy theoretical tool. It is therefore convenient to make a contraction over the variable ω. In the free diffusion case (no orientation potential), this can be done when the friction γ, which makes ω relax (see also Chapter II) is very large compared to $\langle \omega^2 \rangle_{eq}^{1/2}$.

In the presence of an orientation potential, one should also consider the effective frequencies corresponding to the harmonic approximation of this potential. The adiabatic elimination procedure (AEP), as developed in Chapter II, should allow us to take into account the inertial corrections to the standard adiabatic elimination, thereby making it possible to determine the influence of inertia on EPR spectra within the context of a contracted description that retains only the variable Ω. This allows us to arrive at an equation of the form

$$\frac{\partial}{\partial t} p(\Omega, t) = \Gamma_\Omega p(\Omega, t) \tag{2.5}$$

where $p(\Omega, t)$ is the probability of finding the molecule under investigation to have the orientation Ω at time t. Throughout this chapter we shall not take into account the higher-order corrections to the lowest-order result, the standard Favro equation,[47] where

$$\Gamma_\Omega = -\overline{M}(D \cdot \overline{M}V(\Omega)) - \overline{M}D\overline{M} \tag{2.6}$$

D is the molecular diffusion tensor, and $\overline{M}$ is the generator of rotations in the molecular frame. $V(\Omega)$ is the potential whose form has to be made explicit. Note that in the spirit of the RMT, illustrated in several parts of this book, Ω (or in case the pair (Ω, ω)) is the auxiliary variable to be used to get a reliable simulation of the influence of the thermal bath on the set of interest. In the present case this coincides exactly[31] with the well-known SLE theory.[29] This consists of replacing the rigorous operator iL defined by Eq. (2.4) with the dynamical operator

$$\Gamma = i\mathcal{H}_s^x + i\mathcal{H}_I^x(\Omega) + \Gamma_\Omega^+ \tag{2.7}$$

so Eq. (2.3) becomes

$$\frac{d}{dt} S_+(t) = \Gamma S_+(t) = i\left[\mathscr{H}_s + \mathscr{H}_I(\Omega), S_+(t)\right] + \Gamma_\Omega^+ S_+(t) \qquad (2.8)$$

in which the rigorous quantum-mechanical Liouvillian superoperator has been substituted by an effective operator, and Γ_Ω^+ is the operator adjoint to that introduced in Eq. (2.6) (note that if we use the terminology of ref. 47a, we are now in the Heisenberg picture). To apply the Heisenberg method, we have to define a proper scalar product. Let us consider two variables $A(\Omega, I, S)$ and $B(\Omega, I, S)$, which, for the sake of generality, are assumed to depend on molecular orientation and nuclear and electronic spin. Replacing iL with Γ, we are led naturally to define their corresponding scalar product $\langle B|A \rangle$ as follows:

$$\langle B|A \rangle = \text{Tr}_{\{SI\}} \int d\Omega \, B(\Omega, I, S) A(\Omega, I, S) \rho_{\text{eq}}(I, S) w_{\text{eq}}(\Omega) \qquad (2.9)$$

where $\rho_{\text{eq}}(I, S)$ and $w_{\text{eq}}(\Omega)$ denote the equilibrium distribution of the spin system and molecular orientation, respectively. This means that

$$\Gamma_\Omega w_{\text{eq}}(\Omega) = 0$$

$$w_{\text{eq}}(\Omega) = \frac{\exp[-V(\Omega)/KT]}{\int d\Omega \exp[-V(\Omega)/KT]} \qquad (2.10)$$

This definition is made necessary by the fact that in the actual calculations $B(\Omega, I, S)$ will be usually defined as follows

$$B_+(\Omega, I, S) \equiv \phi(\Omega) S_- \qquad (2.11)$$

We shall consider the following two cases:

(i) $\qquad\qquad\qquad\qquad \phi(\Omega) = \text{constant} \qquad\qquad (2.12)$

(ii) $\qquad\qquad\qquad\qquad \phi(\Omega) = g_1^2 \qquad\qquad\qquad (2.13)$

$$g_1^2 = \frac{g_\parallel^2 g_\perp^2}{\tilde{g}^2} \sin^2\alpha + g_\perp^2 \cos^2\alpha$$

and

$$\tilde{g}^2 = g_\parallel^2 \cos^2\beta + g_\perp^2 \sin^2\beta$$

where $g_\parallel$ and $g_\perp$ represent the components of an axial g-splitting tensor and β is the angle between the static magnetic field direction and the cylindrical symmetry axis. Throughout Section III we use case (i) except in Section III.C, where we show how to evaluate the ESR spectra of paramagnetic ions with large anisotropy interactions.

The correlation function $\Phi(t)$ of Eq. (2.1) can be written in terms of the scalar product of Eq. (2.9) as

$$\Phi(t) = \frac{\langle f_0 | f_0(t) \rangle}{\langle f_0 | f_0 \rangle} \tag{2.14}$$

where $|f_0\rangle$ is the first vector of the following biorthogonal set of vectors:

$$|f_0\rangle = |S_+\rangle \qquad\qquad P_0 \equiv |f_0\rangle \langle f_0|f_0\rangle^{-1} \langle f_0|$$

$$|f_1\rangle \equiv (1-P_0)\Gamma|f_0\rangle$$

$$|\tilde{f}_1\rangle \equiv (1-P_0)\Gamma^+|f_0\rangle \qquad P_1 \equiv |f_1\rangle \langle \tilde{f}_1|f_1\rangle^{-1}\langle \tilde{f}_1| \tag{2.15}$$

$$|f_2\rangle \equiv (1-P_0)(1-P_1)\Gamma|f_1\rangle$$

$$|\tilde{f}_2\rangle \equiv (1-P_0)(1-P_1^+)\Gamma^+|\tilde{f}_1\rangle$$

This basis is obtained by the generalized Mori theory (see Chapter I) without requiring $i\Gamma$ to be Hermitian. This leads us to show that the kth-order correlation function

$$\Phi_k(t) = \frac{\langle \tilde{f}_k | f_k(t) \rangle}{\langle \tilde{f}_k | f_k \rangle} \tag{2.16}$$

is related to the $(k+1)$th via the hierarchy relationship

$$\frac{d}{dt}\Phi_k(t) = \lambda_k \Phi_k(t) - \int_0^t \Delta_{k+1}^2 \Phi_{k+1}(t-\tau)\Phi_k(\tau)\,d\tau \tag{2.17}$$

where

$$\lambda_k \equiv \frac{\langle \tilde{f}_k | \Gamma | f_k \rangle}{\langle \tilde{f}_k | f_k \rangle} \tag{2.18}$$

$$\Delta_k^2 \equiv -\frac{\langle \tilde{f}_{k+1} | f_{k+1} \rangle}{\langle \tilde{f}_k | f_k \rangle} \tag{2.18'}$$

By Laplace transforming Eq. (2.14), we obtain

$$\hat{\Phi}_0(z) = \hat{\Phi}(z) = \cfrac{1}{z - \lambda_0 + \cfrac{\Delta_1^2}{z - \lambda_1 + \cfrac{\Delta_2^2}{z - \lambda_2 + \cdots}}} \qquad (2.19)$$

This expression is precisely analogous to that obtained by the original Mori theory,[33] in which the Hermitian nature of Γ makes λ_k purely imaginary and Δ_k^2 purely real. In the present approach, both λ_k and Δ_k^2 are complex quantities, resulting in faster convergence than in ref. 33, where only the last step of the chain has a complex λ_k introduced to simulate (i.e., take account of) the rest of the continued fraction. The real part of λ_n represent the dissipative diffusional terms.

Equation (2.19) is of basic importance for our computer algorithm. This equation will be used to compute the EPR spectra without recourse to diagonalization procedures. The parameters λ_k and Δ_k^2 of Eqs. (2.18) and (2.18′) must be determined in terms of the moments

$$s_n \equiv \frac{\langle f_0 | \Gamma^n | f_0 \rangle}{\langle f_0 | f_0 \rangle}$$

with a procedure which is simpler than that suggested by the projection algebra outlined above. This can be done as follows. Note that $\Phi(t)$ of Eq. (2.2) must be identified with the zeroth order of our hierarchy of correlation functions. Let us develop $\Phi_0(t)$ into a Taylor power series (see also Chapter III):

$$\Phi_0(t) = \sum_{i=0}^{N} s_i^{(0)} \frac{t^i}{i!} \qquad \left(s_i^{(0)} = s_i \right) \qquad (2.20)$$

The number N should be made as large as possible. Then we develop $\Phi_1(t - \tau)$ around $t = \tau$ as follows:

$$\Phi_1(t - \tau) = \sum_{n=0}^{N-2} \frac{1}{n!} s_n^{(1)} (t - \tau)^n = \sum_{n=0}^{N-2} \frac{1}{n!} s_n^{(1)} \sum_{k=0}^{n} \binom{n}{k} t^{n-k} (-\tau)^k$$

$$(2.21)$$

Replacing Eqs. (2.21) and (2.20) into Eq. (2.17) with $k = 0$, we obtain

$$s_m^{(1)} = \frac{s_1^{(0)} s_{m+1}^{(0)} - s_{m+2}^{(0)}}{\left(s_1^{(0)} \right)^2 - s_2^{(0)}} - s_m^{(0)} - \sum_{k=1}^{m-1} s_k^{(1)} s_{m-k}^{(0)} \qquad 0 \leq m \leq N-2 \quad (2.22)$$

This expression allows us to express the first $N-2$ $s_m^{(1)}$ parameters in terms of the first N $s_n^{(0)}$ parameters. In general $s_0^{(i)} = 1$ and

$$s_m^{(i)} = \frac{s_1^{(i-1)}s_{m+1}^{(i-1)} - s_{m+2}^{(i-1)}}{\left(s_1^{(i-1)}\right)^2 - s_2^{(i-1)}} - s_m^{(i-1)} - \sum_{k=1}^{m-1} s_k^{(i)}s_{m-k}^{(i-1)} \qquad 0 \le m \le N-2i$$

$$(2.23)$$

From the definitions of Eqs. (2.15), (2.18) and (2.18′), it is straightforward to get

$$\lambda_i = s_1^{(i)} \tag{2.24}$$

$$\Delta_{i+1}^2 = \left(s_1^{(i)}\right)^2 - s_2^{(i)} \tag{2.25}$$

We have thus shown that $\hat{\Phi}(z)$ can be expressed in a continued fraction form, whose expression can be given in terms of the moments s_n. The only theoretical tool we used to arrive at this important result is the generalized version of the celebrated Mori theory.[33] We now have the problem of evaluating the parameters s_n to obtain the spectra of interest to EPR spectroscopy. This can be done as follows. First, let us define the nth-order state

$$|a_n\rangle \equiv \Gamma^n |f_0\rangle \tag{2.26}$$

where

$$|f_0\rangle = S_+ \Pi_N \tag{2.27}$$

when Π_N is the identity operator in the nuclear space. The general expression for $|a_n\rangle$ reads

$$|a_n\rangle = \sum_{l,p,q,i,j,\alpha,\beta} c(n)_{l,p,q,i,j,\alpha,\beta} D_{p,q}^l(\Omega) A_{i,j} B_{\alpha,\beta} \tag{2.28}$$

We assumed that the magnetic tensors appearing in the spin Hamiltonian and the diffusion tensor have the same principal axis system. The $D_{p,q}^l$ are the Wigner matrices of rank l. The nuclear spin operators are expanded over the set of $(2I+1)^2$ matrices $A_{i,j}$, I being the nuclear angular moment, defined via

$$\left(A_{i,j}\right)_{k,s} = \delta_{i,k}\delta_{j,s} \tag{2.29}$$

$$\Pi_N = \sum_{i=1}^{2I+1} A_{i,i} \tag{2.30}$$

Likewise the electronic spin operators are expanded over the $(2S+1)^2$ matrices $B_{\alpha,\beta}$, which, in turn, are defined in the same way as the $A_{i,j}$ [Eq. (2.29)].

We are now in a position to program the computer to evaluate the multi-dimensional array $c(n)$ with the following iterative expressions:

$$
\begin{aligned}
|a_n\rangle &= \sum_{l,p,q,i,j,\alpha,\beta} c(n)_{l,p,q,i,j,\alpha,\beta} D^l_{p,q}(\Omega) A_{i,j} B_{\alpha,\beta} \\
&= \sum_{l',p',q',i',j',\alpha',\beta'} c(n-1)_{l',p',q',i',j',\alpha',\beta'} \Gamma\left(D^{l'}_{p',q'}(\Omega) A_{i',j'} B_{\alpha',\beta'} \right)
\end{aligned}
$$

$$(2.31)$$

which leads to

$$
c(n)_{l,p,q,i,j,\alpha,\beta} = \sum_{l',p',q',i',j',\alpha',\beta'} R^{l,p,q,i,j,\alpha,\beta}_{l',p',q',i',j',\alpha',\beta'} c(n-1)_{l',p',q',i',j',\alpha',\beta'}
$$

$$(2.32)$$

where

$$
\Gamma\left(D^{l'}_{p',q'}(\Omega) A_{i',j'} B_{\alpha',\beta'} \right) = \sum_{l,p,q,i,j,\alpha,\beta} R^{l,p,q,i,j,\alpha,\beta}_{l',p',q',i',j',\alpha',\beta'} D^l_{p,q}(\Omega) A_{i,j} B_{\alpha,\beta}
$$

Symmetry constraints can significantly reduce the actual number of independent components $c(n)_{l,p,q,i,j,\alpha,\beta}$ to be evaluated. The starting point of our iteration procedure is

$$
c(0)_{l,p,q,i,j,\alpha,\beta} = \delta_{l,0}\delta_{p,0}\delta_{q,0}\delta_{i,j}\delta_{\beta,\alpha+1} c_{\alpha,\alpha+1}
$$

$$(2.33)$$

where $c_{\alpha,\alpha+1}$ is a real number defined by

$$
S_+ = \sum_\alpha c_{\alpha,\alpha+1} B_{\alpha,\alpha+1}
$$

$$(2.34)$$

The parameter s_n can be obtained simply from

$$
s_n = \frac{\langle a_0 | a_n \rangle}{\langle a_0 | a_0 \rangle}
$$

$$(2.35)$$

By using Eq. (2.30) we arrive at

$$s_n = s_n(1) + s_n(2) + \cdots + s_n(2I+1) \tag{2.36}$$

where

$$s_n(i) = \frac{\langle S_- A_{i,i} | \Gamma^n | a_0 \rangle}{\langle a_0 | a_0 \rangle}$$

$$= \frac{\sum\limits_{l',q',p',\alpha'} \overline{D^{l'}_{p'q'}} c_{\alpha',\alpha'+1} c(n)_{l',p',q',i,i,\alpha',\alpha'+1}}{(2I+1)\mathrm{Tr}_{\{S\}}(S_- S_+)} \tag{2.37}$$

and

$$\overline{D^l_{p,q}} = \frac{\int d\Omega\, w_{eq}(\Omega) D^l_{p,q}(\Omega)}{\int d\Omega\, w_{eq}(\Omega)} \tag{2.38}$$

are the order parameters.

Note that Eq. (2.36) suggests an alternative approach to the EPR spectrum consisting in expressing the lineshape as the summation of $2I+1$ continued fractions, one for each of the $2I+1$ contributions to the s_n's of Eq. (2.36). This means that

$$I(\omega) = \sum_{i=1}^{2I+1} I_i(\omega) \tag{2.39}$$

where $I_i(\omega)$ is the real part of the Laplace transform of the correlation function of Eq. (2.14), with the scalar product now given by

$$\langle B | A \rangle^i = \mathrm{Tr}_{\{SI\}} \int d\Omega\, B(\Omega, I, S) A(\Omega, I, S) A_{i,i} \rho_{eq}(I, S) w_{eq}(\Omega) \tag{2.40}$$

To get the results of the next section we have followed both ways. With fairly short microscopic times we have that the latter method leads to a faster convergence. This is because in such a physical condition the spectrum consists of $2I+1$ very distinct lineshapes, one for each continued fraction. A few steps of each continued fraction are then required to clearly identify each component lineshape. The reader who would like to use this approach should keep in mind these two different ways of evaluating spectra and decide which is the most convenient for the actual physical condition. Note that the direct use of Eqs. (2.18) and (2.18′) in the light of the remarks of Chapter IV would be equivalent to the Lanczos method.[40a] This would certainly be preferable to the method here described when convergence problems appear. This does not concern the applications illustrated here. Comparison between the algorithm illustrated here and the result of ref. 40a will be discussed in Section III. In Sections III.B and III.D, we shall compare the present algorithm to the structure of a product of continued fractions (the parameters of which will be determined analytically).

III. APPLICATIONS

In order to highlight the properties of the theory developed in Section II and to make the reader more familiar with this method, we shall describe in this section its application to the evaluation of the EPR spectra of paramagnetic species in the slow-motion regime for physical systems of general experimental interest. As usual in the slow tumbling region, the nonsecular terms in the spin Hamiltonian can be disregarded[48,49]; moreover, without appreciably losing generality, we shall assume the tensors of magnetic interaction in the spin Hamiltonian and the diffusion tensor to be axially symmetric tensors with the same symmetry axis.

In Section III.A, we shall treat the case of a nitroxide spin radical in free diffusion, in III.B we shall evaluate the lineshape of the triplet zero-field splitting and introduce an alternative direct approach. In III.C , we shall obtain the EPR spectra of copper complexes in solution in which the magnetic parameters of the metal ion transition spin Hamiltonian are very anisotropic. These last systems are of very great interest both in the biological field and in the case of vitreous matrices. We shall take into account the effect on the EPR lineshape of different probability transitions for spin packets corresponding to different angles between the principal axis of the magnetic interaction and the static magnetic field. In Section III.D, the lineshape of paramagnetic species dissolved in liquid-crystalline mesophases will be calculated.

We will assume the mesophase director to be parallel to the direction of the static magnetic field. In the last section, III.E, the no-inertia assumption will be rejected and the diffusion operator (2.6) replaced with the complete Hubbard operator.[47b] We shall investigate the spectroscopic effects of molecular inertia in the case of an axially symmetric g-tensor.

The magnetic interactions we shall deal with can be expressed in spherical notation by the Hamiltonian (Appendix B)

$$\mathscr{H} = \sum_{\mu, l, p, q} (-)^q F_\mu^{(l, q)} D_{p, -q}^l (\Omega) T_\mu^{(l, p)}$$

where μ denotes the different interactions, $T_\mu(l, p)$ the corresponding spin

TABLE I

The Irreducible Components of the Zeeman, Hyperfine, and Fine Tensors g, A, δ together with that of the Corresponding Spin Operator[a]

$g^{(0,0)} = \sqrt{3}\, g$	$T_g^{(0,0)} = \beta_e BS_z/\sqrt{3}$
$g^{(2,0)} = \sqrt{\frac{3}{2}}\,(g_{zz} - g)$	$T_g^{(2,0)} = \sqrt{\frac{2}{3}}\,\beta_e BS_z$
$g^{(2,\,\pm 1)} = \mp(g_{xz} \pm ig_{yz})$	$T_g^{(2,\,\pm 1)} = \mp \beta_e BS \pm /2$
$g^{(2,\,\pm 2)} = (g_{xx} - g_{yy} \pm 2ig_{xy})/2$	$T_g^{(2,\,\pm 2)} = 0$
$A^{(0,0)} = \sqrt{3}\, a$	$T_A^{(0,0)} = S \cdot I/\sqrt{3}$
$A^{(2,0)} = \sqrt{\frac{3}{2}}\,(A_{zz} - a)$	$T_A^{(2,0)} = \sqrt{\frac{2}{3}}\,[S_z I_z - (S_+ I_- + S_- I_+)/4]$
$A^{(2,\,\pm 1)} = \mp(A_{xz} \pm iA_{yz})$	$T_A^{(2,\,\pm 1)} = \mp(I_\pm S_z + I_z S_\pm)/2$
$A^{(2,\,\pm 2)} = (A_{xx} - A_{yy} \pm 2iA_{xy})/2$	$T_A^{(2,\,\pm 2)} = I_\pm S_\pm/2$
$\delta^{(0,0)} = 0$	$T_\delta^{(0,0)} = S^2/\sqrt{3}$
$\delta^{(2,0)} = \sqrt{\frac{3}{2}}\,\delta_{zz}$	$T_\delta^{(2,0)} = \sqrt{\frac{2}{3}}\,[S_z^2 - (S_+ S_- + S_- S_+)/4]$
$\delta^{(2,\,\pm 1)} = \mp(\delta_{xz} \pm i\delta_{yz})$	$T_\delta^{(2,\,\pm 1)} = \mp(S_\pm S_z + S_z S_\pm)/2$
$\delta^{(2,\,\pm 2)} = (\delta_{xx} - \delta_{yy} \pm 2i\delta_{xy})/2$	$T_\delta^{(2,\,\pm 2)} = S_\pm^2/2$

[a] It has been assumed that the magnetic field is along the z axis of the laboratory frame.

operators of rank l in the laboratory frame, and $F_\mu^{(l,q)}$ the corresponding component tensors in the molecular frame. In Table I the general expressions for $F_\mu^{(l,q)}$ and $T_\mu^{(l,p)}$ are reported.

A. Hyperfine Structure ($S = 1/2$, $I = 1$)

When dealing with a long microscopic time scale (slow motion), in the high-field approximation, the nonsecular terms are usually disregarded. So the Hamiltonian describing the spin system is

$$\mathcal{H} = \mathcal{H}_s + \mathcal{H}_I \tag{3.1}$$

where

$$\mathcal{H}_s = g\beta_e H_0 S_z + aS_z I_z \tag{3.2}$$

having neglected the nuclear Zeeman Hamiltonian

$$\mathcal{H}_I = \tfrac{2}{3}D_{0,0}^2(\Omega)[b + cI_z]S_z - \left(\tfrac{1}{6}\right)^{1/2} c\left[D_{1,0}^2(\Omega)I_+ - D_{-1,0}^2(\Omega)I_-\right]S_z \tag{3.3}$$

where $g = \frac{1}{3}\mathrm{Tr}\,\mathbf{g} = \frac{1}{3}(g_{\parallel} + 2g_{\perp})$ ($g_{\parallel}, g_{\perp}$ are the components of the g-splitting tensor), β_e is the Bohr magneton, and H_0 is the value of the static magnetic field; $a = \frac{1}{3}\,\mathrm{Tr}\mathbf{A} = \frac{1}{3}(A_{\parallel} + 2A_{\perp})$, where $A_{\parallel}, A_{\perp}$ are the components of the hyperfine $\mathbf{A}$ tensor, $b = (g_{\parallel} - g_{\perp})\beta_e H_0$, $c = A_{\parallel} - A_{\perp}$; the $D_{p,q}^2$ are the Wigner matrices of rank 2 using Rose's notation.[55]

For axially symmetric molecules, (Appendix B)

$$D_{0,0}^{(2)}(\Omega) = \sqrt{\frac{4\pi}{5}}\, Y_{2,0}(\Omega)$$

$$D_{1,0}^{(2)}(\Omega) = \sqrt{\frac{4\pi}{5}}\, Y_{2,1}^*(\Omega) \qquad (3.4)$$

$$D_{-1,0}^{(2)}(\Omega) = \sqrt{\frac{4\pi}{5}}\, Y_{2,-1}^*(\Omega)$$

$$\mathcal{H}_I = \frac{2}{3}\sqrt{\frac{4\pi}{5}}\, Y_{2,0}(\Omega)[b + cI_z]S_z + \sqrt{\frac{4\pi}{30}}\, c\,[Y_{2,-1}(\Omega)I_+ - Y_{2,1}(\Omega)I_-]S_z$$

$$(3.5)$$

where the $Y_{2,m}$ are the spherical harmonics of rank 2. The evolution equation for the S_+ operator is

$$\frac{dS_+}{dt} = \Gamma S_+ \equiv i[\mathcal{H}_s + \mathcal{H}_I, S_+] + \Gamma_{\Omega}^+ S_+ \qquad (3.6)$$

where the diffusional operator of Eq. (3.6), as defined in Eq. (2.6), in the absence of an external potential becomes

$$\Gamma_{\Omega} = -\,\overline{M}\mathbf{D}\overline{M} = -\left[(D_{\parallel} - D_{\perp})M_z^2 + D_{\perp}M^2\right] \qquad (3.7)$$

Using the guidelines of Section II, the states sequence is

$$|a_0\rangle = S_+ \qquad |a_n\rangle = \Gamma^n S_+$$

Using (3.2) and (3.5)–(3.7), it can be shown that the general state $|a_n\rangle$ can be expressed as the following linear combination:

$$|a_n\rangle = \sum_{l,m,i,j} c(n)_{l,m,i,j} Y_{2l,m}(\Omega) A_{i,j} S_+ \qquad (3.8)$$

where $A_{i,j}$ are square matrices of rank 3 and represent a complete operatorial basis set for the nuclear spin $I = 1$. In particular, the operators

I_z, I_+, I_- are expressed by

$$I_z = A_{1,1} - A_{3,3} \qquad I_+ = \sqrt{2}\,(A_{1,2} + A_{2,3}) \qquad I_- = \sqrt{2}\,(A_{2,1} + A_{3,2})$$

$$(3.9)$$

The expansion (3.8), in particular, the proportionality between the state $|a_n\rangle$ and the operator S_+ and the fact that the spherical harmonics $Y_{2l,m}(\Omega)$ represent a complete expansion base, is made possible by the properties of the Hamiltonians (3.2) and (3.5) under consideration. In fact, the Hamiltonian under investigation only depends on S_z, so the proportionality between $|a_n\rangle$ and S_+ is due to

$$[S_z, S_+] = S_+ \qquad (3.10)$$

The second observation follows by simple parity consideration. The introduction of S_+, S_- operators through the nonsecular terms in the Hamiltonian leads to the following expression for $|a_n\rangle$:

$$|a_n\rangle = \sum_{l,m,i,j,\alpha,\beta} c(n)_{l,m,i,j,\alpha,\beta} Y_{2l,m}(\Omega) A_{i,j} B_{\alpha,\beta} \qquad (3.11)$$

with $\alpha, \beta = 1,2$.

The expansion (3.11) can be employed to compute ESR lineshapes for slow-tumbling paramagnetic molecules in the presence of weak static magnetic fields. In such cases the computation of the nitroxide radical ESR spectrum is only a little more time-consuming.

In this section we limit ourselves to considering the expansion (3.8).

The next step, is the calculation of the s_n moments with which it is possible to obtain via Eqs. (2.23)–(2.25) the parameters λ_i, Δ_{i+1}^2 that appear in the expression of the Laplace transform of the correlation function (2.2).

$$s_n = \frac{\langle a_0 | a_n \rangle}{\langle a_0 | a_0 \rangle}$$

$$= \frac{\mathrm{Tr}_{\{SI\}} \int d\Omega\, S_- \Pi_N \left[\displaystyle\sum_{l,m,i,j} c(n)_{l,m,i,j} Y_{2l,m}(\Omega) A_{i,j} S_+ \right] w_{eq}}{(2I+1)\mathrm{Tr}_{\{S\}}(S_- S_+) \int d\Omega\, w_{eq}}$$

$$= \frac{1}{\sqrt{4\pi}\,(2I+1)} \sum_i c(n)_{0,0,i,i} \qquad (3.12)$$

Note in the isotropic phase the distribution function w_{eq} is a constant. So we must use Eq. (2.32), which in the present case can be written

$$c(n+1)_{l,m,i,j} = \sum_{l',m',i',j'} R^{l,m,i,j}_{l',m',i',j'} c(n)_{l',m',i',j'} \qquad (3.13)$$

where $R^{l,m,i,j}_{l',m',i',j'}$ is defined by

$$\Gamma\left(Y_{2l',m'}(\Omega)A_{i',j'}S_+\right) = \sum_{l,m,i,j} R^{l,m,i,j}_{l',m',i',j'} Y_{2l,m}(\Omega)A_{i,j}S_+ \qquad (3.14)$$

The starting point for the iteration is

$$c(0)_{l,m,i,j} = \sqrt{4\pi}\,\delta_{l,0}\delta_{m,0}\delta_{i,j}$$

To obtain the $R^{l,m,i,j}_{l',m',i',j'}$ coefficients for the explicit evaluation of Eq. (3.13) we must calculate the contribution of the following terms of the Liouvillian superoperator:

(i)	$\bar{\omega}_0 S_z$	$\bar{\omega}_0 = ig\beta_e H_0$
(ii)	$AS_z I_z$	$A = ia$
(iii)	$BS_z Y_{2,0}(\Omega)$	$B = \dfrac{2}{3}i\sqrt{\dfrac{4\pi}{5}}\,b$
(iv)	$CY_{2,0}(\Omega)I_z S_z$	$C = \dfrac{2}{3}i\sqrt{\dfrac{4\pi}{5}}\,c$
(v)	$DY_{2,-1}(\Omega)I_+ S_z$	$D = i\sqrt{\dfrac{4\pi}{30}}\,c$
(vi)	$DY_{2,1}(\Omega)I_- S_z$	
(vii)	$-\left[(D_\parallel - D_\perp)M_z^2 + D_\perp M^2\right]$	

$$(3.15)$$

For the sake of simplicity we first consider the spin operator of Eq. (3.15):

1. S_z

$$\left[S_z, A_{i',j'}S_+\right] = A_{i',j'}S_+ = \sum_{i,j}\delta_{i',i}\delta_{j',j}A_{i,j}S_+$$

2. $S_z I_z$

$$\left[S_z I_z, A_{i',j'}S_+\right] = \frac{S_+}{2}\{I_z, A_{i',j'}\}$$

where the $\{\,,\,\}$ represent the anticommutation operator

$$\frac{S_+}{2}\left[\{A_{1,1},A_{i',j'}\}-\{A_{3,3},A_{i',j'}\}\right]$$

$$=\frac{S_+}{2}\left[(A_{1,j'}\delta_{1,i'}+A_{i',1}\delta_{1,j'})-(A_{3,j'}\delta_{i',3}+A_{i',3}\delta_{j',3})\right]$$

$$=\sum_{i,j}\frac{1}{2}\left[(\delta_{1,i}\delta_{j',j}\delta_{i',1}+\delta_{i',i}\delta_{1,j'}\delta_{1,j})\right.$$

$$\left.-(\delta_{3,i}\delta_{j,j'}\delta_{i',3}+\delta_{i,i'}\delta_{3,j}\delta_{j',3})\right]A_{i,j}S_+$$

3. I_+S_z

$$\left[I_+S_z,A_{i',j'}S_+\right]=\frac{S_+}{2}\{I_+,A_{i',j'}\}$$

$$=\frac{\sqrt{2}}{2}\left[S_+\{A_{1,2},A_{i',j'}\}+S_+\{A_{2,3},A_{i',j'}\}\right]$$

$$=\frac{\sqrt{2}}{2}S_+\left(A_{1,j'}\delta_{2,i'}+A_{i',2}\delta_{j',1}+A_{2,j'}\delta_{3,i'}+A_{i',3}\delta_{j',2}\right)$$

$$-\sum_{i,j}\frac{\sqrt{2}}{2}\left[\delta_{1,i}\delta_{j,j'}\delta_{2,i'}+\delta_{i,i'}\delta_{2,j}\delta_{j',1}\right.$$

$$\left.+\delta_{2,i}\delta_{j,j'}\delta_{3,i'}+\delta_{i,i'}\delta_{3,j}\delta_{j',2}\right]A_{i,j}S_+$$

4. I_-S_z

$$\left[I_-S_z,A_{i,j}S_+\right]=\frac{S_+}{2}\{I_-,A_{i',j'}\}$$

$$=\frac{\sqrt{2}}{2}S_+\left(A_{2,j'}\delta_{1,i'}+A_{i',1}\delta_{j',2}+A_{3,j'}\delta_{2,i'}+A_{i',2}\delta_{j',3}\right)$$

$$=\sum_{i,j}\frac{\sqrt{2}}{2}\left[\delta_{2,i}\delta_{j,j'}\delta_{1,i'}+\delta_{i,i'}\delta_{1,j}\delta_{j',2}\right.$$

$$\left.+\delta_{3,i}\delta_{j,j'}\delta_{2,i'}+\delta_{i,i'}\delta_{2,j}\delta_{j'3}\right]A_{i,j}S_+ \qquad (3.16)$$

having used

$$A_{i,j}A_{l,k}=\delta_{j,l}A_{i,k}$$

The contribution to the $c(n+1)_{l,m,i,j}$ of Eq. (3.13) due to the terms of Eq. (3.15) are, respectively,

1. $\bar{\omega}_0 c(n)_{l,m,i,j}$

2. $\dfrac{A}{2}\left[\delta_{i,1}c(n)_{l,m,1,j}+\delta_{j,1}c(n)_{l,m,i,1}-\delta_{i,3}c(n)_{l,m,3,j}-\delta_{j,3}c(n)_{l,m,i,3}\right]\equiv$
   $\dfrac{A}{2}M(l,m,i,j)$

3. $B\left[C_{0P}c(n)_{l-2,m,i,j}+C_{00}c(n)_{l,m,i,j}+C_{0M}c(n)_{l+2,m,i,j}\right]$

   where

   $$C_{00}\equiv C(l,2,l,m,0)C(l,2,l,0,0)\sqrt{\dfrac{5}{4\pi}}$$

   $$C_{0M}\equiv C(l+2,2,l,m,0)C(l+2,2,l,0,0)\left[\dfrac{2(l+2)+1}{2l+1}\right]^{1/2}\sqrt{\dfrac{5}{4\pi}}$$

   $$C_{0P}\equiv C(l-2,2,l,m,0)C(l-2,2,l,0,0)\left[\dfrac{2(l-2)+1}{2l+1}\right]^{1/2}\sqrt{\dfrac{5}{4\pi}}$$

   $$(3.17)$$

   $C(l_1,l_2,l_3,m_1,m_2)$ denotes the Clebsch-Gordan coefficients according to Rose's notation.[55]

4. $C\left[C_{0P}M(l-2,m,i,j)+C_{00}M(l,m,i,j)+C_{0M}M(l+2,m,i,j)\right]/2$

5. $\dfrac{D}{\sqrt{2}}\left[C_{MP}\left(\delta_{1,i}c(n)_{l-2,m+1,2,j}\right.\right.$

   $+\delta_{j,2}c(n)_{l-2,m+1,i,1}+\delta_{i,2}c(n)_{l-2,m+1,3,j}+\delta_{j,3}c(n)_{l-2,m+1,i,2}\big)$

   $+C_{M0}\big(\delta_{1,i}c(n)_{l,m+1,2,j}+\delta_{j,2}c(n)_{l,m+1,i,1}$

   $+\delta_{i,2}c(n)_{l,m+1,3,j}+\delta_{j,3}c(n)_{l,m+1,i,2}\big)$

   $+C_{MM}\big(\delta_{1,i}c(n)_{l+2,m+1,2,j}+\delta_{j,2}c(n)_{l+2,m+1,i,1}$

   $\left.+\delta_{i,2}c(n)_{l+2,m+1,3,j}+\delta_{j,3}c(n)_{l+2,m+1,i,2}\big)\right]$

be using (3.13) (3.14) where

$$C_{M0} \equiv C(l,2,l,m+1,-1)C(l,2,l,0,0)\sqrt{\frac{5}{4\pi}}$$

$$C_{MM} \equiv C(l+2,2,l,m+1,-1)C(l+2,2,l,0,0)\left[\frac{2(l+2)+1}{2l+1}\right]^{1/2}\sqrt{\frac{5}{4\pi}}$$

$$C_{MP} \equiv C(l-2,2,l,m+1,-1)C(l-2,2,l,0,0)\left[\frac{2(l-2)+1}{2l+1}\right]^{1/2}\sqrt{\frac{5}{4\pi}}$$

6.

$$\frac{D}{\sqrt{2}}\Big\{C_{1P}\big[\delta_{i2}c(n)_{l-2,m-1,1,j}$$

$$+ \delta_{j,1}c(n)_{l-2,m-1,i,2}+\delta_{i,3}c(n)_{l-2,m-1,2,j}+\delta_{2,j}c(n)_{l-2,m-1,i,3}\big]$$

$$+ C_{10}\big[\delta_{i,2}c(n)_{l,m-1,1,j}+\delta_{j,1}c(n)_{l,m-1,i,2}$$

$$+ \delta_{i,3}c(n)_{l,m-1,2,j}+\delta_{2,j}c(n)_{l,m-1,i,3}\big]$$

$$+ C_{1M}\big[\delta_{i,2}c(n)_{l+2,m-1,1,j}+\delta_{j,1}c(n)_{l+2,m-1,i,2}+\delta_{i,3}c(n)_{l+2,m-1,2,j}$$

$$+ \delta_{2,j}c(n)_{l+2,m-1,i,3}\big]\Big\}$$

where

$$C_{10} \equiv C(l,2,l,m-1,1)C(l,2,l,0,0)\sqrt{\frac{5}{4\pi}}$$

$$C_{1M} \equiv C(l+2,2,l,m-1,1)C(l+2,2,l,0,0)\left[\frac{2(l+2)+1}{2l+1}\right]^{1/2}\sqrt{\frac{5}{4\pi}}$$

$$C_{1P} \equiv C(l-2,2,l,m-1,1)C(l-2,2,l,0,0)\left[\frac{2(l-1)+1}{2l+1}\right]^{1/2}\sqrt{\frac{5}{4\pi}}$$

7. $-\big[(D_\| - D_\perp)m^2 + D_\perp l(l+1)\big]c(n)_{l,m,i,j}$

For a fixed value of n, the index l explores all the even values ranging from 0 to $2n$, and m should range from l to $-l$, making the dimension of $c(n)$ very large. However, as remarked in Section II, symmetry constraints significantly reduce the actual number to be evaluated. In fact, it is possible to show that:

1. m ranges from -2 to 2 because the index m is made to change by the terms $Y_{2,-1}(\Omega)I_+$ and $Y_{2,1}(\Omega)I_-$ of Eqs. (3.15) and $(I_+)^3 = (I_-)^3 = 0$ since $I = 1$.

2. $c(n)_{l,m,i,j} = (-)^m c(n)_{l,-m,j,i}$. This useful relationship can be obtained by using the operator T defined as follows:

$$T = KP\Pi_s \qquad (3.18)$$

where

$$KY_{l,m}(\Omega) = (-)^m Y_{l,-m}(\Omega) \qquad (3.19)$$

$$PA_{i,j} = A_{j,i} \qquad (3.20)$$

and Π_s is the identity operator of the electronic spin space.

Since $\mathcal{H} = \mathcal{H}^+$, T commutes with $\mathcal{H}$. This leads us to $T(\Gamma^n S_+)T^{-1} = \Gamma^n T S_+ T^{-1} = \Gamma^n S_+$.[56] So, by applying this transformation to the right-hand side of the equation

$$\Gamma^n S_+ = \sum_{l,m,i,j} c(n)_{l,m,i,j} Y_{2l,m}(\Omega) A_{i,j} S_+ \qquad (3.21)$$

we have

$$\sum_{i,j,l,m} c(n)_{l,m,i,j} T\left[A_{i,j} Y_{2l,m}(\Omega) S_+\right] T^{-1}$$

$$= \sum_{i,j,l,m} c(n)_{l,m,i,j} A_{j,i} (-)^m Y_{2l,-m}(\Omega) S_+$$

$$= \sum_{i,j,l,m} (-)^m c(n)_{l,-m,j,i} A_{i,j} Y_{2l,m}(\Omega) S_+$$

$$\qquad (3.22)$$

From Eqs. (3.21)–(3.22), we obtain the useful property 2. In consequence, we are allowed to choice as independent terms $c(n)_{l,0,1,1}$, $c(n)_{l,0,2,2}$, $c(n)_{l,0,3,3}$, $c(n)_{l,1,3,2}$, $c(n)_{l,1,2,1}$, and $c(n)_{l,2,3,1}$. In Appendix A we provide the recursion formulas for these coefficients.

We are now in a position to calculate the s_n parameters of Eq. (3.12) and the λ_i, Δ_{i+1}^2 necessary to construct the continued fraction providing the ESR lineshape. We will now apply this algorithm to the ESR spectra of a nitroxide radical in free diffusion for different values of the diffusion coefficient while varying the anisotropy of the corresponding diffusion process. The ratio of the parallel diffusion coefficient to the orthogonal is shown in Figs. 1 to 4 to vary from $\frac{1}{2}$ to 100.

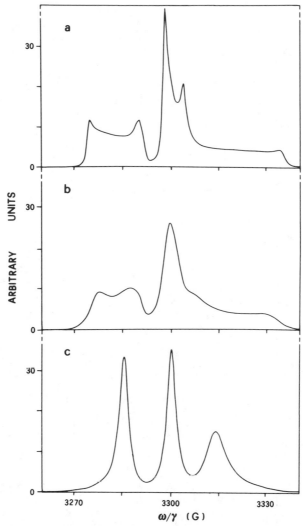

Figure 1. Plot of the absorption spectrum for a nitroxide spin probe dissolved in an isotropic medium. Magnetic parameters are $g_{\parallel} = 2.0027$, $g_{\perp} = 2.0075$, $A_{\parallel} = 32$ G, $A_{\perp} = 6$ G, $H_0 = 3300$ G. The values of the diffusion tensor components give $D_{\parallel}/D_{\perp} = 0.5$. (a) $r = 100$, (b) $r = 10$, (c) $r = 1$.

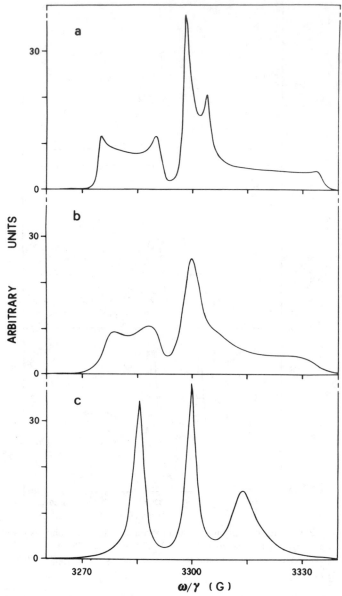

Figure 2. Plot of the absorption spectrum for a nitroxide spin probe dissolved in a iso-tropic medium. Magnetic parameters for the spin probe are the same as in Fig. 1. $D_{\parallel}/D_{\perp} = 1$. (a) $r = 100$, (b) $r = 10$, (c) $r = 1$.

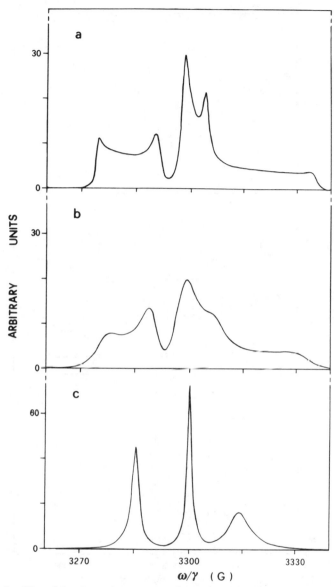

Figure 3. Plot of the absorption spectrum for a nitroxide spin probe dissolved in a iso-tropic medium. Magnetic parameters for the spin probe are the same as in Fig. 1. $D_{\parallel}/D_{\perp} = 10$. (a) $r = 100$, (b) $r = 10$, (c) $r = 1$.

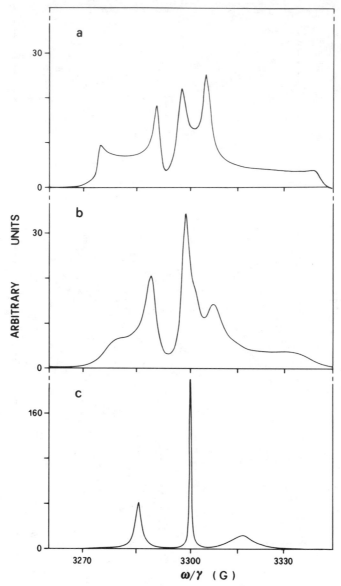

Figure 4. Plot of the absorption spectrum for a nitroxide spin probe dissolved in a iso-tropic medium. Magnetic parameters for the spin probe are the same as in Fig. 1. $D_{\parallel}/D_{\perp} = 100$. (a) $r = 100$, (b) $r = 10$, (c) $r = 1$.

From Eq. (6.18) of Chapter II,

$$D_\perp = \frac{\langle \omega_\perp^2 \rangle I_\perp}{\mathscr{F}_\perp} \qquad D_\| = \frac{\langle \omega_\|^2 \rangle I_\|}{\mathscr{F}_\|} \qquad (3.23)$$

where $\mathscr{F}_\|$ indicates the friction affecting the rotation around the molecular symmetry axis and $\mathscr{F}_\perp$ that affecting the rotation of the axis itself. Therefore this is equivalent to making the ratio $\mathscr{F}_\perp / \mathscr{F}_\|$ vary from $\frac{1}{2}$ to 100.

Moreover, for a given $D_\|/D_\perp$, the ESR lineshapes are related (parts a, b, c of each figure) to different values of the parameter

$$r = \frac{2}{3} |g_\| - g_\perp| \frac{\beta_e H_0}{D_\perp} \qquad (3.24)$$

which, in a sense, indicates the degree to which our system is non-Markovian. Parts a, b, c, of each figure, at a given value of $D_\|/D_\perp$, correspond to an increasing diffusional mobility, as a quick look to the ESR spectra suggests.

Moving from Fig. 1 to Fig. 4, the anisotropy of the diffusion motion increases due to a faster diffusion process around the symmetry axis of the molecule. In the highly rigid regime (Figs. 1a, 2a, 3a, 4a), the ESR spectrum is noted to depend weakly on the r parameter for low values of $D_\|$ while showing a drastic change at the highest value of $D_\|$. The modification of ESR lineshape appears to be more marked in the case of fast diffusional motion. Such behavior can be ascribed to the pseudosecular terms of the Hamiltonian (3.3), which become more important when the local field components at the Larmor nuclear frequencies increase. In fact, analysis of the spin Hamiltonian (3.3) shows that while the secular terms are modulated only by the motions involving the $D_\perp$ diffusion tensor component, the pseudosecu-

TABLE II

Minimum Number of Steps n_s Needed to Achieve the Convergence of the C.F. and (in brackets) Elapsed Computing Times (s) with an IBM 3033 Computer to Obtain the ESR Spectra Reported in Figs. 1–4

r \ $D_\|/D_\perp$	0.5	1	10	100
1	12(2.1)	15(2.7)	15(2.7)	20(37*)
10	23(4.8)	23(4.8)	23(4.8)	30(66*)
100	32(8.7)	32(8.7)	32(72*)	34(80*)

lar terms are also modulated by the reorientational processes around the molecular symmetry axis. Thus, in the slow-motion regime the spin relaxation of the paramagnetic probe becomes unaffected by molecular reorientational motion around its own symmetry axis, while the characteristic features of the ESR lineshape depend on the value of the $D_\perp$ coefficient.

In Table II are reported the relevant data on computational convergence and execution times corresponding to the case illustrated in Figs. 1–4. Asterisks denote some extreme case where extended precision has been used in the computation in order to overcome numerical stability. In these cases the alternative procedure of projection operator (see, for instance, Chapter IV) could be followed.

B. Zero Field Splitting ($S = 1$)

For an $S = 1$ spin we now consider the presence of a zero-field term along with an axially symmetric g tensor leading in the secular approximation to the following Hamiltonian:

$$\mathcal{H} = \omega_0 S_z + Y_{2,0}(\Omega) S_z [\eta + \delta S_z] \qquad (3.25)$$

where $\omega_0 = g\beta_e H_0$, $\eta = \frac{2}{3}\sqrt{4\pi/5} \; (g_\parallel - g_\perp)\beta_e H_0$, $\delta = \sqrt{4\pi/5} \, (\frac{2}{3})(\delta_\parallel - \delta_\perp)$, and $\delta_\parallel, \delta_\perp$ are the components of the fine-structure tensor. It is well known that the evolution equation for the variable of interest S_+ is given by

$$\frac{d}{dt} S_+ = i[\mathcal{H}, S_+] + \Gamma_\Omega^+ S_+ \equiv \Gamma S_+ \qquad (3.26)$$

where $\Gamma_\Omega = -[(D_\parallel - D_\perp) M_z^2 + D_\perp M^2]$. This example is an important illustration of our algorithm, which is versatile and of general validity. However, in some special cases alternative methods can be faster.

It is straightforward to show that the special case of a spin $\frac{1}{2}$ having an axially symmetric g tensor (widely used as a tutorial example for the theory behind ESR spectroscopy[43,48]), when direct use is made of Eqs. (2.18) and (2.18′), leads to a continued fraction of the same kind as Eq. (2.19) with the expansion parameters

$$\lambda_j = +i\frac{\eta}{2\pi} \int d\Omega \, Y_{2l,0}^*(\Omega) Y_{2l,0}(\Omega) Y_{2l,0}(\Omega) - 2l(2l+1)D_\perp \qquad (3.27)$$

$$\Delta_j^2 = \frac{\eta^2}{4\pi^2} \left[\int d\Omega \, Y_{2l-2,0}^* Y_{2l,0}(\Omega) Y_{2l,0}(\Omega) \right]^2 \qquad (3.28)$$

This result coincides with the result of Knak Jensen.[48a]

In this section we shall detail the analytical derivation of the absorption spectrum for a magnetic species in a triplet spin state tumbling in a viscous disordered liquid. This analytical structure consists of the sum of product of continued fractions. The standard Mori structure of Eq. (2.19) is recovered in the absence of orienting potentials.

This will serve the main purpose of checking the general algorithm illustrated in Section II. This approach also renders straightforward a clearer physical interpretation of the main features of the corresponding EPR spectra.

In Section III.B.1 we shall use the CFP illustrated in Section III.A, whereas to the latter method we shall devote Section III.B.2.

1. Evaluation of the Moments

According to Section III.A, the general expression for the state $|a_n\rangle$ is

$$|a_n\rangle = \Gamma^n|a_0\rangle = \sum_{l,\alpha,\beta} c(n)_{l,0,\alpha,\beta} Y_{2l,0}(\Omega) B_{\alpha,\beta} \qquad (3.29)$$

where $|a_0\rangle = S_+$.

The spherical components of the spin angular momentum ($S = 1$) on the base $B_{\alpha\beta}$ are given by

$$S_z = B_{1,1} - B_{3,3}$$
$$S_+ = \sqrt{2}\,(B_{1,2} + B_{2,3}) \qquad (3.30)$$
$$S_- = \sqrt{2}\,(B_{2,1} + B_{3,2})$$

Since the Hamiltonian of Eq. (3.25) commutes with the M_z operator, it is useful to denote the $c(n)_{l,0,\alpha,\beta}$ with the shorter symbol $c(n)_{l,\alpha,\beta}$. As in Section III.A, we must find the recursive expression for the $c(n)_{l,\alpha,\beta}$ coefficients. We shall follow the same method as that illustrated by Eqs. (3.13)–(3.16). Thus we consider the evolution of the operator $B_{\alpha,\beta}$ due to the spin operator part of the Hamiltonian (3.25). We have

(i) $\quad i\left[S_z, B_{\alpha',\beta'}\right] = i\left\{\delta_{1,\alpha'}B_{1,\beta'} - \delta_{1,\beta'}B_{\alpha',1} - \delta_{3,\alpha'}B_{3,\beta'} + \delta_{3,\beta'}B_{\alpha',3}\right\}$

$\qquad = i\sum_{\alpha,\beta}\left[\delta_{1,\alpha'}\delta_{1,\alpha}\delta_{\beta,\beta'} - \delta_{1,\beta'}\delta_{\alpha,\alpha'}\delta_{1,\beta}\right.$

$\qquad\qquad \left. - \delta_{3,\alpha'}\delta_{3,\alpha}\delta_{\beta,\beta'} + \delta_{3,\beta'}\delta_{\alpha,\alpha'}\delta_{3,\beta}\right] B_{\alpha,\beta}$

(ii) $\quad i\left[S_z^2, B_{\alpha',\beta'}\right] = i\left\{\delta_{1,\alpha'}B_{1,\beta'} - \delta_{1,\beta'}B_{\alpha',1} + \delta_{3,\alpha'}B_{3,\beta'} - \delta_{3,\beta'}B_{\alpha',3}\right\}$

$\qquad = i\sum_{\alpha,\beta}\left[\delta_{1,\alpha'}\delta_{1,\alpha}\delta_{\beta,\beta'} - \delta_{1,\beta'}\delta_{\alpha,\alpha'}\delta_{1,\beta}\right.$

$\qquad\qquad \left. + \delta_{3,\alpha'}\delta_{3,\alpha}\delta_{\beta,\beta'} - \delta_{3,\beta'}\delta_{\alpha,\alpha'}\delta_{3,\beta}\right] B_{\alpha,\beta}$

The contributions to the $c(n+1)_{l,\alpha,\beta}$ due to the superoperator of Eq. (3.26) are

(i') $i\omega_0 M_+^{(n)}(l,\alpha,\beta)$

(i'') $i\eta\left[C_{OP}M_+^{(n)}(l-2,\alpha,\beta)+C_{00}M_+^{(n)}(l,\alpha,\beta)+C_{0M}M_+^{(n)}(l+2,\alpha,\beta)\right]$

(ii) $i\delta\left[C_{OP}M_-^{(n)}(l-2,\alpha,\beta)+C_{00}M_-^{(n)}(l,\alpha,\beta)+C_{0M}M_-^{(n)}(l+2,\alpha,\beta)\right]$

(iii) $-\left[(D_{\|}-D_{\perp})m^2+D_{\perp}l(l+1)\right]c(n)_{l,\alpha,\beta}$ (3.31)

Where the following definitions are used:

$$\begin{aligned}
M_+^{(n)}(l,\alpha,\beta) &\equiv \left[\delta_{1,\alpha}c(n)_{l,1,\beta}-\delta_{1,\beta}c(n)_{l,\alpha,1}\right.\\
&\quad\left.-\delta_{3,\alpha}c(n)_{l,3,\beta}+\delta_{3,\beta}c(n)_{l,\alpha,3}\right]\\
M_-^{(n)}(l,\alpha,\beta) &\equiv \left[\delta_{1,\alpha}c(n)_{l,1,\beta}-\delta_{1,\beta}c(n)_{l,\alpha,1}\right.\\
&\quad\left.+\delta_{3,\alpha}c(n)_{l,3,\beta}-\delta_{3,\beta}c(n)_{l,\alpha,3}\right]
\end{aligned}$$
       (3.32)

When using Eq. (2.34) the initial condition for the $c(n)$ coefficients becomes

$$c(0)_{l,\alpha,\beta}=\sqrt{8\pi}\,\delta_{l,0}\delta_{\beta,\alpha+1} \qquad (3.33)$$

As a consequence, we can give to the s_n moments the following expression:

$$s_n=\frac{\langle a_0|a_n\rangle}{\langle a_0|a_0\rangle}=\mathrm{Tr}_{\{S\}}\frac{\int d\Omega\, S_-\left[\sum_{l,\alpha,\beta}c(n)_{l,\alpha,\beta}Y_{2l,0}(\Omega)B_{\alpha,\beta}\right]w_{eq}}{\mathrm{Tr}_{\{S\}}\{S_-S_+\}\int w_{eq}\,d\Omega}$$

$$=\frac{\sum_{\alpha}c(n)_{0,\alpha,\alpha+1}}{\sqrt{32\pi}} \qquad (3.34)$$

Equations (2.23)–(2.25) provide the expansion parameters λ_i, Δ_{i+1}^2 and therefore the ESR lineshape.

2. Analytical Structure

We will now develop the algorithm which makes it possible in this special case to obtain the ESR spectrum more directly. It is easy to verify that the autooperators of the Liouvillian superoperator (3.26) are given by

$$
\begin{aligned}
|+\rangle &\equiv [S_+ + (S_z S_+ + S_+ S_z)]\frac{1}{2\sqrt{2}} \\
|-\rangle &\equiv [S_+ - (S_z S_+ + S_+ S_z)]\frac{1}{2\sqrt{2}}
\end{aligned}
\tag{3.35}
$$

so that

$$
S_+ = \sqrt{2}\,(|+\rangle + |-\rangle)
\tag{3.36}
$$

Let the expansion bases for the corresponding autospaces of Γ be

$$
|l_+\rangle \equiv |+\rangle Y_{2l,0}(\Omega)/\int d\Omega\, |Y_{2l,0}(\Omega)|^2 w_{eq}(\Omega)
\tag{3.37}
$$

$$
|l_-\rangle \equiv |-\rangle Y_{2l,0}(\Omega)/\int d\Omega\, |Y_{2l,0}(\Omega)|^2 w_{eq}(\Omega)
\tag{3.38}
$$

Upon reapplication of the scalar product defined in Eq. (2.9), Γ turns out to be tridiagonal on each base. Restricted to the first autospace, using the base (3.37), the matrix elements of Γ read

$$
\begin{aligned}
\gamma_l &\equiv \langle l_+|\Gamma|l_+\rangle = i\omega_0 + i(\eta+\delta)\langle l_+|Y_{2,0}(\Omega)|l_+\rangle - 2l(2l+1)D_\perp \\
c_l^{l+1} &\equiv \langle (l+1)_+|\Gamma|l_+\rangle = i(\eta+\delta)\langle (l+1)_+|Y_{2,0}(\Omega)|l_+\rangle \\
c_l^{l-1} &\equiv \langle (l-1)_+|\Gamma|l_+\rangle = i(\eta+\delta)\langle (l-1)_+|Y_{2,0}(\Omega)|l_+\rangle
\end{aligned}
\tag{3.39}
$$

Analogously using the base (3.38), we have

$$
\begin{aligned}
\bar\gamma_l &\equiv \langle l_-|\Gamma|l_-\rangle = i\omega_0 + i(\eta-\delta)\langle l_-|Y_{2,0}(\Omega)|l_-\rangle - 2l(2l+1)D_\perp \\
\bar c_l^{l+1} &\equiv \langle (l+1)_-|\Gamma|l_-\rangle = i(\eta-\delta)\langle (l+1)_-|Y_{2,0}(\Omega)|l_-\rangle \\
\bar c_l^{l-1} &\equiv \langle (l-1)_-|\Gamma|l_-\rangle = i(\eta-\delta)\langle (l-1)_-|Y_{2,0}(\Omega)|l_-\rangle
\end{aligned}
\tag{3.40}
$$

Thus, the time evolution equation (3.26) can be replaced by a proper combination of the following sets of infinite linear differential equations:

$$\frac{d}{dt}|0_+\rangle = c_0^1|1_+\rangle$$

$$\frac{d}{dt}|1_+\rangle = c_1^0|0_+\rangle + \gamma_1|1_+\rangle + c_1^2|2_+\rangle$$

$$\vdots$$

$$\frac{d|l_+\rangle}{dt} = c_l^{l-1}|(l-1)_+\rangle + \gamma_l|l_+\rangle + c_l^{l+1}|(l+1)_+\rangle \tag{3.41}$$

$$\frac{d|0_-\rangle}{dt} = \bar{c}_0^1|1_-\rangle$$

$$\frac{d|1_-\rangle}{dt} = \bar{c}_1^0|0_-\rangle + \bar{\gamma}_1|1_-\rangle + \bar{c}_1^2|2_-\rangle$$

$$\vdots$$

$$\frac{d|l_-\rangle}{dt} = \bar{c}_l^{l-1}|(l-1)_-\rangle + \bar{\gamma}_l|l_-\rangle + \bar{c}_l^{l+1}|(l+1)_-\rangle \tag{3.42}$$

By Laplace transforming Eqs. (3.41) and (3.42), we get

$$z|\hat{0}_+\rangle = |\hat{0}_+\rangle + c_0^1|\hat{1}_+\rangle$$

$$(z-\gamma_1)|\hat{1}_+\rangle = c_1^0|\hat{0}_+\rangle + |\hat{1}_+\rangle + c_1^2|\hat{2}_+\rangle$$

$$\vdots$$

$$(z-\gamma_l)|\hat{l}_+\rangle = c_l^{l-1}|\widehat{(l-1)}_+\rangle + |\hat{l}_+\rangle + c_l^{l+1}|\widehat{(l+1)}_+\rangle \tag{3.43}$$

and

$$z|\hat{0}_-\rangle = |\hat{0}_-\rangle + \bar{c}_0^1|\hat{1}_-\rangle$$

$$(z-\bar{\gamma}_1)|\hat{1}_-\rangle = \bar{c}_1^0|\hat{0}_-\rangle + |\hat{1}_-\rangle + \bar{c}_1^2|\hat{2}_-\rangle$$

$$\vdots$$

$$(z-\bar{\gamma}_l)|\hat{l}_-\rangle = \bar{c}_l^{l-1}|\widehat{(l-1)}_-\rangle + |\hat{l}_-\rangle + \bar{c}_l^{l+1}|\widehat{(l+1)}_-\rangle \tag{3.44}$$

Here the $|\hat{l}_+\rangle$ denotes the Laplace transform of the state $|l_+\rangle$ and $|\hat{l}_\pm\rangle$ are the initial conditions for $|l_\pm\rangle$. We truncate these sets at the $(l+1)$th order (i.e., we assume $c_l^{l+1} = \bar{c}_l^{l+1} = 0$). This allows us to define, from the last equations of (3.43) and (3.44),

$$|\hat{l}_+\rangle = \frac{1}{z - \gamma_l}|\hat{l}_+\rangle + \frac{c_l^{l-1}}{z - \gamma_l}|\overline{(l-1)}_+\rangle \qquad (3.45a)$$

$$|\hat{l}_-\rangle = \frac{1}{z - \bar{\gamma}_l}|\hat{l}_-\rangle + \frac{\bar{c}_l^{l-1}}{z - \bar{\gamma}_l}|\overline{(l-1)}_-\rangle \qquad (3.45b)$$

respectively.

By substituting (3.45a) into the last equation of (3.43), and (3.45b) into the last one of (3.44), we get

$$(z - \gamma_{l-1})|\overline{(l-1)}_+\rangle = c_{l-1}^{l-2}|\overline{(l-2)}_+\rangle + |\widehat{(l-1)}_+\rangle$$

$$+ \frac{c_{l-1}^l c_l^{l-1}}{z - \gamma_l}|\overline{(l-1)}_+\rangle + \frac{c_{l-1}^l}{z - \gamma_l}|\hat{l}_+\rangle$$

thereby providing

$$|\overline{(l-1)}_+\rangle = \frac{1}{z - \gamma_{l-1} - c_{l-1}^l c_l^{l-1}/(z - \gamma_l)}$$

$$\times \left[c_{l-1}^{l-2}|\overline{(l-2)}_+\rangle + |\widehat{(l-1)}_+\rangle + \frac{c_{l-1}^l}{z - \gamma_l}|\hat{l}_+\rangle \right]$$

and

$$(z - \bar{\gamma}_{l-1})|\overline{(l-1)}_-\rangle = \bar{c}_{l-1}^{l-2}|\overline{(l-2)}_-\rangle + |\widehat{(l-1)}_-\rangle$$

$$+ \frac{\bar{c}_{l-1}^l \bar{c}_l^{l-1}}{z - \bar{\gamma}_l}|\overline{(l-1)}_-\rangle + \frac{\bar{c}_{l-1}^l}{z - \bar{\gamma}_l}|\hat{l}_-\rangle$$

which, in turn, results in

$$|\overline{(l-1)}_-\rangle = \frac{1}{z - \bar{\gamma}_{l-1} - \bar{c}_{l-1}^l \bar{c}_l^{l-1}/(z - \bar{\gamma}_l)}$$

$$\times \left[\bar{c}_{l-1}^{l-2}|\overline{(l-2)}_-\rangle + |\widehat{(l-1)}_-\rangle + \frac{\bar{c}_{l-1}^l}{z - \bar{\gamma}_l}|\hat{l}_-\rangle \right]$$

Proceeding in this way we shall ultimately obtain for $|\hat{0}_+\rangle$ and $|\hat{0}_-\rangle$ the following expressions (having set $z = i\omega$):

$$|\hat{0}_+\rangle = F_0(\omega)\Big(|\hat{0}_+\rangle + F_1(\omega)c_0^1\big(|\hat{1}_+\rangle + F_2(\omega)c_1^2\big(|\hat{2}_+\rangle + F_3(\omega)c_2^3\big(|\hat{3}_+\rangle$$
$$+ \cdots + F_l(\omega)c_{l-1}^l|\hat{l}_+\rangle\big) \cdots \Big) \tag{3.46}$$

$$|\hat{0}_-\rangle = \bar{F}_0(\omega)\Big(|\hat{0}_-\rangle + \bar{F}_1(\omega)\bar{c}_0^1\big(|\hat{1}_-\rangle + \bar{F}_2(\omega)\bar{c}_1^2\big(|\hat{2}_-\rangle + \bar{F}_3(\omega)\bar{c}_2^3\big(|\hat{3}_-\rangle$$
$$+ \cdots + \bar{F}_l(\omega)\bar{c}_{l-1}^l|\hat{l}_-\rangle\big) \cdots \Big) \tag{3.47}$$

where

$$F_0(\omega) = \cfrac{1}{i\omega - \cfrac{c_0^1 c_1^0}{i\omega - \gamma_1 - \cfrac{c_1^2 c_2^1}{i\omega - \gamma_2 \cdots}}}$$

$$F_1(\omega) = \cfrac{1}{i\omega - \gamma_1 - \cfrac{c_1^2 c_2^1}{i\omega - \gamma_2 \cdots}} \tag{3.48}$$

$$\vdots$$

$$F_l(\omega) = \cfrac{1}{i\omega - \gamma_l}$$

Note that the continued fraction F_k has one step more than the F_{k+1} one. The $\bar{F}_l(\omega)$ can be obtained from the $F_l(\omega)$ by substituting the γ_l, c_{l-t}^l with the $\bar{\gamma}_l, \bar{c}_{l-1}^l$, respectively. The ESR spectrum $I(\omega)$ can be expressed by

$$I(\omega) = \mathrm{ReL} \frac{\langle S_-|S_+(t)\rangle}{\langle S_- S_+\rangle} = \mathrm{Re} \frac{\big((\langle \hat{0}_+| + \langle \hat{0}_-|)(|\hat{0}_+\rangle + |\hat{0}_-\rangle\big)}{2} \tag{3.49}$$

In the absence of an orienting potential we are therefore allowed to write

$$I(\omega) = \mathrm{Re} \frac{F_0(\omega) + \bar{F}_0(\omega)}{2} \tag{3.50}$$

An inspection of Eqs. (3.50) and (3.25) suggests that the ESR spectrum for the present system can be seen as a superimposition of two spins $\frac{1}{2}$ ESR spectra with the anisotropy factor $\Delta \bar{g}$ proportional to $(\eta + \delta)$ and $(\eta - \delta)$, respectively. This is illustrated in Fig. 5, showing two typical ESR spectra for $\eta > \delta > 0$ and $\delta > \eta > 0$, respectively. The ESR lineshapes for some values of the ratio between the anisotropy amplitude of the fine structure tensor

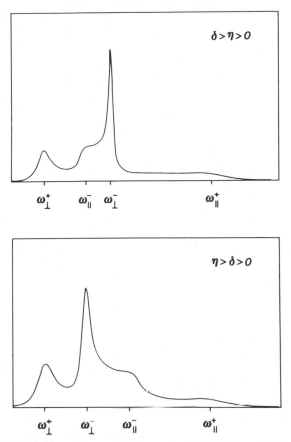

Figure 5. Typical absorption spectra for a magnetic species in a triplet spin state tumbling in a viscous disordered liquid. $\omega_\parallel^\pm = g_\parallel \beta_e H_0 \pm \delta_\parallel$, $\omega_\perp^\pm = g_\perp \beta_e H_0 \pm \delta_\perp$, where β_e is the Bohr magneton, and H_0 is the value of the static magnetic field.

and that of the splitting tensor are reported in Fig. 6. This figure also illustrates the ESR spectra of two different values of the $D_\perp$ diffusion tensor component. The expression for the Hamiltonian (3.25), in the secular approximation, shows that the spin relaxation processes are driven by reorientational diffusion of the molecular symmetry axis. Table III provides the number of steps needed to reach the convergence of the continued fractions (3.50) calculations for different D values ranging through four orders of magnitude. Both the simple analytical relations between physical parameters and expansion coefficients (3.39), (3.40) and the fast convergence allow this alternative algorithm to provide a deeper insight into the physical contributions to the relaxation processes.

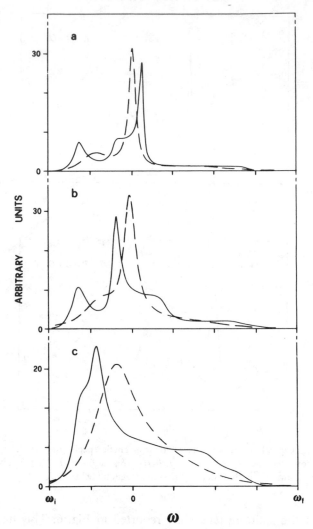

Figure 6. Absorption lineshape for a molecule in a triplet spin state dissolved in a disordered liquid. $\omega_0 = 0$. Frequency sweeps from $\omega_i = -\frac{2}{3}(\eta + \delta)$ to $\omega_f = \frac{4}{3}(\eta + \delta)$. $D_\perp = 10^{-2}\eta$ (continuous line), $D_\perp = 10^{-1}\eta$ (dashed line). (a) $\delta = 1.5\eta$, $(b)\delta = 0.5\eta$, (c) $\delta = 0.15\eta$.

TABLE III

Minimum Number of Steps n_s to Reach Convergence
of the C.F. Describing the Absorption of a Molecule
in a Triplet Spin State with Isotropic
Reorientation $(\bar{\eta} = \max\{|\eta - \delta|, |\eta + \delta|\})$

$D_\perp / \bar{\eta}$	10	1	10^{-1}	10^{-2}	$5 \cdot 10^{-3}$
n_s	2	2	2	5	6

C. Isolated Paramagnetic Ions with Large Anisotropy $(Cu^{2+}: S = \frac{1}{2}, I = 3/2)$

Electron spin resonance studies have many uses—for example, to investigate a large number of substances (e.g., polycrystalline material), to characterize the dynamics of molecules of biological interest, and to study the diffusion of paramagnetic ions in glassy matrices.

In this subsection we deal with the ESR spectrum of Cu^{2+} in a physical system the mobility of which varies very greatly. We calculate the ESR lineshape of this metal ion in axial symmetry sites for physical systems whose mobility ranges from rigid to high diffusion regimes.

Henceforth we assume that the dominant process driving the magnetic relaxation is reorientational diffusion. To calculate the ESR spectrum for ions of high magnetic anisotropy in viscous systems, which in general show an absorption line which covers some hundreds of gauss, it is necessary to take into account some special aspects of the problem.

First, for a given value of the components of the diffusion tensor, the lineshape is calculated by varying the frequency of the radiation field while keeping fixed the value of the static magnetic field. Thus the ratio between the values of magnetic terms and the diffusion tensor components is kept constant. However, this procedure is nearly equivalent to irradiating with a constant frequency wave while sweeping the static magnetic field across the resonant values, as in magnetic resonance experiments, provided that the anisotropy of the magnetic parameters is low. In the case dealt with here, to calculate the lineshape by varying the frequency (for values of the diffusion tensor components near or lower than splitting tensor anisotropy) implies that the intensity of the absorption line at low fields is underestimated. On the other hand, it is necessary to remember that the parameters λ_i, Δ_i^2 which appear in the expression of the continued fraction depend on the value of the static magnetic field. In consequence, to obtain a theoretical ESR spectrum comparable with the experimental one, it would be necessary to employ a matching technique between continued fractions covering the required magnetic field interval, in which the static magnetic field value is changed in a discrete way and the field of frequency change is suitably limited. However, this is beyond the scope of this article, so keeping in mind our previous remarks, we will reproduce ESR spectra calculated by keeping the value of the static magnetic field constant.

Second, we point out that the intensity of the absorption spectrum depends (as noted in the introduction to this section) on the angle that the quantization axis of the single spin packet makes with the oscillating magnetic field direction. It has been proved that neglecting this dependence significantly changes the ion's transition ESR spectrum.[50a] This dependence will be taken into account by adopting the scalar product (2.9) the weighting of Eq. (2.13).

The explicit form of the Hamiltonian describing the spin system studied in this section and the evolution equation for the variable of interest S_+ are formally the same as those of Eqs. (3.2), (3.3), and (3.6), respectively. Here the expansion base $A_{i,j}$ for the nuclear spin operators is formed by matrices of rank 4 to provide a suitable expansion for the components of the angular momentum

$$I_z = \tfrac{3}{2}A_{11} + \tfrac{1}{2}A_{22} - \tfrac{1}{2}A_{33} - \tfrac{3}{2}A_{44}$$
$$I_+ = \sqrt{3}\,A_{12} + 2A_{23} + \sqrt{3}\,A_{34} \tag{3.51}$$
$$I_- = \sqrt{3}\,A_{21} + 2A_{32} + \sqrt{3}\,A_{43}$$

The expression of the nth state $|a_n\rangle$ and the implicit form of the moment s_n are, except for the dimensions, the same of those of Eqs. (3.8) and (3.12), respectively. The explicit forms of the $c(n)_{l,m,i,j}$ coefficients required to determine the s_n moments are given in Appendix A. Here we limit ourselves to noting that there are analogous constraints limiting the dimension of $\mathbf{c}(n)$. Thus a set of possible independent terms to be calculated is

$$\begin{array}{llll} c(n)_{l,0,11} & c(n)_{l,0,2,2} & c(n)_{l,0,3,3} & c(n)_{l,0,4,4} \\ c(n)_{l,1,2,1} & c(n)_{l,1,3,2} & c(n)_{l,1,4,3} & c(n)_{l,2,3,1} \\ c(n)_{l,2,4,2} & c(n)_{l,3,4,1} \end{array}$$

In Appendix A the recursion formulas for the coefficients of interest are reported for convenience.

As in the case of Section III.A, we are now in a position to evaluate the ESR spectrum of a copper complex in solution. As an illustrative example we consider the case where the copper ion occupies a molecular site with axial symmetry. The magnetic parameters characterizing the spin Hamiltonian of the metal ion under study are of the same order of magnitude as those drawn from the literature for the $CuCl_2$-morpholine complex.[50b]

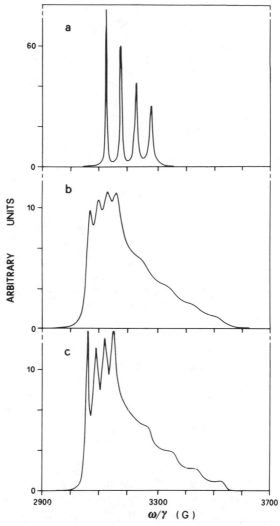

Figure 7. Absorption lineshape for a copper complex dissolved in a disordered liquid of different viscosity. The diffusion process is assumed to be spherically symmetric. Magnetic parameters are $g_{\parallel} = 2.2$, $g_{\perp} = 2$, $A_{\parallel} = 90$ G, $A_{\perp} = 30$ G. (a) $\bar{r} = 10^{-1}$, (b) $\bar{r} = 20$, (c) $\bar{r} = 100$.

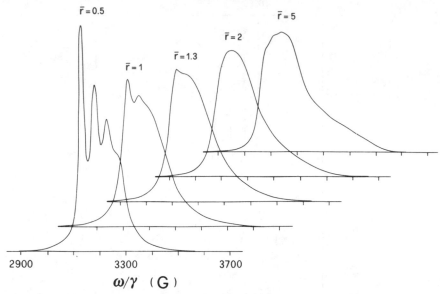

Figure 8. Absorption profiles for a copper complex dissolved in a disordered liquid. Molecular reorientation in this intermediate region of motion is isotropic. Magnetic parameters are the same as in Fig. 7.

To evaluate the corresponding ESR spectrum, we assumed the diffusion process to be spherically symmetric. The spectra illustrated in Fig. 7 concern different values of the parameter

$$\bar{r} = \frac{\frac{2}{3}|A_{\parallel} - A_{\perp}|}{\underline{D}} \tag{3.52}$$

where $\underline{D}$ is the corresponding diffusion coefficient. The opposite limits of the range of the parameter $\bar{r}$ explored in Fig. 7 are concerned with typical structures of rapid motion and powderlike spectra, respectively.

In the intermediate region ($0.5 \leq \bar{r} \leq 5$), Fig. 8, the transition from the second regime to the first takes place at $\bar{r}$ about 2.

In this region we observe that the lineshape features depending on the components of the g-splitting and hyperfine tensors are gradually replaced by ones depending on the trace of A, g, whose anisotropy is averaged out by the rapid motion of the molecules. The information provided by the set of curves of Fig. 8 are of special relevance for experimentalists. The diffusion constant of paramagnetic species can indeed be derived via a comparison of experimental and theoretical lineshapes in this transition region. This em-

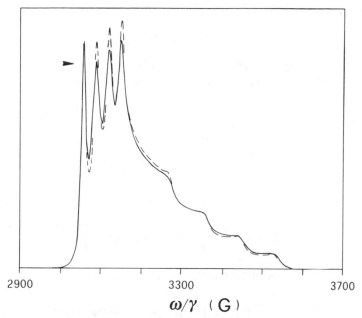

2900   3300   3700

$$\omega/\gamma \; (\; G \;)$$

Figure 9. Influence on the lineshape of the pseudosecular terms in the spin Hamiltonian for $\bar{r} = 100$. Magnetic parameters are the same as in Fig. 7. The continuous curve is drawn using the complete spin Hamiltonian. If we delete the pseudosecular contributions, the dashed curve is obtained. The arrow indicates a peak of the dashed curve.

phasizes the importance of a fast and efficient algorithm such as that illustrated in this chapter. This allows a straightforward comparison to be made.

Figure 9 shows the influence of the pseudosecular terms on ESR lineshape calculated in the very slow regime ($\bar{r} = 100$). It can be observed that in general the spectral features tend to depend weakly on the pseudosecular contribution except in the relative intensity of the high-frequency lines.

D.   ESR Lineshape in the Presence of an Orienting Potential

The two algorithms already developed and used to reproduce ESR lineshapes of paramagnetic species in free diffusion are applied in this subsection to the case of spin probes dissolved in liquid crystalline mesophases. The main point of difference with the previously examined cases is due to the introduction of an orienting potential whose nature is directly reflected in the structure of the Fokker-Planck operator, which in the diffusional assumption is given by Eq. (2.6). The explicit form of the potential we use in this subsection is given by[51]

$$V(\Omega) = \lambda P_2(\cos\beta) \tag{3.53}$$

where β denotes the second Euler angle and $P_2(\cos\beta)$ is the second-order Legendre polynomial. The expression for the diffusional operator is now given by

$$\Gamma_v = -\left[(D_{\|} - D_{\perp})M_z^2 + D_{\perp}M^2\right]$$
$$- \lambda \frac{D_{\perp}}{2}\sqrt{\frac{4\pi}{5}}\left[M_+(M_- Y_{2,0}(\Omega)) + M_-(M_+ Y_{2,0}(\Omega))\right] \tag{3.54}$$

and the equilibrium distribution of the molecular orientation takes the form

$$W_{eq}(\Omega) = \exp\left[\frac{-V(\beta)}{KT}\right]\bigg/ \int d\Omega \exp\left[\frac{-V(\beta)}{KT}\right] \tag{3.55}$$

The expression for the order parameters defined by Eq. (2.38) now becomes

$$\overline{P}_l = \int d\Omega\, W_{eq}(\Omega)\, P_l(\cos\beta) \bigg/ \int d\Omega\, W_{eq}(\Omega) \tag{3.56}$$

We first show that the completely analytical approach developed in Section III.B.2 also holds in the presence of an orienting potential. Later on we apply the general CFP method already applied in Sections III.A and III.C.

It has been assumed that the director axis defining the orientational order of the mesophase has the same direction as the static magnetic field.

1. Axial g-Splitting Tensor ($S = 1/2$)

The Hamiltonian describing this spin system is

$$\mathcal{H} = \omega_0 S_z + \eta Y_{2,0}(\Omega) S_z$$

so that the Liouvillian superoperator is given by

$$\Gamma = i(\omega_0 + \eta Y_{2,0}(\Omega))S_z^x + \Gamma_v^+$$

where η has the same meaning as in Section III.B. Since S_+ is a self-operator of Γ, it is quite natural to take as an expansion base

$$|l\rangle \equiv \tfrac{1}{2}S_+ Y_{2l,0}(\Omega)\bigg/ \int d\Omega\, |Y_{2l,0}(\Omega)|^2 W_{eq}(\Omega)$$

on which Γ is tridiagonal.

Analogously to Section III.B.2 we now construct a single system of infinite differential equations whose coefficients are given by

$$\gamma_l = \langle l | \Gamma | l \rangle = i\omega_0 + i\eta \langle l | Y_{2,0}(\Omega) | l \rangle + \langle l | \Gamma_v^+ | l \rangle$$

$$c_l^{l+1} = \langle l+1 | \Gamma | l \rangle = i\eta \langle (l+1) | Y_{2,0}(\Omega) | l \rangle + \langle (l+1) | \Gamma_v^+ | l \rangle$$

$$c_l^{l-1} = \langle l-1 | \Gamma | l \rangle = i\eta \langle (l-1) | Y_{2,0}(\Omega) | l \rangle + \langle (l-1) | \Gamma_v^+ | l \rangle$$

By Laplace transforming the differential equation obtained by the expansion of Γ on the base $|l\rangle$ and making a truncation which neglects a given $\overline{|l+1\rangle}$ regarded as being irrelevant to the evolution of S_+ by using the iterative procedure developed in Section III.B.2, we get for the Laplace transform of $|0\rangle$, $|\hat{0}\rangle$ the expression

$$|\hat{0}\rangle = F_0(\omega)\Big(|\hat{0}\rangle + F_1(\omega)c_0^1\big(|\hat{1}\rangle + F_2(\omega)c_1^2\big(|\hat{2}\rangle$$

$$+ F_3(\omega)c_2^3\big(|\hat{3}\rangle + \cdots + F_l(\omega)c_{l-1}^l|\hat{l}\rangle\big)\cdots\Big) \qquad (3.59)$$

The ESR spectrum is given by

$$I(\omega) = \text{Re}\,\frac{\langle \hat{0} | \hat{0} \rangle}{\langle \hat{0} | \hat{0} \rangle} \qquad (3.60)$$

thereby implying that we must calculate the scalar products with weighting function given by the equilibrium distribution $e^{\lambda \Gamma_2}$. By defining the quantities

$$P_l' \equiv \sqrt{4l+1}\,\overline{P}_{2l} \qquad (3.61)$$

the ESR lineshape proves to have the form

$$I(\omega) = F_0(\omega)\big(1 + F_1(\omega)c_0^1\big(P_1' + F_2(\omega)c_1^2\big(P_2' + \cdots$$

where the $F_l(\omega)$ are formally the same of those of Eqs. (3.48).

Figure 10 shows the spectra at two different values of the diffusion coefficient as a function of the order parameter $\overline{P}_2$ in the slow-motion regime. The curves establish a comparison between the lineshapes of hypothetical free diffusion paramagnetic species with different values of the order parameter $\overline{P}_2$. In the high-rigidity case the lineshape intensity at a certain field value B is proportional to the number of molecules, the principal axis of which forms an angle between Ω and $\Omega + d\Omega$ with respect to the static magnetic field. In

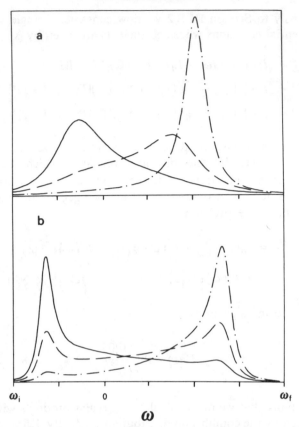

Figure 10. Absorption lineshape for an axial g-splitting interaction for a spin $1/2$ system in the presence of an orienting potential. Vertical scales are expressed in the same units. $\omega_0 = 0$. (a) $D_\perp = 10^{-1}\eta$, (b) $D_\perp = 5 \times 10^{-3}\eta$. The frequency sweep ranges from $\omega_i = -\frac{2}{3}\eta$ to $\omega_f = \frac{4}{3}\eta$. (——) $\bar{P}_2 = 0$; (---) $\bar{P}_2 = 0.3$; (−·−) $\bar{P}_2 = 0.6$.

the free diffusion case this results in the spectrum illustrated by the continuous curve of Fig. 10b, which is well known in the literature. In the presence of an orienting potential, as an effect of the symmetry of the magnetic parameters (note that we have assumed that the molecular principal axis is parallel to the principal axis of the **g** tensor), the lineshape turns out to be increasingly peaked around the value of the field corresponding to the parallel component of the **g** tensor with increasing $\bar{P}_2$. In Fig. 10a (dealing with a diffusion process faster than that of Fig. 10b), a different behavior is exhibited: as a partial averaging on the **g** tensor anisotropy is made, the line-

TABLE IV

Number of Steps n_s Needed to Reach the Convergence
of the C.F. Describing the Absorption of a Molecule
in $S = 1/2$ Spin State as a Function of Parameters
$\bar{P}_2$ and $D_\perp / \eta$

$D_\perp / \eta$ \ $\bar{P}_2$	0	0.3	0.6	0.9
10	2	2	4	15
1	2	2	4	14
10^{-1}	2	3	5	12
10^{-2}	5	5	5	10
$5 \, 10^{-3}$	6	7	8	10

shapes are no longer peaked around the values of the magnetic fields corresponding to the components of the g-splitting tensor.

Table IV indicates the number of steps of the continued fraction $F_0(\omega)$ necessary to attain convergence when aiming at evaluating the lineshape as a function of the order parameter $\bar{P}_2$ and the diffusion coefficient. With low-order parameters and high values of the diffusion coefficient, it is possible to directly study the lineshape in the region intermediate between the fast-motion and high-viscosity dominions.

The information provided by Table IV leads us to the following conclusions:

1. For low values of the order parameter, as memory strength increases (i.e., D decreases), a larger number of excited states of the operator Γ_v are involved, and thus a larger number of steps are required for convergence.

2. For constant D, as the order parameter increases, a larger number of spherical harmonics are involved in producing a more peaked equilibrium distribution. This effect also means that more steps are needed for convergence.

3. Surprisingly, at $\bar{P}_2 = 0.9$, the number of steps required for convergence decreases with increasing memory strength. We believe that this effect is reminiscent of the decoupling phenomenon induced by external field.[52] In other words, it seems that strong external fields reduce the intensity of the local fluctuating fields. This effect indeed becomes more pronounced as the memory strength is increased, thereby resulting in more rapid convergence.

2.   Zero Field Splitting (S = 1)

The Hamiltonian describing this spin system in the secular approximation is given by Eq. (3.25) of Section III.B. The evolution equation of the operator S_+ is given by

$$\frac{dS_+}{dt} = i[\mathcal{H}, S_+] + \Gamma_v^+ S_+ \equiv \Gamma S_+ \tag{3.62}$$

In this section we shall apply the algorithm developed in the second part of Section III.B. This algorithm makes it possible to obtain directly the continued fraction parameters. As in Section III.B, we shall choose for Γ the expansion basis provided by Eqs. (3.37) and (3.38), which in this case results in the following coefficients:

$$\gamma_l \equiv \langle l_+ | \Gamma | l_+ \rangle = i\omega_0 + i(\eta + \delta)\langle l_+ | Y_{2,0}(\Omega) | l_+ \rangle + \langle l_+ | \Gamma_v^+ | l_+ \rangle$$

$$c_l^{l+1} \equiv \langle (l+1)_+ | \Gamma | l_+ \rangle$$

$$= i(\eta + \delta)\langle (l+1)_+ | Y_{2,0}(\Omega) | l_+ \rangle + \langle (l+1)_+ | \Gamma_v^+ | l_+ \rangle \tag{3.63}$$

$$c_l^{l-1} \equiv \langle (l-1)_+ | \Gamma | l_+ \rangle$$

$$= i(\eta + \delta)\langle (l-1)_+ | Y_{2,0}(\Omega) | l_+ \rangle + \langle (l-1)_+ | \Gamma_v^+ | l_+ \rangle$$

and

$$\bar{\gamma}_l \equiv \langle l_- | \Gamma | l_- \rangle = i\omega_0 + i(\eta - \delta)\langle l_- | Y_{2,0}(\Omega) | l_- \rangle + \langle l_- | \Gamma_v^+ | l_- \rangle$$

$$\bar{c}_l^{l+1} \equiv \langle (l+1)_- | \Gamma | l_- \rangle$$

$$= +i(\eta - \delta)\langle (l+1)_- | Y_{2,0}(\Omega) | l_- \rangle + \langle (l+1)_- | \Gamma_v^+ | l_- \rangle \tag{3.63'}$$

$$\bar{c}_l^{l-1} \equiv \langle (l-1)_- | \Gamma | l_- \rangle$$

$$= +i(\eta - \delta)\langle (l-1)_- | Y_{2,0}(\Omega) | l_- \rangle + \langle (l-1)_- | \Gamma_v^+ | l_- \rangle$$

Note that the last contribution to each off-diagonal coefficient of Eqs. (3.63) depends on the orienting potential. This contribution vanishes when the potential is turned off.

Now it is possible to obtain sets of equations analogous to (3.41)–(3.44). Then by applying the usual truncation technique and the iterative procedure it is possible to obtain for $|\hat{0}_+\rangle$ and $|\hat{0}_-\rangle$ expressions reminiscent of Eqs. (3.46) and (3.47) respectively, with the coefficients provided by Eqs. (3.63).

In these cases we obtain

$$I(\omega) = \mathrm{Re}\frac{\left(\langle\hat{0}_+| + \langle\hat{0}_-|\right)\left(|\hat{0}_+\rangle + |\hat{0}_-\rangle\right)}{2}$$

$$= \mathrm{Re}\left(\frac{F_0(\omega)}{2}\left\{1 + F_1(\omega)c_0^1\left[P_1' + F_2(\omega)c_1^2(P_2' + \cdots)\right]\right\}\right.$$

$$\left. + \frac{\bar{F}_0(\omega)}{2}\left\{1 + \bar{F}_1(\omega)\bar{c}_0^1\left[P_1' + \bar{F}_2(\omega)\bar{c}_1^2(P_2' + \cdots)\right]\right\}\right) \quad (3.64)$$

where the P_l' are defined by Eq. (3.61).

As an example we illustrate in Fig. 11 the ESR spectra evaluated by Eq. (3.64). For given values of g-tensor anisotropy and diffusion coefficient we study the dependence on the parameters $\bar{P}_2$ and on the variation of the parameter δ around η. Figure 12 shows ESR spectra under the same physical conditions as Fig. 11 except for a diffusion coefficient value corresponding to a faster motion (10 times greater than that of Fig. 11). It is worth noting that a consequence of the secular approximation to the spin Hamiltonian is that only the diffusion tensor component orthogonal to the molecular principal axis turns out to be relevant to relaxation processes. Also, in the presence of an orienting potential (Figs. 11 and 12), the absorption spectrum can be thought of as being the superposition of two single lineshapes such as those of the spin $\frac{1}{2}$ systems in free diffusion dealt with in Section III.B. Note that the convergence of component with the larger effective g-splitting anisotropy implies that of triplet lineshape. So data concerning convergence with varying both order parameters and diffusion coefficient may be derived by Table IV (as an example compare Table III and first column of Table IV).

3. Hyperfine Interaction ($S = \frac{1}{2}$, $I = 1$; $S = \frac{1}{2}$, $I = \frac{3}{2}$)

In this subsection we show the effect of the orienting potential on the ESR lineshape of paramagnetic species of particular practical interest, namely, the nitroxide stearic spin probe and Cu^{2+} ion complex dissolved in a nematic liquid crystal.

As mentioned in the introduction, the director potential of the mesophase is assumed to have the same direction as the external static magnetic field. The Hamiltonian describing these spin systems is given by Eq. (3.5). To adapt the CFP already applied to these systems in the free diffusion case to the present case, some small modifications have to be made. These minor changes concern the step where the $c(n+1)_{l,m,i,j}$ coefficients and the s_n moments are evaluated. In fact, as can be seen in Appendix A, Eq. (A.2), the presence in the diffusion operator of a term depending on the orienting potential

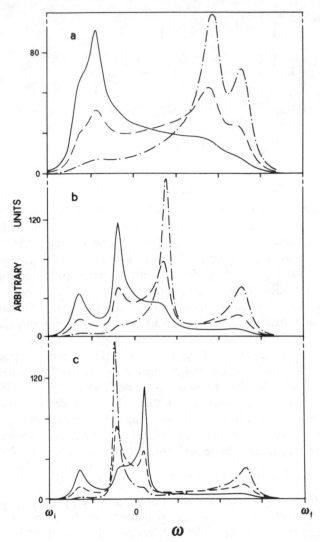

Figure 11.   Absorption lineshape for a molecule in a triplet spin state subjected to an orienting potential. $\omega_0 = 0$. $D_\perp = 10^{-2}\eta$. The frequency sweep ranges from $\omega_i = -\frac{2}{3}(\eta + \delta)$ to $\omega_f = \frac{4}{3}(\eta + \delta)$. (——) $\bar{P}_2 = 0$; (- - -) $\bar{P}_2 = 0.3$; ($\cdot - \cdot - \cdot$) $\bar{P}_2 = 0.6$. (a) $\delta = 0.15\eta$, (b) $\delta = 0.5\eta$, (c) $\delta = 1.5\eta$.

368

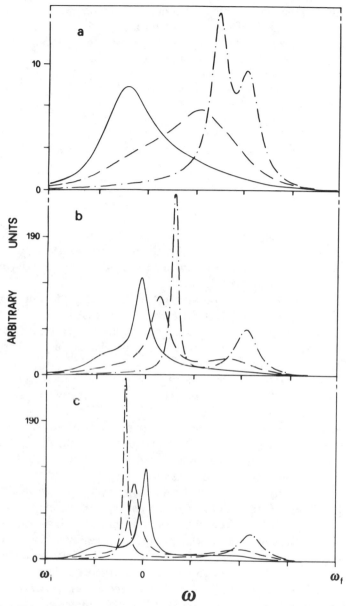

Figure 12. Absorption lineshape for a molecule in a triplet spin state subjected to an orienting potential. The parameters are the same as in Fig. 11, except for $D_\perp = 10^{-1}\eta$. The frequency sweep ranges from $\omega_i = -\frac{2}{3}(\eta + \delta)$ to $\omega_f = \frac{4}{3}(\eta + \delta)$. (——) $\overline{P}_2 = 0$; (---) $\overline{P}_2 = 0.3$; (—·—) $\overline{P}_2 = 0.6$. (a) $\delta = 0.15\eta$, (b) $\delta = 0.5\eta$, (c) $\delta = 1.5\eta$.

369

modifies the C_0 diagonal element of Smolukowsky operator and makes the off-diagonal ones C_M, C_P different from zero.

When evaluating moments, which now have to be expressed in terms of the order parameters (2.38), Eq. (2.36) becomes

$$
s_n = \sum_{i=1}^{2I+1} s_n(i) = \sum_{i=1}^{2I+1} \sum_{l=0}^{2n} \frac{(4l+1)^{1/2}\overline{P}_{2l}c(n)_{l,0,i,i}}{2I+1}
\qquad (3.65)
$$

where the initial condition for the recursive relation reported in Appendix A for the $c(n)_{l,m,i,j}$ coefficients is

$$
c(n)_{l,m,i,j} = \delta_{l,0}\delta_{m,0}\delta_{i,j}
\qquad (3.66)
$$

Via the same procedure we can calculate the continued fraction parameters and then obtain the ESR lineshape. In the following we show the effect on the lineshape in the presence of the orienting potential of varying the r parameter defined in Section III.A.

Figure 13 shows the absorption spectra for a nitroxide spin probe and compares the case of free diffusion ($\overline{P}_2 = 0$) with those obtained for various values of the orienting potential. Such spectra have been calculated for an isotropic diffusion coefficient. Note the drastic change in lineshape at the onset of molecular order.

This effect increases as the value of the r parameter increases. This is in line with the generally accepted opinion that a non-Markovian system is more sensitive to the details of microscopic dynamics. Note also that the lineshapes significantly narrow as the order parameter increases. This can be traced back to the fact that the mean squared value of the fluctuation of the local fields "felt" by the resonating electron spins, decreases as the intensity of the orienting field increases. The number of steps needed for convergence of the continued fraction relevant to the ESR spectra shown in Fig. 13 are reported in Table V. In addition to data concerning the curves of Fig. 13, Table V also provides information on the number of steps needed for convergence in the further case concerning an extremely narrow and intense lineshape obtained at high order parameter ($\overline{P}_2 = 0.97$). It can be noted that the number of steps necessary to achieve convergence is always larger than in the corresponding cases involved in the preceding discussion (Table IV). This is obviously due to the presence of hyperfine interaction, which imply an increase in the number of states necessary for the description of the time evolution of the variable of interest. Note that now the number of convergence steps decreases with increasing values of the order parameter. This is precisely the reverse of the behavior illustrated in Table IV. There, the num-

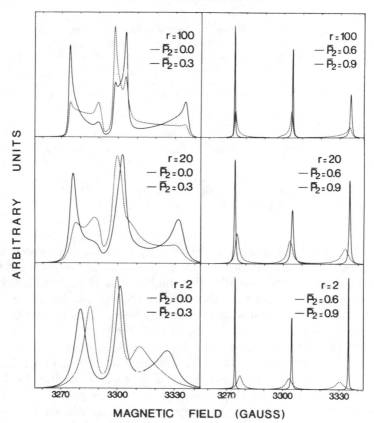

Figure 13. Absorption spectra simulations for a nitroxide probe with axially symmetric g tensor and hyperfine tensor. $D = D_\perp = D_{\parallel}$, $r = \frac{2}{3}|g_{\parallel} - g_\perp|\beta_e H_0/D$: β_e is the Bohr magneton, $H_0 = 3300$ G, the static magnetic field. Magnetic parameters are $g_{\parallel} = 2.0027$, $g_\perp = 2.0075$, $A_{\parallel} = 32$ G, and $A_\perp = 6$ G. $\bar{P}_2$ is the most significant order parameter. Each pair of curves involves the same area, whereas for graphical reasons different pairs can involve different areas.

ber of steps was an increasing function of the order parameter (at a fixed value of the diffusion coefficient) as a probable consequence of the fact that a larger number of spherical harmonics is needed to reproduce an equilibrium distribution function caused to peak sharply by a strong orienting potential.

In the present cases, in contrast, when the order parameter increases, the ESR lineshapes are narrowed, because the relaxation processes are less effective. Thus, narrowing in turn involves a reduction in the number of excited states necessary to obtain convergence and thereby a faster attainment of the result.

TABLE V

Minimum Number of Steps n_s Needed to Reach the Convergence
of the C.F. and (in brackets) Elapsed Computing Times (s)
to Obtain ESR Spectra[a] of a Nitroxide Spin Radical
as a Function of Parameters $\bar{P}_2$ and r.

$\bar{P}_2$ $\quad$ r	0	0.3	0.6	0.9	0.97
2	21(4.3)	18(3.4)	18(3.4)	16(3.1)	11(2.7)
10	25(6)	24(6.1)	21(4.3)	16(3.1)	10(2.7)
100	32(8.7)	28(6.8)	25(6)	19(3.7)	7(2.5)

[a] The data for the spectra corresponding to the values of $\bar{P}_2$ between 0 and 0.9 refer to the curves reported in Fig. 15.

On the other hand, increasing the memory strength, as indicated by Table IV, means that a larger number of excited states of the operator Γ_v are involved, thereby again increasing the number of steps required for convergence. The orienting potential significantly reduces the intensities of the local fields and compensates for the increase in the memory strength to reverse this trend again in the last column of Table V.

Figure 14 illustrates the spectra of a stearic spin probe for two greatly different values of the ratio $D_{\parallel}/D_{\perp}$. For very different diffusion limits in the presence of an orienting potential, this shows how sensitive to changes of the order parameter the dynamics of the spin probe are. The slightly different features of the two series of spectra can be ascribed to the modulation of secular terms of the Hamiltonian of Eq. (3.3). The two series of spectra exhibit virtually the same behavior, except that the series of curves reproduced in Fig. 14 by the continuous lines seem to be more rigid.

Finally, as in the case of free isotropic diffusion for a Cu^{2+} ion complex in a highly viscous solvent, we plan to establish how important the pseudosecular terms are in the presence of an orienting potential at different values of the order parameter. Figure 15 shows also that in the high-rigidity case the pseudosecular terms can still affect the relative intensities of the components of the spectrum as in the free diffusion case.

E. Inertial Effects on the ESR Lineshape

In this section we treat the problem of evaluating an orientational correlational function without the inertial approximation (which assumes the molecular velocity relaxed to thermal equilibrium) and determining the spectroscopic effects of molecular inertia on a spin system $S = \frac{1}{2}$ whose Hamiltonian is described by an axially anisotropic Zeeman interaction.

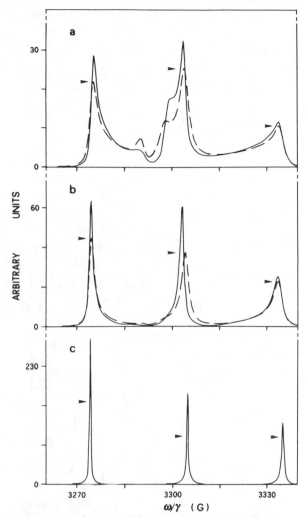

Figure 14. Absorption lineshape for a nitroxide radical dissolved in an ordered liquid. Magnetic parameters are the same as in Fig. 1. $H_0 = 3300$ G. (a) $\overline{P}_2 = 0.3$, (b) $\overline{P}_2 = 0.6$, (c) $\overline{P}_2 = 0.9$. The spectra for $D_{\parallel}/D_{\perp} = 0.5$, $r = 50$ are indicated by the continuous lines. The spectra for $D_{\parallel}/D_{\perp} = 50$, $r = 50$ are denoted by the dashed lines and are not reproduced when they nearly coincide with the curves in the previous case. The heights of peaks in the latter case are denoted by arrows.

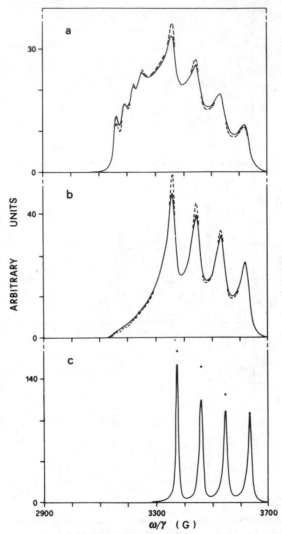

Figure 15. Absorption lineshape for copper complex dissolved in an ordered liquid. Magnetic parameters are the same as defined in Fig. 7. (a) $\overline{P}_2 = 0.3$, (b) $\overline{P}_2 = 0.6$, (c) $\overline{P}_2 = 0.9$. The continuous line is the absorption spectrum obtained via the complete spin Hamiltonian. If the pseudosecular terms are neglected, the dashed line is obtained. In (c), dots indicate the position of the peaks when the approximate Hamiltonian is used.

The expression of the spin Hamiltonian, in the secular approximation, is given by

$$\mathcal{H} = \omega_0 S_z + \eta Y_{2,0}(\Omega)S_z$$

The evolution equation of the variable of interest S_+ is well known to be

$$\frac{dS_+}{dt} = i[\mathcal{H}, S_+] + \Gamma_\omega^+ S_+ = \Gamma S_+ \tag{3.67}$$

where Γ_ω is the Hubbard operator for the spherical top[47b]

$$\Gamma_\omega = -\sum_{j=1}^{3}\left(i\omega_j M_j - \frac{\mathcal{F}}{I_0}\frac{\partial}{\partial\omega_j}\omega_j - D\left(\frac{\mathcal{F}}{I_0}\right)^2\frac{\partial^2}{\partial\omega_j^2}\right) \tag{3.68}$$

ω_i are the components of angular velocity in the molecular frame, $\mathcal{F}$ is the molecular friction coefficient, $\overline{M}$ is the molecular rotation generator, and $\underline{D}$ satisfies the fluctuation-dissipation relationship

$$\underline{D} = \frac{KT}{\mathcal{F}} \tag{3.69}$$

I_0 being the inertial moment of the spherical molecule. With such a choice for Γ_ω, it is no longer possible to obtain directly the analytical expression for the parameters of the continued fraction, and we must have recourse to the general approach employed in Section II.

Now the general state $|a_n\rangle$ is given by $(|a_0\rangle = S_+)$

$$|a_n\rangle = \sum_{l,m,i,j,k} c(n)_{l,m,i,j,k} Y_{2l,m}(\Omega) H_i(\overline{\omega}_1) H_j(\overline{\omega}_2) H_k(\overline{\omega}_3) S_+ \tag{3.70}$$

where $H_j(\overline{\omega}_i)$, the Hermite polynomial of jth order, is a function of the ith component of the reduced angular velocity $\overline{\omega} \equiv (I_0/KT)^{1/2}\omega$. Let us modify the scalar product (2.9), introducing $\overline{\omega}$ as a stochastic variable with a suitable distribution function $W_0(\overline{\omega})$. As a consequence the moment s_n reads:

$$s_n = \frac{\langle a_0|a_n\rangle}{\langle a_0|a_0\rangle}$$

$$= \left[\text{Tr}_{\{S\}}\int d\Omega\, d\overline{\omega}\, w_{eq}(\Omega)W_0(\overline{\omega})S_-\right.$$

$$\times\left(\sum_{l,m,i,j,k} c(n)_{l,m,i,j,k} Y_{2l,m}(\Omega)H_i(\overline{\omega}_1)H_j(\overline{\omega}_2)H_k(\overline{\omega}_3)S_+\right)\right]$$

$$\times\left[\text{Tr}_{\{S\}}(S_-S_+)\int d\Omega\, w_{eq}(\Omega)\int d\overline{\omega}\, W_0(\overline{\omega})\right]^{-1}$$

$$= \frac{1}{\sqrt{4\pi}}c(n)_{0,0,0,0,0} \tag{3.71}$$

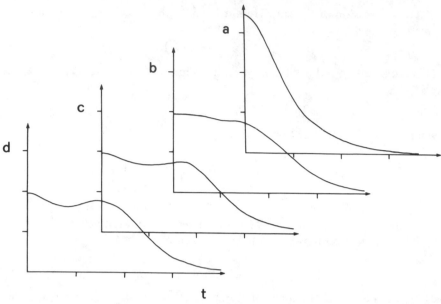

Figure 16. Real part of Laplace transform of $\langle \omega_s \omega_s(t) \rangle$. (a) $\langle \omega^2 \rangle^{1/2}/\underline{D} = 5$, (b) $\langle \omega^2 \rangle^{1/2}/\underline{D} = 1$, (c) $\langle \omega^2 \rangle^{1/2}/\underline{D} = 0.6$, (d) $\langle \omega^2 \rangle^{1/2}/\underline{D} = 0.5$. Vertical and horizontal scales are in arbitrary units.

The equilibrium distribution function $W_0(\overline{\omega})$ for the variable ω is given by

$$W_0(\overline{\omega}) = \Pi_i \exp\left(\frac{-\overline{\omega}_i^2}{2} \right) \tag{3.72}$$

As usual, following the procedure outlined in Section II, algebraic iterative equations, relating $c(n+1)$ to $c(n)$ can be derived. Assuming as a starting point:

$$c(0)_{l,m,i,j,k} = \sqrt{4\pi}\, \delta_{l,0} \delta_{m,0} \delta_{i,0} \delta_{j,0} \delta_{k,0}$$

the problem is reduced to one of computation only.

In Figs. 16, and 17 are shown the correlation function of the angular velocity and the ESR lineshape for various values of the diffusion coefficient $\underline{D}$.

Note that the use of the operator of Eq. (3.68) instead of the diffusional limit 3.7, which allows us to evaluate the spectroscopic effects due to molecular inertia, shows the emergence of strong correlation effects when the dif-

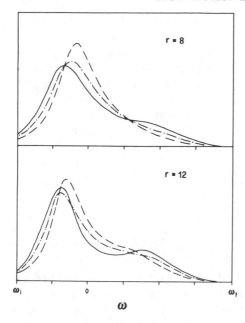

Figure 17. Absorption lineshape for an axial g-splitting interaction for a spin $1/2$ system. Vertical scales are expressed in the same units. $\omega_0 = 0$. The frequency sweep ranges from $\omega_i = -\frac{2}{3}\eta$ to $\omega_f = \frac{4}{3}\eta$. $\langle \omega^2 \rangle^{1/2}/\underline{D} = 2$ (continuous line) $\langle \omega^2 \rangle^{1/2}/\underline{D} = 3.162$ (dot dashed line), $\langle \omega^2 \rangle^{1/2}/\underline{D} = 31.62$ (dashed line). r parameter is defined in Eq. (3.24), setting $D_{\perp} = \underline{D}$.

fusion coefficient $\underline{D}$ assumes values near the square root of the mean quadratic value of the angular velocity of the tagged molecule.

In particular, the multiplicative stochastic frequency $\omega_s = \eta Y_{2,0}(\Omega)$ shows a correlation function quite different from that obtained when inertial effects are neglected. In fact, a direct calculation shows the correlation function of ω_s to be

$$\langle \omega_s \omega_s(t) \rangle = \frac{\eta^2}{4\pi} \exp(-6D_{\perp}|t|)$$

whose Laplace transform is a Lorentzian curve.

IV. FINAL REMARKS

It appears clear from Chapters I, III, and IV that the Mori theory is the major theoretical tool behind the algorithm illustrated in Section II, which derives the expansion parameters λ_i and Δ_i^2 from the moments s_n. This theory also affords us with a second straightforward way of determining these parameters: that of deriving them directly from the biorthogonal basis set of states $|f_i\rangle$ and $|\bar{f}_i\rangle$ (Eqs. 2.15). As discussed at length in Chapter IV, this is an especially stable way of building up λ_i and Δ_i^2. The Lanczos method followed by the authors of ref. 40, being completely equivalent to this method, shares this special numerical stability. The former approach has been used in this paper without meeting any relevant problem of numerical stability (of special relevance in this sense was the comparison made in Section III.B.2 between this method and a completely analytical one of proven stability). We feel that the former approach is faster then the latter (when no simple direct analytical expressions for λ_i and Δ_i^2 are available).

The influence of inertia can be taken into account by using directly the Hubbard diffusion operator. However, a general program of calculation should consider a physical condition more general than that explored in ref. 35. We believe that this will be made possible by combined use of AEP and CFP. Indeed it has been shown in Chapter II that the AEP allows us to correct the Favro equation without enlarging the system to be studied via inclusion of the variable ω. This is left to future work.

Note that our approach is a sort of happy compromise between the Kubo[29,53] and Mori[33,54] theories. The Kubo theory of the stochastic Liouville equation[29] has been widely applied to EPR spectroscopy by Freed and co-workers,[53] whereas Kivelson and Ogan[54] illustrate how to apply the Mori theory to the same field of investigation. The present approach uses the generalized Langevin equation method as a pure calculation technique to solve the stochastic Liouville equation of Kubo. This affords both computational advantages and a deeper understanding of the involved physical phenomena.

APPENDIX A

In this appendix we shall provide explicit expressions for the recursion formulas of Eq. (3.13) for the case of a nitroxide spin probe and a copper complex in solution. To this end we need these preliminary definitions:

$$A = i\frac{A_{\|} + 2A_{\perp}}{3}$$

$$B = i\frac{2}{3}\sqrt{\frac{4\pi}{5}}\,(g_{\|} - g_{\perp})\beta_e H_0$$

$$C = i\frac{2}{3}\sqrt{\frac{4\pi}{5}}\,(A_{\|} - A_{\perp})$$

$$D = i\sqrt{\frac{4\pi}{30}}\,(A_{\|} - A_{\perp}) \tag{A.1}$$

$$C_M(m) \equiv \int d\cos\theta \, d\phi \, Y_{l,m}^*(\theta,\phi) \Gamma_\Omega^+ Y_{l-2,m}(\theta,\phi)$$

$$C_0(m) \equiv \int d\cos\theta \, d\phi \, Y_{l,m}^*(\theta,\phi) \Gamma_\Omega^+ Y_{l,m}(\theta,\phi)$$

$$C_P(m) \equiv \int d\cos\theta \, d\phi \, Y_{l,m}^*(\theta,\phi) \Gamma_\Omega^+ Y_{l+2,m}(\theta,\phi) \qquad (A.2)$$

$$C_{10}(m) \equiv \sqrt{\frac{5}{4\pi}} \, C(l,2,l,m-1,1) C(l,2,l,0,0)$$

$$C_{1M}(m) \equiv \sqrt{\frac{5}{4\pi}} \, C(l+2,2,l,m-1,1) C(l+2,2,l,0,0) \left[\frac{2(l+2)+1}{2l+1} \right]^{1/2}$$

$$C_{1P}(m) \equiv \sqrt{\frac{5}{4\pi}} \, C(l-2,2,l,m-1,1) C(l-2,2,l,0,0) \left[\frac{2(l-2)+1}{2l+1} \right]^{1/2}$$

$$C_{00}(m) \equiv \sqrt{\frac{5}{4\pi}} \, C(l,2,l,m,0) C(l,2,l,0,0)$$

$$C_{0M}(m) \equiv \sqrt{\frac{5}{4\pi}} \, C(l+2,2,l,m,0) C(l+2,2,l,0,0) \left[\frac{2(l+2)+1}{2l+1} \right]^{1/2}$$

$$C_{0P}(m) \equiv \sqrt{\frac{5}{4\pi}} \, C(l-2,2,l,m,0) C(l-2,2,l,0,0) \left[\frac{2(l-2)+1}{2l+1} \right]^{1/2}$$

$$C_{M0}(m) \equiv \sqrt{\frac{5}{4\pi}} \, C(l,2,l,m+1,-1) C(l,2,l,0,0)$$

$$C_{MM}(m) \equiv \sqrt{\frac{5}{4\pi}} \, C(l+2,2,l,m+1,-1) C(l+2,2,l,0,0) \left[\frac{2(l+2)+1}{2l+1} \right]^{1/2}$$

$$C_{MP}(m) \equiv \sqrt{\frac{5}{4\pi}} \, C(l-2,2,l,m+1,-1) C(l-2,2,l,0,0) \left[\frac{2(l-2)+1}{2l+1} \right]^{1/2}$$

$$(A.3)$$

β_e is the Bohr magneton, $i = \sqrt{-1}$, H_0 is the magnetic field. The symbol $Y_{l,m}(\theta,\phi)$ denotes the spherical harmonics. Note that we are considering only even l, because for the sake of conciseness we have changed the notation $Y_{2l',m}$ used throughout the text into $Y_{l,m}$ ($l = 2l'$). Furthermore, we follow Rose's notation,[55] according to which $C(l_1, l_2, l_3, m_1, m_2)$ denotes the Clebsch-Gordan coefficients. Note that quantities defined by (A.2) and (A.3) refer to the same (general) pair of indices l and m and are independent of i and j.

The set of expressions (A.3) is redundant, since these are not independent of each other,[56] but the present formulation allows for a clear understanding of recursion formulas.

We are now in a position to provide relationships connecting the $c(n)_{l,m,i,j}$ [here renamed $\overline{W}(l, m, i, j)$] to the $c(n-1)_{l,m,i,j}$ [here renamed $W(l, m, i, j)$]. When exploring the independent triads of indices m, i, and j, we get for a nitroxide spin probe:

$$\overline{W}(l,0,1,1) = \frac{D}{\sqrt{2}}\left\{[C_{MM}(0) + C_{1M}(0)] W(l+2,1,2,1)\right.$$

$$+ [C_{M0}(0) + C_{10}(0)] W(l,1,2,1)$$

$$+ [C_{MP}(0) + C_{1P}(0)] W(l-2,1,2,1)\}$$

$$+ [(C+B)C_{0P}(0) + C_M(0)] W(l-2,0,1,1)$$

$$+ [(C+B)C_{00}(0) + C_0(0) + A] W(l,0,1,1)$$

$$+ [(C+B)C_{0M}(0) + C_P(0)] W(l+2,0,1,1)$$

$$\overline{W}(l,0,2,2) = \frac{D}{\sqrt{2}}\left\{[C_{MM}(0) + C_{1M}(0)][W(l+2,1,3,2) + W(l+2,1,2,1)]\right.$$

$$+ [C_{M0}(0) + C_{10}(0)][W(l,1,3,2) + W(l,1,2,1)]$$

$$+ [C_{MP}(0) + C_{1P}(0)][W(l-2,1,3,2) + W(l-2,1,2,1)]\}$$

$$+ [BC_{0P}(0) + C_M(0)] W(l-2,0,2,2)$$

$$+ [BC_{00}(0) + C_0(0)] W(l,0,2,2)$$

$$+ [BC_{0M}(0) + C_P(0)] W(l+2,0,2,2)$$

$$\overline{W}(l,0,3,3) = \frac{D}{\sqrt{2}}\left\{[C_{MM}(0) + C_{1M}(0)] W(l+2,1,3,2)\right.$$

$$+ [C_{M0}(0) + C_{10}(0)] W(l,1,3,2)$$

$$+ [C_{MP}(0) + C_{1P}(0)] W(l-2,1,3,2)\}$$

$$+ [(-C+B)C_{0P}(0) + C_M(0)] W(l-2,0,3,3)$$

$$+ [(-C+B)C_{00}(0) + C_0(0) - A] W(l,0,3,3)$$

$$+ [(-C+B)C_{0M}(0) + C_P(0)] W(l+2,0,3,3)$$

$$\overline{W}(l,1,3,2) = \frac{D}{\sqrt{2}}\left\{W(l+2,2,3,1)C_{MM}(1) + W(l,2,3,1)C_{M0}(1)\right.$$

$$+ W(l-2,2,3,1)C_{MP}(1)$$

$$- \{[W(l+2,0,2,2) + W(l+2,0,3,3)]C_{1M}(1)$$

$$+ [W(l,0,2,2) + W(l,0,3,3)]C_{10}(1)$$

$$+ [W(l-2,0,2,2) + W(l-2,0,3,3)]C_{1P}(1)\}\}$$

$$+\left[\left(-\frac{C}{2}+B\right)C_{0P}(1)+C_M(1)\right]W(l-2,1,3,2)$$

$$+\left[\left(-\frac{C}{2}+B\right)C_{00}(1)+C_0(1)-\frac{A}{2}\right]W(l,1,3,2)$$

$$+\left[\left(-\frac{C}{2}+B\right)C_{0M}(1)+C_P(1)\right]W(l+2,1,3,2)$$

$$\overline{W}(l,1,2,1)=\frac{D}{\sqrt{2}}\{W(l+2,2,3,1)C_{MM}(1)+W(l,2,3,1)C_{M0}(1)$$

$$+W(l-2,2,3,1)C_{MP}(1)$$

$$-\{[W(l+2,0,2,2)+W(l+2,0,1,1)]C_{1M}(1)$$

$$+[W(l,0,2,2)+W(l,0,1,1)]C_{10}(1)$$

$$+[W(l-2,0,2,2)+W(l-2,0,1,1)]C_{1P}(1)\}\}$$

$$+\left[\left(\frac{C}{2}+B\right)C_{0P}(1)+C_M(1)\right]W(l-2,1,2,1)$$

$$+\left[\left(\frac{C}{2}+B\right)C_{00}(1)+C_0(1)+\frac{A}{2}\right]W(l,1,2,1)$$

$$+\left[\left(\frac{C}{2}+B\right)C_{0M}(1)+C_P(1)\right]W(l+2,1,2,1)$$

$$\overline{W}(l,2,3,1)=\frac{(-D)}{\sqrt{2}}\{[W(l+2,1,2,1)+W(l+2,1,3,2)]C_{1M}(2)$$

$$+[W(l,1,2,1)+W(l,1,3,2)]C_{10}(2)$$

$$+[W(l-2,1,2,1)+W(l-2,1,3,2)]C_{1P}(2)\}$$

$$+[BC_{0P}(2)+C_M(2)]W(l-2,2,3,1)$$

$$+[BC_{00}(2)+C_0(2)]W(l,2,3,1)$$

$$+[BC_{0M}(2)+C_P(2)]W(l+2,2,3,1)$$

In the case of a copper complex dissolved in solution, when exploring the independent triads of indices m, i, and j, we get

$$\overline{W}(l,0,1,1)=\frac{\sqrt{3}}{2}D\{[C_{MP}(0)+C_{1P}(0)]W(l-2,1,2,1)$$

$$+[C_{M0}(0)+C_{10}(0)]W(l,1,2,1)$$

$$+[C_{MM}(0)+C_{1M}(0)]W(l+2,1,2,1)\}$$

$$+\left[C_{0P}(0)\left(\frac{3C}{2}+B\right)+C_M(0)\right]W(l-2,0,1,1)$$

$$+\left[C_{00}(0)\left(\frac{3C}{2}+B\right)+C_0(0)+\frac{3A}{2}\right]W(l,0,1,1)$$

$$+\left[C_{0M}(0)\left(\frac{3C}{2}+B\right)+C_P(0)\right]W(l+2,0,1,1)$$

$$\overline{W}(l,0,2,2)=\frac{D}{2}\left\{\left[C_{MP}(0)+C_{1P}(0)\right]\right.$$

$$\cdot\left[\sqrt{3}\,W(l-2,1,2,1)+2W(l-2,1,3,2)\right]$$

$$+\left[C_{M0}(0)+C_{10}(0)\right]\left[\sqrt{3}\,W(l,1,2,1)+2W(l,1,3,2)\right]$$

$$+\left[C_{MM}(0)+C_{1M}(0)\right]\left[\sqrt{3}\,W(l+2,1,2,1)+2W(l+2,1,3,2)\right]\right\}$$

$$+\left[C_{0P}(0)\left(\frac{C}{2}+B\right)+C_M(0)\right]W(l-2,0,2,2)$$

$$+\left[C_{00}(0)\left(\frac{C}{2}+B\right)+C_0(0)+\frac{A}{2}\right]W(l,0,2,2)$$

$$+\left[C_{0M}(0)\left(\frac{C}{2}+B\right)+C_P(0)\right]W(l+2,0,2,2)$$

$$\overline{W}(l,0,3,3)=\frac{D}{2}\left\{\left[C_{MP}(0)+C_{1P}(0)\right]\right.$$

$$\cdot\left[2W(l-2,1,3,2)+\sqrt{3}\,W(l-2,1,4,3)\right]$$

$$+\left[C_{M0}(0)+C_{10}(0)\right]\left[2W(l,1,3,2)+\sqrt{3}\,W(l,1,4,3)\right]$$

$$+\left[C_{MM}(0)+C_{1M}(0)\right]$$

$$\cdot\left[2W(l+2,1,3,2)+\sqrt{3}\,W(l+2,1,4,3)\right]\right\}$$

$$+\left[C_{0P}(0)\left(-\frac{C}{2}+B\right)+C_M(0)\right]W(l-2,0,3,3)$$

$$+\left[C_{00}(0)\left(-\frac{C}{2}+B\right)+C_0(0)-\frac{A}{2}\right]W(l,0,3,3)$$

$$+\left[C_{0M}(0)\left(-\frac{C}{2}+B\right)+C_P(0)\right]W(l+2,0,3,3)$$

$$\overline{W}(l,0,4,4)=\frac{\sqrt{3}}{2}D\left\{\left[C_{MP}(0)+C_{1P}(0)\right]W(l-2,1,4,3)\right.$$

$$+\left[C_{M0}(0)+C_{10}(0)\right]W(l,1,4,3)$$

$$+\left[C_{MM}(0)+C_{1M}(0)\right]W(l+2,1,4,3)\right\}$$

$$+\left[C_{0P}(0)\left(-\frac{3C}{2}+B\right)+C_M(0)\right]W(l-2,0,4,4)$$

$$+\left[C_{00}(0)\left(-\frac{3C}{2}+B\right)+C_0(0)-\frac{3A}{2}\right]W(l,0,4,4)$$

$$+\left[C_{0M}(0)\left(-\frac{3C}{2}+B\right)+C_P(0)\right]W(l+2,0,4,4)$$

$$\overline{W}(l,1,2,1)=D\Bigg(C_{MP}(1)W(l-2,2,3,1)+C_{M0}(1)W(l,2,3,1)$$

$$+C_{MM}(1)W(l+2,2,3,1)$$

$$-\frac{\sqrt{3}}{2}\{C_{1P}(1)[W(l-2,0,1,1)+W(l-2,0,2,2)]$$

$$+C_{10}(1)[W(l,0,1,1)+W(l,0,2,2)]$$

$$+C_{1M}(1)[W(l+2,0,1,1)+W(l+2,0,2,2)]\}\Bigg)$$

$$+[C_{0P}(1)(C+B)+C_M(1)]W(l-2,1,2,1)$$

$$+[C_{0M}(1)(C+B)+C_P(1)]W(l+2,1,2,1)$$

$$+[C_{00}(1)(C+B)+C_0(1)+A]W(l,1,2,1)$$

$$\overline{W}(l,1,3,2)=D\Bigg(\frac{\sqrt{3}}{2}\{C_{MP}(1)[W(l-2,2,3,1)+W(l-2,2,4,2)]$$

$$+C_{M0}(1)[W(l,2,3,1)+W(l,2,4,2)]$$

$$+C_{MM}(1)[W(l+2,2,3,1)+W(l+2,2,4,2)]\}$$

$$-\{C_{1P}(1)[W(l-2,0,3,3)+W(l-2,0,2,2)]$$

$$+C_{10}(1)[W(l,0,3,3)+W(l,0,2,2)]$$

$$+C_{1M}(1)[W(l+2,0,3,3)+W(l+2,0,2,2)]\}\Bigg)$$

$$+[C_{0P}(1)B+C_M(1)]W(l-2,1,3,2)$$

$$+[C_{00}(1)B+C_0(1)]W(l,1,3,2)$$

$$+[C_{0M}(1)B+C_P(1)]W(l+2,1,3,2)$$

$$\overline{W}(l,1,4,3) = D\bigg(C_{MP}(1)W(l-2,2,4,2)$$

$$+ C_{M0}(1)W(l,2,4,2) + C_{MM}(1)W(l+2,2,4,2)$$

$$- \frac{\sqrt{3}}{2}\{ C_{1P}(1)[W(l-2,0,4,4) + W(l-2,0,3,3)]$$

$$+ C_{10}(1)[W(l,0,4,4) + W(l,0,3,3)]$$

$$+ C_{1M}(1)[W(l+2,0,4,4) + W(l+2,0,3,3)]\}\bigg)$$

$$+ [C_{0P}(1)(-C+B) + C_M(1)]W(l-2,1,4,3)$$

$$+ [C_{00}(1)(-C+B) + C_0(1) - A]W(l,1,4,3)$$

$$+ [C_{0M}(1)(-C+B) + C_P(1)]W(l+2,1,4,3)$$

$$\overline{W}(l,2,3,1) = \frac{D}{2}\bigg(\sqrt{3}\,[C_{MP}(2)W(l-2,3,4,1)$$

$$+ C_{M0}(2)W(l,3,4,1) + C_{MM}(2)W(l+2,3,4,1)]$$

$$- \{ C_{1P}(2)[\sqrt{3}\,W(l-2,1,3,2) + 2W(l-2,1,2,1)]$$

$$+ C_{10}(2)[\sqrt{3}\,W(l,1,3,2) + 2W(l,1,2,1)]$$

$$+ C_{1M}(2)[\sqrt{3}\,W(l+2,1,3,2) + 2W(l+2,1,2,1)]\}\bigg)$$

$$+ \bigg[C_{0P}(2)\Big(\frac{C}{2}+B\Big) + C_M(2)\bigg]W(l-2,2,3,1)$$

$$+ \bigg[C_{00}(2)\Big(\frac{C}{2}+B\Big) + C_0(2) + \frac{A}{2}\bigg]W(l,2,3,1)$$

$$+ \bigg[C_{0M}(2)\Big(\frac{C}{2}+B\Big) + C_P(2)\bigg]W(l+2,2,3,1)$$

$$\overline{W}(l,2,4,2) = \frac{D}{2}\bigg(\sqrt{3}\,[C_{MP}(2)W(l-2,3,4,1)$$

$$+ C_{M0}(2)W(l,3,4,1) + C_{MM}(2)W(l+2,3,4,1)]$$

$$- \{ C_{1P}(2)[2W(l-2,1,4,3) + \sqrt{3}\,W(l-2,1,3,2)]$$

$$+ C_{10}(2)[2W(l,1,4,3) + \sqrt{3}\,W(l,1,3,2)]$$

$$+ C_{1M}(2)[2W(l+2,1,4,3) + \sqrt{3}\,W(l+2,1,3,2)]\}\bigg)$$

$$+ \left[C_{0P}(2)\left(-\frac{C}{2} + B \right) + C_M(2) \right] W(l-2,2,4,2)$$

$$+ \left[C_{00}(2)\left(-\frac{C}{2} + B \right) + C_0(2) - \frac{A}{2} \right] W(l,2,4,2)$$

$$+ \left[C_{0M}(2)\left(-\frac{C}{2} + B \right) + C_P(2) \right] W(l+2,2,4,2)$$

$$\overline{W}(l,3,4,1) = -\frac{\sqrt{3}}{2} D \{ C_{1P}(3)[W(l-2,2,4,2) + W(l-2,2,3,1)]$$

$$+ C_{10}(3)[W(l,2,4,2) + W(l,2,3,1)]$$

$$+ C_{1M}(3)[W(l+2,2,4,2) + W(l+2,2,3,1)] \}$$

$$+ [C_{0P}(3)B + C_M(3)] W(l-2,3,4,1)$$

$$+ [C_{00}(3)B + C_0(3)] W(l,3,4,1)$$

$$+ [C_{0M}(3)B + C_P(3)] W(l+2,3,4,1)$$

APPENDIX B

In our description of spin reorientational relaxation processes, tensorial quantities are used for which it is necessary to know the transformation properties concerning rotation. A clear and compact formulation is obtained by replacing the cartesian components with a representation in terms of irreducible spherical components.[55] It is known that any representation of the group of rotations can be developed into a sum of irreducible representations D^l of dimension $2l+1$. If for the description of general rotation $R(\Omega)$ we use the Euler angles $\Omega \equiv (\alpha, \beta, \gamma)$, this rotation will be defined by

$$R(\Omega) = e^{-i\alpha M_z} e^{-i\beta M_y} e^{-i\gamma M_z} \tag{B.1}$$

The choice of a base where M^2 and M_z turn out to be diagonal allows us to obtain the matrices representing D^l:

$$D^l_{m',m}(\Omega) = \langle l, m' | R(\Omega) | l, m \rangle$$
$$= e^{-im'\alpha} \langle l, m' | e^{-i\beta M_y} | l, m \rangle e^{-im\gamma} \tag{B.2}$$

The matrix element of $e^{-i\beta M_y}$ is the scalar quantity

$$d^l_{m',m}(\beta) \equiv \langle l, m' | e^{-i\beta M_y} | l, m \rangle \tag{B.3}$$

which is termed the reduced Wigner matrix.

An irreducible tensor of rank l T^l will be defined as a set of $2l + 1$ operators $T^{(l,m)}$ $(m = -l, -l+1, \ldots, l)$, which is transformed according to the $(2l + 1)$-dimensional representation of the rotation group D^l:

$$T^{(l,m)'} \equiv R(\Omega) T^{(l,m)} R^{-1}(\Omega) = \sum_{m'} D^l_{m',m}(\Omega) T^{(l,m')} \qquad (B.4)$$

Definition (B.4) states that T^l is subjected to the same transformation law as the spherical harmonics of rank l.[55]

This transformation law is quite simple, and on it relies the main advantages of using spherical tensors in problems involving rotations. The Wigner matrices defined by Eq. (B.2) provide a set complete and orthogonal in the space of Euler angles, thereby making it possible to use them as a suitable expansion basis set.

Due to unitarity properties of rotations:

$$\sum_m D^{l*}_{m',m}(\Omega) D^l_{m'',m}(\Omega) = \delta_{m',m''}$$

$$\sum_m D^{l*}_{m,m'}(\Omega) D^l_{m,m''}(\Omega) = \delta_{m',m''} \qquad (B.5)$$

which comes directly from definition (B.2). Furthermore, the matrices $D^l_{m',m}(\Omega)$ satisfy the orthogonal property on the complete solid angle:

$$\int d\Omega \, D^{l*}_{m,n}(\Omega) D^{l'}_{m',n'}(\Omega) = \frac{8\pi^2}{2j+1} \delta_{l,l'} \delta_{m,m'} \delta_{n,n'} \qquad (B.6)$$

where $d\Omega = d\alpha \, d\cos\beta \, d\gamma$. The product of Wigner matrices can be expanded over their basis set as

$$D^{l'}_{m',n'}(\Omega) D^{l''}_{m'',n''}(\Omega) = \sum_{l=|l'-l''|}^{l=l'+l''} C(l', l'', l; m', m'') C(l', l'', l; n', n'')$$

$$\times D^l_{m'+m'', n'+n''}(\Omega) \qquad (B.7)$$

where $C(l_1, l_2, l_3; m_1, m_2)$ is a Clebsch-Gordan coefficient. Of frequent use are the symmetry properties:

$$D^{l*}_{m,n}(\alpha, \beta, \gamma) = (-)^{m+n} D^l_{-m,-n}(\alpha, \beta, \gamma) = D^l_{n,m}(-\gamma, -\beta, -\alpha) \qquad (B.8)$$

Let us note also that for special values of their indices the Wigner matrices reduce to important special functions:

$$D^l_{0,m}(\alpha,\beta,\gamma) = (-)^m \left(\frac{4\pi}{2l+1} \right)^{1/2} Y^*_{l,m}(\beta,\gamma)$$

$$D^l_{m,0}(\alpha,\beta,\gamma) = \left(\frac{4\pi}{2l+1} \right)^{1/2} Y^*_{l,m}(\alpha,\beta) \qquad \text{(B.9)}$$

$$D^l_{0,0}(\alpha,\beta,\gamma) = d^l_{0,0}(\beta) = P_l(\cos\beta)$$

where $Y_{l,m}$ is a spherical harmonic and P_l is a Legendre polynomial.

References

1. J. H. Van Vleck, *Phys. Rev.*, **74**, 1168 (1948).
2. M. H. L. Pryce and K. W. H. Stevens, *Proc. Phys. Soc. (London)*, **63**, 36 (1950).
3. A. Abragam and K. Kambe, *Phys. Rev.*, **91**, 894 (1953).
4. A. Abragam, *Principles of Nuclear Magnetism*, (Clarendon, Oxford, England, 1961, Ch. 4.
5. R. G. Gordon, *J. Chem. Phys.*, **40**, 1973 (1964).
6. R. G. Gordon, *J. Chem. Phys.*, **41**, 1819 (1964).
7. R. G. Gordon, *J. Chem. Phys.*, **39**, 2788 (1963).
8. R. G. Gordon, *J. Math. Phys.*, **9**, 655 (1968).
9. R. G. Gordon, *J. Math. Phys.*, **9**, 1087 (1968).
10. J. C. Wheeler and R. G. Gordon, *J. Chem. Phys.*, **51**, 5566 (1969).
11. P. Mansfield, *Phys. Rev.*, **151**, 199 (1966).
12. (a) W. T. P. Choy, *J. Chem. Phys.* **46**, 1578 (1967); (b) S. Vega, *Adv. Magn. Resonance*, **6**, 259, Academic, N.Y, 1973; (c) P. Grigolini, *J. Chem. Phys.*, **56**, 5930 (1972).
13. J. C. Gill, *J. Phys. C: Solid State Phys.*, **4**, 1420 (1971).
14. A. Lonke, *J. Math. Phys.*, **12**, 2422 (1971).
15. F. Lado, J. D. Memory, and G. W. Parker, *Phys. Rev.*, **B4**, 1406 (1971).
16. G. W. Parker, *J. Chem. Phys.*, **58**, 3274 (1973).
17. E. T. Cheng and J. D. Memory, *Phys. Rev.*, **B6**, 1714 (1972).
18. W. F. Wurzbach and S. Gade, *Phys. Rev.*, **B6**, 1724, (1972).
19. S. J. Knak Jensen and E. K. Hansen, *Phys. Rev.*, **B7**, 2910 (1973).
20. K. Tomita and H. Mashiyama, *Progr. Theor. Phys.*, **51**, 1312 (1974).
21. M. Engelsberg and I. J. Lowe, *Phys. Rev.*, **B10**, 822 (1974).
22. M. Engelsberg and I. J. Lowe, *Phys. Rev.*, **B12**, 3547 (1975).
23. M. Engelsberg, I. J. Lowe, and J. L. Carolan, *Phys. Rev.*, **B7**, 924 (1973).
24. (a) A. Sur and I. J. Lowe, *Phys. Rev.*, **B11**, 1980 (1975); (b) A. Sur and I. J. Lowe, *Phys. Rev.*, **B12**, 4597 (1975).
25. C. W. Myles, C. Ebner, and P. A. Fedders, *Phys. Rev.*, **B14**, 1 (1976).
26. S. J. Knak Jensen and E. K. Hansen, *Phys. Rev.*, **B13**, 1903 (1976).
27. T. Y. Hwang and I. J. Lowe, *Phys. Rev.*, **B17**, 2845 (1978).
28. I. Waller, *Z. Phys.*, **79**, 380 (1932).

29. R. Kubo, in *Advances in Chemical Physics*, **15**, 101, Wiley, N.Y, 1969.

30. M. Ferrario and P. Grigolini, *Chem. Phys. Lett.*, **62**, 100 (1979).

31. P. Grigolini, *Chem. Phys.*, **38**, 389 (1979).

32. R. Kubo, *J. Phys. Soc. Japan*, **12**, 570 (1957).

33. H. Mori, *Prog. Theor. Phys.*, **33**, 423 (1965); **34**, 399 (1965).

34. N. I. Akhiezer, *The Classical Moment Problem*, Oliver and Boyd, Edinburgh, 1976.

35. M. Giordano, P. Grigolini, and P. Marin, *Chem. Phys. Lett.*, **83**, 554 (1981).

36. M. Dupuis, *Prog. Theor. Phys.*, **37**, 502 (1967).

37. H. S. Wall, *Analytical Theory of Continued Fractions*, Van Nostrand, New York, 1948.

38. (a) R. R. Whitehead and W. Watt, *J. Phys.*, **A14**, 1887 (1981); (b) R. R. Whitehead, A. Watt, B. J. Cole, and I. Morrison, in *Advances in Nuclear Physics*, **9**, 123, Plenum, N.Y, 1977; (c) R. R. Whitehead and A. Watt, *J. Phys.*, **G4**, 835 (1978).

39. C. Lanczos, *J. Res. Nat. Bur. Stand.*, **45**, 255 (1950).

40. (a) G. Moro and J. H. Freed, *J. Phys. Chem.*, **84**, 2837 (1980); (b) G. Moro and J. H. Freed, *J. Chem. Phys.*, **74**, 3757 (1981); (c) M. H. Lee, I. M. Kim, and R. Dekeyser, *Phys. Rev. Lett.*, **52**, 1579 (1984) and references therein.

41. A. J. Dammers, Y. K. Levine, and J. A. Tyon, *Chem. Phys. Lett.*, **88**, 198 (1982).

42. G. Sauermann, (a) *Physica*, **66**, 331 (1973); (b) *Lett. Nuovo Cimento*, **3**, 489 (1970).

43. (a) A. Baram, *Mol. Phys.*, **41**, 823 (1980); (b) A. Baram, *ibid.*, **44**, 1099 (1981); (c) A. Baram, *ibid.*, **45**, 309 (1982); (d) A. Baram, *J. Phys. Chem.*, **87**, 1676 (1983).

44. J. G. Powles and B. Carazza, in *Magnetic Resonance*, G. K. Coogan et al., eds., Plenum, New York, 1970.

45. M. W. Evans and J. Jarwood, *Adv. Mol. Rel. Int. Proc.*, **21**, 1 (1981).

46. N. G. van Kampen, *Stochastic Processes in Physics and Chemistry*, North-Holland, Amsterdam, 1981.

47. (a) L. Dale Favro, *Fluctuation Phenomena in Solids*, R. Burgess ed., Academic New York, 1965; (b) P. S. Hubbard, *Phys. Rev.*, **A6**, 2421 (1972).

48. (a) S. J. Knak Jensen, in *Electron Spin Relaxation in Liquids*, L. T. Muus and P. W. Atkins, eds., Plenum, New York, 1972; (b) P. W. Atkins and B. P. Hill, *Mol. Phys.*, **29**, 761 (1975); (c) G. L. Monroe and H. L. Friedman, *J. Chem. Phys.*, **66**, 955 (1977); (d) H. L. Friedman, L. Blum and G. Yue, *J. Chem. Phys.* **65**, 4396 (1976). Note that in these papers the reader can find approximated techniques of calculation to be added to the set of refs. 15–44.

49. L. T. Muus and P. W. Atkins, eds., *Electron Spin Relaxation in Liquids*, Plenum, New York, 1972.

50. (a) B. Bleaney, *Proc. Phys. Soc.* (*London*), **75**, 621 (1960); (b) H. R. Gershmann and J. D. Swalen, *J. Chem. Phys.*, **36**, 3221 (1962).

51. C. Zannoni, in *The Molecular Physics of Liquid Crystals*, G. R. Luckhurst and G. W. Gray, eds., Academic, London, 1979, p. 51.

52. P. Grigolini, *Mol. Phys.*, **31**, 1717 (1976).

53. J. H. Freed, in *Electron Spin Relaxation in Liquids*, L. T. Muus and P. W. Atkins, eds., Plenum, New York, 1972, and references therein.

54. D. Kivelson and K. Ogan, *Adv. Magn. Reson.*, **7**, 71 (1974).

55. M. E. Rose, *Elementary Theory of Angular Momentum*, Wiley, New York, 1957.

56. M. Giordano, P. Grigolini, D. Leporini and P. Marin, *Phys. Rev.*, **A28**, 2474 (1983).

IX

THE THEORY OF CHEMICAL REACTION RATES*

T. FONSECA, J. A. N. F. GOMES, P. GRIGOLINI, and
F. MARCHESONI

CONTENTS

I. MOTIVATION

The theoretical study of the rates of condensed phase chemical reactions is, by comparison with the gas phase, in an early state of development. However, the last few years have seen renewed interest in the subject, with important new results being obtained, and this in turn has spurred novel experimental activity.

When compared with reactions in the gas phase, the theory of condensed phase processes meets the additional difficulty of having to deal with solvent interactions, but the detailed understanding of their effects is crucial in many

*Work supported by INIC, Lisbon, and CNR, Rome.

areas of chemistry, biology, and related sciences. Among the interactions between the reactive species and the solvent, we should distinguish the specific interactions (e.g., electrostatic or those originating from hydrogen bonding) from the stochastic forces generated by the thermal motions of the solvent particles. In this chapter we shall be concerned only with interactions of the latter type and discuss some recent developments.

The strength of the stochastic interactions between the solvent and the reactant may be described (in the Markovian case) by a parameter γ, the damping rate or friction coefficient, which is some measure of the coupling at the microscopic level and is related to the macroscopic viscosity, density, or pressure. Viscosity effects on reaction rates have been known for a long time, but the understanding of the mechanism of these interactions and of the most general form of their effects was very poor. The so-called cage effect is commonly used in discussing the observed viscosity dependence of the reaction rates.[1-3] The motion of the reactant(s) into the reacting position, the "encounter" in the language of collision theory, is hindered by higher viscosities η, and in the diffusive regime the number of these encounters is $Z \propto \eta^{-1}$. However, this same factor increases the difficulty for the reactant(s) to move out of position, providing a sort of cage wall, and the time spent inside this cage, or the number of collisions following the initial encounter, is $n \propto \eta$. Any reaction rate that is proportional to the total number of collisions, $Z \times n$, will be independent of the viscosity. Under certain circumstances, however, the rate will be proportional to $Z \propto \eta^{-1}$. One class of processes where this is well known to happen is the quenching of fluorescence, where the chemical process is so fast that the rate is controlled by the diffusion of the quencher to the excited molecule.[4] Similar behavior is observed in very fast proton-transfer reactions.[5] For very high viscosities, the quantity n will be very large, the chemical reaction will always occur at an early stage of the encounter, and the rate will be proportional to $Z \propto \eta^{-1}$. Then, we should expect that the rate of a chemical reaction would always go as η^{-1} for sufficiently high viscosities.

In the other extreme case, when the coupling of the reactant(s) to the solvent is very weak, the reaction rate will also decrease. In fact, once the higher energy molecules have reacted, the replenishment of this top energy layer will be too slow to maintain thermal equilibrium and the rate will slow down.

We shall show that this behavior is predicted in all stochastic theories, the major effort being directed to understanding the conditions when such extreme regimes fail and to predicting the detailed general form of the rate constant.

The most widely used theoretical tool for the understanding of chemical kinetics is still the transition state theory (TST) in its original form,[6] or in one of its modern versions.[7] Because it is used throughout this chapter as a major reference for comparison of the results obtained with the stochastic theories, it is useful to recall its basic principles and final expression. Conventional transition state theory depends on the following general assumptions (for a detailed discussion of the theory, see, e.g., ref. 8):

1. The rate of a chemical reaction may be calculated by focusing attention on the "transition state," the region near the col or saddle point of the potential energy surface that must be crossed in the process of converting the reactants into products.
2. The transition state is in quasi-thermodynamic equilibrium with the reactants, and the removal of the products does not affect the reactants' equilibrium up to the transition state.
3. In the region around the col, the motion along the reaction coordinate can be treated as free translational motion.

The rate k_{TST} is calculated as the product of the population at the transition state and the frequency at which one such species will go into products. The final result may be cast in the form

$$k_{TST} = \kappa \frac{K_B T}{h} \frac{Q_{\ddagger}}{Q_A} \exp\left(-\frac{E_b}{K_B T}\right) \qquad (1)$$

where $Q_{\ddagger}$ and Q_A are the partition functions associated with the transition state and the reactant, respectively, E_b is the activation energy (the potential energy of the col above the ground state of the reactants), and h, K_B, and T have the usual meaning. κ is the so-called transmission coefficient, an ad hoc factor usually taken close to unity, measuring the fraction of the forward moving transition state molecules that actually become products and are not reflected.

If one considers a system with a single degree of freedom, $Q_{\ddagger} = 1$ and $Q_A = (1 - \hbar\omega_0/K_B T)^{-1} \approx K_B T/\hbar\omega_0$ (for $K_B T \gg \hbar\omega_0$), the partition function of the harmonic vibration of the reactant, the TST rate is given by

$$K_{TST} = \frac{\omega_0}{2\pi} \exp\left(-\frac{E_b}{K_B T}\right) \qquad (2)$$

if the transmission coefficient is assumed unity. This is the TST rate expression that we will always consider in later sections.

The most interesting applications of transition state theory have been, perhaps, in solution chemistry, and a number of detailed improvements have been made to bring in some of the effects of the solvent. This has been done mostly within the framework of thermodynamics by introducing in Eq. (1)

the solvent dependence of the assumed equilibrium between reactants and transition state. The activation energy is then solvent-dependent, and quantities like activation entropy and activation volume are used in the discussion. (This is thoroughly treated in standard textbooks, for example, in ref. 1.) Other effects originate in the intrinsically dynamic interactions between solvent and solute and are thus not amenable to this kind of thermodynamic treatment. The stochastic theories that have expanded so much in these last few years attempt to deal with these more complicated interactions.

The plan of this chapter is as follows. In Section II, the basic ideas of the method of Kramers are reviewed, and recent generalizations, especially the progress made in bridging the two Kramers limits, are discussed. The remainder of the chapter is devoted to discuss two lines of current development of the theory that seem very promising for the interpretation of chemical rate processes in condensed media. Section III deals with the problem of the interaction of the reactive coordinate with other nonreactive modes and establishes a connection with the field of nonequilibrium nonlinear statistical thermodynamics. The difficulties arising from the breakdown of the hypothesis of time-scale separation (non-Markovian effects) which may be very relevant in condensed phase processes are considered in Section IV. The improved physical interpretation that may be achieved by the general strategy that is the subject of this volume is discussed in Section V. We should note that the three cornerstone techniques of the delta-like strategy proposed in Chapter I are used in Sections III and IV.

II.   THE KRAMERS MODEL AND ITS EXTENSIONS

For our purposes, a chemical reaction is viewed as the passage over a barrier of a particle under the influence of random forces originating in its environment. It was Marcelin[9] who first represented a chemical reaction by the motion of a point in phase space, thus using for the first time the rigorous methods of statistical mechanics. He suggested that the course of a chemical reaction could be followed by the trajectory of a point in the $2n$-dimensional space defined by the n position coordinates necessary to describe the reacting system together with the corresponding conjugate momenta.

Inspired by Christiansen's[10] treatment of a chemical reaction as a diffusional problem, Kramers[11] studied the model of a particle in Brownian motion in a one-dimensional force field and predicted the existence of three fundamental kinetic regimes, depending on the magnitude of the friction. The basic hypothesis and results of this work will be summarized below, as many of the results most recently obtained using more sophisticated models are still best described by reference to Kramers' original model and reduce to Kramers models when the appropriate limits are taken.

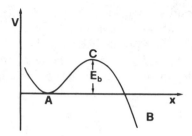

Figure 1.   The Kramers potential.

A.   The Kramers Model

Consider an ensemble of noninteracting particles—the reactant—under the influence of (1) a force derived from an external one-dimensional potential $V(x)$ consisting of a well A and an adjacent barrier C (see Fig. 1) and (2) an irregular force resulting from random collisions between the reactant particles and solvent particles at a given temperature T.

Kramers[11] identified the chemical reaction with the escape over the barrier of the reactant particles initially located in the potential well. The irregular force simulates the interaction with the solvent, which is thus treated as a heat bath.

The motion of a particle (mass M) in the Kramers model may be described by the following Langevin equation:

$$\dot{x} = v \tag{3a}$$

$$\dot{v} = -\gamma v - \frac{1}{M}\frac{\partial V}{\partial x} + \frac{1}{M}F(t) \tag{3b}$$

where γ is the friction coefficient (or damping rate) and $F(t)$ is the irregular force associated with the coupling to the heat bath. This force is assumed to be Markovian, that is, the forces at different times are assumed to be uncorrelated. It may then be defined by

$$\langle F(t) \rangle = 0 \tag{4a}$$

$$\langle F(0)F(t) \rangle = 2\gamma M K_B T \delta(t) \tag{4b}$$

where Eq. (4b) is an expression of the fluctuation-dissipation theorem,[12] which relates the friction γ with the magnitude of the irregular forces acting on the particle.

The Langevin equation (3a, b) is equivalent to the following Fokker-Planck equation which drives the probability distribution in phase space:

$$\frac{\partial}{\partial t}\rho(x,v,t) = \left\{ -v\frac{\partial}{\partial x} + \frac{1}{M}\frac{\partial V}{\partial x}\frac{\partial}{\partial v} + \gamma\left[\frac{\partial}{\partial v}v + \frac{K_B T}{M}\frac{\partial^2}{\partial v^2}\right] \right\}\rho(x,v,t)$$

$$\tag{5}$$

In order to obtain simple analytical results from this equation, Kramers assumed further that (3) the potential is parabolic near A, $V(x) = \frac{1}{2}M\omega_0^2 x^2$, and near C, $V(x) = Q - \frac{1}{2}M\omega_b^2(x - x_c)^2$, and that (4) the height of the barrier is much larger than the thermal energy, $E_b \gg K_B T$, so that the reaction process is slow and quasi-stationary. Under these conditions he was able to obtain the following simple expression for the rate of particle flow over the barrier:

$$k_K = \frac{\gamma \omega_0}{4\pi \omega_b}\left[\left(1 + \frac{4\omega_b^2}{\gamma^2}\right)^{1/2} - 1\right]\exp\left(-\frac{E_b}{K_B T}\right) \qquad (6)$$

It is important to consider two limiting cases where this general expression may be simplified. For small frictions, $\gamma \ll 2\omega_b$, Eq. (6) gives the same expression as that obtained earlier in transition state theory,

$$k_{\text{TST}} = \frac{\omega_0}{2\pi}\exp\left(-\frac{E_b}{K_B T}\right) \qquad \gamma \ll 2\omega_b \qquad (7)$$

The condition of validity of this expression is easily understood. If the time scale of the damping, $1/\gamma$, is much larger than the time scale of the motion atop the barrier, $1/\omega_b$, then the particle will have an effectively free motion in its downhill path out of the well. It should be kept in mind that this is exactly one of the fundamental hypotheses of transition state theory. It would be wrong, however, to conclude that there is no lower limit on the friction for the correct applicability of the TST expression. For extremely low frictions, the coupling to the heat bath is no longer able to maintain the quasi-thermodynamic equilibrium in the well, thus invalidating an assumption made by Kramers to derive Eq. (6) and also the underlying conventional TST. For this extreme low-friction region, Kramers[11] was able to calculate the rate by converting the Fokker-Planck equation [Eq. (5)] into a diffusion equation for the energy; the exchange of energy between the heat bath and the particle is the rate-limiting step in these conditions. The following approximate rate equation was obtained:

$$k_{\text{low}} = \gamma \frac{E_b}{K_B T}\exp\left(-\frac{E_b}{K_B T}\right) \qquad (8)$$

This energy diffusion process should apply when the characteristic time of damping, $1/\gamma$, is much larger than the time of equilibrium escape of a particle from the well, $1/k_{\text{TST}}$.

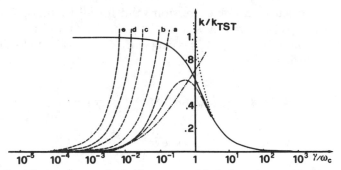

Figure 2.   The reaction rate in the Kramers model relative to its transition state value. The low friction rate is plotted for $(\omega_b/\omega_0)(E_b/K_BT) = 1$ (a), 2 (b), 5 (c), 10 (d), and 20 (e). Also plotted is the interpolated rate of Eq. (13) and that calculated by Buttiker, Harris, and Landauer.[13] (——) k_K/k_{TST}; ($\cdots$) k_{high}/k_{TST}; (---) k_{low}/k_{TST}; (———) k_{int}/k_{TST}; (—·—) k_{BHL}/k_{TST}.

Kramers[11] suggested that transition state theory should apply in the range of frictions $(K_BT\omega_0/2\pi E_b) \lesssim \gamma \lesssim \frac{1}{5}\omega_b$, the lowest limit corresponding to the point where expressions (7) and (8) give the same rate value.

The general Kramers expression (6) may be simplified in the region of very high friction, $\gamma \gg 2\omega_b$, where a purely diffusive regime is attained,

$$ k_{high} = \frac{\omega_0\omega_b}{2\pi\gamma} \exp\left(-\frac{E_b}{K_BT} \right) \tag{9} $$

The plots in Fig. 2 suggest that the limits of validity of transition state theory may be fairly narrow or altogether nonexistent, contrary to the prediction made by Kramers. The nonequilibrium effects duly treated for extremely low friction may start being felt before the TST plateau is approached. This is more likely to occur for the lower barriers and larger ratios ω_0/ω_b.

The following argument may help in understanding the connection between the extreme low-friction regime and the diffusive one. Kramers identified very clearly the two processes that determine the rate of escape: the thermal escape out of the bottom of the well and the actual diffusive crossing of the barrier. The slowest of the two becomes the limiting step and determines the overall rate given by Eqs. (8) and (6), respectively. If we assume that these are successive processes according to the scheme

$$ A \underset{k_l^E}{\overset{k_r^E}{\rightleftharpoons}} A^* \overset{k^d}{\rightarrow} B \tag{10} $$

we can easily calculate a rate expression valid for all frictions. The overall rate of scheme (10) is given by $k_{\text{int}} = k^d k_r^E / (k_l^E + k^d)$ and the Boltzmann equilibrium requires that $(k_r^E/k_l^E) = \exp(-E_b/K_B T)$. The following limits are clearly satisfied:

$$k^d \gg k_l^E \qquad k_{\text{int}} = k_l^E \exp\left(-\frac{E_b}{K_B T}\right) = k_{\text{low}} \tag{11}$$

$$k^d \ll k_l^E \qquad k_{\text{int}} = k^d \exp\left(-\frac{E_b}{K_B T}\right) = k_K \tag{12}$$

and the general expression for the overall rate may be written as

$$k_{\text{int}}^{-1} = k_{\text{low}}^{-1} + k_K^{-1} \tag{13}$$

The results of this two-step model are also shown in Fig. 2.

Büttiker, Harris, and Landauer[13] refined the Kramers treatment of the low-friction case, allowing for a nonzero density of particles at the energy of the barrier, and obtained an expression for the rate k_{BHL} which may be cast in the form

$$k_{\text{BHL}} = \frac{[1+4k_{\text{TST}}/k_{\text{low}}]^{1/2}-1}{[1+4k_{\text{TST}}/k_{\text{low}}]^{1/2}+1} k_{\text{low}} \tag{14}$$

This expression converges to k_{TST} for high friction and starts correcting k_{low} according to

$$k_{\text{BHL}} \sim \left\{1-\left(\frac{k_{\text{low}}}{k_{\text{TST}}}\right)^{1/2}+\frac{1}{2}\left(\frac{k_{\text{low}}}{k_{\text{TST}}}\right)-\frac{1}{8}\left(\frac{k_{\text{low}}}{k_{\text{TST}}}\right)^{3/2}+\frac{1}{8}\left(\frac{k_{\text{low}}}{k_{\text{TST}}}\right)^2 \cdots\right\} k_{\text{low}} \tag{15}$$

while Eq. (13) introduces a correction of the form

$$k_{\text{int}} \sim \left\{1-\left(\frac{k_{\text{low}}}{k_K}\right)+\left(\frac{k_{\text{low}}}{k_K}\right)^2 \cdots\right\} k_{\text{low}} \tag{16}$$

As may be seen by comparing Eqs. (15) and (16) and also by inspection of the plots in Fig. 2, our interpolating expression gives a rate higher than that calculated by the method of Büttiker, Harris, and Landauer[13], but for higher frictions it approaches the Kramers function in the correct way. (See Section II.B for further discussion of this point.)

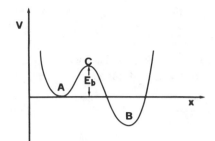

Figure 3.   The unidimensional bistable potential.

B.   The Bistable Model and Other Generalizations
of the Kramers Method

In the model studied by Kramers,[11] the particles are assumed to be initially at the well around A and to be lost as they escape above the barrier. Many physical processes, however, are more realistically modeled by a bistable potential (see Fig. 3), namely when two states A and B may be interconverted. In the original Kramers model no back-crossings from B to A were considered; the particles were somehow absorbed as they arrived at B.

$$A \overset{k_r}{\underset{k_l}{\rightleftharpoons}} B \qquad (17)$$

It should be noted that states A and B are not well-defined states but rather probability distributions around the potential minima A and B. For high barriers like those assumed originally by Kramers, there should be no ambiguities, but one should be careful when dealing with small barriers. One way to deal with this problem rigorously is to work with the eigenvalues of the operator driving the probability distribution in time. For simplicity, consider the case of a symmetric potential, and let $\varphi_n(x)$ be the eigenfunctions associated with eigenvalues λ_n. The following interpretation emerges from an interesting paper by van Kampen[14]: $\varphi_0(x)$ ($\lambda_0 = 0$) is the equilibrium distribution; the lowest nonzero eigenvalue, λ_1, is usually the one defining the chemical relaxation rate, as it corresponds to the slowest time scale, and its associated eigenfunction $\varphi_1(x)$ is antisymmetric. To this first level of approximation, the probability distribution is given by $\varphi_0(x) +$ $\varphi_1(x)\exp(-\lambda_1 t)$ and describes the evolution from $t = 0$, when the two functions may cancel each other in the right-hand well, up to the final equilibrium distribution $\varphi_0(x)$. It is easy to see that $\lambda_1 = k_r + k_l$, and the individual one-way rate constants may be determined if the equilibrium constant is known as well. Other methods of avoiding this ambiguity consist of calculating the expectation value of the position, $\langle x \rangle$, (see, e.g., ref. 15) or the total population in a well defined by $n_A(t) = \int_{-\infty}^{x_c} dx\, \rho(x, t)$ (see, e.g., ref. 16) and look for the time evolution of these variables.

Brinkman[17] considered the bistable potential problem and showed that the diffusive, very-high-friction regime of Kramers was still correct,

$$k^r_{high} = \frac{\omega_A \omega_c}{2\pi\gamma} \exp\left(- \frac{V_c - V_A}{K_B T} \right) \tag{18}$$

$$k^l_{high} = \frac{\omega_B \omega_c}{2\pi\gamma} \exp\left(- \frac{V_c - V_B}{K_B T} \right) \tag{19}$$

Instead of the quasi-stationary state assumption of Kramers, he assumed only that the density of particles in the vicinity of the top of the barrier was essentially constant. Visscher[18] included in the Fokker-Planck equation a source term to account for the injection of particles so as to compensate those escaping and evaluated the rate constant in the extreme low-friction limit. Blomberg[19] considered a symmetric, piecewise parabolic bistable potential and obtained a partial solution of the Fokker-Planck equation in terms of tabulated functions; by requiring this piecewise analytical solution to be continuous, the rate constant is obtained. The result differs from that of Kramers only when the potential has a sharp, nonharmonic barrier.

Brinkman,[17] Landauer and Swanson,[20] and Donnelly and Roberts[21] made important progress in extending Kramers' method to models with several spatial dimensions. For the relatively simple models that were worked out, the major conclusions attained by Kramers do hold well. (A more detailed discussion of this point is given in the next section.)

Van Kampen[14] presented a detailed analysis of a specialized one-dimensional, symmetric double-well potential and obtained expressions for the eigenfunctions and eigenvalues of the associated Smoluchowski equation. He was able to reproduce and correct the Kramers result in the diffusional limit and clarified the various relaxation processes that occur in the different time scales of a reaction process with a high barrier. Taking as initial distribution a delta-like function placed at the bottom of one well, which is equivalent to considering a linear combination of infinitely many excited states, he showed that a quasi-equilibrium is attained after an initial fast relaxation process; this quasi-equilibrium consists of an equally weighted linear combination of the ground state (the equilibrium distribution) and the first excited state. The eigenvalue of the first excited state corresponds to the Kramers rate of escape. This shows how the Kramers theory gives a satisfactory description of the slow escape process, while an accurate picture of the faster processes, consisting mainly in the initial relaxation inside the well, would require the evaluation of an overwhelming number of excited states.

Larson and Kostin[22] considered a symmetric double-well potential and solved the Fokker-Planck equation driving the probability distribution in phase space, assuming the barrier to be high. Three cases were considered: the diffusional limit, where the Smoluchowski equation may be used; an intermediate range of the friction coefficient γ; and the limit of very low friction. A variational approach was used to calculate the eigenvalues, and the eigenfunctions were obtained by a perturbational technique. For high and intermediate values of the friction, asymptotic formulas for the rate are given, their accuracy being tested against numerical calculations. Starting from the limit $\gamma \to 0$, they proposed a semiempirical expression apparently valid for all frictions and suggested that the lower limit of validity of TST would be far higher than predicted by Kramers.[11] (See also the discussion at the end of Section II.A.)

Several other attempts have been made to derive general expressions for the chemical rate, valid from the extreme low-friction regime to the moderate and high friction ones. Earlier, Visscher[23] had performed numerical calculations in the transition region between low and intermediate friction regimes and fitted a one-parameter expression which appeared to cover the entire range of frictions. Skinner and Wolynes[16] constructed a sequence of Padé approximants from the analytical results known for small friction and large friction; although some theoretical difficulties may arise with this use of the approximants, the results obtained seem very satisfactory. The same technique was applied very recently by Garrity and Skinner.[24] Montgomery, Chandler, and Berne[25] used a stochastic dynamics trajectory method to solve the bistable either piecewise harmonic or piecewise constant potential and found that the actual rate was always below 50% of the TST value. (For comparison with Fig. 2, we note that the parameters taken correspond to $E_b/K_B T = 4.9$ and, for the piecewise harmonic potential, $\omega_0/\omega_b = 3.05$.)

Büttiker, Harris, and Landauer[13] extended the treatment made by Kramers[11] for the extreme low-friction regime to take into account the effect that the flow of particles out of the well has on their distribution inside the well. They obtained the rate expression of Eq. (14), valid from the extreme low-friction region up to intermediate friction but converging to the TST value (see Fig. 2). Carmeli and Nitzan[26] proposed a new approach based on a division of the particle phase space into two overlapping regions. In the first, for the lower energies deep inside the wells, the variation of phase is assumed to be much faster than that of the energy, and a diffusion equation for the energy will hold. The second corresponds to the higher energy region near the top of the barrier, where a spatial diffusion of the particles may be assumed. The final expression for the rate, k_{CN}, may be written in the form

$$k_{CN}^{-1} = \tau_1 + s k_K^{-1} \tag{20}$$

400   T. FONSECA ET AL.

where τ_1 is the mean first passage time for the particle to reach the boundary between the two regions referred above, k_K is the Kramers rate given by Eq. (6), and s is a complicated factor assuming values between $\frac{1}{2}$ (for $\gamma \to 0$) and 1 (for large γ). This method and the final result, Eq. (20), should be compared with the very simplistic two-step model discussed at the end of Section II.A. The factor s now introduced makes the rate k_{CN} always larger than k_{int} but closely related to it. Moving away from the smallest frictions, the corrections introduced are of the form

$$k_{CN} \sim \left\{ 1 - s\left(\frac{k_{low}}{k_K}\right) + s^2\left(\frac{k_{low}}{k_K}\right)^2 \cdots \right\} k_{low} \qquad (21)$$

if we identify $\tau_1^{-1} \approx k_{low}$. As the factor s is close to $\frac{1}{2}$ in this region, it is clear how k_{CN} is closer to k_{low} (or more rigorously, to τ_1^{-1}) than k_{int}. A factor $s = \frac{1}{2}$ for the first-order correction in the form of Eq. (21) had been proposed earlier by Visscher[23] to fit the numerical results of a Kramers-type model that includes a source term. This should be contrasted with the refined Kramers treatment of Büttiker, Harris, and Landauer,[13] which overcorrects k_{low} by comparison with the two-step model. [See Eqs. (14) and (15) and Fig. 2.] The approach of Carmeli and Nitzan has been generalized by the same authors to the non-Markovian case, but this is the subject of Section IV.

Very recently, Lavenda[27] devised an interesting method of solution of the Kramers problem in the extreme low-friction limit. He was able to show that it could be reduced to a formal Schrödinger equation for the radial part of the hydrogen atom and thus be solved exactly. One particular form of the long-time behavior of the rigorous rate equation coincides with that obtained by Kramers with the quasi-stationary hypothesis and may thus clarify the implications of this hypothesis. The method of Lavenda is reminiscent of that used by van Kampen[14] but applied to a Smoluchowski equation for the diffusion of the energy.

III. MULTIMODAL THEORIES

In the Kramers[11] model, the reaction process is described by the motion of a particle along a single coordinate. This is what in the jargon of chemical kinetics is called the reaction coordinate, a concept lacking rigorous definition in most cases. In actual problems of chemical interest, the barrier may be fairly wide near the saddle point and, besides, the normal mode separation may break down in that region. Real systems do usually require a many-coordinate description, and the coupling among these modes may play an important role in the rate process. Landauer and Swanson[20] extended Kramers' work to the general multidimensional case to find that in the diffusive regime (high friction) the rate expression showed the same deviation from the TST value as that found in one dimension. In the other extreme case, for very low frictions, however, there appeared to be an effect of dimensionality. It is the aim of this section to evaluate the results obtained with multimodal theories, and we start by discussing in Section III.A two interesting attempts to deal with more detailed models, one to bring in the effects of the solvent, the other to deal directly with a two-dimensional coupled system. Later, in Section III.B, another detailed model is presented which aims to supplement the results of these two works.

A. Two Detailed Models

The two particular models of mode coupling that we shall briefly discuss in this subsection are illuminating about the many different mechanisms that are involved and the difficulty in establishing a general simple pattern.

Grote and Hynes[28] studied a model for an exchange reaction in solution,

$$A + BC \rightarrow AB + C \tag{22}$$

assuming that the motion in the saddle region is separable into reactive and nonreactive normal modes. The solvent dynamics act on the motion on each mode and may also induce a dynamical coupling among them. In the particular case of Eq. (22), the reactive mode is the antisymmetric stretch of the molecular system ABC. For example, it is easy to see that the solvent reaction forces upon the translational mode (one of the nonreactive normal modes) will couple this one into the reactive mode. This coupling may have three sources: (1) the different masses of the atoms, (2) the different friction on the central atom relative to the more exposed external atoms, and (3) the cross correlation between the atomic forces. Grote and Hynes described the motion on each coordinate q_i by a generalized Langevin equation of the type

$$\ddot{q}_i(t) = -\omega_i^2 q_i(t) - \sum_j \int_0^t d\tau\, \gamma_{ij}(\tau)\dot{q}_j(t-\tau) + F_i(t) \tag{23}$$

where the frequency ω_i is imaginary for the reactive mode. They found that, except for the limiting cases of very high and very low friction, the rate of the reaction would depend very markedly on the assumed friction kernels $\gamma_{ij}(\tau)$. [It should be kept in mind that these are related to the correlation functions of the solvent forces, $\gamma_{ij}(\tau) = \langle F_i(0)F_j(\tau)\rangle/K_B T$.] Moreover, the mode coupling reduced the effective friction that was "felt" on the reactive mode. This shows how important and complex may be the role played by the solvent in determining the reaction rate.

Another source of coupling between the reactive and the other modes may result from the shape of the potential of the (solvent free) reacting system. A particular case of this class was studied by Christoffel and Bowman,[29] who considered a two-dimensional potential based on that of ammonia,

$$V(x, y) = \left[\tfrac{1}{2}ax^2 + \tfrac{1}{2}bx^4 + V_0\exp(-cx^2)\right] + \tfrac{1}{2}m\left[\omega_y(x)\right]^2 y^2 \quad (24a)$$

with

$$\omega_y(x) = \omega_0\left[1 - \lambda\exp(-\alpha x^2)\right] \quad (24b)$$

This has the form of a double-well oscillator coupled to a transverse harmonic mode. The adiabatic approximation was discussed in great detail from a number of quantum-mechanical calculations, and it was shown how the two-dimensional problem could be reduced to a one-dimensional model with an effective potential where the barrier top is lowered and a third well is created at the center as more energy is pumped into the transverse mode. From this change in the reactive potential follows a marked increase in the reaction rate. Classical trajectory calculations were also performed to identify certain specifically quantal effects. For the higher energies, both classical and quantum calculations give parallel results.

B.   The Coupled Double-Well Oscillator

In this subsection we extend Christoffel and Bowman's investigation to the condensed phase. This is done within a classical context reminiscent of the work of Grote and Hynes,[28] and we make extensive use of both AEP and CFP (see the first four chapters of this volume). A more detailed account is given by Fonseca et al.[30]

Consider a bidimensional model potential,

$$V(x, y) = \phi(x) + \omega_{\text{eff}}^2(x)y^2 \quad (25)$$

where x is the reaction coordinate and y is some transverse normal mode, $\phi(x)$ is a symmetric double-well potential modeling the chemical reaction, and

$$\omega_{\text{eff}}(x) = \left[\tfrac{1}{2}\omega_0^2 + \psi(x)\right]^{1/2} \quad (26)$$

with

$$\psi(x) = -\frac{\omega_0^2}{2}\lambda_{\text{int}}\exp\left(-\frac{x^2}{r^2}\right) \quad (27)$$

λ_{int} and r may be regarded as measures of the intensity and the range, respectively, of the coupling of the transverse mode onto the reactive motion. For Grote and Hynes' assumption on the mode separability in the saddle region to be valid, a fairly large value of r is required. In fact, when $r \gg a$ ($2a$ is the distance between the two minima of the reactive potential), the effect of the deterministic coupling can be viewed as a simple upward translation of the double-well potential on the energy axis; for $r \lesssim a$, however, the reaction coordinate is driven by an effective potential which has a smaller barrier and, in some cases, a third well, an effect already found in Christoffel and Bowman's work.

The classical motion of a stochastic particle in the potential defined by Eq. (25) may be described by the following set of equations:

$$
\begin{aligned}
\dot{x} &= v \\
\dot{v} &= -\phi'(x) - \gamma v - \psi'(x)y^2 + f(t) \\
\dot{y} &= w \\
\dot{w} &= -\lambda w - \omega_0^2 y - 2y\psi(x) + f'(t)
\end{aligned}
\tag{28}
$$

The stochastic forces $f(t)$ and $f'(t)$ are assumed to be of the form of Gaussian white noises and to be statistically uncorrelated; this means that the coupling between reactive and nonreactive modes via the solvent is completely neglected. However, the noise affecting the nonreactive mode is transmitted into the reactive one originating the appearance of multiplicative noise effects. Although the Fokker-Planck equation corresponding to the set of Eqs. (28) may be written straightforwardly, its explicit solution involves some technical difficulties. In order to avoid these difficulties we shall make a set of assumptions similar to those of Christoffel and Bowman. The pair of variables (y, w) is assumed to be much faster than the pair (x, v); if this condition applies, the AEP can be applied to obtain a simpler Fokker-Planck equation depending only on the slow variables, and the CFP can be used to determine the time evolution of the observables driven by that equation. This kind of approach allows us to determine the rate constant for the chemical process under investigation in the following two different physical situations:

1. *System in Thermal Equilibrium.* The two modes have available the same thermal energy, and in this case, we study the whole range of values of the friction γ on the reactive mode.
2. *System Being Excited.* We assume that the nonreactive mode can be continuously heated by an external source without affecting the reactive one, thereby creating a physical situation where a canonical equilibrium does not exist.

In both of these physical situations we assume the nonreactive mode to be overdamped, with the friction λ so large (with respect to ω_0) as to allow the set of Eqs. (28) to be replaced by

$$\dot{x} = v$$

$$\dot{v} = -\phi'(x) - \gamma v - \psi'(x) y^2 + f(t) \tag{29}$$

$$\dot{y} = -\frac{\omega_0^2}{\lambda} y - \frac{2}{\lambda} \psi(x) y + \frac{f'(t)}{\lambda}$$

The AEP is applied to the set of Eqs. (29) in order to obtain from its equivalent Fokker-Planck equation,

$$\frac{\partial}{\partial t} \rho(x, v, y, t) = \left\{ -v \frac{\partial}{\partial x} + \phi'(x) \frac{\partial}{\partial v} + \psi'(x) y^2 \frac{\partial}{\partial v} \right.$$

$$+ \gamma \left[\frac{\partial}{\partial v} v + \langle v^2 \rangle \frac{\partial^2}{\partial v^2} \right] + \frac{2}{\lambda} \psi(x) \frac{\partial}{\partial y}$$

$$\left. + \frac{\omega_0^2}{\lambda} \left[\frac{\partial}{\partial y} y + \langle y^2 \rangle \frac{\partial^2}{\partial y^2} \right] \right\} \rho(x, v, y, t) \tag{30}$$

the one describing the time evolution of the probability distribution of the slow variables, $\sigma(x, v, t)$:

$$\frac{\partial}{\partial t} \sigma(x, v, t) = \left\{ -v \frac{\partial}{\partial x} + \frac{\partial}{\partial v} \phi'(x) + \gamma \frac{\partial}{\partial v} v \right.$$

$$+ \frac{\partial}{\partial v} \psi'(x) \langle y^2 \rangle - \frac{2}{\omega_0^2} \frac{\partial}{\partial v} \langle y^2 \rangle \psi'(x) \psi(x)$$

$$\left. + \gamma \langle v^2 \rangle \frac{\partial^2}{\partial v^2} + \frac{\partial^2}{\partial v^2} \langle y^2 \rangle^2 \frac{\lambda}{\omega_0^2} [\psi'(x)]^2 \right\} \sigma(x, v, t) \tag{31}$$

(This result is obtained by using corrections up to the second perturbational order; for a detailed discussion of how the perturbation parameter is defined, see Chapter I.)

It is illuminating to briefly discuss the significance and importance of each term in Eq. (31). The first three terms are trivial, as they are nothing but a description of the deterministic evolution in the reactive mode in the absence of coupling with the nonreactive mode; the sixth term is also trivial and is a diffusional term corresponding to the effect of the stochastic force

$f(t)$ over the reactive mode. More interesting are the remaining terms: in fact, the fourth term is equivalent to the standard adiabatic correction to the reactive potential found by Christoffel and Bowman and arises as a consequence of the deterministic coupling of the reactive mode to the nonreactive mode. When one applies the AEP to the set of Eqs. (29), this term appears as the first-order correction. The fifth and seventh terms appear as the second-order correction provided by the AEP; the fifth is a nonstandard adiabatic correction to the reactive potential, and the seventh is a multiplicative diffusional term that transmits to the reactive mode the effect of the thermal fluctuations acting on the nonreactive mode. It can also be proved that higher order corrections provided by the AEP will generate the true effective potential "felt" by the reactive mode.

Equation (31) is valid when the characteristic times of the position x and velocity v are similar; when both v and y are assumed to be fast variables, the AEP applied to the set of Eqs. (29) leads to

$$
\frac{\partial}{\partial t}\sigma(x,t) = \left\{ \frac{1}{\gamma}\frac{\partial}{\partial x}\left[\phi'(x) + \langle y^2 \rangle \psi'(x) \right.\right.
$$

$$
\left. -2\frac{\langle y^2 \rangle}{\omega_0^2}\psi'(x)\psi(x) - \frac{\langle y^2 \rangle^2}{\gamma^2 R_1}\psi'(x)\psi''(x) \right]
$$

$$
+ \frac{1}{\gamma}\frac{\partial^2}{\partial x^2}\left[\langle v^2 \rangle + \frac{\langle y^2 \rangle^2}{\gamma^2 R_1}\left[\psi'(x)\right]^2 \right]
$$

$$
+ \frac{1}{\gamma}\frac{\partial}{\partial x}\left[\frac{2\langle y^2 \rangle^2}{\gamma^2(1+2R_1)}\psi'(x)\psi''(x) \right.
$$

$$
\left.\left. + 4\frac{\langle y^2 \rangle}{(\omega_0^2)^2}\psi'(x)\psi(x)^2 \right]\right\}\sigma(x,t) \tag{32}
$$

where $R_1 = (\omega_0^2/\lambda)/\gamma \equiv \tau_v/\tau_y$. This equation was obtained taking into account corrections up to the fourth order on the AEP and considering v as a fast variable but not infinitely fast when compared to x. It must be noted that the last term in this equation is, with respect to Eq. (31), the next non-standard adiabatic correction to the reactive potential. It is interesting to study how Eq. (32) behaves with γ or more directly with R_1; when R_1 assumes large values, which is equivalent to taking small values of γ and therefore making the system more inertial, the fourth and seventh terms of Eq. (32) cancel each other out, and the resulting equation is equivalent to that obtained starting from Eq. (31) and eliminating (with the AEP) the velocity v. When R_1 tends to zero, γ tends to infinity, and the diffusional limit is approached; in this limit, Eq. (32) can be rewritten as

$$
\frac{\partial}{\partial t}\sigma(x,t) = \left\{ \frac{1}{\gamma}\frac{\partial}{\partial x}\left[\phi'(x) + \psi'(x)\langle y^2\rangle \right.\right.
$$

$$
\left. -2\frac{\langle y^2\rangle}{\omega_0^2}\psi'(x)\psi(x) + 4\frac{\langle y^2\rangle}{(\omega_0^2)^2}\psi'(x)\psi(x)^2 \right]
$$

$$
-\frac{1}{\gamma}\frac{\partial}{\partial x}\frac{\langle y^2\rangle^2}{\gamma^2 R_1}\psi'(x)\psi''(x)
$$

$$
\left. +\frac{1}{\gamma}\frac{\partial^2}{\partial x^2}\left[\langle v^2\rangle + \frac{\langle y^2\rangle^2}{\gamma^2 R_1}[\psi'(x)]^2 \right] \right\}\sigma(x,t) \qquad (33)
$$

This same equation can be obtained from the set of Eqs. (29) assuming v infinitely faster than x. This touches the Itô-Stratonovich controversy discussed by Faetti et al. in Chapter X (note that R_1 is to be identified with their parameter R^{-1}). In line with their remarks, we are led to the conclusion that when the system becomes inertial the Itô description is valid [see Eq. (31)] and that when inertia is completely absent [see Eq. (33)] the Stratonovich description is attained.

The results obtained considering that the system is thermalized can be summarized in Fig. 4, where the chemical reaction rate k is displayed as a function of R_1 ($= \tau_v/\tau_y$). τ_y was kept constant, and therefore this figure exhibits the same kind of k dependence on γ (γ is the friction acting on the reactive mode) as that already discussed in Section II.A. Note, however, that new effects originating from the coupling between reactive and nonreactive modes appear in this case, as will be discussed later on. When $R_1 \to 0$, the high-friction region is attained and a linear dependence of k on $1/\gamma$ is obtained, in agreement with the classical Kramers result. As R_1 increases, the system becomes more inertial, and it is also interesting to note that as λ_{int} increases, straight lines of increasing slope are obtained. This is a manifestation of the role played by inertia: the sensitivity of the reaction rate k to the intensity of the coupling increases as the reactive system becomes more iner-

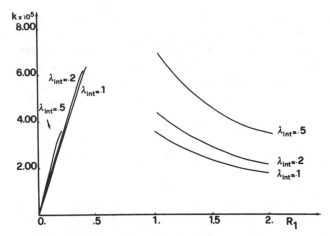

Figure 4. Synergism of inertia and multiplicative fluctuation. Variation of the reaction rate k throughout the whole range of values of the friction γ [$R_1 = (\omega_0^2/\lambda)/\gamma$, and ω_0^2/λ was kept constant]. The curves on the left-hand side were obtained using the CFP, and k was identified with the inverse (at the origin) of the Laplace transform describing the time evolution of the average value of the position $\langle x \rangle$. Those on the right-hand side were obtained by the first-passage time technique, and k was identified with the inverse of that time. The parameters were given the values $E_b = 2 \times 10^{-7}$, $a = 0.5$, $r = 2a$, and $\langle y^2 \rangle \omega_0^2 = \langle v^2 \rangle = \langle w^2 \rangle = 10^{-7}$. Atomic units are used throughout. [Taken from T. Fonseca et al., J. Chem. Phys., **80**, 1826 (1984).]

tial. Above a certain value of R_1, the lines start bending down, a sign that the intermediate friction regime is being approached (see introduction). Unfortunately, we had difficulty with the convergence of the CFP in this region, and therefore there is little reliance to be placed on results provided by those computer calculations.

As R_1 tends to infinity, the energy-controlled regime is approached and the important role played by the interaction between reactive and nonreactive modes can be assessed by some remarks on Eq. (31). Let us consider the case where $\langle y^2 \rangle = 0$. If $\gamma \langle v^2 \rangle$ is also assumed to vanish, Eq. (31) describes a purely deterministic process, and the overcoming of the barrier is rigorously forbidden when the total energy of the reactant is lower than the barrier height. However, when the coupling between reactive and nonreactive modes is restored, the reactant undergoes the influence of the fluctuations acting on the nonreactive mode, and this can supply enough energy for the reactant to overcome the barrier. Fluctuations become ineffective near the top of the barrier, where their intensity vanishes as implied by $\psi'(0) = 0$. This means that inertia is absolutely necessary for the barrier to be really overcome. As a result of such a synergism between inertia and multiplicative fluctuations, the chemical reaction can take place even when Kramers' theory predicts vanishingly small rates. This is an interesting property, a quantitative discussion of which requires a point of view completely different from the one considered until now. To derive a Fokker-Planck equation for the energy, we follow Lindenberg and Seshadri,[31] who used energy and displacement as independent variables. We define the energy as

$$E = \frac{v^2}{2} + \Theta(x) \tag{34a}$$

$$\Theta(x) = \phi(x) + \psi(x)\langle y^2 \rangle - [\psi(x)]^2 \frac{\langle y^2 \rangle}{\omega_0^2} \tag{34b}$$

In the absence of the additive and multiplicative stochastic forces, E would be a constant of motion, rigorously independent of time. Under the influence of these fluctuations, E becomes time-dependent, but its dynamics will certainly be very slow compared to the dynamics of the variable x, thus allowing us to also eliminate the space variable. Starting from Eq. (31), we rewrite it in terms of the new pair of variables, x and E, and after eliminating x with a procedure introduced by Stratonovich,[32] the following final equation is obtained:

$$\frac{\partial}{\partial t}\sigma(E,t) = \left\{ -\frac{\partial}{\partial E}\left[-\gamma\frac{\varphi(E)}{\varphi'(E)} + \gamma\langle v^2 \rangle + \frac{\lambda_{int}^2}{r^4}\langle y^2 \rangle^2 \frac{\lambda}{\omega_0^2}\frac{\chi'(E)}{\varphi'(E)} \right] \right.$$
$$\left. + \frac{1}{2}\frac{\partial^2}{\partial E^2}\left[2\gamma\langle v^2 \rangle \frac{\varphi(E)}{\varphi'(E)} + 2\frac{\lambda_{int}^2}{r^4}\langle y^2 \rangle^2 \frac{\lambda}{\omega_0^2}\frac{\chi(E)}{\varphi'(E)} \right] \right\}\sigma(E,t) \tag{35}$$

where

$$\varphi(E) = \int dx\, [E - \Theta(x)]^{1/2} \tag{36a}$$

$$\chi(E) = \int dx\, x^2 e^{-x^2/r^2}[E - \Theta(x)]^{1/2} \tag{36b}$$

with the integration extending over a domain that includes all values of x for which $E \geq \Theta(x)$.

To evaluate the chemical reaction rate via Eq. (35), we adopt the first-passage time method,[32,33] identifying k with the inverse of the mean first-passage time. The results are displayed on the right-hand side of Fig. 4 and in Fig. 5.

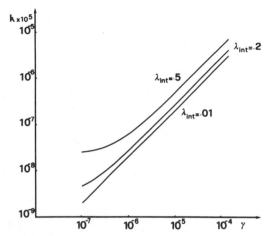

Figure 5. Variation of the reaction rate in the low-friction region, $\gamma \to 0$. These results were obtained using the first-passage time technique. Parameters were given the same values as in Fig. 4. [Taken from T. Fonseca et al., J. Chem. Phys., **80**, 1826 (1984).]

The results illustrated on the right-hand side of Fig. 4 show that in this region the increase of k is much more sensitive to the increases in λ_{int} than it is in the high-friction region, thereby corroborating our statements about the role of inertia. This trend is especially emphasized in the limit $\gamma \to 0$ and is better seen in Fig. 5. As remarked above, the reaction rate stays finite in this zero-friction limit, counter to Kramers' prediction.

Until now we have limited ourselves to study the thermalized system, that in physical condition (1) cited at the very beginning of this subsection. When we assume that the nonreactive mode may be continuously heated by an external source, the system ceases to be thermalized, and interesting new effects can occur as a consequence of the coupling between reactive and nonreactive modes. Returning to Eq. (32) we can guess what really happens when $\omega_0^2 \langle y^2 \rangle$ is increased: on the one side, the deterministic effect over the reactive potential increases and consists of lowering the barrier to be overcome. However, and in addition to this effect, the intensity of the multiplicative fluctuations is increased with respect to the intensity of the additive fluctuations; this creates a gradient of temperature inside the reactant well that pushes the reactant particles to the region near the barrier while supplying them with energy. This effect vanishes at $x = 0$ (the barrier top), but due to the presence of the additive fluctuations the reaction occurs with a velocity that is much faster than in the absence of this effect. If we continue to increase the energy of the nonreactive mode, a threshold region is attained when the deterministic counterpart of the multiplicative diffusional term

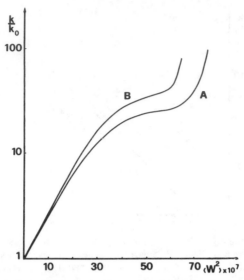

Figure 6. Variation of the reaction rate while increasing the energy in the nonreactive mode. Curve A corresponds to results obtained when the role of inertia was completely neglected. Curve B, in turn, corresponds to results obtained when these effects were present. The parameters were given the values $\lambda_{int} = 0.5$, $E_b = 2 \times 10^{-7}$, $\langle v^2 \rangle = 1 \times 10^{-7}$, $R_1 = \frac{1}{3}$, $a = 0.5$, and $r = 2a$. $k_0 = 1.2 \times 10^{-5}$ is the k value when $\langle v^2 \rangle = \langle w^2 \rangle$. [Taken from T. Fonseca et al., J. Chem. Phys., **80**, 1826 (1984).]

equals the frequency corresponding to the harmonic expansion of the effective potential around the top of the barrier. In the absence of additive fluctuations, it is well known after the work of Schenzle et al.[34,35] that this threshold corresponds to centering the probability distribution at the top of the barrier, and in chemical language we can roughly identify this with an activation process. When the threshold is passed and we continue to pump energy into the nonreactive mode, the probability distribution tends to become still more concentrated on the top of the barrier, rendering the chemical reaction even faster. The results obtained in this particular physical condition are displayed in Fig. 6.

Curve A was obtained using Eq. (33), that is, completely neglecting the role of inertia; curve B was obtained using Eq. (32), where these effects are present. The increase in the reaction rate is very clear; the threshold region corresponds to the plateau, and the increase of k after this region is much more marked. Once again the role of inertia is to speed up the chemical reaction, and this seems to imply that the threshold condition would be attained at lower values of the energy given to the nonreactive mode.

IV. NON-MARKOVIAN EFFECTS ON THE RATE

In this section we shall explore a different kind of generalization of the Kramers theory to take into account the problems resulting from the breakdown of the time-scale separation between the reactive mode and its thermal bath. This problem may also be found in the multimodal theories in Section III when the nonreactive modes are not much faster than the motion along the reaction coordinate.

Computer simulations of the molecular dynamics of the liquid state[36-39] (see also Chapter VI) clearly show that the correlation function of the velocity variable is not exponential; rather it usually exhibits a sort of damped oscillatory behavior. This means that the Markovian assumption is often invalid. This makes it necessary, when studying a chemical reaction in a liquid phase, to replace the standard Kramers condition [see Eq. (4b)] with a more realistic correlation function having a finite lifetime. Recall the rate expression obtained by Kramers for moderate to high frictions, Eq. (6). This may be cast into the form $k_K = k_{TST} f(\omega_b, \gamma)$, where k_{TST}, given by Eq. (7), is essentially an equilibrium property depending on the thermodynamic equilibrium inside the well. As a canonical equilibrium property, it is not affected by whether or not the system is Markovian. The calculation of the factor $f(\omega_b, \gamma)$ depends, however, on the dynamics of the system and will thus be modified when non-Markovian behavior is allowed for.

Another problem of interest concerns the effect of external radiation fields. In the overdamped regime this will be shown to be reminiscent of the effect of the nonreactive modes. These problems will be the major topics of the present section.

This section is organized as follows: in subsection A the approaches based on the assumption of heat bath statistical equilibrium and those which use the generalized Langevin equation are reviewed for the case of a bounded one-dimensional Brownian particle. A detailed analysis of the activation dynamics in both schemes is carried out by adopting AEP and CFP techniques. In subsection B we shall consider a case where the non-Markovian character of the variable velocity stems from the finite duration of the coherence time of the light used to activate the chemical reaction process itself.

A. Noise-Activated Escape Rate in the Presence of Memory Effects

To discuss the idea of noise-activated reactions we begin by noting that the random forces which occur in the Langevin equation related with the process under investigation may have quite different origins. In an ordinary microscopic derivation of a Langevin equation (or the corresponding Fokker-Planck equation), the random term is interpreted as associated with the thermal fluctuations of the system. This thermal or internal noise scales with the size of the system (except near instability points).[40-43] A different interpretation of such a contribution to a Langevin equation is necessary, however, when this is thought to model what can be defined as an external noise. In this latter case, one considers a system which experiences fluctuations that are not "self-originating." These fluctuations can be due to a fluctuating environment or can be the result of an externally applied random force. The mathematical modeling of these fluctuations is done by considering a deterministic equation appropriate in the absence of external fluctuations and then considering the external parameter which undergoes fluctuations to be a stochastic variable. The noise term of the stochastic differential equation so obtained is usually multiplicative in nature, that is, it depends on the instantaneous value of the variable of the system. It does not scale with the system size and is not necessarily small. We can regard the external noise as an external force field which drives the system, always maintaining its statistical equilibrium. Among the experimental situations in the presence of external noises so far considered, the example of illuminated chemical reactions[43] may be of particular interest for our readers.

In Chapters I, X, and XI it is stressed that the "microscopic" derivation of equations such as some of those used here should be discussed carefully. This is to avoid some ambiguous features of a purely phenomenological treatment. However, as these are widely used in the literature of stochastic processes, we shall show how to approach the problem of their solution while avoiding those difficulties by using a more rigorously founded "microscopic" derivation (see Chapters X and XI).

1. Examples of Non-Markovian External Noises

Let us focus on the one-dimensional dynamics of an order parameter x exhibiting bistability, that is,

$$\dot{x} = f(x, \mathbf{a}) \tag{37}$$

where $\mathbf{a}$ denotes an external control parameter. The flow $f(x, \mathbf{a})$ is assumed to possess three real roots $\{x_1, x_u, x_2\}$. We choose $x_1 < x_2$, where x_1 and x_2 denote locally stable steady states and x_u is an intermediate, locally unstable, steady state. In the presence of a fluctuating control parameter $\mathbf{a}$, the deterministic flow in Eq. (37) should be replaced by a stochastic one:

$$\dot{x} = f(x, \mathbf{a}) + g(x)\xi(t) \tag{38}$$

where the multiplicative noise (state-dependent coupling) represents the linear coupling of $\mathbf{a}$ to the order parameter x in the dynamical flow, Eq. (37). A common example of Eq. (38) is provided by the Smoluchowski approximation of the random walk of a Brownian particle bounded into a symmetrical double-well potential,

$$V(x) = -\frac{ax^2}{2} + \frac{bx^4}{4} \tag{39}$$

In such a case $g(x)$ is assumed to be 1, $x_u = 0$, and $x_{1,2} = \pm(a/b)^{1/2}$.

The problem may be formulated as follows. Given random noises $\xi(t)$ with different correlation parameters τ_1 and τ_2 but possessing identical spectral densities S_ξ ($\omega = 0$) at zero frequency, that is,

$$S_\xi(0) = \int \langle \xi_1(t)\xi_1(0) \rangle \, dt = \int \langle \xi_2(t)\xi_2(0) \rangle \, dt = 2D \tag{40}$$

what is the relationship between the corresponding activation rates of the metastable states?

Hänggi and Riseborough[44] carried out an exact calculation of the activation rates for the bistable flow of Eq. (38) for the case when the noise of the control parameter can be modeled by a telegraphic noise of vanishing mean,

$$\xi(t) = d(-1)^{n(t)} \tag{41a}$$

$$\langle \xi(t)\xi(s) \rangle = \frac{D}{\tau} \exp\left(-\frac{|t-s|}{\tau}\right) \tag{41b}$$

where $n(t)$ is a Poisson counting process with parameter $(2\tau)^{-1}$ and d denotes a random step with density

$$\rho_d = \frac{1}{2}\left\{ \delta\left[d - \left(\frac{D}{\tau}\right)^{1/2}\right] + \delta\left[d + \left(\frac{D}{\tau}\right)^{1/2}\right]\right\}. \tag{42}$$

We may now elaborate on the problem posed above: the system with a smaller correlation time τ is subject to random forces with larger amplitude (see Fig. 7), and this might lead to the conclusion that the rate was enhanced.

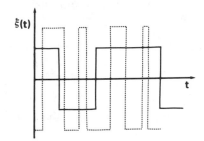

$\xi(t)$

t

Figure 7. Sketch of possible realizations of a telegraphic noise, $\xi(t)$, for differing correlation times τ, Eq. (41). The solid curve is for τ_1 and the dashed curve for τ_2, where $\tau_2 < \tau_1$. [Taken from P. Hänggi and P. Riseborough, *Phys. Rev.*, **A27**, 3329 (1983).]

However, the time interval over which the force is constant decreases; since the random force changes sign more rapidly, one might now expect that the system did not have enough time to reach the point of instability, and consequently the rate would be suppressed for a smaller correlation time. Thus, it is not obvious a priori which of the two random forces, $\xi_1(t)$ or $\xi_2(t)$, yields a smaller rate, that is, a larger escape time.

The analysis made in ref. 44 is based on the discussion of the related exact non-Markovian master equation[45,46] and allows us to conclude that when the noise intensity $S_\xi(0)$, Eq. (40), is constant the rates are exponentially enhanced with decreasing correlation time τ and this is independent of the specific form of the nonlinear bistable flow $f(x,\mathbf{a})$ and also of whether the random noise is additive or multiplicative. [The only condition imposed is $g(x) \neq 0$ in $\{x_1, x_u, x_2\}$.]

An important property of the telegraphic noise, Eq. (41), is the approach to a Gaussian white noise in the limit $\tau \to 0$.[47] With

$$\lim_{\tau \to 0} \frac{1}{2\tau} e^{-|t|/\tau} = \delta(t)$$

Eq. (41) reduces to

$$\langle \xi(t)\xi(s) \rangle = 2D\delta(t-s) \tag{43}$$

From now on we consider the stochastic differential equation (38) with $\xi(t)$ a random force associated with a zero-mean Gaussian process and an autocorrelation function given by Eq. (41). This system has been thoroughly investigated by Sancho et al.[48-50] The use of such Gaussian noises is justified by the central limit theorem.[11b,41] For a Gaussian noise with exponential autocorrelation, Eq. (41), the so-called Ornstein-Uhlenbeck noise, one is unable to derive exact expressions for activation rates.[48a,50,51] In ref. 50 an approximate Fokker-Planck equation is obtained for the probability distribution of the variable x by applying functional methods. These methods provide an alternative to the more often used cumulant techniques[49,52,53] and may be shown to lead to consistent results.[49,50] The same approximate Fokker-Planck equation, however, can be recovered with the AEP technique. The AEP can be applied by introducing an equivalent formulation of the process under investigation, Eq. (38):

$$\dot{x} = f(x,\mathbf{a}) + g(x)\xi$$
$$\dot{\xi} = -\frac{1}{\tau}\xi + \eta(t) \tag{44}$$

The exact equivalence of these formulations may be proved,[54] for the case where:

1. $\eta(t)$ is a white Gaussian noise with

$$\langle \eta(t) \rangle = 0 \quad \text{and} \quad \langle \eta(t)\eta(s) \rangle = \frac{2D}{\tau^2}\delta(t-s) \quad (45)$$

2. A fluctuation-dissipation relationship for the auxiliary variable ξ is understood, and it is initially prepared at its Gaussian equilibrium with

$$\langle \xi^2(0) \rangle = \langle \xi^2 \rangle_{eq} = \frac{D}{\tau} \quad (46)$$

The perturbation reduction of the corresponding Markovian Fokker-Planck equation for the two-variable process $(x(t), \xi(t))$ to an approximate one in $x(t)$ has been carried out in Section V.A of Chapter II. For brevity we report only the approximate time-evolution equation for $\sigma(x, t)$ up to order D;

$$\frac{\partial}{\partial t}\sigma(x, t) = \left\{ -\frac{\partial}{\partial x}f(x, \mathbf{a}) + D\frac{\partial}{\partial x}g(x)\frac{\partial}{\partial x}[g(x) - \tau M(x)] \right\}\sigma(x, t)$$

$$(47a)$$

where

$$M(x) = f(x)g'(x) - f'(x)g(x) \quad (47b)$$

where the prime denotes the derivative. [The reader can find a detailed discussion of some technical properties of Eq. (47) in Chapter II.]

The problem we are addressing now is the same one posed in ref. 44 for a case of a non-Markovian telegraphic noise: Given Gaussian noises with different autocorrelation times τ_1 and τ_2 but identical intensities $2D$, Eq. (40), which of them will provide a smaller rate (larger escape time)? Since detailed balance does not hold for Eq. (44), the standard methods[11a,20,55] fail in evaluating the activation rate of the non-Markovian process under investigation, and the more general method of refs. 56 and 57 is rather cumbersome because the stationary probability $\sigma_{st}(x, \xi)$ should first be determined perturbatively. If T denotes the mean first passage time[32,33] to reach the barrier top, the activation rate can be estimated as

$$K = \tfrac{1}{2}T^{-1} \quad (48)$$

where the factor $\frac{1}{2}$ takes into account that the random walker has equal chance to either continue to the adjacent stable state or return to the old one. Without loss of generality, we consider the particular case of the Smoluchowski approximation of the random walk of a Brownian particle bound into a symmetrical double-well potential, that is,

$$f(x,a) = V'(x) \quad \text{and} \quad g(x) = 1$$

where $V(x)$ is given in Eq. (39). The chemical meaning of this model has been discussed at length in the preceding sections. If $x = -\infty$ is a natural reflecting boundary and $x = x_u = 0$ an absorbing state, one finds,[32,58] for the mean first passage time $T(x)$ of a walker which started out at $x(0) = x_{st} < 0$,

$$T(x) = \int_{x_{st}}^{0} \frac{dy}{\sigma_{st}(y)D(y)} \int_{-\infty}^{y} \sigma_{st}(z)\, dz \tag{49}$$

$\sigma_{st}(x)$ denotes the stationary probability of the approximate Fokker-Planck equation, Eq. (47). $D(x)$ is the corresponding diffusion coefficient, that is, $D(x) = D(1 - \tau M(x))$. With the assumptions of (1) small enough autocorrelation time τ and (2) weak noise such that $D < a^2/b$, we can evaluate $T(x)$ by applying the method of steepest descent to Eq. (49). From Eqs. (47) and (39), we obtain

$$T(x) = \frac{\pi}{a\sqrt{2}} \left(\frac{1+2a\tau}{1-a\tau} \right)^{1/2} \exp\left(\frac{\Delta\phi}{D} \right) \tag{50}$$

with

$$\Delta\phi = \int_{0}^{x_1} \frac{f(y,a)}{1+\tau f(y,a)} dy = \frac{a^2}{4b}\left(1 - \frac{a^2\tau^2}{2}\right) + O(\tau^3) \tag{51}$$

Since Eq. (49) takes into account only the term of order $D\tau$, the term of order τ^2 in Eq. (51) is meaningless and the term linear in τ in $\Delta\phi$ vanishes exactly. For $\tau = 0$, our result equals the well-known Smoluchowski rate.[11a]

The main conclusion we can draw is that the activation rates for non-Markovian processes like Eq. (44) decrease as τ increases; the exact result of ref. 44 can thus be extended to the case of Gaussian random forces of finite correlation time as well. However, if we take Eq. (50) seriously, we obtain an Arrhenius factor, $\exp(\Delta\phi/D)$, of $T(x)$ which does not exhibit a dependence on τ. This is in contrast to the result found for telegraphic noises, where the Arrhenius factor increases with increasing autocorrelation time τ (see ref. 44). The result of a numerical simulation for $T(x)$ based on the bi-

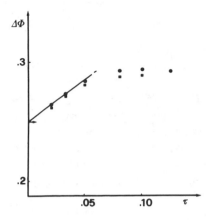

Figure 8. $\Delta\Phi$, defined in Eq. (51), versus the noise autocorrelation time τ. The computer simulation of Eqs. (44) has been carried out by applying the numerical algorithm of ref. 50 with an integration step of 0.01. The values of the parameters are $a = b = 1$, $D = 0.1$ ($\bullet$) and $D = 0.05$ ($\blacksquare$). T is the average over 1000 first passage times considering the initial conditions (46) for ξ and $\sigma(x,0) = \delta(x - x_1)$ for x. The maximum error bar in our numerical simulation is estimated to be about 10%. The arrow denotes the white noise limit $\Delta\Phi(\tau = 0)$. [Taken from P. Hänggi et al., Z. Phys. **B56**, 333 (1985).]

stable flow is given in Fig. 8. In contrast with our prediction in Eq. (50), $\Delta\phi$ increases with increasing autocorrelation time τ. The increase is proportional to the first order in τ and is not dependent on the small noise parameter D. The origin of the disagreement must be ascribed to the expansion in a Taylor series of $\exp\{\mathscr{L}_a s\}$ appearing in the memory kernel of Eq. (3.15) of Chapter II. This expansion is proven[60] virtually equivalent to considering only $\mathscr{L}_b$ (see Chapter II) as being the unperturbed part of the total operator $\mathscr{L}$. When $\mathscr{L}_a$ is replaced with an equivalent linear operator so as to avoid this perturbative expansion, then a complete agreement with the predictions of the master equation method[59] is recovered.[60,61]

2. Chemical Reactions Driven by Bona Fide Non-Markovian Fluctuation-Dissipation Processes

When the chemical reaction process takes place in condensed phase (for example, in a liquid), a reliable description of it seems to be achieved[49] simply by replacing Eqs. (3) with

$$\dot{x} = v \tag{52a}$$

$$\dot{v} = -\frac{1}{M}V'(x) - \int_0^t \varphi(t - \tau)v(\tau)\,d\tau + f(t) \tag{52b}$$

where the kernel $\varphi(t)$ and the stochastic force are related to each other via

$$\langle v^2 \rangle \varphi(t) = \langle f(0)f(t) \rangle \tag{53}$$

This takes into account the fact that the stochastic force $f(t)$ can have a finite correlation time, for example,

$$\varphi(t) = \frac{\gamma}{\tau_c} \exp\left(-\frac{t}{\tau_c}\right) \qquad (54)$$

In the absence of the external potential V, Eqs. (52) can be given a rigorous derivation from a microscopic Liouville equation (see Chapter I). We make the naive assumption that when an external potential driving the reaction coordinate is present, the two contributions (the deterministic motion resulting from the external potential and the fluctuation-dissipation process described by the standard generalized Langevin equation) can simply be added to each other.

A more realistic and more general treatment would presumably lead to a set of equations like Eqs. (52), with the potential $V(x)$ fluctuating as a consequence of couplings with nonreactive modes (see Section III). For the sake of simplicity, we study separately the two different aspects. While Section III was devoted to pointing out the role of multiplicative fluctuations (derived from nonlinear microscopic Liouvillians) in the presence of additive white noise, this subsection is focused on the effects of a non-Markovian fluctuation-dissipation process (with a time convolution term provided by a rigorous derivation from a hypothetical microscopic Liouvillian) in the presence of a time-independent external potential.

A more general expression for $\varphi(t)$ can be derived from the continued fraction expansion, ref.[62]

$$\hat{\varphi}(z) = \cfrac{\Delta_1^2}{z + \lambda_1 + \cfrac{\Delta_2^2}{z + \lambda_2 + \Delta_3^2 \cfrac{}{\ddots} \quad + \cfrac{\Delta_n^2}{z + \hat{\varphi}_n(z)}}} \qquad (55)$$

defining its Laplace transform. In the explicit calculations presented in this subsection, we shall limit ourselves to considering the case of Eq. (54) which corresponds to truncating Eq. (55) at the first order ($\Delta_2^2 = 0$) while assuming $\lambda_1 = 1/\tau_c$ and $\Delta_1^2 = \gamma/\tau_c$. A truly rigorous derivation from a microscopic Liouvillian would lead to $\lambda_i = 0$ unless coherent oscillatory motions have to be simulated (in that case λ_i would be purely imaginary numbers). The chain of Eq. (55) is often truncated at the nth order by assuming $\hat{\varphi}_n(z) = \gamma_n$. When this is done, the dissipative term γ_n simulates the infinite remainder of the chain. In most cases (see, for example, Grote and Hynes[28a,37]) $\varphi(t)$ is given a certain analytical expression without taking into account the formal constraints provided by the derivation from a hypothetical microscopic Liouvillian. In such a case the parameters λ_i can be real numbers. If we adopt the basic ideas of the RMT (which in the present linear case to which the standard generalized Langevin equation applies is virtually equivalent to the methods described by Ferrario and Grigolini[63]), we find that the set of Eqs. (52) is equivalent to

$$\dot{x} = v \tag{56a}$$

$$\dot{v} = -V'(x) + A_1 \tag{56b}$$

$$\dot{A}_1 = -\Delta_1^2 v - \lambda_1 A_1 + A_2 + \xi_1(t) \tag{56c}$$

$$\dot{A}_2 = -\Delta_2^2 A_1 - \lambda_2 A_2 + A_3 + \xi_2(t) \tag{56d}$$

$$\vdots$$

$$\dot{A}_n(t) = -\Delta_n^2 A_{n-1} - \lambda_n A_n + \xi_n(t) \tag{56n}$$

The random forces $\xi_1(t), \ldots, \xi_n(t)$ are Gaussian white noises of zero mean and correlations

$$\langle \xi_i(t)\xi_j(s) \rangle = 2\delta_{ij} K_B T \lambda_i (\Delta_1^2 \cdots \Delta_i^2) \delta(t-s) \tag{57}$$

These forces are introduced[64] so as to supplement the frictions λ_i with the corresponding noise term and guarantee the attainment of a canonical equilibrium. The Fokker-Planck equation associated with the set of Eqs. (56) can be written as

$$\frac{\partial}{\partial t}\rho(q,t) = \frac{\partial}{\partial q_\mu}D_{\mu\nu}\left[(K_B T)^{-1}\left(\frac{\partial}{\partial q_\nu}U(q)\right) + \frac{\partial}{\partial q_\nu}\right]\rho(q,t) \tag{58}$$

where a summation over repeated indices is implicit, $\mu, \nu = 1, \ldots, n+2$, and $q = x, v, A_1, \ldots, A_n$. The generalized potential U is

$$U(q) = \frac{V(x)}{M} + \frac{v^2}{2} + \frac{A_1^2}{2\Delta_1^2} + \cdots + \frac{A_n^2}{2\Delta_1^2 \cdots \Delta_n^2} \tag{59}$$

and the kinetic matrix $D_{\mu\nu}$ is

$$
D_{\mu\nu} = K_B T
\begin{bmatrix}
0 & -1 & 0 & 0 & \cdots & \cdots & & 0 \\
1 & 0 & -\Delta_1^2 & 0 & \cdots & \cdots & & 0 \\
0 & \Delta_1^2 & \lambda_1\Delta_1^2 & -\Delta_1^2\Delta_2^2 & \cdots & \cdots & & 0 \\
0 & 0 & \Delta_1^2\Delta_2^2 & & \cdots & \cdots & & \\
\vdots & \vdots & \vdots & \vdots & \vdots & \vdots & & -\Delta_1^2\cdots\Delta_n^2 \\
0 & & & & & & \Delta_1^2\cdots\Delta_n^2 & \lambda_n\Delta_1^2\cdots\Delta_n^2
\end{bmatrix}
$$

$$(60)$$

The equilibrium stationary solution of Eq. (58) is

$$
\rho_{st}(q) = N\exp\left[-\frac{U(q)}{K_B T}\right]
$$

$$(61)$$

where N is a normalization constant.

As mentioned above, in the explicit calculations of this subsection we shall consider $\lambda_0 = 0$, $\lambda_1 \neq 0$, $\Delta_2^2 = 0$. This is the simplest case satisfying the requirements of a rigorous derivation from a microscopic Liouvillian. Of course, for the non-Markovian nature of the variable velocity v to result in observable effects, the effective friction term

$$
\gamma_{eff} \equiv \left[\int_0^\infty \langle v(0)v(t)\rangle \, dt\right]^{-1}
$$

cannot be infinitely large compared with the frequency ω_0, the harmonic approximation around the bottom of the reactant well. This means that inertial effects cannot be disregarded. An interesting discussion of the influence of inertia on the escape over the potential barrier (variational in nature) can be found in a paper by Larson and Kostin.[22b] Their results are valid in the limit of white noise and provide a reliable check of our approach. Furthermore, in an earlier paper,[22a] the same authors improved the Kramers result for the diffusional case by evaluating corrections to the linearization of the Brownian motion within the barrier region. Such an assumption is usually[28a,37,65] at the basis of any approximate analytical calculation of the activation rates. However, as shown by Fonseca et al., when using the CFP this approximation can be avoided. Therefore we shall apply the CFP in the most advanced form reviewed by Grosso and Pastori in Chapter III to the Fokker-Planck equation [Eq. (58)] with $n = 1$ and $\lambda_1 \neq 0$, which hopefully should account also for the corrections of ref. 22b.

In Larson and Kostin[22b] notation we change variables as follows:

$$x \rightarrow x\tilde{a}$$
$$v \rightarrow v\gamma\tilde{a}$$
$$t \rightarrow \gamma^2\tilde{a}^2\frac{t}{q} \tag{62}$$
$$A_1 \rightarrow A_1\Delta_1^2\tilde{a}$$

where

$$\tilde{a} = \left(\frac{a}{b}\right)^{1/2} \qquad \gamma = \frac{\Delta_1^2}{\lambda_1} \qquad q = \frac{\gamma}{\langle v^2 \rangle} \tag{63}$$

The reduced Fokker-Planck equation now reads

$$\frac{\partial}{\partial t}\sigma(x, v, A_1, t) = \left[-\alpha v\frac{\partial}{\partial x} + 4c(x^3 - x)\frac{\partial}{\partial v} - \alpha' A_1\frac{\partial}{\partial v} \right.$$

$$\left. + \alpha v\frac{\partial}{\partial A_1} + \alpha'\frac{\partial}{\partial A_1}A_1 + \frac{\partial^2}{\partial A_1^2} \right]\sigma(x, v, A_1, t) \tag{64}$$

where

$$\alpha = \gamma^3\frac{\tilde{a}^2}{q} \qquad \alpha' = \lambda_1\gamma^2\frac{\tilde{a}^2}{q}$$

$$c = \frac{\gamma}{V_0} \qquad V_0 = \frac{a^2}{4b} \tag{65}$$

Let us note that in these dimensionless variables c plays the role of barrier height, while $\gamma = \Delta_1^2/\lambda_1$ is the effective friction constant. This can be shown by following the heuristic argument of Section I of Chapter II. Let us assume that A_1 relaxes so fast that $\dot{A}_1$, Eq. (56c), is approximately zero; the system of Eqs. (3) will be recovered provided $\gamma = \Delta_1^2/\lambda_1$. Moreover, in Section V.C of Chapter II it is shown that the AEP corrections to the trivial Markovian approximation of Eq. (58), $n = 1$, are perturbation terms in the parameter

$$g = 2\left(\frac{\alpha}{\alpha'}\right)^{1/2} = \frac{2\Delta_1}{\gamma} \tag{66}$$

In other words, if we keep γ fixed and vary g, we explore situations with different "memory strength." Following the prescription of ref. 66, we define the escape time from the reactant well to the product well as the area below the curve $\langle x(t) \rangle / \langle x(0) \rangle$. For fairly high values of the barrier c, this curve is mostly one exponential throughout the entire time range but for a narrow region close to $t = 0$. This fast relaxation depends significantly on the starting point distribution, $\sigma(x,0)$.[15a] Let us assume $\sigma(x,0)$ to be given by a delta of Dirac placed at the bottom of one well. This choice may enhance the effect of the short time relaxation on our definition of escape time,

$$\tau_K \equiv \hat{\Phi}(0) \tag{67}$$

where $\hat{\Phi}(0)$ is the Laplace transform of $\langle x(t) \rangle / \langle x(0) \rangle$ at zero frequency. However, for large enough values of c, $k = \tau_K^{-1}$ can be relied on as a sensible estimate of the activation rate of the process.

Figure 9 describes the results obtained by applying the CFP. The most remarkable feature of these results, is the increase in the rate k as the parameter g increases. A further remarkable finding is that for $g \to 0$ (Markovian limit) the accurate value of Larson and Kostin[22b] is attained within a precision of a few percent.

Figure 9 is the main result of the present discussion. However, we can attempt to arrive at an analytical expression for the rate of escape over the

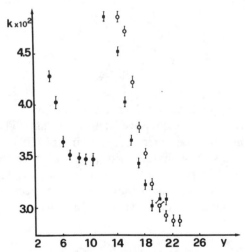

Figure 9. The escape rate k [see Eq. (67)] as a function of $y = 4/g^2 = \alpha'/\alpha$ when $c = 5$. The approach used to calculate k is the "exact" one described in Chapter II of this volume. The values of the parameter α are as follows: (O) $\alpha = 10$, (■) $\alpha = 12$, and (●) $\alpha = 20$. [Taken from F. Marcheroni and P. Grigolini, J. Chem. Phys., 78, 6287 (1983).]

barrier by using the generalization of Kramers' ideas to systems with many variables.[20,55,67,68,69] Let us come back to the multidimensional potential $U(q)$, Eq. (59): It has two metastable minima at $q_{1,2} = (x_{1,2}, 0, \cdots, 0)$ and a saddle point $q_0 = (x_0, 0, \cdots, 0)$. This generalization essentially consists in the following. One first looks for a quasi-stationary state of Eq. (58). In this state there exists a nonvanishing probability current from one metastable minimum to the other. The nonequilibrium stationary state and the probability current are calculated by linearizing around the saddle point q_0. The escape rate is given by the flux of probability current through a surface containing the point q_0. The calculation of k has been discussed in detail by Langer[20] for a general Fokker-Planck equation with the form of Eq. (58). The final result is

$$k = \frac{|\kappa|}{2\pi} \left(\frac{\det M^1}{|\det M^0|} \right)^{1/2} \exp\left(-\frac{\Delta U}{K_B T} \right) \tag{68a}$$

where

$$\Delta U = U(q_0) - U(q_1) \qquad M_{\mu\nu}^0 = \left. \frac{\partial^2 U}{\partial q_\mu \partial q_\nu} \right|_{q=q_0} \qquad M_{\mu\nu}^1 = \left. \frac{\partial^2 U}{\partial q_\mu \partial q_\nu} \right|_{q=q_1}$$

$$\tag{68b}$$

Note that $\det M^0$ is taken in absolute value in Eq. (68). This corresponds to replacing the negative eigenvalue M_{11}^0, which indicates the single direction of instability, by its absolute value. The dynamical factor κ is defined as the negative eigenvalue of the matrix $M^0 D / K_B T$. It is important to note that the dynamics of the system only enters into Eq. (68) through κ. This factor depends on the kinetic coefficients $D_{\mu\nu}$, Eq. (60), while the remaining terms in Eq. (68) are completely determined by the potential U of the stationary solution. For the case under study, Eqs. (58) and (59), Eq. (68) reduces to

$$k = \frac{|\kappa|}{2\pi} \left(\frac{U''(x_1)}{|U''(x_0)|} \right)^{1/2} \exp\left[-\frac{U(x_0) - U(x_1)}{K_B T} \right] \tag{69}$$

In this particular case,

$$\frac{M^0 D}{K_B T} = \begin{bmatrix} 0 & -U''(x_0) & 0 & \cdots & \vdots & & 0 \\ 1 & 0 & -\Delta_1^2 & 0 & \vdots & & 0 \\ 0 & 1 & \lambda_1 & -\Delta_2^2 & \vdots & & 0 \\ 0 & 0 & 1 & \lambda_2 & \vdots & & \\ \cdots & \cdots & \cdots & \cdots & \vdots & \cdots & \\ & \cdots & & \cdots & \vdots & & -\Delta_n^2 \\ 0 & \cdots & & \cdots & \vdots & 1 & \lambda_n \end{bmatrix} \tag{70}$$

The eigenvalues of this matrix admit a continued fraction expansion

$$-\kappa = -\cfrac{U''(x_0)}{-\kappa + \cdots}\cfrac{\Delta_1^2}{-\kappa + \lambda_1 + \cdots} \cdots \cfrac{\Delta_n^2}{-\kappa + \lambda_n} \tag{71}$$

From this expression it is clear that κ is the negative solution of the implicit relation

$$-\kappa = \frac{-U''(x_0)}{-\kappa + \hat{\varphi}(-\kappa)} \tag{72}$$

Equation (72) coincides with the analytical result of Grote and Hynes.[37] In the Markovian limit, $\hat{\varphi}(t) = 2\gamma\delta(t)$,

$$\kappa = \frac{\gamma}{2} - \left(\frac{\gamma^2}{4} - U''(x_0) \right)^{1/2}$$

and we recover the Kramers result, Eq. (6).

The whole effect of the non-Markovian dynamics is contained in κ. As long as the fluctuation-dissipation relation, Eq. (57), is satisfied, the existence of the non-Markovian kernel modifies the dynamics but not the equilibrium solution, Eq. (61), and, on the other hand, a change in the dynamics of the system only changes the value of κ in Eq. (69). The general form of the non-Markovian effects on κ have also been obtained by Hänggi and Mojtabai.[65] Their elegant derivation is based on a non-Markovian master equation first established by Adelman[70] for the probability density of the process which is solved by using the main basic assumption of Kramers. Their results are again proven to agree with those of Grote and Hynes.[37]

As a particular example of Eq. (72) we can consider the case of an Ornstein-Uhlenbeck noise,[37,65] where $n = 1$, $\lambda_1 = \tau_c^{-1}$ and $\Delta_1^2 = \gamma/\tau_c$. In this case κ is the negative solution of

$$-\kappa^3 - \lambda_1\kappa^2 - \left[\Delta_1^2 + U''(x_0) \right]\kappa + U''(x_0)\lambda_1 = 0 \tag{73}$$

The Markovian limit corresponds to $\lambda_1 \to \infty$. By solving Eq. (73) to the lowest order in λ_1^{-1} it is easy to see that in this case the non-Markovian dynamics leads to an enhancement of the decay rate k. In the notation of ref. 22b, an approximate expression for Eq. (69) may then be written as

$$k = k(g = 0)\left[1 - \frac{4c}{\alpha}\left(1 - \frac{g^2}{4} \right) \right] \tag{74}$$

where $k(g = 0)$ is the Kramers escape rate (in the diffusional limit) and g is the parameter of "memory strength" defined in Eq. (66). The same result has been obtained in ref. 66 by adopting the variational method of ref. 22b. Equation (74) is the analytical counterpart of the exact results reported in Fig. 9: As shown in ref. 66, the agreement with numerical results is only qualitative. Before concluding this subsection, we would like to mention a further way to explore the effects of non-Markovian statistics on the rate of escape from a well. This consists in applying the AEP of Chapter II to the Fokker-Planck equation, Eq. (58), so as to build up a reduced diffusion-like equation for the variable x alone. As the chance of proper simulating these effects relies on a faithful simulation of inertia, we quote here the interesting result of Gardiner,[71] which shows that this actually happens. He considered a corrected Smoluchowski equation which is a particular case of the more general reduced equation mentioned above. By using a first-passage time technique he could explore the whole region going from low- to high-friction regime and obtained results in agreement with those of a computer simulation. It therefore seems possible to explore also the effects of a non-white noise by applying the same procedure to the more general reduced equation mentioned above.

B.   Activation of a Chemical Reaction Process via Electromagnetic Excitation

The subject of this subsection is closely related to that of Section III. Indeed, we shall show that the effect of a radiation field on an overdamped reacting system produces activated states which are reminiscent and formally similar to those arrived at by the coupling between reactive and non-reactive modes.

Hänggi[72] studied the model potential

$$V(x) = -\frac{d(t)}{2}x^2 + \frac{b}{4}x^4 \tag{75}$$

where the frequency $d(t)$ is a stochastic parameter such as

$$d(t) = d + \eta(t) \tag{76}$$

and $\eta(t)$ is a Gaussian white noise. The main result of this study is that the presence of a multiplicative coupling with the heat bath makes the activation rate increase with respect to that in the Kramers model, where a purely additive noise is considered. In the following we give a detailed discussion of the interplay of additive and multiplicative noises on the basis of a phenomenological model for a photoactivated chemical reaction. De Kepper and

Horsthemke[43] have already used a radiation field as a source of noise. As in refs. 73 and 74, we model the action of a radiation field $h(t)$ with a finite coherence time $1/\lambda$, in terms of the following set of stochastic differential equations:

$$\dot{x} = v$$
$$\dot{v} = -V'(x) - \gamma v + E(x)h(t) + f(t) \tag{77}$$

where $f(t)$ is a Gaussian white noise with zero mean and autocorrelation function

$$\langle f(t)f(0)\rangle = 2D\delta(t) \equiv 2\gamma\langle v^2\rangle\delta(t) \tag{78}$$

$V(x)$ is assumed to be the usual symmetrical double-well potential $[V(x) = -\delta x^2/2 + \beta x^4/4]$; the third term on the right-hand side of Eq. (77) is the coupling between the Brownian particle and the external radiation field, which is characterized through its autocorrelation function

$$\langle h(t)h(0)\rangle = 2\langle w^2\rangle\exp(-\lambda t)\cos\omega t \tag{79}$$

Equation (79) has the physical meaning that the coherence of the electromagnetic field is lost in a time $1/\lambda$. Models of this kind are frequently used to depict laser light.[75] The electrical dipole of the system interacting with the external field is assumed to have the simple form

$$E(x) = \mu(x - x_0)^n \tag{80}$$

In order to relate the system of Eqs. (77) to a time-independent Fokker-Planck formalism, we replace that set of stochastic differential equations with the equivalent one,

$$\dot{x} = v$$
$$\dot{v} = -V'(x) - \gamma v + E(x)(y + z) + f(t)$$
$$\dot{y} = -i\omega y - \lambda y + \eta_y(t) \tag{81}$$
$$\dot{z} = i\omega z - \lambda z + \eta_z(t)$$

The AEP allows us to simplify the discussion of this model provided that we can choose a (slowly relaxing) variable of interest. For that reason we shall focus on an electromagnetic field of frequency comparable to the frequency corresponding to the harmonic expansion of the reactant well, ω_0. The diffusional assumption implies $\gamma \gg 2\omega_0$. Furthermore, we shall assume that our experimental apparatus allows us to observe only long-time regions corresponding to $t \gtrsim \gamma/\omega_0^2$ so that the dynamics induced by the radiation field belongs to the short-time region if $\lambda \gg 2\omega_0$. When it is further assumed that the stochastic forces $\eta_y(t)$ and $\eta_z(t)$ are independent of each other and related to the field intensity by

$$\langle \eta_y(t)\eta_y(0)\rangle = 2D_y\delta(t)$$
$$\langle \eta_z(t)\eta_z(0)\rangle = 2D_z\delta(t)$$
(82)

where

$$D_y = D_z^* = (\lambda + i\omega)\langle w^2\rangle \equiv \xi\langle w^2\rangle$$
(83)

the current problem takes a form resembling that of the model studied in Section V.C of Chapter II. Let us focus our attention on the case $\gamma \gg \lambda$, that is, one in which non-Markovian effects due to light statistics are more relevant than inertial corrections. The perturbation expansion of Eq. (5.31) of Chapter II can then be rewritten as

$$\frac{\partial}{\partial t}\sigma(x,t) = \left\{ \frac{1}{\gamma}\frac{\partial}{\partial x}j(x) + \frac{\langle w^2\rangle}{\gamma^2(\gamma + \xi)}\frac{\partial^2}{\partial x^2}E^2(x) \right.$$

$$\left. + \frac{\langle w^2\rangle}{\xi(\gamma + \xi)\gamma}\frac{\partial}{\partial x}E(x)\frac{\partial}{\partial x}E(x) + \text{c.c.} \right\}\sigma(x,t)$$
(84)

where $j(x) = \langle v^2\rangle\partial/\partial x + V'(x)$ and γ^{-3} or higher order terms have been neglected.

Let us study in detail the case where the particle dipole $E(x)$, Eq. (80), is given by

$$E(x) = \mu x$$
(85)

Equation (84) can be put in a simpler form:

$$\frac{\partial}{\partial t}\sigma(x,t) = \left\{ -\frac{\partial}{\partial x}(-d_\varrho x + bx^3) + D\frac{\partial^2}{\partial x^2} + Q\frac{\partial}{\partial x}x\frac{\partial}{\partial x}x \right\}\sigma(x,t)$$
(86)

where

$$D = \frac{\langle v^2 \rangle}{\gamma} \tag{87}$$

$$d = \frac{\delta}{\gamma} \qquad b = \frac{\beta}{\gamma} \tag{88}$$

$$Q = \frac{2\langle w^2 \rangle}{\gamma^2 \lambda (1 + \omega^2/\lambda^2)} \tag{89}$$

$$d_Q = d - Q \frac{1 + \lambda/\gamma + \omega^2/\lambda\gamma}{(1 + \lambda/\gamma)^2 + \omega^2/\lambda\gamma} \tag{90}$$

As a result of AEP, the initial system of the set of Eqs. (81) is reduced to the equation describing the diffusional motion of a Brownian particle which undergoes the action of an additive and a multiplicative noise (with intensities D and Q, respectively) in the presence of a renormalized bounding potential, Eq. (90). The Markovian limit corresponds to $\lambda \to \infty$. If we take such a limit at a fixed value of γ, $d_Q = d$, and the case studied by Hänggi[72] is recovered. Of course, having neglected the condition $\lambda \ll \gamma$ we have reduced the problem to a trivial diffusional (lowest-order) approximation.

The escape rate for the process described by the Fokker-Planck equation, Eq. (86), has been studied in ref. 73. We choose $\Phi(t) = \langle x(t) \rangle / \langle x(0) \rangle$ as the observable of interest, $\langle x(\infty) \rangle = 0$. Then we apply the approach described in the Section IV.A to evaluate the escape rate k as the area below the curve $\Phi(t)$: $k = \hat{\Phi}(0)^{-1}$, where $\hat{\Phi}(0)$ is the Laplace transform of $\Phi(t)$ at zero frequency. To make the convergence of the computer calculations faster, the CFP algorithm has been applied by taking

$$\sigma(x,0) = N|x|^{-1 + d_Q/Q} \exp\left(-\frac{d_Q}{2Q} x^2 \right) \tag{91}$$

as the initial distribution (N is a normalization constant). This is the stationary distribution in the absence of additive noise. The most remarkable results are reported in Fig. 10. When $Q = 0$, k exactly coincides with the corresponding result of Larson and Kostin.[22b] For small values of Q, k is a linear function of Q. A first change in the slope of $k(Q)$ is exhibited at those values of Q corresponding to the onset of the continuum in the spectrum of the purely multiplicative Fokker-Planck operator,[34,35] that is, the Fokker-Planck operator of Eq. (51) with $D = 0$. A second is found when the threshold of the phase transition (see ref. 34) is reached. The main conclusion is that

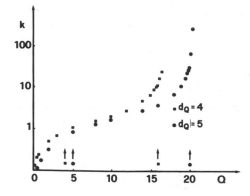

Figure 10. k as a function of the intensity of the multiplicative noise. k is defined as $\dot{\Phi}(0)^{-1}$ with $\Phi(t) = \langle x(t)\rangle/\langle x(0)\rangle$. The two arrows on the left denote the point where the discrete branch of the eigenvalue spectrum disappears (see Schenzle and Brand[34]). The two arrows on the right denote the phase transition threshold. [Taken from S. Faetti et al., *Z. Phys.*, **B47**, 353 (1982).]

the cooperative presence of a multiplicative noise produces a marked increase of the thermal (i.e., additive) activation rate.

The question raised at the beginning of the present section is still unanswered. A simple argument, however, can provide information on the role played by non-Markovian dynamics in the problem under investigation. When $\lambda < \infty$, $d_Q < d$, so that at a fixed value of Q the rate of escape will be larger than in the Markovian limit, making $k(Q, d_Q)$ a decreasing function of d_Q (see Fig. 10). We are in the presence of a striking effect due to the synergism of different—and statistically unrelated—noise sources, nonlinearity and inertia. We showed in the preceding subsection that the effect of an external additive noise, non-Markovian in its nature, would be to lessen the activation rate of the process with respect to the Markovian case first studied by Kramers. Furthermore, it has been found,[50] by means of a numerical simulation, that the non-Markovian dynamics affects the diffusional relaxation in the presence of external multiplicative noises in a similar way. Contrary to these findings, our major conclusion in this subsection is that when both additive and multiplicative external noises act on the system, a finite correlation time of the multiplicative noise determines an increase of the activation rate. This subject is discussed further in Chapter X.

V. DISCUSSION AND GENERAL PERSPECTIVE

In order to get a satisfactory perspective of the state of the art in the limited sector of the theory of chemical reactions which is explored in this chapter, we shall devote this final section to the following basic aspects:

1. The relation between current theories and selected experiments and discussion of the extent to which the details of theoretical predictions have been confirmed so far.

2.   The relation between chemical reaction rate theories and some recent advances in the field of nonequilibrium statistical thermodynamics.

A.   Supporting Experimental Evidence

Experimental confirmation of the theoretical predictions discussed in this chapter is still far from completely satisfactory. It may be expected that in the near future fresh experimental results will come to motivate new developments in the theory and greatly improve the understanding of the actual experimental conditions where the theoretically predicted effects are relevant. The aim of this subsection is not that of giving a comprehensive review of the already very sizable mass of relevant experimental studies (for more comprehensive reviews see refs. 7b and 76); we shall draw attention to certain difficulties in the interpretation of experimental results in relation to the theory developed here and refer to a few representative pieces of experimental work.

One difficulty of connecting theory and experiment arises from the fact that the relation between the microscopic coupling parameter between the reaction coordinate and the medium (the friction coefficient) and the macroscopic observables is not well understood. The usual rule of thumb follows Stokes's law and states that the friction is proportional to the macroscopic bulk viscosity; however, this may be grossly incorrect. It would be advantageous to use a local viscosity obtained from the measurement of some sort of molecular relaxation phenomenon, but this is not always available.

An alternative strategy is to look at the activation volumes $V^{\ddagger}$ related to the pressure dependence of the rate constant by the thermodynamic relation

$$V^{\ddagger} = - K_B T \left(\frac{\partial \ln k}{\partial P} \right)_T \tag{92}$$

The volume of activation should be formed by an equilibrium (quasi-thermodynamic) part related to the TST rate, $V_{\mathrm{TST}}^{\ddagger}$, plus an extra part, $V_D^{\ddagger}$, originating from the dynamic interaction with the solvent. $V_{\mathrm{TST}}^{\ddagger}$ may be estimated with reasonable assumptions about the transition-state conformation, and thus access is gained to $V_D^{\ddagger}$, that is, the pressure dependence of k / k_{TST}. This pressure dependence is felt through the friction,

$$V_D^{\ddagger} = - K_B T \left[\frac{d}{d\gamma} \frac{K}{K_{\mathrm{TST}}} \right] \left(\frac{\partial \gamma}{\partial P} \right)_T \tag{93}$$

This method was proposed by Montgomery, Chandler, and Berne,[25] who suggested that $(\partial \gamma / \partial P)_T$ could be estimated from the equation of state of the solvent, together with a hard-sphere collision expression, and thus the friction dependence of k / k_{TST} could be assessed.

One would like to know what experimental conditions lead to the energy transfer controlled regime or to the diffusive regime and whether the plateau of transition between these two regimes approaches the TST rate. The experimental evidence to answer this sort of question is still very fragmentary. The first series of very interesting experiments appearing to cover the whole range of friction-dependent kinetic regimes has only very recently been performed. Hasha, Eguchi, and Jonas[77] did a high-pressure NMR study of the conformational isomerization of cyclohexane in several solvents so as to cover a viscosity range of about 50 times. They found a clear transition from the rate-increasing low-friction regime to the rate-decreasing high-friction region, but this decrease does not exceed 7.5% of the maximum for a friction 10 times higher.

Fleming et al.,[78] in a series of studies of the solvent viscosity dependence of the rate of isomerization of several organic molecules (e.g., diphenylbutadiene) in alkane and alcohol solvents, found a similar deviation: For the higher viscosities, the observed rate is lower than that predicted by a fitted Kramers expression. This effect has been explained as coming from the non-Markovian nature of the coupling to the heat bath by Velsko, Waldek, and Fleming;[78c] by Bagchi and Oxtoby,[79] using Grote and Hynes[28] formalism; and also by Carmeli and Nitzan[26b] within their generalized theory.

Other reactions have been studied that appear to also require consideration of non-Markovian effects. For example, in a recent study of the photoisomerization of *trans*-stilbene and *trans*-1,1'-biindanylidene, Rothenberger, Negus, and Hochstrasser[80] found deviations from the Kramers rate in the case of *trans*-stilbene. These discrepancies were tentatively related to the larger flexibility of this molecule but appeared to be well simulated by the non-Markovian theory of Grote and Hynes.[28]

The fitting of the theoretical models to experimental data does normally require adjustment of the frequency parameters (ω_0, ω_b) related to the molecular potential, since the latter is frequently unknown. It has been noted by several authors[26b,80] that the values obtained appeared to be unrealistic, which sheds doubt as to the validity of the interpretation given to the data.

An explanation of the enhancement and other anomalies of the catalytic reaction rates on metals and certain insulators associated with the large fluctuations of the internal degrees of freedom that occur near a phase transition or by alloying has been attempted by d'Agliano, Schaich, Kumar, and Suhl[81] within the framework of stochastic theories.

To sum up the current position of the experimental evidence on the viscosity effect on condensed phase reaction rates, we may say that the most commonly observed effect is the inverse proportionality associated with the diffusive (high-friction) regime. In some cases, deviations are observed for lower viscosities which fit well with Kramers intermediate friction regime predictions. (See, for example, the analysis made by McCaskill and Gilbert[82] of data of Shank et al.[83] for the optically induced conformational changes in 1,1'-binaphthyl in several solvents.) Furthermore, there is now enough experimental evidence to show that in more particular conditions the energy transfer controlled (very low friction) regime will set in and may be accompanied by a wealth of finer effects that are discussed in this chapter.

Some of the theoretical results discussed here may also be checked by analogous computer simulation, a topic discussed by Faetti et al. in Chapter X.

B. Settled and Unsettled Problems in the Field of Chemical Reaction Rate Theory

The current attempts at generalizing the Kramers theory of chemical reactions touch two major problems: The fluctuations of the potential driving the reaction coordinate, including the fluctuations driven by external radiation fields, and the non-Markovian character of the relaxation process affecting the velocity variable associated to the reaction coordinate. When the second problem is dealt with within the context of the celebrated generalized Langevin equation

$$\dot{v} = - \int_0^t \varphi(t - \tau)v(\tau)\, d\tau + f(t) \tag{94}$$

supplemented by the fluctuation-dissipation relationship

$$\varphi(t) = \frac{\langle f(0)f(t)\rangle_{eq}}{\langle v^2\rangle_{eq}} \tag{95}$$

these topics seem now to be at a fully developed level of understanding. As already illustrated in the foregoing sections, the chemical relaxation process is then described by

$$\dot{v} = - \frac{\partial V}{\partial x} - \int_0^t \varphi(t - \tau)v(\tau)\, d\tau + f(t) \tag{96}$$

where V is the external potential driving the reaction coordinate x.

Carmeli and Nitzan[84] have provided a complete treatment of this problem. They assumed the memory kernel φ to be given the following analytical expression:

$$\varphi(t) = \Omega^2 \exp(-\Gamma t) \tag{97}$$

As already stressed in the foregoing sections, the standard case studied by Kramers is recovered by assuming Γ to be infinitely large. In such a case, $\varphi(t)$ can be replaced by

$$\varphi(t) = 2\gamma\delta(t) \equiv 2\frac{\Omega^2}{\Gamma}\delta(t) \qquad (98)$$

which, when replaced into Eq. (96), results in the standard set of equations studied by Kramers [see Eq. (6)].

The parameter

$$\gamma = \frac{\Omega^2}{\Gamma} \qquad (99)$$

can be thought of as a measurement of the friction intensity in the strong memory region also.

As we have already discussed, further parameters of interest are the frequencies ω_0 and ω_b deriving from the harmonic approximation at the bottom of the reactant well and the top of the barrier, respectively. Carmeli and Nitzan[84a] evaluated the reaction rate throughout the entire friction dominion ranging from the low-friction regime ($\omega_b \gg \gamma$) to the high-friction regime. (This has also been commented on in Section II.B.) They also studied the dependence of the reaction rate on the correlation time

$$\tau_c = \frac{1}{\Gamma} \qquad (100)$$

Their interesting results are shown in Figs. 11 and 12. We learn from these

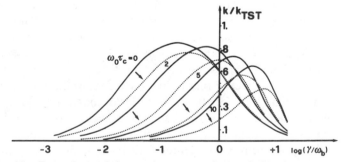

Figure 11.   Dependence of the reaction rate on the correlation time τ_c. A comparison is made between the analytical results provided by (---) the two-step model [Eq. (13)], and (—) those obtained by Carmeli and Nitzan.[84a] The results of the two-step model were obtained using an effective friction in order to simulate the non-Markovian character of the chemical process [see Eqs. (116) and (124)]. $\omega_0/\omega_b = 5$.

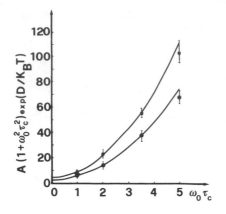

Figure 12. Numerical results provided by Carmeli and Nitzan.[84b] (■) $E_b = 5K_BT$; (●) $E_b = 2.5K_BT$. These numerical results were reproduced by the analytical result provided by Eq. (117) using $A = K_BT/\gamma E_b$ as a fitting parameter. (■) $A = 0.0292$ for $E_b = 5K_BT$; (●) $A = 0.236$ for $E_b = 2.5K_BT$.

results that the effect of increasing the correlation time τ_c is equivalent to shifting the curve corresponding to the case of white noise to the right. The height of the curves also changes as τ_c is varied, the form of this change depending on the ratio ω_0/ω_b.

These results are of very special interest, as they provide a definite answer to questions such as: (1) Can the rate be increased beyond any limit by adjusting the value of τ_c? (2) On another region of the friction, can an increase in τ_c make the reaction time infinitely large? By inspection of Carmeli and Nitzan's results, we conclude that an unbounded growth of τ_c makes the reaction rate vanishingly small; however, when an initial increase in the rate (as a consequence of the growth of τ_c) is observed, it is bound to reach a maximum value and then decrease to a vanishing rate for $\tau_c \to \infty$. This is not merely a problem of academic interest. Considerable attention is currently being devoted to enzyme chemistry,[85] where the enigma to be solved concerns how the activation process takes place. Enzymes succeed in increasing the reaction rates by about six orders of magnitude. A possible mechanism could be the presence of cooperative effects which make τ_c tend to infinity. However, Carmeli and Nitzan's results[84] show that in the case of a barrier as high as $18K_BT$, the effect of increasing τ_c cannot produce an increase in the chemical reaction rate larger than one order of magnitude. This suggests that the enigma of enzyme chemistry has to be solved by other mechanisms —for example, the interaction with nonreactive modes. This is another aspect concerning the generalization of Kramers theory and it touches problems which seem to be still the subject of controversy, such as the validity of the AEP itself.

Concerning the first aspect, on the other hand, we are already in a position to get a fairly definite view, which is clearly illustrated by the results of Carmeli and Nitzan (see Figs. 11 and 12).

What about the role played within this context by the general strategy of this book (as symbolized by the delta-like diagram of Chapter I)? We shall devote a large part of this final section to showing how this strategy may contribute to clarifying the physical meaning of these results. A calculation completely satisfactory from a quantitative point of view should, however, rely largely on the methods developed by other authors (those of Carmeli and Nitzan seem to be of special interest). For the sake of clarity we shall recall some of the key results of the foregoing sections.

When considering the special case studied by Carmeli and Nitzan, the RMT replaces Eq. (96) with

$$\dot{x} = v$$

$$\dot{v} = -\frac{\partial V}{\partial x} + w \tag{101}$$

$$\dot{w} = -\Omega^2 v + \Gamma w + F(t)$$

where $F(t)$ is a white Gaussian noise defined by

$$\langle F(t) \rangle = 0$$

$$\langle F(0)F(t) \rangle = 2\Gamma \langle w^2 \rangle_{eq} \delta(t) \tag{102}$$

The physical meaning of Eq. (101) has already been discussed in the foregoing sections, as well as in Chapter I.

First of all we shall apply Eq. (101) to study the low-friction regime:

$$\gamma \ll \omega_0 \tag{103}$$

and we assume ω_0 and ω_b to be of the same order of magnitude. The standard Kramers theory corresponds to

$$\Omega \ll \omega_0 \ll \Gamma \tag{104}$$

To take into account the fact that $f(t)$ of Eq. (94) is not rigorously white, we should explore also the region where

$$\Omega \ll \Gamma \ll \omega_0 \tag{105}$$

which is precisely that explored by Carmeli and Nitzan.[84b] Their latest results,[84a] however, seem to apply also to

$$\Gamma \lesssim \Omega \ll \omega_0 \tag{106}$$

Since, in the low-friction regime, the escape is largely determined by the behavior of the Brownian particle in the well, we shall focus our attention on that. When considering barriers of large intensity, we are allowed to replace Eq. (101) with its harmonic approximation

$$\dot{x} = v$$
$$\dot{v} = -\omega_0^2 x + w \qquad (107)$$
$$\dot{w} = -\Omega^2 v - \Gamma w + F(t)$$

We assumed the origin of the space coordinate to be at the bottom of the reactant well. By adopting the method of the stochastic normal modes (see Chapter II), Eq. (107) is replaced by

$$\dot{\xi}_+ = -\Lambda_+ \xi_+ + F_+(t)$$
$$\dot{\xi}_- = -\Lambda_- \xi_- + F_-(t) \qquad (108)$$
$$\dot{\xi}_0 = -\Lambda_0 \xi_0 + F_0(t)$$

To determine these normal modes one has to diagonalize the matrix

$$A = \begin{bmatrix} 0 & i\omega_0 & 0 \\ -i\omega_0 & 0 & i\Omega \\ 0 & -i\Omega & -\Gamma \end{bmatrix} \qquad (109)$$

This antisymmetric form can easily be derived from Eq. (107) by multiplying the variables v and w by suitable constants. Note the similarity of this matrix with that of ref. 39.

We may exploit the fact that ω_0 is much larger than the other parameters, Γ and Ω. First of all, let us rewrite the matrix A in the basis set, where it can be given the form

$$A = \begin{bmatrix} i\omega_0 & 0 & \dfrac{i\Omega}{\sqrt{2}} \\ 0 & -i\omega_0 & -\dfrac{i\Omega}{\sqrt{2}} \\ -\dfrac{i\Omega}{\sqrt{2}} & \dfrac{i\Omega}{\sqrt{2}} & -\Gamma \end{bmatrix} \qquad (110)$$

Then, by a perturbation calculation, we obtain

$$\Lambda_+ \cong i\omega_0 + \frac{\Omega^2}{2(i\omega_0 + \Gamma)}$$

$$\Lambda_- \cong -i\omega_0 + \frac{\Omega^2}{2(i\omega_0 - \Gamma)} \qquad (111)$$

$$\Lambda_0 \cong -\Gamma$$

This means that the normal modes ξ_+ and ξ_- are characterized by frequencies Ω_+ and Ω_- given by

$$i\Omega_+ = i\left[\omega_0 + \frac{\Omega^2\omega_0}{2(\omega_0^2 + \Gamma^2)}\right]$$

$$i\Omega_- = -i\left[\omega_0 - \frac{\Omega^2\omega_0}{2(\omega_0^2 + \Gamma^2)}\right] \qquad (112)$$

both with the same damping,

$$\frac{\Gamma_{\text{eff}}}{2} = \frac{\Omega^2\Gamma}{2(\omega_0^2 + \Gamma^2)} = \frac{\gamma}{2(1 + \omega_0^2\tau_c^2)} \qquad (113)$$

If we focus our attention on the damping while neglecting the less important effect on the frequencies, we have that the same result could be obtained from the Markovian system

$$\dot{x} = v$$

$$\dot{v} = -\omega_0^2 x - \Gamma_{\text{eff}} v + f^*(t) \qquad (114)$$

with the Gaussian white stochastic force $f^*(t)$ defined by

$$\langle f^*(0)f^*(t)\rangle = 2\Gamma_{\text{eff}}\langle v^2\rangle_{\text{eq}}\delta(t) \qquad (115)$$

Note that in the non-Markovian case ($\omega_0\tau_c \gg 1$) the effective damping

$$\Gamma_{\text{eff}} = \frac{\gamma}{1 + \omega_0^2\tau_c^2} \qquad (116)$$

turns out to be much smaller than the damping in the absence of the external field. This is a well-understood effect, discussed at length by Grigolini.[86]

A strong external field acting on a non-Markovian system tends to decouple that system from its thermal bath, thereby rendering smaller its effective damping.

In other words, if we are exploring the low-friction regime, the interplay of non-Markovian statistics and external field renders the system still more inertial, thereby widening the range of validity of the formula provided by Kramers for the low-friction regime provided that γ be replaced by $\gamma/(1 + \omega_0^2\tau_c^2)$.

Figure 12 shows that this simple expression agrees fairly well with both the theory of Carmeli and Nitzan and the result of their purely numerical calculations. The plots in Fig. 12 show how well the non-Markovian effects on the rate may be simulated by a simple multiplicative factor $(1 + \omega_0^2\tau_c^2)^{-1}$. For the sake of comparison, we fitted an expression with this factor to Carmeli and Nitzan's results so as to include their accurate Markovian rate.

Using Eq. (113), the Markovian low-friction expression of Kramers [Eq. (8)] may be generalized to the non-Markovian case,

$$k_{\text{Low}}(\tau_c) = \frac{\gamma}{1 + \omega_0^2\tau_c^2}\left(\frac{E_b}{K_BT}\right)\exp\left(-\frac{E_b}{K_BT}\right) \qquad (117)$$

The discrepancies between the rate given by this expression and that calculated by Carmeli and Nitzan are mostly due to their improved Markovian part.

We believe that the arguments above should convince the reader that the interesting phenomenon detected by Carmeli and Nitzan is another manifestation of the decoupling effect, well understood at least since 1976 (see ref. 86). The only physical systems, the dissipative properties of which are completely independent of whether or not an external field is present, are the purely ideal Markovian ones. Non-Markovian systems in the presence of a strong external field provoking them to exhibit fast oscillations are characterized by field-dependent dissipation properties. These decoupling effects have also been found in the field of molecular dynamics in the liquid state studied via computer simulation (see Evans, Chapter V in this volume).

The region ranging from $\gamma = \infty$ to $\gamma \sim \omega_b$ can also be explored using the RMT. In Section IV we showed that the basic ideas of the RMT supplemented by the generalization of the Kramers theory to the multidimensional case allows us to recover the simple expression first derived by Grote and Hynes.[37] This quite interesting formula reads

$$k_H = k_{\text{TST}}\frac{\lambda_r}{\omega_b} \qquad (118)$$

where

$$k_{TST} = \frac{\omega_0}{2\pi} \exp\left(\frac{-E_b}{K_B T}\right)$$

and

$$\lambda_r = \frac{\omega_b^2}{\lambda_r + \hat{\varphi}(\lambda_r)} \tag{119}$$

with

$$\hat{\varphi}(\lambda_r) = \int_0^\infty dt \exp(-\lambda_r t)\varphi(t) \tag{120}$$

In the case considered by Carmeli and Nitzan,[84b] we obtain

$$\lambda_r = \frac{\omega_b^2}{\lambda_r + \Omega^2/(\lambda_r + \Gamma)} \tag{121}$$

In the high-friction region when the additive noise is almost white, we have

$$\frac{\omega_b^2}{\gamma} \ll \omega_b \ll \gamma \ll \Gamma \tag{122}$$

This means that ω_b^2/γ can be disregarded compared to both Γ and $\Omega^2/\Gamma = \gamma$. In other words,

$$\lambda_r = \frac{\omega_b^2}{\gamma} \tag{123}$$

which is the Kramers high-friction result.

As Γ decreases, this simple formula results in an increase of λ_r in qualitative agreement with Fig. 11 from the Carmeli and Nitzan work.

We are thus in a position to state that the RMT provides quite simple formulas which give a clear and simple picture of the chemical reaction rate as a function of $\omega_b \tau_c$ (or $\omega_0 \tau_c$). The two-step model introduced in Section II.A naturally leads to

$$k_{int}^{-1}(\tau_c) = k_{high}^{-1}(\tau_c) + k_{low}^{-1}(\tau_c) \tag{124}$$

This expression gives a useful semiquantitative description of chemical rate in the most general case, Markovian or non-Markovian and in any range of friction. Figure 11 illustrates the corresponding results, which are qualitatively the same as those of Carmeli and Nitzan.

The generalization of the Kramers theory involving the problem of multiplicative fluctuations is still an open field of investigation. A large part of the discrepancies between the AEP and the other approaches do certainly derive from the fact that this theory is applied to a set of differential equations, the formal expression of which seems to be not completely legitimate. For instance, a rigorous microscopic derivation certainly cannot result in formal expressions such as those of Eqs. (38) and (44).

The physical reasons for the acceleration of the chemical reaction rate as a result of a coupling with nonreactive modes is the subject of interesting investigations which seem to support the point of view according to which the AEP can successfully be applied to the study of transient dynamics even when failing in reproducing the correct equilibrium distributions which are attained at a much larger time scale. The results of Chapter X show that when the time necessary to get the final equilibrium state (of Gaussian type) is virtually infinitely large, a quasi-equilibrium state is predicted by the AEP which is found to be in surprisingly good agreement with the results of analog simulation. This opens a fertile new field of investigation, which could be of significant relevance for the theory of chemical reactions and especially that of enzyme chemistry.

References

1. J. E. Lefler and E. Grunwald, *Rates and Equilibria of Organic Reactions*, Wiley, New York, 1963.

2. R. W. Weston, Jr., and H. A. Schwarz, *Chemical Kinetics*, Prentice-Hall, Englewood Cliffs, New York, 1972.

3. H. Eyring, S. H. Lin, and S. M. Lin, *Basic Chemical Kinetics*, Wiley, New York, 1980.

4. B. Williamson and V. K. La Mer, *J. Amer. Chem. Soc.*, **70**, 547 (1960).

5. M. T. Emerson, E. Grunwald, and R. A. Kromhout, *J. Chem. Phys.*, **33**, 547 (1960).

6. (a) M. G. Evans and M. Polanyi, *Trans. Faraday Soc.*, **31**, 875 (1935); (b) H. Eyring, *J. Chem. Phys.*, **3**, 107 (1935); (c) H. Eyring and W. F. K. Wynne-Jones, *J. Chem. Phys.*, **3**, 492 (1935).

7. (a) P. Pechukas, *Ber. Bunsenges. Phys. Chem.*, **86**, 372 (1982); (b) D. G. Trulhar, W. L. Hase, and J. T. Hynes, *J. Phys. Chem.*, **87**, 2664 (1983).

8. (a) K. J. Laidler, *Theories of Chemical Reactions*, McGraw-Hill, New York, 1969; (b) I. W. M. Smith, *Kinetics and Dynamics of Elementary Gas Reactions*, Butterworths, London, 1980.

9. R. Marcelin, *C. R. Hebd. Seances Acad. Sci. (Paris)*, **158**, 116, 407 (1914).

10. J. A. Christiansen, *Z. Phys. Chem.*, **B33**, 145 (1936).

11. (a) H. A. Kramers, *Physica*, **7**, 284 (1940); (b) S. Chandrasekhar, *Rev. Mod. Phys.*, **15**, 1 (1943).

12. R. Kubo, *Rep. Prog. Phys.*, **29**, 255 (1966).

13. M. Büttiker, E. P. Harris, and R. Landauer, *Phys. Rev.*, **B28**, 1268 (1983).

14. (a) N. G. van Kampen, *J. Stat. Phys.*, **17**, 71 (1977); (b) N. G. van Kampen, *Suppl. Prog. Theor. Phys.*, **64**, 389 (1978).

15. (a) T. Fonseca, P. Grigolini, and P. Marin, *Phys. Lett.*, **A88**, 117 (1982). (b) T. Fonseca, J. A. N. F. Gomes, and P. Grigolini, *Int. J. Quantum Chem.*, **23**, 473 (1983).

16. J. L. Skinner and P. G. Wolynes, *J. Chem. Phys.*, **72**, 4913 (1980).

17. H. C. Brinkman, *Physica*, **22**, 29 (1956).

18. P. B. Visscher, *Phys. Rev.*, **B13**, 3272 (1976).

19. C. Blomberg, *Physica*, **A86**, 49 (1977).

20. R. Landauer and J. A. Swanson, *Phys. Rev.*, **121**, 1668 (1961).

21. R. J. Donnelly and P. H. Roberts, *Proc. Roy. Soc.*, **A312**, 519 (1969).

22. R. S. Larson and M. D. Kostin, (a) *J. Chem. Phys.*, **69**, 4821 (1978); (b) *ibid.*, **72**, 1392 (1980).

23. P. B. Visscher, *Phys. Rev.*, **B14**, 347 (1976).

24. D. K. Garrity and J. L. Skinner, *Chem. Phys. Lett.*, **95**, 46 (1980).

25. J. A. Montgomery, Jr., D. Chandler, and B. J. Berne, *J. Chem. Phys.*, **70**, 4056 (1979).

26. B. Carmeli and A. Nitzan, (a) *Phys. Rev. Lett.*, **51**, 233 (1983); (b) *Phys. Rev.*, **A29**, 1481 (1984).

27. B. H. Lavenda, *Lett. Nuovo Cimento.*, **37**, 20 (1983).

28. R. F. Grote and J. T. Hynes, (a) *J. Chem. Phys.*, **74**, 4465 (1980); (b) *ibid.*, **75**, 2191 (1981).

29. K. M. Christoffel and J. M. Bowman, *J. Chem. Phys.*, **74**, 5057 (1981).

30. T. Fonseca, J. A. N. F. Gomes, P. Grigolini, and F. Marchesoni, (a) *J. Chem. Phys.*, **79**, 3320 (1983); (b) *ibid.*, **80**, 1826 (1984).

31. K. Lindenberg and V. Seshadri, *Physica*, **A109**, 483 (1981).

32. R. L. Stratonovisch, *Topics in the Theory of Random Noise*, Gordon and Breach, New York, 1967, Vol. I, p. 115.

33. G. H. Weiss, in *Stochastic Processes in Chemical Physics: The Master Equation*, I. Oppenheim, K. E. Schuder, and G. H. Weiss, ed., MIT Press, Cambridge, 1977, p. 361.

34. A. Schenzle and H. Brand, *Phys. Rev.*, **A20**, 1628 (1979).

35. R. Graham and A. Schenzle, *Phys. Rev.*, **A26**, 1676 (1982).

36. S. H. Northrup and J. T. Hynes, *J. Chem. Phys.*, **73**, 2700 (1980).

37. R. F. Grote and J. T. Hynes, *J. Chem. Phys.*, **73**, 2715 (1980).

38. M. W. Evans, *J. Chem. Phys.*, **76**, 5480 (1982).

39. M. W. Evans, G. T. Evans, W. T. Coffey, and P. Grigolini, *Molecular Dynamics*, Wiley, New York, 1982.

40. N. G. van Kampen, *J. Stat. Phys.*, **24**, 175 (1981).

41. N. G. van Kampen, *Stochastic Processes in Physics and Chemistry*, North-Holland, Amsterdam, 1981.

42. M. Suzuki, *Adv. Chem. Phys.*, **46**, 195 (1980).

43. P. De Kepper and W. Horsthemke, *C. R. Acad. Sci. Paris Ser. C*, **287**, 251 (1978).

44. P. Hänggi and P. Riseborough, *Phys. Rev.*, **A27**, 3329 (1983).

45. K. Kitahara, W. Horsthemke, R. Lefever, and Y. Inaba, *Prog. Theor. Phys.*, **64**, 1233 (1980).

46. P. Hänggi, *Phys. Rev.*, **A26**, 2996 (1982).

47. V. I. Klyotskin, *Radiophys. Quantum Electron.*, **20**, 382 (1977).

48. J. M. Sancho and M. San Miguel, (a) *Z. Phys.*, **B36**, 357 (1980); (b) *ibid.*, **43**, 361 (1981).

49. L. Garrido and J. M. Sancho, *Physica*, **A115**, 479 (1982).

50. J. M. Sancho, M. San Miguel, S. L. Katz, and J. D. Gunton, *Phys. Rev.*, **A26**, 1589 (1982).

51. P. Hänggi, *Z. Phys.*, **B31**, 407 (1978).

52. R. Kubo, *J. Math. Phys.*, **4**, 174 (1963).

53. N. G. van Kampen, *Phys. Rev.*, **24**, 171 (1976).

54. M. San Miguel and J. M. Sancho, *J. Stat. Phys.*, **22**, 605 (1980).

55. J. S. Langer, *Ann. Phys.* (*N.Y.*), **54**, 258 (1969).

56. P. Talkner and P. Hanggi, *Phys. Rev.*, **A29**, 768 (1984).

´7. Z. Schuss, *S.I.A.M. Rev.*, **22**, 119 (1980).

58. K. Schulter, Z. Schulton, and Z. Szabo, *J. Chem. Phys.*, **74**, 4426 (1981).

59. P. Hänggi, F. Marchesoni, and P. Grigolini, *Z. Phys.* **B56**, 333 (1984).

60. T. Fonseca, P. Grigolini, and D. Pareo, *J. Chem. Phys.* (in press); S. Faetti, L. Fronzoni, and P. Grigolini, *Phys. Rev. A* (in press); P. Grigolini, Proceeding of the Florence Meeting on Dynamical Systems, January 1985.

61. P. Hanggi, T. J. Mroczkowski, F. Moss, and P. V. E. McClintock *Phys. Rev. A* (in press).

62. H. Mori, (a) *Prog. Theor. Phys.*, **34**, 399 (1965); (b) *ibid.*, **34**, 423 (1965).

63. M. Ferrario and P. Grigolini, *J. Math. Phys.*, **20**, 2567 (1979).

64. P. Grigolini, *J. Stat. Phys.*, **27**, 283 (1982).

65. P. Hänggi and E. Mogtabai, *Phys. Rev.*, **A26**, 1168 (1982).

66. F. Marchesoni and P. Grigolini, *J. Chem. Phys.*, **78**, 6287 (1983).

67. E. Helfand, *J. Chem. Phys.*, **54**, 4651 (1971).

68. C. Blomberg, *Chem. Phys.*, **37**, 219 (1979).

69. E. Guardia, F. Marchesoni, and M. San Miguel, *Phys. Lett.*, **A100**, 15 (1984).

70. S. A. Adelman, *J. Chem. Phys.*, **69**, 124 (1976).

71. C. W. Gardiner, *J. Stat. Phys.*, **30**, 157 (1983).

72. P. Hänggi, *Phys. Lett.*, **A78**, 304 (1980).

73. S. Faetti, P. Grigolini, and F. Marchesoni, *Z. Phys.*, **B47**, 353 (1982).

74. F. Marchesoni and P. Grigolini, *Physica*, **A121**, 269 (1983).

75. H. Haken, *Handbuch der Physik*, vol. 25/2C, Springer, Berlin, 1970.

76. J. T. Hynes, in *The Theory of Chemical Reaction Dynamics*, M. Baer, ed., C R C Press, Boca Raton, FL, in press, (1984).

77. D. L. Hasha, T. Eguchi, and J. Jonas, (a) *J. Chem. Phys.*, **75**, 1573 (1981); (b) *ibid.*, *J. Am. Chem. Soc.*, **104**, 2290 (1982).

78. (a) S. P. Velsko and G. R. Fleming, *J. Chem. Phys.*, **76**, 3553 (1982); (b) S. P. Velsko and G. R. Fleming, *Chem. Phys.*, **65**, 59 (1982); (c) S. P. Velsko, D. H. Waldek, and G. R. Fleming, *J. Chem. Phys.*, **78**, 249 (1983); (d) S. P. Velsko, D. H. Waldek, and G. R. Fleming, *Chem. Phys. Lett.*, **93**, 322 (1983).

79. B. Bagchi and D. W. Oxtoby, *J. Chem. Phys.*, **78**, 2735 (1983).

80. G. Rothenberger, D. K. Negus, and R. M. Hochstrasser, *J. Chem. Phys.*, **79**, 5360 (1983).

81. E. G. d'Agliano, W. L. Schaich, P. Kumar, and H. Suhl, in *Nobel Symposium 24, Collective Properties of Physical Systems*, B. Lundquist, ed., Academic Press, New York, 1973, p. 200.

82. J. S. McCaskill and R. G. Gilbert, *Chem. Phys.*, **44**, 389 (1979).

83. C. V. Shank, E. P. Ippen, O. Teschke, and K. B. Eisenthal, *J. Chem. Phys.*, **67**, 5547 (1977).

84. (a) B. Carmeli and A. Nitzan, *J. Chem. Phys.* **79**, 393 (1983); (b) *Phys. Rev. Lett.*, **49**, 423 (1982).

85. (a) G. R. Welch, B. Somogyi, and S. Damjanovich, *Prog. Biophys. Mol. Biol.*, **39**, 109 (1982); (b) J. A. McCammon, *Rep. Prog. Phys.*, **47**, 1 (1984).

86. P. Grigolini, *Mol. Phys.*, **31**, 1717 (1976)

X

EXPERIMENTAL INVESTIGATION ON THE EFFECT OF MULTIPLICATIVE NOISE BY MEANS OF ELECTRIC CIRCUITS

S. FAETTI, C. FESTA, L. FRONZONI, and P. GRIGOLINI

CONTENTS

I. INTRODUCTION

The major aim of the present chapter is to utilize the theoretical tools of Chapter I to face a subject which is becoming of increasing interest: the problem of noise-induced phase transitions.[1] The theoretical starting point to investigate these phenomena is usually typified by stochastic differential equations such as

$$\dot{x} = f(x) + g(x)F(t) \qquad (1.1)$$

where $F(t)$ is a Gaussian noise defined by

$$\langle F(0)F(t) \rangle = 2Q\,\delta(t) \qquad (1.2)$$

$F(t)$ is usually named multiplicative stochastic noise, since it multiplies the function $g(x)$, which depends on the variable of interest x. Due to the multiplicative nature of the stochastic force $F(t)$, it is not immediately

445

clear which is the Fokker-Planck form to be associated with Eq. (1.1). Two major proposals have found large applications in the literature of stochastic processes. The first, especially popular among physicists, is the Stratonovich[2] algorithm (see, for example, Schenzle and Brand[3]):

$$\frac{\partial}{\partial t}\sigma(x;t) = \left\{ -\frac{\partial}{\partial x}f(x) + Q\frac{\partial}{\partial x}g(x)\frac{\partial}{\partial x}g(x) \right\}\sigma(x;t) \qquad (1.3)$$

The second one is the Itô form[4]

$$\frac{\partial}{\partial t}\sigma(x;t) = \left\{ -\frac{\partial}{\partial x}f(x) + Q\frac{\partial^2}{\partial x^2}g^2(x) \right\}\sigma(x;t) \qquad (1.4)$$

It is evident that when $g(x) = 0$ the two rules lead to different Fokker-Planck equations. This raised some controversy among physicists as to which should be applied. West et al.[5] studied the case of a stochastic linear oscillator (driven by an additive stochastic force). They pointed out that the variable energy turns out to be driven by a Langevin equation of the same form as Eq. (1.1). This was proved to lead to a Fokker-Planck equation consistent with the original one (affected by no ambiguity since it concerns a purely additive case) only when the Stratonovich rule is applied. Van Kampen[6] argued that this is not a decisive argument in favor of the Stratonovich rule, as this only proves that the Stratonovich rule has to be invoked when transforming the Langevin equation within the Stratonovich algorithm. According to van Kampen[6] the controversy between the Itô and Stratonovich rules stems from the improper use of nonlinear Langevin equations such as Eq. (1.1). This controversy could be completely bypassed if use would be made of his master equation approach. Smythe et al.[7,8] made an analog experiment to assess whether Itô or Stratonovich provides an accurate description of the physical world. They concluded that the Stratonovich rule accurately describes what actually happens in nature.

We believe that a definite conclusion on such an interesting issue cannot be reached without facing the problem of how to derive Eq. (1.1) from a real system involving virtually infinite degrees of freedom. A rigorous treatment of this problem is certainly beyond our current means. Nevertheless, the reduced model theory (RMT), an important corner of our delta-like theoretical background (see Chapter I), allows us to replace the rigorous, ideal, procedure with a faithful simulation. In a sense we share van Kampen's point of view: According to the prescription of the RMT, we never have direct recourse to equations of the same kind as Eq. (1.1). The interaction of x with the stochastic environment always takes place via intermediate auxiliary (or " virtual") variables which interact with their thermal baths in a standard unambiguous manner.

We shall take the opportunity to shed further light on the nature of this approach by illustrating how to derive an unambiguous picture of the phenomenon which Eq. (1.1) would aim at describing. A rigorous equation of motion for the variable x is

$$\frac{d}{dt}x = \Gamma x \tag{1.5}$$

where $\Gamma \equiv \mathscr{L}^+$, that is Γ has to be identified with the operator adjoint to the dynamical operator which drives the corresponding Schrödinger picture [see Eq. (2.1) of Chapter I]:

$$\frac{\partial}{\partial t}\rho(t) = \mathscr{L}\rho(t) \tag{1.6}$$

where $\rho(t)$ represents the probability distribution of all variables of the system. When dealing with a physical problem, $\mathscr{L}$ should be identified with the operator iL denoting, in the classical case, the classical Poisson brackets. To make contact with the notation used in Chapter I, let us note that the set of variables of interest $\mathbf{a}$ is assumed to be reduced now to the monodimensional variable x. As noted already in that introductory chapter, we have to define a set of auxiliary variables $\mathbf{b}_V$, the main role of which is to mimic the influence of the real irrelevant variables $\mathbf{b}_R$. This obliges us to replace the rigorous operator iL with an effective Liouvillian iL_{eff}, where dissipative and diffusional terms are present. If x is the space coordinate of a Brownian particle, it seems to us natural to regard Eq. (1.1) as the inertialess contraction of a more complete set of equations where the corresponding velocity is present. Furthermore, the assumption that $F(t)$ is a white Gaussian noise is certainly too drastic.

These remarks naturally lead us to replace Eq. (1.1) with

$$\begin{aligned}
\dot{x}(t) &= v(t) \\
\dot{v}(t) &= -\gamma v(t) + G(x) + g(x)\xi(t) + f_a(t) \\
\dot{\xi}(t) &= -\lambda\xi(t) + f_s(t)
\end{aligned} \tag{1.7}$$

where $G(x)$ is assumed to be the derivative of a potential V,

$$G(x) \equiv -\frac{d}{dx}V \tag{1.8}$$

The additive stochastic force $f_a(t)$, assumed to be a white Gaussian noise, is defined by

$$\langle f_a(0)f_a(t)\rangle = 2D_a\delta(t) \tag{1.9}$$

and is related to the friction γ via the standard fluctuation-dissipation relationship

$$\gamma\langle v^2 \rangle_{eq} = D_a \qquad (1.10)$$

To recover the ideal case of Eq. (1.1) we would have to assume that $\langle v^2 \rangle_{eq}$ vanishes. The analog simulation of Section III, however, will involve additive stochastic forces, which are an unavoidable characteristic of any electric circuit. It is therefore convenient to regard D_a as a parameter the value of which will be determined so as to fit the experimental results. In the absence of the coupling with the variable ξ, Eq. (1.7) would describe the standard motion of a Brownian particle in an external potential field $G(x)$. This potential is modulated by a fluctuating field ξ. The stochastic motion of ξ, in turn, is driven by the last equation of the set of Eq. (1.7), which is a standard Langevin equation with a white Gaussian noise $f_s(t)$ defined by

$$\langle f_s(0)f_s(t) \rangle = 2D_\xi \delta(t) \qquad (1.11)$$

$$D_\xi \equiv \lambda \langle \xi^2 \rangle_{eq} \qquad (1.11')$$

When the time $1/\lambda$ or $1/\gamma$ becomes very small, the rough approaches consist in assuming that v and ξ relax almost instantaneously to their equilibrium values; that is, one puts $\dot{v} = 0$ and $\dot{\xi} = 0$ in the second and the third of the set (1.7). The expressions for $v(t)$ and $\xi(t)$ so obtained are then replaced in the first equation of this set. This allows Eq. (1.1) to be recovered [if $f_a(t) = 0$] provided that

$$f(x) \equiv \frac{G(x)}{\gamma} \qquad (1.12)$$

and

$$F(t) \equiv \frac{f_s(t)}{\lambda\gamma} \qquad (1.13)$$

Equations (1.13), (1.11), and (1.11') imply that $F(t)$ is a white Gaussian noise defined by

$$\langle F(0)F(t) \rangle = 2Q \delta(t) \qquad (1.14)$$

where

$$Q \equiv \frac{\langle \xi^2 \rangle_{eq}}{\lambda\gamma^2} \qquad (1.15)$$

Now the question arises whether Eq. (1.3) or Eq. (1.4) should be used.

The Fokker-Planck equation corresponding to the system of Eq. (1.7) is found unambiguously to be

$$\frac{\partial}{\partial t}\rho(x,v,\xi;t) = \left\{-v\frac{\partial}{\partial x} + \gamma\left[\frac{\partial}{\partial v}v + \langle v^2\rangle_{eq}\frac{\partial^2}{\partial v^2}\right] - g(x)\xi\frac{\partial}{\partial v}\right.$$

$$\left. + \lambda\left[\frac{\partial}{\partial \xi}\xi + \langle \xi^2\rangle_{eq}\frac{\partial^2}{\partial \xi^2}\right]\right\}\rho(x,v,\xi;t) \qquad (1.16)$$

Analytic solutions of Eq. (1.16) cannot be obtained, and thus one must make some approximations. In many physical systems, γ and λ are very large ($\gamma \to \infty$, $\lambda \to \infty$); therefore both v and ξ can be considered as fast relaxing variables and eliminated by an adiabatic elimination procedure (AEP). The AEP of Chapter II allows us to derive from Eq. (1.16) the equation of motion of $\sigma(x;t)$ regarded as being the "contracted distribution":

$$\sigma(x;t) = \int dv\,d\xi\,\rho(x,v,\xi;t) \qquad (1.17)$$

When the fourth-order interaction terms are retained, we obtain

$$\frac{\partial}{\partial t}\sigma(x;t) = \left\{\frac{\partial}{\partial x}\frac{1}{\gamma}V' + \frac{1}{\gamma^3}\frac{\partial}{\partial x}V'V'' + \frac{\langle \xi^2\rangle_{eq}}{\lambda\gamma^2}\frac{\partial}{\partial x}g(x)\frac{\partial}{\partial x}g(x)\right.$$

$$\left. + \frac{\langle \xi^2\rangle_{eq}}{\lambda\gamma^2}\frac{1}{1+R}\frac{\partial}{\partial x}g(x)g'(x) + \frac{D_a}{\gamma^2}\frac{\partial^2}{\partial x^2} + \frac{D_a}{\gamma^4}\frac{\partial}{\partial x}V''\frac{\partial}{\partial x}\right\}\sigma(x;t)$$

$$(1.18)$$

where

$$R = \frac{\gamma}{\lambda} \qquad (1.19)$$

R is an important parameter of this theory which expresses the ratio of the correlation time of the multiplicative stochastic noise ξ to that of the velocity v. The second term on the right-hand side of Eq. (1.18) is a potential term independent of the intensity of the noise $\langle \xi^2\rangle_{eq}$, the role of which becomes increasingly irrelevant as $\gamma \to \infty$. We cannot, however, neglect the last two terms, which express the lowest-order contribution coming from the noise ξ.

We see that $R = 0$ leads us to the Itô form, whereas $R = \infty$ means that the Stratonovich rule is followed. Therefore the AEP allows us to shed light on the Itô-Stratonovich controversy. In particular, the validity of either approach is found to be strictly dependent on the relative time scales of the v and ξ variables.

The remarks above can be extended from physical to "nonphysical" systems, that is, cooperative systems concerning biology, economics, and so on. For example, the genetic model studied in the next chapter of this volume aims at determining the effects of environmental fluctuations on gene frequency. However, a more realistic model should take into account that the direct influence of environment is exerted on the phenotypic characteristics rather than the genotypic ones. This means that a well-defined lag time characterizes the influence of environment on the genotype frequency.

In this chapter we discuss to what extent this point of view is supported by the results of analog simulation, while keeping in mind the special case where

$$V \equiv \frac{V_0(x^2 - a^2)^2}{a^4} \tag{1.20}$$

which is a double-well potential.

Section II is devoted to reviewing accurately the investigative work done so far in the field of analog simulation. After assessing that no relevant attention has been devoted yet to the point of view stressed above, we describe in Section III our experimental apparatus. Section IV will be devoted to a comparison of experimental and theoretical results. Concluding remarks and a perspective of further investigative work are given in Section V.

II. REVIEW OF EXPERIMENTAL RESEARCH

Stratonovich[2] first studied the influence of external fluctuations via a vacuum tube oscillator. He noticed a phenomenological behavior reminiscent of that of physical systems far from equilibrium. His pioneer work showed that the use of electric circuits is a simple means of shedding light on general problems, thereby stimulating further experimental work of this kind.

Landauer[9] studied the behavior of a bistable system consisting of a tunnel diode and focused his attention on the activation jump between the two states resulting from thermal fluctuations. Later, Matsuno[10] analyzed the power spectrum of the noise affecting the electric conductivity of a Gunn diode. Diode oscillations were proven to be accompanied by modulation of both amplitude and frequency. The noise spectrum was shown to exhibit typical $1/f$ characteristics, and fluctuations were found to appear in the region close to the critical threshold with an increasing relaxation time as the threshold was approached. This behavior is largely reminiscent of that exhibited by systems undergoing first- and second-order phase transitions. This striking analogy has been stressed by Woo and Landauer[11] by means of experimental investigation on a parametrically excited subharmonic oscillator. They determined the physical conditions necessary to get a source or pump excitation almost completely void of fluctuations. In accordance with the fluctuation-dissipation theory, they traced back the presence of fluctuations to circuit resistances.

In 1973, Kawakubo et al.[12] did a more detailed investigation on the behavior of a Gunn diode. They proved that the distribution of the fluctuations near threshold is broadened and deviates from the Gaussian type. The non-Gaussian distribution near threshold comes from an emergence of nonlinearity which is due to the enhancement of fluctuations associated with the instability of the system.

A Gunn diode again[13] was used to study the low-frequency spectrum. In this frequency domain a critical region was detected where the system alternates between oscillatory and nonoscillatory behavior with a random time period.

Drosdziok[14] used a self-sustained oscillator to study the transition to the state of self-sustained oscillations, and he pointed out the strong analogy between this transition and that occurring in a laser.

Experiments on Wien-bridge oscillators[15] showed such anomalous fluctuations to exhibit a variance proportional to $(\beta_c - \beta)^{-1}$, where β is the reaction factor and β_c the corresponding threshold value. The relaxation time of the fluctuation correlation function, too, exhibited a $(\beta_c - \beta)^{-1}$ behavior, thereby proving the existence of a critical slowing down.

The fluctuations of a self-excited oscillator have been studied via a model based on the Van der Pol equation.[16] Such an equation has been used to account for the amplitude and phase fluctuations. A growth of the time coherence of phase above threshold is attributed to a decrease in an apparent diffusion coefficient for phase fluctuations.

The time behavior as experimentally detected on a Wien-bridge oscillator[17] has been accounted for by using the theories of Kubo et al.[18]

Morton and Corrsin[19] used an electric analog device to simulate the Fokker-Planck equation of a linear and cubic spring.

Kawakubo et al.[20] studied the behavior of a self-excited oscillator under the influence of an external noise. They found that the self-oscillations are suppressed by the external noise with a threshold behavior reminiscent of the phase transitions of a ferroelectric sample.

Horn et al.[21] simulated phase transitions of the first and second kind, and a three-critical point, via a Wien-bridge oscillator. Such behavior is fairly well accounted for by the Landau theory except for very close to the critical point. Furthermore, Horn et al. observed that the phase of their oscillator had a

short coherence time. Kabashima et al.[22] built up a parametric oscillator with
ferromagnetic nuclei. With this tool, they found a novel kind of transition
from an oscillatory to a nonoscillatory state induced by an external noise.
The oscillator is pumped with a sinusoidal current upon which an almost
white noise is superimposed. Such an external noise changes the oscillation
time decay, and a slowing down appears in proximity to the critical threshold.
The behavior of variance contrasts with that of the system above, in that at
the critical threshold no divergence is exhibited.

The analogy between these phenomena and phase-transition processes
stimulated the development of a new approach based on electric circuits that
had no oscillating behavior but nevertheless exhibited state-transition phenomena. Morimoto[23] built a device consisting of two transistors coupled to
an emitter via a feedback resistance. As the feedback resistance varies, phase
transitions of the first kind are found, with corresponding fluctuations close
to the threshold. He also found a three-state phase diagram as a function of
feedback resistance. These states are characterized by different values of
voltage and current in the different branches of the circuit and exhibit
different stochastic properties.

The two-transistor circuit is then replaced by a more complicated one with
three transistors.[24] This system shows a still wider variety of physical conditions, in that five different states are observed. Several oscillations with decay rates, changing from one state to another, are detected. Furthermore, the
author detected critical fluctuations and slowing-down phenomena. These
devices are characterized by metastable and unstable states. This distinctive
feature makes them very interesting compared to the preceding devices.

The role played by an external noise on a self-sustained oscillator has been
studied by Harada et al.[25] Their device exhibits two kinds of instability,
termed hard and soft mode. These authors focused their attention on charge
and current fluctuations. Charge fluctuations exhibit an anomalous increase
in the variance of noise and a slowing down in both the hard and soft modes,
while the variance of the current increases only in the hard mode. A new
physical observable termed "irreversible circulation of the fluctuations" is
introduced. This provides a measure of the deviation from detailed balance.
Good agreement is obtained between experimental and theoretical results.

An interesting investigation on the influence of multiplicative non-white
noise in an analog circuit simulating a Langevin equation of a Brownian
particle in a double-well potential has been carried out by Sancho et al.[26]
This device allowed them to study the stationary properties as a function of
the noise correlation time. Theory in a white-noise limit cannot provide a
satisfactory explanation for experimental results such as a relative maximum
of the probability distribution and the maximum position in the stationary
distribution for noises of weak intensity.

We would like to recall the experimental work of Smythe et al.[7,8] already mentioned in Section I, who tried to decide whether the Stratonovich or the Itô prescription should be used.

Finally, we would like to mention the work of Fauve and Heslot,[27] who studied the contemporary action of a periodic and stochastic force on a bistable electric system. They found that applied force and noise exhibit resonant behavior. Resonance takes place when the external force time period is equal to the Kramers rate of escape.[28]

None of the experiments done so far seems to deal explicitly with the problems posed in Section I. In the next section we shall give a detailed account of a further analog experiment inspired by those reviewed in this section, the main aim of which is to assess whether or not the point of view of Section I is correct. The interpretation of the corresponding results (Section IV) will involve a joint use of the theoretical tools mentioned in Chapter I.

III. EXPERIMENTAL APPARATUS

A. The Analog Circuit

In this section we describe the electronic apparatus which allows us to simulate the set of Eq. (1.7) when the potential of Eq. (1.8) is given the form of Eq. (1.20).

A block diagram of the electric circuit we used is shown in Fig. 1. To minimize the effects of thermal drift, offset voltages, linearity errors, and internal noise, we used the minimum possible number of active components. The first integration, therefore, is simply done by using passive components (R_1, R_2, C_1). This is followed by a Miller integrator (utilizing a Texas Instruments TL080), the output V_2 of which is applied to the first multiplier/adder. This is a four-quadrant multiplier (Analog Devices AD534) with three inputs (X, Y, Z), all differential, and providing a high

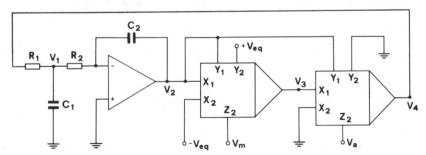

Figure 1.   A block diagram of the electronic circuit simulating a double-well potential.

degree of flexibility, with the transfer function $(X_1 - X_2)(Y_1 - Y_2)/V_r + Z_2$ (V_r is a reference voltage internal to the device).

We apply V_2 to both X_1 and Y_1, and the voltages $-V_{eq}$ and $+V_{eq}$ to X_2 and Y_2, respectively. As shown later on, $-V_{eq}$ and $+V_{eq}$ have to be interpreted as being the equilibrium positions of the double-well potential. The input Z_2 is used to apply to the circuit the multiplicative noise V_m. The output V_3 of the first multiplier is multiplied by V_2 via the second multiplier, which is identical to the first. This also performs the function of adding the additive noise V_a, via Z_2, to the product $V_2 V_3$. The output V_4 of this multiplier is connected to the first integrator, thereby closing the feedback loop of the Miller integrator. If no noise is applied to the circuit ($V_m = 0$, $V_a = 0$), the value attained by V_2 should have to be equal to either $-V_{eq}$ or $+V_{eq}$. Nevertheless, because of various offset voltages, as a rule this does not happen, thereby rendering it necessary that these be balanced.

The four functions of the electric circuit of Fig. 1 are expressed by the following equations:

$$\frac{V_4 - V_1}{R_1} = \frac{V_1}{R_2} + C_1 \dot{V}_1 \quad \text{(first integrator)} \tag{3.1}$$

$$\frac{V_1}{R_2} = -\dot{V}_2 C_2 \quad \text{(second integrator)} \tag{3.2}$$

$$V_3 = \frac{(V_2 + V_{eq})(V_2 - V_{eq})}{V_r} + V_m \quad \text{(first multiplier)} \tag{3.3}$$

$$V_4 = \frac{V_3 V_2}{V_r} + V_a \quad \text{(second multiplier)} \tag{3.4}$$

The combined use of these equations results in the circuit equation:

$$\ddot{V}_2 = -\frac{R_1 + R_2}{R_1 R_2 C_1} \dot{V}_2 - \frac{V_2^3}{R_1 R_2 C_1 C_2 V_r^2} + \frac{V_{eq}^2}{R_1 R_2 C_1 C_2 V_r^2} V_2$$

$$- \frac{V_m}{R_1 R_2 C_1 C_2 V_r} V_2 - \frac{V_a}{R_1 R_2 C_1 C_2} \tag{3.5}$$

This equation is equivalent to the one describing the system of Eq. (1.7) when the potential V [eq. (1.8)] is given the double-well expression of Eq. (1.20), that is,

$$\ddot{x}(t) = -\gamma \dot{x}(t) - \beta x^3(t) + \alpha_0 x(t) + \xi(t) x(t) + f_a(t) \tag{3.6}$$

where $\alpha_0 = 4V_0/a^2$ and $\beta = 4V_0/a^4$.

TABLE I

Relation between the system and the analog simulator

x	V_2	α_0	$\dfrac{V_{eq}^2}{R_1 R_2 C_1 C_2 V_r^2}$
$v \equiv \dot{x}$	$-\dfrac{V_1}{R_2 C_2}$	ξ	$-\dfrac{V_m}{R_1 R_2 C_1 C_2 V_r}$
γ	$\dfrac{R_1 + R_2}{R_1 R_2 C_1}$	f_a	$-\dfrac{V_a}{R_1 R_2 C_1 C_2}$
β	$\dfrac{1}{R_1 R_2 C_1 C_2 V_r^2}$	a	V_{eq}

This table indicates how the variables and parameters of the system of eq. (1.7) are related to these of the electric circuit of eq. (3.5)

Table I relates the parameters of the analog circuit to those of the system of Eq. (1.7). As shown in Table I, the parameters defining the main properties of the system, except the noise $(\gamma, \alpha_0, \beta)$ are R_1, R_2, C_1, C_2, V_r, and V_{eq}. When aiming at assigning values to these parameters, we have to take into account the electrical characteristics of the active components (input and output impedances, frequency response, etc.).

B.   The Noise Source

To give $f_s(t)$, the noise driving the variable $\xi(t)$ via

$$\dot{\xi}(t) = -\lambda \xi(t) + f_s(t) \qquad (3.7)$$

[i.e., the last one of the set of equations of (1.7)], the required properties of a white Gaussian noise at a high level of accuracy, we applied the operating principle of a linear-feedback shift register (LFSR)[29,30] supplemented by significant improvements, which, to the best of our knowledge, are not yet available in the literature on this subject.

The main flaw of the noise obtained by filtered maximal sequences is that the amplitude distributions are usually affected by a skewness.[31] This problem has been faced by Tomlison and Galvin.[32] Their method of eliminating this effect does not seem to be suitable for our purposes, since we need to get completely rid of skewness. Furthermore, for our work program to be pursued, the cutoff frequency of the low-pass filter has to be selectable in a wide interval of values (as a function of λ).

Our method is based on the property that the m sequence generated by an LFSR with exclusive-OR feedback function is complementary to the one generated by the same LFSR with inverted feedback function (exclusive-NOR), that is, one sequence differs from the other in replacing the 0's by 1's

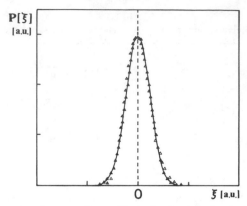

Figure 2.   Comparison between the amplitude distribution of noise in (Δ) the presence of skewness and ($\blacktriangle$) in the absence of skewness, respectively. The solid curve indicates the theoretical Gaussian curve.

and vice versa. As a consequence, the two sequences are affected by antisymmetric skewings, which are exactly compensated by alternating the two feedback complementary functions. If this is made at random time intervals, the time duration of which is, however approximately, as large as the time transit across the shift register, and if filter cutoff frequency is lower than /150 of the clock frequency, the resulting noise is fairly Gaussian (above a value five times the standard deviation). Moreover, from Fourier analysis the noise spectrum is proved to be virtually flat below the cutoff frequency. In the present experiment the cutoff frequency is 120 kHz. The commutation of the two feedback functions is driven by the sequence generated by an auxiliary LFSR. This is based on the same method as before, that of alternating the feedback functions, but with commutation being driven, at equal time intervals, by a flip-flop cascade. No starting circuit is required.

Figure 2 shows the distribution of noise amplitude in two different cases: (a) without alternating ex-or/ex-nor functions (open triangles), (b) with alternating ones (full triangles).

The electronic circuit generating the Gaussian noise is shown in Fig. 3. The auxiliary 17-stage LFSR, with taps at stages 5 and 17 of the shift register SR_1, is driven by stage 4 of the flip-flop cascade FF and in turn drives the main 17-stage LFSR, which also works with taps at stages 5 and 17 of the shift register SR_2. All shift registers use CMOS 4006 I.C. The low-pass filtering is performed by the feedback loop of the output amplifier. The λ parameter in Eq. (3.7) is related to the R_2C_2 time constant of the filter:

$$\lambda = \frac{1}{R_2C_2}$$

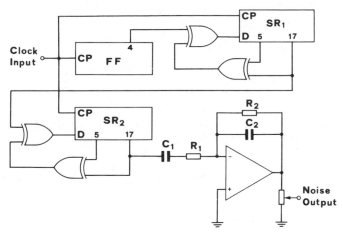

Figure 3.   A block diagram of the electronic circuit which generates the Gaussian noise.

The AC coupling of the output amplifier sets the lower cutoff frequency of the noise spectrum (time constant is R_1C_1). In our experiment, this frequency is below 1 Hz.

The original improvements described above allow us to build up a simple and inexpensive apparatus, which is nevertheless especially suitable for the particular purposes of this paper. Indeed, we limited ourselves to using a few standard monolithic components.

Note that even after turning off both the multiplicative and additive noise, residual multiplicative and additive noises are still present in the circuit. Throughout the applications of this chapter, we shall keep the additive noise turned off while providing an experimental estimate of the residual additive noise (see Section IV), which will turn out to be extremely weak. For the sake of simplicity we assumed that $f_a(t)$ of Eq. (1.7) is white and Gaussian even when it simulates such residual noise.

IV.   FLUCTUATING DOUBLE-WELL POTENTIAL: A COMPARISON OF THEORY AND EXPERIMENT

In this section we shall compare theory with the results of the experimental apparatus described in Section III. To make this comparison easier, let us consider the system of Eq. (1.7) when the potential of Eq. (1.20) is used. In this case,

$$V'(x) = -\alpha_0 x + \beta x^3 \qquad (4.1)$$

where

$$\alpha_0 \equiv \frac{4V_0}{a^2} \tag{4.2}$$

$$\beta \equiv \frac{4V_0}{a^4} \tag{4.2'}$$

The AEP developed in Chapter II allows us to obtain Eq. (1.18), which, when explicitly related to the potential of Eq. (1.20), reads

$$\frac{\partial}{\partial t}\sigma(x;t) = \left[-\frac{\partial}{\partial x}(dx - bx^3) + Q\frac{\partial}{\partial x}x\frac{\partial}{\partial x}x + D\frac{\partial}{\partial x^2} \right]\sigma(x;t) \tag{4.3}$$

where we are considering the case $g(x) = x$ and where

$$Q \equiv \frac{\langle \xi^2 \rangle_{eq}}{\lambda \gamma^2} \tag{4.3a}$$

$$D \equiv \frac{D_a}{\gamma^2} \tag{4.3b}$$

$$d_0 \equiv \frac{\alpha_0}{\gamma} \tag{4.3c}$$

$$b \equiv \frac{\beta}{\gamma} \tag{4.3d}$$

$$d \equiv d_0 - \frac{Q}{1+R} \tag{4.3e}$$

Equations with the same structure as Eqs. (4.3) have been studied extensively by Schenzle and Brandt[3] and other authors.[33,34,37] We shall summarize here the most interesting results of their investigation.

First, it can be assessed straightforwardly that Eq. (4.3) admits the steady state distribution

$$\sigma_{ss}(x) \equiv \sigma(x;\infty) = A\left(D + Qx^2\right)^{1/2[d_0/Q + bD/Q^2 - (2+R)/(1+R)]}\exp\left(-\frac{bx^2}{2Q}\right) \tag{4.4}$$

where A is a normalization constant defined by

$$A^{-1} \equiv \int dx \left(D + Qx^2\right)^{1/2[d_0/Q + bD/Q^2 - (2+R)/(1+R)]}\exp\left(-\frac{bx^2}{2Q}\right) \tag{4.5}$$

Note that $\sigma_{ss}(x)$ of Eq. (4.4) is the result of a sort of coarse-grained observation, the scarce resolution of which makes it impossible to observe the details of the competition between energy pumping and dissipation. When considering the whole system of Eq. (1.7), as will be shown later, the system reaches a compromise between the two processes, that is, a steady state which is not to be confused with an ordinary equilibrium state. This is the reason why we use the symbol $\sigma_{ss}(x)$ rather than $\sigma_{eq}(x)$ which is usually used to denote standard canonical equilibria. For noises of small intensity, $\sigma_{ss}(x)$ of Eq. (4.4) is virtually equivalent to two Gaussian distributions with center at $x = \pm a$ which are the minima of the double-well potential of Eq. (1.20). Adopting the symbols of the present section, we can also write

$$a = \left(\frac{d_0}{b}\right)^{1/2} \tag{4.6}$$

As Q increases, the two Gaussians shift toward $x = 0$ and are merged into a single distribution which peaks at $x = 0$ when Q reaches the threshold value Q_1:

$$Q_1 \equiv \frac{d_0(1+R)}{4+2R}\left[1 + \sqrt{1 + \frac{4bD}{d_0^2}\left(\frac{2+R}{1+R}\right)}\right] \tag{4.7}$$

Henceforth this will be referred to as the first threshold. This effect has often been termed "noise-induced phase transition." Figure 4 shows how $\sigma_{ss}(x)$ is

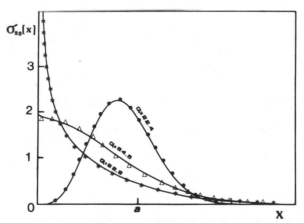

Figure 4.   Steady state probability distribution $\sigma_{ss}(x)$ for $R = 0.4$ and for some values of the intensity Q of the multiplicative stochastic noise. Curves correspond to the theoretical predictions of the AEP for $D = 0.0001$ at different values of Q, while symbols denote the corresponding experimental results obtained by using the electronic circuit. The parameters of the double-well potential are $d_0 = 109$ and $b = 122$ [Eq. (4.3)].

modified with increasing Q for a given value of the parameter R. The points correspond to the experimental values obtained by means of the electronic circuit of Section III. The full curves correspond to the theoretical predictions of the AEP.

A. The Case of a Purely Multiplicative Noise: Steady State Distribution

The limiting case $D = 0$ in Eq. (4.4) is pathological. In such a case the probability flux vanishes at $x = 0$, thereby making the solution dependent on the initial position x_0. If $x_0 > 0$, the steady state distribution is entirely bounded to the positive semiaxis. In this semiaxis region the solution is still described by Eq. (4.3) with $D = 0$. Precisely the reverse takes place when $x_0 < 0$. Henceforth we shall always refer ourselves to the case $x_0 > 0$. For $Q > Q_1$, the equilibrium distribution diverges at $x = 0$:

$$\lim_{x \to 0^+} \sigma_{ss}(x) \propto x^{-(d_0/Q)[(Q-Q_1)/Q_1]} \tag{4.8}$$

where, in this case,

$$Q_1 \equiv d_0\left(\frac{1+R}{2+R}\right) \tag{4.8'}$$

A second critical event takes place when

$$Q = Q_2 \equiv d_0(1+R) \tag{4.9}$$

(Henceforth this will be termed the second threshold.) When Q exceeds the second threshold, the solution of Eq. (4.4) loses meaning because it diverges more rapidly than x^{-1} and therefore proves to be nonnormalizable. In such a case the steady state solution is expressed via

$$\sigma_{ss}(x) = 0 \qquad \text{for } x < 0 \tag{4.10a}$$
$$\sigma_{ss}(x) = 2\delta(x) \qquad \text{for } x > 0 \tag{4.10b}$$

When $D = 0$ and $Q < Q_2$, the moments of the distribution of Eq. (4.4) are given by

$$\langle x^m \rangle_{ss} = \left(\frac{b}{2Q}\right)^{-m/2} \frac{\Gamma(d/2Q + m/2)}{\Gamma(d/2Q)} \tag{4.11}$$

where Γ stands for the gamma function. For even moments ($m = 2n$), this is reduced to

$$\langle x^m \rangle_{ss} = \left(\frac{b}{2Q}\right)^{-n}\left(\frac{d}{2Q} + n - 1\right) \cdots \left(\frac{d}{2Q}\right) \tag{4.12}$$

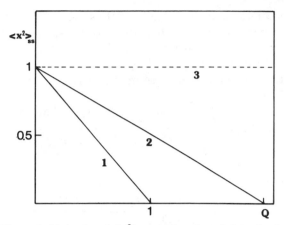

Figure 5.   Theoretical behavior of $\langle x^2 \rangle_{ss}$ as a function of Q in the absence of additive noise ($D = 0$). Curves 1, 2, and 3 correspond to $R = 0$ (Itô), $R = 1$, and $R = \infty$ (Stratonovich), respectively. The parameters of the double well are $d_0 = 1$ and $b = 1$ [Eqs. (4.3)].

Let us note the special case

$$\langle x^2 \rangle_{ss} = \frac{d}{b} = \frac{d_0 - Q/(1+R)}{b} \tag{4.13}$$

The theoretical behavior of $\langle x^2 \rangle_{ss}$ as a function of Q for several values of R is shown in Fig. 5.

B.   The Influence of Additive Noise and Inertia on Equilibrium Properties

One has to emphasize that the behavior of the system at $D = 0$ is pathological. The divergence of $\sigma_{ss}(x)$ at $x = 0$ has to be traced back mainly to the fact that this is the space region where the intensity of the multiplicative stochastic force vanishes. Recall that we are considering the case where the term $g(x)\xi(t)$ is given the form $x\xi(t)$. An extremely weak stochastic force of an additive nature is therefore enough to remove the divergence of $\sigma_{ss}(x)$ at $x = 0$. This property can be derived directly from Eq. (4.4). As a consequence of that, the second threshold disappears and the even moments of the steady state distribution never vanish, whereas the odd moments always vanish as a consequence of the fact that, however weak it is, an additive stochastic force ultimately allows the barrier to be overcome. The time necessary for this to happen can be overwhelmingly large if the intensity of the additive stochastic force is weak. However, this is enough to guarantee the steady state distribution to be symmetric around $x = 0$. The effect of additive noise becomes increasingly significant as the order of the moment increases. This depends on the fact that the additive stochastic force depopulates the region around $x = 0$ while favoring the large-x regions, thereby proving large effects on $\langle x^n \rangle_{ss}$ with large n.

The analytical expression for these moments is

$$\langle x^m \rangle_{ss} = \frac{\int_{-\infty}^{+\infty} x^m (D + Qx^2)^\delta \exp(-Kx^2)\, dx}{\int_{-\infty}^{\infty} (D + Qx^2)^\delta \exp(-Kx^2)\, dx} \tag{4.14}$$

where

$$\delta \equiv \frac{1}{2}\left(\frac{d_0}{Q} + \frac{bD}{Q^2} - \frac{2+R}{1+R} \right) \tag{4.15a}$$

$$K \equiv \frac{b}{2Q} \tag{4.15b}$$

Further information on moments stems from the work of Fujisaka and Grossman,[33] who showed that all the even moments can be expressed in terms of the second moment $\langle x^2 \rangle_{ss}$ via the following recursion relationships:

$$\langle x^{2(n+2)} \rangle_{ss} = \left\{ \frac{d_0 + Q[2n + (2R+1)/(R+1)]}{b} \right\} \langle x^{2(n+1)} \rangle_{ss}$$
$$+ \frac{D}{b}(2n+1)\langle x^{2n} \rangle_{ss} \tag{4.16}$$

Unfortunately, no exact analytical expression for the second moment is available except in the case where $D = 0$.

The behavior of $\langle x^2 \rangle_{ss}$ as a function of Q while keeping D fixed at several values of R is shown in Fig. 6, whereas Fig. 7 shows the behavior of $\langle x^2 \rangle_{ss}$ as a function of Q at two different values of D, while keeping R fixed. The analysis of these results shows that in the large Q region the behavior of the system is significantly influenced by the presence of an additive stochastic force. We are also in a position to state that in the special case where $0 < R < \infty$,

$$\lim_{Q \to \infty} \langle x^2 \rangle_{ss}$$

$$= - \frac{(2)^{-1/2(1+R)} \Gamma(-1/2(1+R)) \Gamma[(2+R)/2(1+R)](D/b)^{1/2(1+R)}(Q/b)^{R/1+R}}{(1+R)\Gamma(\frac{1}{2})\Gamma(1/2(1+R))} \tag{4.17}$$

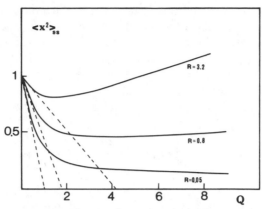

Figure 6. Theoretical behavior of $\langle x^2 \rangle_{ss}$ as a function of Q at several values of the parameter R. The intensity of the additive noise is $D = 0.01$, while the parameters of the double well are $d_0 = 1$ and $b = 1$ [Eqs. (4.3)]. The dashed lines correspond to the theoretical predictions in the case $D = 0$.

From Eq. (4.17) it appears evident that $\langle x^2 \rangle_{ss}$ tends to infinity with increasing Q. The increase of $\langle x^2 \rangle_{ss}$ with increasing Q is characterized by a power law with an exponent which increases with R. In consequence, the additive stochastic force produces increasingly significant effects with increasing Q.

In several respects the finite inertia of the system of Eq. (1.7) produces effects similar to those of a weak additive stochastic force. In particular, as a consequence of finite inertia, the escape from the well is also possible in the absence of the additive stochastic noise. These effects are not taken into

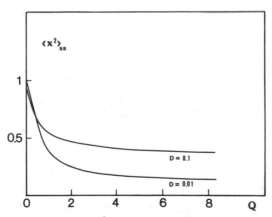

Figure 7. Theoretical behavior of $\langle x^2 \rangle_{ss}$ as a function of Q at two different values of the parameter D. The other parameters are $R = 0.05$, $d_0 = 1$, and $b = 1$ [Eq. (4.3)].

account by the AEP, which leads to Eq. (1.18), which in turn does not admit any escape from the well in the absence of an additive stochastic force. To prove this, let us make the adiabatic elimination of the variable ξ alone from the system of Eq. (1.16). Note that this implies that ξ is the fastest variable of the system which, in a sense, means that $R = 0$. In this case we obtain

$$\frac{\partial}{\partial t}\rho(x,v;t) = \left\{-v\frac{\partial}{\partial x} + V'\frac{\partial}{\partial v} + \frac{\langle\xi^2\rangle_{\text{eq}}}{\lambda}x^2\frac{\partial^2}{\partial v^2}\right.$$

$$\left. + \gamma\langle v^2\rangle_{\text{free}}\frac{\partial^2}{\partial v^2} + \gamma\frac{\partial}{\partial v}v\right\}\rho(x,v;t) \qquad (4.18)$$

where $\langle v^2\rangle_{\text{free}}$ indicates the weak equilibrium value of v^2 in the presence of the purely additive stochastic force alone. Equation (4.18) can be interpreted as the diffusion of a Brownian particle in a double-well potential in the presence of a thermal gradient which makes the temperature vanish at $x = 0$ (if $\langle v^2\rangle_{\text{free}} = 0$). We shall show that the AEP does not take into account the effects of this gradient, which, on the contrary, produces a flux of energy toward $x = 0$, which ultimately will force the Brownian particle to escape from the well where it is initially found (see Fig. 8).

To show this, let us consider the case of an extremely weak Q. This means that the system has enough time to reach a steady state condition within the initial well before escaping. This steady state, in turn, can be found by making the linearization assumption indicated in Fig. 9. We are thus in a position to replace Eq. (4.18) with

$$\frac{\partial}{\partial t}\rho(x,v;t) = \left\{-v\frac{\partial}{\partial x} + \Omega_{\text{eff}}^2 x\frac{\partial}{\partial v} + \frac{\langle\xi^2\rangle_{\text{eq}}}{\lambda}(x-a)^2\frac{\partial^2}{\partial v^2}\right.$$

$$\left. + \gamma\left(\frac{\partial}{\partial v}v + \langle v^2\rangle_{\text{free}}\frac{\partial^2}{\partial v^2}\right)\right\}\rho(x,v;t) \qquad (4.19)$$

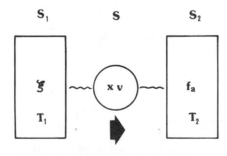

flow of energy

Figure 8. An intuitive representation of the multiplicative stochastic process under study. The arrow indicates that energy flows continuously from left to right. Energy flows only from S_1 into S witout admitting any reverse process as the dynamics of ξ is assumed to be completely independent of the dynamics of S. Then the system S tries to dissipate this energy into the thermal bath S_2. The whole situation is interpreted as a flow of energy from a hot to a cold heat bath, i.e., $T_1 > T_2$, where T_1 and T_2 are the temperature of S_1 and S_2, respectively.

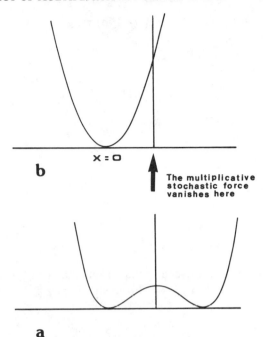

Figure 9. (a) Double-well potential. (b) Linearized potential close to the minimum of the double well.

where Ω_{eff} is the effective frequency corresponding to the linearization of the potential V of Eq. (1.20) around the bottom of the well,

$$\Omega_{\text{eff}} = \left(\frac{8V_0}{a^2} \right)^{1/2} \tag{4.19a}$$

and where the origin of the x axis has been located in the bottom of the well.

The time evolution of momenta of the $\rho(x, v; t)$ distribution can be obtained by writing

$$\frac{d}{dt} \langle x^n v^m \rangle = \langle \Gamma^+ x^n v^m \rangle \tag{4.19b}$$

where Γ^+ represents the adjoint of the operator of Eq. (4.19) and n and m

are two arbitrary integers. From Eq. (4.19b) we get

$$\frac{d}{dt}\langle v^2 \rangle = -2\Omega_{\text{eff}}^2 \langle xv \rangle + \frac{2\langle \xi^2 \rangle_{\text{eq}}}{\lambda}\langle x^2 \rangle + \frac{2\langle \xi^2 \rangle_{\text{eq}} a^2}{\lambda} - 2\gamma\langle v^2 \rangle + 2\gamma\langle v^2 \rangle_{\text{free}}$$

$$(4.20a)$$

$$\frac{d}{dt}\langle xv \rangle = \langle v^2 \rangle - \Omega_{\text{eff}}^2 \langle x^2 \rangle - \gamma\langle xv \rangle \tag{4.20b}$$

$$\frac{d}{dt}\langle x^2 \rangle = 2\langle xv \rangle \tag{4.20c}$$

$$\frac{d}{dt}\langle v \rangle = \Omega_{\text{eff}}^2 \langle x \rangle - \gamma\langle v \rangle \tag{4.20d}$$

$$\frac{d}{dt}\langle x \rangle = \langle v \rangle \tag{4.20e}$$

The latter pair of equations means that when the steady state is attained, $\langle xv \rangle_{\text{ss}} = 0$, $\langle x \rangle_{\text{ss}} = 0$, and $\langle v \rangle_{\text{ss}} = 0$. From the former pair we get

$$\langle x^2 \rangle_{\text{ss}} = \frac{\gamma\langle v^2 \rangle_{\text{free}} + \langle \xi^2 \rangle_{\text{eq}} a^2 / \lambda}{\gamma\Omega_{\text{eff}}^2 - \langle \xi^2 \rangle_{\text{eq}} / \lambda} \tag{4.21}$$

This equation produces a divergence when the external noise is so intense as to get the condition

$$\gamma = \frac{\langle \xi^2 \rangle_{\text{eq}}}{\lambda\Omega_{\text{eff}}^2} \tag{4.22}$$

Note that this divergence does not disappear when the additive stochastic force vanishes. This is a consequence of the fact that $a \neq 0$. Earlier analysis of divergences of this kind showed[35-40] that these can be accounted for as follows: A purely multiplicative stochastic force means a continuous flux of energy into the system. The Brownian particle can dissipate part of its energy into the thermal bath responsible for the damping γ (note that the temperature of this thermal bath has to be considered as vanishing when the additive stochastic force vanishes). To react against this continuous flux of energy, the Brownian particle is also obliged to explore regions of higher potential energy. This simple physical argument accounts for the first threshold predicted via use of the AEP. When obliged by the external flux of energy to explore regions of higher potential energy, the Brownian particle between the two available regions of higher potential energy prefers to ex-

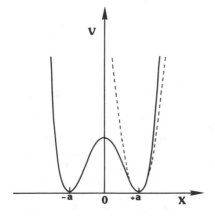

Figure 10. The potential V in the presence of which the motion of the Brownian particle takes place. The right well is considered. The dashed curve expresses the harmonic expansion of the potential V around $x = a$. The regions $0 < x < a$ and $a < x < \infty$ are softer and harder than the harmonic expansion, respectively. In the hard-potential region the local frequency becomes larger and larger as x increases.

plore the region around $x = 0$, since there the multiplicative stochastic force vanishes and the potential is softer than the harmonic approximation (see Fig. 10). Note that this interpretation is quite satisfactory from a physical point of view. There is no chance of obtaining an interpretation of this kind from reduced equations such as Eq. (1.3). This is further evidence for the need to relate equations of this kind to their microscopic model. The advantages coming from this approach are twofold: in addition to shedding light on the Itô-Stratonovich controversy, we also account for the physical reasons behind striking phenomena such as the appearance of the first threshold.

Note, however, that Eq. (4.21) loses its validity in the case of strong multiplicative noise, as it is based on a linearization assumption. The remarks above have to be related to a purely qualitative level of interpretation.

When no additive stochastic force is present and when $\langle \xi^2 \rangle_{eq} \ll \lambda \gamma \Omega_{eff}^2$, Eq. (4.21) results in

$$\langle x^2 \rangle_{ss} \approx \frac{\langle \xi^2 \rangle_{eq}^2 a^2}{\lambda \gamma \Omega_{eff}^2} \tag{4.23}$$

Let us consider the case where $\langle x^2 \rangle_{ss}$ is so small as to render Eq. (4.19) indistinguishable from

$$\frac{\partial}{\partial t} \rho(x, v; t) = \left\{ -v \frac{\partial}{\partial x} + V' \frac{\partial}{\partial v} + \frac{\langle \xi^2 \rangle_{eq}}{\lambda} a^2 \frac{\partial^2}{\partial v^2} + \gamma \frac{\partial}{\partial v} v \right\} \rho(x, v; t)$$

$$\tag{4.24}$$

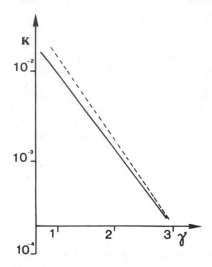

Figure 11. The rate of escape k as a function of the friction coefficient. The dashed curve is obtained from Eq. (4.25). The solid line denotes the result obtained by evaluating the time evolution of $A = x - \langle x \rangle_{ss}$ by using the CFP. Here we have considered the initial distribution $\sigma(x) = \delta(x - x_0)$. The parameters used are $V_0 = 0.25$, $a = 1$, $x_0 = 0.9a$, $\langle x \rangle_{ss} = 0$, $\langle \xi^2 \rangle_{eq} = 1$, and $\lambda = 10$.

Note that Eq. (4.24) provides the same stationary second moment as Eq. (4.19). Equation (4.24), via the well-known Kramers approach[30] (see also Fonseca et al., Chapter IX) leads us to the rate of escape from the well:

$$K = \frac{1}{4\pi} \left[\left(\frac{\gamma^2}{4} + d_0 \right)^{1/2} - \frac{\gamma}{2} \right]^{-1} \exp\left[-\frac{V_0}{a^2} \frac{\lambda\gamma}{\langle \xi^2 \rangle_{eq}} \right] \qquad (4.25)$$

In ref. 39 a computer calculation of the rate of escape was done based on the CFP of Chapter III. The agreement between this calculation and Eq. (4.25) is good, as shown in Fig. 11. Note that the combination of weak inertia with multiplicative noise results in a finite rate of escape which is rigorously forbidden by the AEP when no additive noise is present.

We are now in a position to test with our analog circuit the predictions of the AEP concerning the Itô-Stratonovich controversy. We believe that inertia and a weak residual circuit force have similar effects, so that some caution is need in interpreting the predictions of the AEP.

Figures 12a and b show the dependence of $\langle x^2 \rangle_{ss}$ on Q for some values of R as obtained by using the experimental method of Section III. We would like to stress again that a great deal of attention has been devoted to limiting the effects of spurious additive noise from the circuit and that if the well is not exactly symmetric the multiplicative noise can itself produce a spread of the variable x. In spite of our efforts, a weak additive stochastic force proves to be present in our electrical circuit.

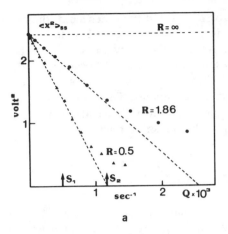

a

Figure 12a. $\langle x^2 \rangle_{ss}$ as a function of Q (experimental results). (●) $R = 0.5$; (▲) $R = 1.86$; the corresponding dashed lines indicate the theoretical results as provided by Eq. (4.13), which in turn is derived from the theory of Graham and Schenzle.[42] The horizontal dashed line denotes the Stratonovich behavior ($R = \infty$). The parameters d_0 and b of Eq. (4.13) are $d_0 = 902$ s^{-1} and $b = 373$ s^{-1} V^{-2}. Arrow S_1 denotes the value of Q at which the first threshold is reached when the case $R = 0.5$ is considered. Arrow S_2 denotes the second threshold obtained as the intersection of the corresponding dashed line with the abscissa axis. Note that $S_1/S_2 = 2.35$ is very close to the theoretical value $S_1/S_2 = 2 + R = 2.5$.

We can see, therefore, that the agreement between the AEP and the experimental results is satisfactory as long as the weak Q region is explored (see Fig. 12a). Such agreement is lost when the region $Q > Q_2$ is attained.

One cannot expect that such an agreement covers also the large-Q region. Two main reasons have already been singled out: inertia and weak residual stochastic forces of additive kind. In particular, as shown in Figs. 6 and 7, the additive stochastic noise forces $\langle x^2 \rangle_{ss}$ to increase when the large-Q region is attained. However, the effect of the additive noise is relevant for

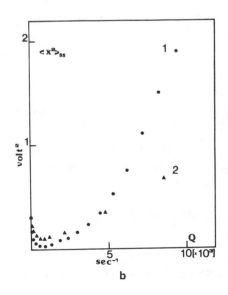

b

Figure 12b. $\langle x^2 \rangle_{ss}$ versus Q (experimental results) for two different values of R. (▲) $R = 1.4$, (●) $R = 0.1$. The parameters of the electric circuit have been modified in such a way as to increase the range of allowable values of the intensity Q of the multiplicative stochastic noise. Note that in the large Q region, $\langle x^2 \rangle_{ss}$ deviates largely from the behavior predicted by the AEP. In fact, $\langle x^2 \rangle_{ss}$ increases as a linear function of Q. The parameters of the system are $d_0 = 222$ s^{-1}, $b = 739$ s^{-1} V^{-2}, $j = 5070$ s^{-1}, and $D = 0.054$ V^2 s^{-1} [Eq. (4.3)].

$R \gg 1$, whereas it is almost negligible for $R \ll 1$. On the other hand, the experimental results clearly indicate that deviations from Eq. (4.13) mainly occur for $R \to 0$. In Section V we shall show that these deviations have to be traced back to the finite inertia still present in our system. We can conclude for the moment, however, that the Itô-Stratonovich controversy is settled clearly by our theoretical and experimental arguments. This supports again our research line based on the RMT, which, as shown above, also permits a straightforward physical interpretation of the threshold phenomena.

V. CONCLUDING REMARKS AND PERSPECTIVE ON FURTHER INVESTIGATIVE WORK

To support our belief that inertia is the main reason for the deviations from the predictions of the AEP, let us study the dynamical properties of the system of Eq. (4.3). To shed light on that, let us consider again, as in Section IV, the ideal case $D = 0$. The spectrum of the eigenvalues of the operator of Eq. (4.3) has been studied by Schenzle and Brand[3] and by Graham and Schenzle,[42] who showed that this consists of the superposition of two clearly distinct contributions: (1) the discrete spectrum of the eigenvalues

$$\varepsilon_m = 2m\left(d_0 - \frac{2m+1+2R}{1+R}Q\right) \tag{5.1}$$

supplemented by the condition

$$m < \frac{1}{4}\left(\frac{d_0}{Q} - \frac{1}{1+R}\right) \tag{5.2}$$

(2) A continuous spectrum of eigenvalues for

$$\varepsilon > \frac{[d_0 - Q/(1+R)]^2}{4Q} \tag{5.3}$$

From Eq. (5.2) we derive that for $Q > Q_0 = d_0(1+R)/(5+4R)$ the spectrum consists only of the continuous contribution. In the case where $Q < Q_0$, the long-time behavior of the autocorrelation function

$$\Phi(t) \equiv \frac{\langle(x(t)-\langle x\rangle_{ss})(x(0)-\langle x\rangle_{ss})\rangle}{\langle x^2\rangle_{ss} - \langle x\rangle_{ss}^2} \tag{5.4}$$

is mainly determined by the lowest eigenvalue, thereby resulting in an exponential decay. In the case where $Q > Q_0$, on the other hand, the spectrum, consisting only of a continuum kind with a complicated density of states, generates a time decay difficult to be described with simple analytical expressions. For $t \to \infty$, Graham and Schenzle[42] found that its time behavior is of the type

$$\Phi(t) \propto t^{-3/2} \exp\left[-\left(d_0 - \frac{Q}{1+R}\right)^2 \frac{t}{Q}\right] \tag{5.4'}$$

In consequence, when Q gets closer to the second threshold Q_2 [Eq. (4.9)], the term $t^{-3/2}$ becomes dominant.

Let us define the relaxation time of the variable x as

$$T \equiv \int_0^\infty \Phi(t)\, dt \tag{5.5}$$

We obtain thus

$$\lim_{Q \to Q_2} T^{-1} = 0 \tag{5.6}$$

In consequence, the AEP predicts that a slowing down takes place around the second threshold.

This analysis is, however, a little bit partial. We would like to notice that a model like that of the system of Eq. (1.7) can also be interpreted as a model of classical activation.[43] The relaxation time defined by Eq. (5.5) would then be interpreted as a sort of activation time, that is, the time required for a starting point condition extremely favorable for the escape from a well. An experimenter aiming at activating a chemical reaction process would be greatly disappointed if the activation time were infinite!

A different view can be gained if the relaxation time T is evaluated by using the CFP of Chapter III. Earlier work of this kind has been done by Fujisaka and Grossman,[33] who basically limited themselves to using the second-order truncation of the corresponding continued fraction. By using the CFP of Chapter III, it is possible to extend such a kind of calculation up to the 22nd order (Faetti et al.[43,44]). The results of this calculation are described in Fig. 13 together with the results of the numerical experiment of ref. 45 (we are now considering the case $R = \infty$).

The disagreement between the CFP and the results of Graham and Schenzle (and the computer simulation of ref. 45 as well) can be accounted for by remarking that the decay of the function $\Phi(t)$ consists of a fast relaxation regime followed by a slow regime characterized by the behavior $t^{-3/2}$. The CFP technique describes in a fairly accurate way the fast relaxa-

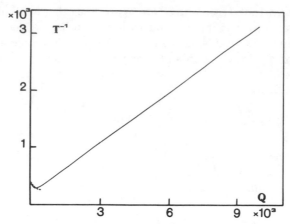

Figure 13. $1/T$ as a function of Q for $R = \infty$ (theoretical results obtained by the CFP). The other parameters characterizing the system are $D = 0$, $d_0 = 222$, and $b = 739$ [Eq. (4.3)]. The dashed line at the lower left-corresponds to the results of the numerical experiment of ref. 45 normalized to our unities.

tion process, while failing to describe the slow tail. We believe that by taking proper care of this problem, one could also successfully evaluate this long-time region by using the CFP. However, our current versions of the CFP computer program are not yet suited for that. As a consequence, the relaxation time T provided by the CFP has to be regarded as the lifetime of the fast early decay process.

Note, however, that the slow tail is difficult to detect experimentally because this slow tail concerns a residual weak part of the original signal which almost completely disappears. As a matter of fact, Faetti et al.,[44] by using a version of the circuit of Section III which completely neglects inertia (by disregarding the variable v), obtained results which from a qualitative point of view agree with the predictions of the CFP.

Further results have been obtained by using the electric circuit of Section III. This experiment shows that after the first threshold, T^{-1} exhibits a rapid increase with increasing Q. This increase is still more rapid than predicted by the CFP. The experimental behavior of T^{-1} as a function of Q at several values of R is shown in Fig. 14. These results show that the basic assumptions of the AEP are invalidated with increasing Q. This points out also the importance of a technique such as the current version of the CFP which allows us to get information on the short-time region, which turns out to be the most significant region within the present context.

We shall argue indeed that this fast increase of T^{-1} with increasing Q is the main reason behind the experimental deviations from the theoretical

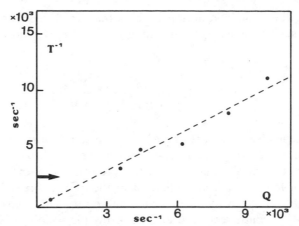

Figure 14. $1/T$ as a function of Q for $R = 0.1$ (experimental results). The parameters characterizing the system are $d_0 = 222$ s^{-1}, $b = 739$ s^{-1} V^{-2}, $j = 5070$ s^{-1}, and $D = 0.054$ V^2 s^{-1}. Note that the abscissa scale is so greatly contracted as to make it impossible to draw the decreasing behavior of $1/T$ with increasing Q which would be exhibited in the weak-Q region.

provisions of the AEP. Since the fast relaxation exhibited by this variable prevents us from still considering this variable as slow, we have to look for a new slow variable.

If we assume that the variable x is fast, we can consider as slow the energy variable defined by

$$E \equiv \frac{v^2}{2} - \frac{\alpha_0 x^2}{2} + \frac{\beta x^4}{4} \qquad (5.7)$$

Note that this variable can be regarded as being characterized by the relaxation time[46]

$$T_E \equiv \frac{3}{2\gamma} \qquad (5.8)$$

This shows that the energy variable can be regarded as being slow when exploring the large-Q region where the relaxation time of the x variable is much smaller than γ^{-1}.

In the large-Q region, both the effects coming from the term $\alpha_0 x^2/2$ and those of the weak circuit noise can be disregarded. By following the Stratonovich approach as reviewed by Seshadri et al.,[46] we can show that

$$\langle E \rangle_{ss} = 15 \left(\frac{12\pi^2 \langle \xi^2 \rangle_{eq}}{5\lambda\gamma\beta^{1/2}\Gamma^4\left(\frac{1}{4}\right)} \right) \qquad (5.9)$$

By repeating the same kind of approach as that which led us to Eq. (4.20) we can find such relationships concerning $\langle x^2 \rangle_{ss}$, $\langle x^4 \rangle_{ss}$, and $\langle v^2 \rangle_{ss}$ to obtain

$$\langle x^2 \rangle_{ss} = 20 \left(\frac{12\pi^2}{5\Gamma^4(1/4)} \right)^2 \frac{\langle \xi^2 \rangle_{eq}}{\gamma\lambda\beta} \tag{5.10}$$

This linear dependence of $\langle x^2 \rangle_{ss}$ on $\langle \xi^2 \rangle_{eq}$ is in satisfactory agreement with the experimental results in the large-Q region (see Fig. 12b).

In conclusion, the deviations from the predictions of the AEP can be accounted for fairly well. The predictions of the AEP in the small-Q region are supported completely by the experimental results. This sheds light on the Itô-Stratonovich controversy and also confirms the validity of our general strategy based on using suitable auxiliary variables (RMT).

References

1. W. Horsthemke and R. Lefever, *Phys. Lett.*, **64A**, 19 (1977); R. Lefever, in *Stochastic, Nonlinear Systems in Physics, Chemistry and Biology*, L. Arnold and R. Lefever, eds., Springer, Berlin, 1981, p. 127; W. Horsthemke, *ibid.*, p. 116.

2. R. L. Stratonovich, *Topics in the Theory of Random Noise*, Gordon and Breach, New York, Vol. 1, 1963, Vol. 2, 1967.

3. A. Schenzle and H. Brand, *Phys. Rev.*, **20A**, 1628 (1979).

4. K. Itô, *Mem. Math. Soc.*, **4**, 1 (1951).

5. B. J. West, A. R. Bulsara, K. Lindemberg, V. Seshadri, and K. E. Schuler, *Physica*, **97A**, 211 (1979).

6. N. G. van Kampen, *J. Stat. Phys.*, **24**, 175 (1981).

7. I. Smythe, F. Moss, and P. V. E. McClintock, *Phys. Lett.*, **97A**, 95 (1983).

8. I. Smythe, F. Moss, and P. V. E. McClintock, *Phys. Rev. Lett.*, **51**, 1062 (1983).

9. R. Landauer, *J. Appl. Phys.*, **33**, 2209 (1962).

10. K. Matsuno, *Appl. Phys. Lett.*, **12**, 404 (1968); K. Matsuno, *Phys. Lett.*, **31A**, 6, 335 (1970).

11. J. W. F. Woo and R. Landauer, *IEEE J. Quantum Electronics*, **7**, 435 (1971).

12. T. Kawakubo, S. Kabashima, and K. Nishimura, *J. Phys. Soc. Japan*, **34**, 1460 (1973).

13. S. Kabashima, H. Yamazaki, and T. Kawakubo, *J. Phys. Soc. Japan*, **40**, 921 (1976).

14. S. Drosdziok, *Z. Physik*, **261**, 431 (1973).

15. T. Kawakubo, S. Kabashima, and M. Ogishima, *J. Phys. Soc. Japan*, **34**, 1149 (1973).

16. T. Kawakubo and S. Kabashima, *J. Phys. Soc. Japan*, **37**, 1199 (1974).

17. S. Kabashima, M. Itsumi, T. Kawakubo, and T. Nagashima, *J. Phys. Soc. Japan*, **39**, 1183 (1975).

18. R. Kubo, K. Matsuno, and K. Kitahara, *J. Stat. Phys.*, **9**, 51 (1973).

19. J. B. Morton and S. Corrsin, *J. Math. Phys.*, **10**, 361 (1969).

20. T. Kawakubo, S. Kabashima, and M. Itsumi, *J. Phys. Soc. Japan*, **41**, 699 (1976).

21. P. M. Horn, T. Carruthers, and M. T. Long, *Phys. Rev.*, **A14**, 833 (1976).

22. S. Kabashima, S. Kogure, T. Kawakubo, and T. Okada, *J. Appl. Phys.*, **50**, 6296 (1979).

23. Y. Morimoto, *J. Phys. Soc. Japan*, **50**, 28 (1981).

24. Y. Morimoto, *J. Phys. Soc. Japan*, **50**, 2459 (1981); **52**, 1086 (1983).

25. K. Harada, S. Kuhura, and K. Hirakawa, *J. Phys. Soc. Japan*, **50**, 2450 (1981).

26. J. M. Sancho, M. San Miguel, H. Yamazaki, and T. Kawakubo, *Physica*, **116A**, 560 (1982).

27. S. Fauve and F. Heslot, *Phys. Lett.*, **A97**, 5 (1983).

28. H. A. Kramers, *Physica*, **7**, 284 (1940).

29. S. Golomb, *Shift Register Sequences*, Holden-Day, San Francisco, 1967.

30. Madhu S. Gupta, ed., *Electrical Noise: Fundamentals and Sources*, IEEE Press, New York, 1977.

31. B. M. Smith, *Proc. IEEE*, **54**, 793 (1966).

32. G. H. Tomlison and P. Galvin, *Electronic Lett.*, **11**, 77 (1975).

33. H. Fujisaka and S. Grossman, *Z. Phys.*, **B43**, 69 (1981).

34. M. Suzuki, K. Kaneko, and F. Sasagawa, *Prog. Theor. Phys.*, **65**, 829 (1981).

35. B. J. West, K. Lindemberg, and V. Seshadri, *Physica*, **102A**, 470 (1980).

36. K. Lindemberg, V. Seshadri, K. E. Shulerand, and B. West, *J. Stat. Phys.*, **23**, 755 (1980).

37. K. Lindemberg, V. Seshadri, and B. J. West, *Phys. Rev.*, **22A**, 2171 (1980).

38. N. G. van Kampen, *Physica*, **102A**, 489 (1980).

39. P. Hänggi, *Phys. Lett.*, **78A**, 489 (1980).

40. P. Grigolini, *Phys. Lett.*, **84A**, 301 (1981).

41. C. Festa, L. Fronzoni, P. Grigolini, and F. Marchesoni, *Phys. Lett.*, **102A**, 95 (1984).

42. R. Graham and S. Schenzle, *Phys. Rev.*, **A25**, 1731 (1982).

43. S. Faetti, P. Grigolini, and F. Marchesoni, *Z. Phys.*, **B47**, 353 (1982).

44. S. Faetti, C. Festa, L. Fronzoni, P. Grigolini, F. Marchesoni, and V. Palleschi, *Phys. Lett.*, **99A**, 25 (1983).

45 J. M. Sancho, M. S. Miguel, S. L. Katz, and J. D. Gunton, *Phys. Rev.*, **26A**, 1589 (1982).

46. V. Seshadri, B. J. West, and K. Lindemberg, *Physica*, **102A**, 470 (1980).

XI

INTERDISCIPLINARY SUBJECTS (POPULATION GENETICS): THE TIME PROPERTIES OF A MODEL OF RANDOM FLUCTUATING SELECTION

P. GRIGOLINI, F. MARCHESONI, and S. PRESCIUTTINI

CONTENTS

I. INTRODUCTION

The Fokker-Planck equation, better known among biologists as the Kolmogorov forward equation, was introduced into the field of population genetics by Wright[1] in 1945. Since then a variety of stochastic processes in the change of gene frequencies have been treated by means of this approach, from random sampling of gametes in finite populations to random fluctuations in systematic evolutionary pressures, of which fluctuation of selection intensity is especially relevant. It is clear, therefore, that this field could be investigated with the strategy of the reduced model theory (RMT) outlined in Chapter I, the main aim of which is to provide a general foundation for models of this class. For instance, the Itô-Stratonovich controversy discussed in Chapter X is, among genetists, a controversy about the progress of selection when selection coefficients vary at random: Kimura[2] has adopted the Itô prescription, whereas Gillespie[3] has followed Stratonovich. It seems that in this scientific environment it is not yet clear which prescription is preferable.[4] The results of Chapter X make it clear, however, that this dilemma can be restated as a question of whether or not some kind of inertia is exerting its influence in the systems under study. If we find that inertia is irrelevant to the stochastic selection models, then it will become clear that the results of Gillespie are to be preferred to those of Kimura.

This type of problem has recently become of great interest in the field of physics,[5-20] being directly related to basic problems such as noise-induced phase transitions[5-10] and the dependence of the relaxation time of the variable of interest on the intensity of external noise terms.[13-16] As in Chapter X, the reduced equation of motion itself can be assumed to be reliable only when this relaxation time decreases with increasing noise intensity. Further problems of interest are whether or not the threshold of a noise-induced phase transition is characterized by a slowing down[13-16,19,20] and what is the analytical form of the long-time decay.[17,18]

Another very important feature of the stochastic equations considered here, when they are subjected to RMT analysis, is their resemblance to the general formalism arrived at in the thermodynamics of nonequilibrium processes: this suggests an analogy between the effects of multiplicative noise and the continuous flux of energy which maintains the systems far from equilibrium. This is considered the main characteristic of self-organizing living systems[21-23] and means that multiplicative stochastic models could take on a new and fundamentally important role.

For all these reasons it is clear that the "microscopic" foundation of the equations under study, interpreted as in Chapter I, is of particular importance for a deeper understanding of the various outcomes of its application. Therefore, we will pay special attention in this chapter to the construction of a model for fluctuating selection: We will consider an optimum model in which the maximum fitness fluctuates at random with time.

The basic formula on which most of the work of population genetics is founded is

$$\dot{x} = x(1-x)s \tag{1}$$

The simplest model from which this equation can be derived describes a population which consists of two distinguishable types of independently self-reproducing entities whose change in number with time (respectively, N_1 and N_2) is determined by the laws $\dot{N}_1 = w_1 N_1$ and $\dot{N}_2 = w_2 N_2$. If the frequency x of individuals of type 1 is defined as $x = N_1/(N_1 + N_2)$, then the change of frequency with time will be described by Eq. (1), where $s = w_1 - w_2$. For macroscopic biological systems, Eq. (1) is considered a good approximation, valid for small values of s, for the change in time of the frequency of an allele (say A_1) in a diploid population in which two alleles exist at a given gene and there is no dominance [i.e., a population in which three genotypes, A_1A_1, A_1A_2, and A_2A_2 are distinguishable with respective fitnesses w_{11}, $(w_{11} + w_{22})/2$, and w_{22}]; s, the selection coefficient, is in this case $(w_{22} - w_{11})/2$.

Fluctuations of selection intensity have received much attention in population genetics because it is quite reasonable to assume that some relevant environmental variables change the fitness value of different genotypes over generations. Deterministic treatments derive from the model of Haldane and Jayakar[34]; stochastic fluctuations were first considered by Kimura[2] and Dempster.[35] Recent reviews are those of Felstein[36] and Maynard-Smith and Hoekstra.[37]

Generally in stochastic models the randomness is added to Eq. (1) by rewriting s as a random process $s(t)$; thus Eq. (1) becomes

$$\dot{x} = x(1 - x)s(t) \tag{2}$$

Kimura[2] arrived at a Fokker-Planck type of equation for the gene frequency probability density and showed its solution in the symmetric particular case (i.e., when the selective difference between the alleles was 0 in the long-term average) by using the formal similarity with the heat conduction equation. Another way of solving Eq. (2) has been given by Gillespie[24] by means of the change-of-variable technique, and a third one also by Gillespie,[3] who arrived at an equation similar to Kimura's[2] but with an additional second-order term deriving from the Stratonovich prescription.

A general problem of theoretical population genetics is the maintenance of stable polymorphisms (maxima asymptotic densities at $0 < x < 1$-about 30% of the genes are polymorphic, i.e., with more than one allele at frequencies $> .01$, in natural populations). For this problem the treatments of Eq. (2) lead to the same result: Random fluctuations of selection, even in the symmetric case, eventually cause one or the other of the alleles to be fixed (or, better, "quasi-fixed," because generally the population is thought of as being of infinite size). However, as regards the time at which the process of quasi-fixation occurs, there are no unique results, different approaches giving different answers. On the other hand, the question is not a secondary one; clearly, the more time that is necessary to change a gene frequency of a given amount under selection, the more time is allowed for mutation and other forces to restore a given level of heterozygosity (which is the average proportion of heterozygote genes per individual in a given population).

In this chapter we shall show how the RMT tackles the problem of the time of gene frequency decay under stochastic selection. To do so we will build up a detailed model (Section II) which provides a mechanism with which fluctuations of selection can arise. This is in line with the strategy underlying the RMT to look for a microscopic foundation for the equation under study. It will be shown that this model leads to a standard diffusion-like equation, for which (Section III) we shall investigate the corresponding transient properties.

II.   A DETAILED MODEL FOR STOCHASTIC SELECTION

In quantitative genetics the macroscopic properties of biological organisms, the "characters," are represented as if they are determined by a number (very often unknown) of "microscopic" variables, the *genes*, which can be found in different states (*alleles*), and which generally show additivity of effects. Every individual has, therefore, as regards the measure of any character, a *genotypic value*: the sum, over all genes affecting that character, of the effect of each allele which is actually present. It may happen, however, that the genotypic value does not correspond to the one which is in fact observed (the *phenotypic value*); this is because a certain amount of external noise (*environmental deviation*) is always unavoidably associated with onto-genetic development; it causes an uncontrollable shift of the macroscopic measure from the score which would be anticipated on the basis of the particular allelic arrangement (*genotype*) characteristic of the individual. The randomness of the noise explains, for example, the normal distribution of the phenotypic values usually exhibited by the populations of individuals sharing the same genotype.

In population genetics, the *fitness* (w) is defined generally as a property of the genotype and measures its rate of self-reproduction in comparison with the rates of other genotypes of the same population. However, it is often treated in mathematical models as a property of one allele in comparison with another. In this case it is the average fitness of the genotypes which contain that allele compared with the average fitness of the genotypes which contain the other. Obviously this is a gross simplification of what occurs in nature, because in the real world natural selection acts on phenotypes and then determines the fitness of their genotypes, so that the selection coefficient ($s = 1 - w$) for a given gene must be considered a function of both the entire system of gene frequencies and the relationships connecting the genotypes to the phenotypes. Nevertheless, as a good approximation, which simply makes it possible to tackle the questions of population genetics, one may assume that a given selection coefficient is applied directly to a given gene. Note that this choice automatically excludes any lag between the application of the forces which act on gene frequencies and the actual change ("inertia"), thus making the Gillespie prescription preferable to that of Kimura in the controversy mentioned earlier. We return to this point later.

However, a line of research of population genetics, which deals with the so-called optimum models, makes an attempt to give a microscopic foundation to the determination of the fitness of the genotypes. It is based on the assumption that there is a relationship between the phenotypic score of each individual at a given character and its fitness, independently of the genotypes which, together with the environment, determine that score. The function describing such a connection is usually a quadratic curve, $w_i = 1 - k(f_i - m)^2$, where w_i and f_i, respectively, are the fitness and phenotypic value of the individual i, k is an arbitrary constant determining the strength of selection, and m is the optimal phenotype. This assumption is based on the widespread idea that natural selection is in most cases "stabilizing" (Haldane), or "centripetal" (Simpson), or "normalizing" (Waddington), that is, that it favors the intermediate phenotypic values and acts against both extremes (see, for instance, ref. 33). Alternatively, the Gaussian function [$w_i = \exp\{-(f_i - m)^2/k\}$ is often used: this is more suitable for analytical treatments and is not distinguishable from the quadratic curve when k is large. Wright[25] was perhaps the first who studied extensively such a "quadratic deviation model" applied to a quantitative character. He showed that in the more general cases the involved genes could not be maintained heterozygous (i.e., with more than one allele); eventually half the genes would be fixed on the "plus" allele (the one which increases the phenotypic value) and half on the "minus." Other contributions were made by Kojima,[26] Lewontin,[27] Latter,[28] Bulmer,[29] Lande,[30] and Kimura.[31] In no case have the stochastic fluctuations of the optimum phenotype been considered.

Let us now consider the model depicted in Fig. 1. The abscissa axis is the individual phenotypic value f for a measurable character (body size, weight, or other) controlled by a single biallelic gene and affected by environmental deviation in a haploid infinite population. In this way the analysis is greatly simplified, but one can easily extend the model to a diploid population for which a quantitative character, contributed by a number of additively combining genes, is considered. In this case one may think to examine phenotypic differences which are due in average to only one gene, the others being in mutation/selection equilibrium, and the heterozygotes having the fitness exactly intermediate between the fitnesses of the opposite homozygotes; also, selection is so weak that the genetic variance of the background does not vary in time. This explains why in the model the phenotypic distributions P_1 and P_2 of the two genotypes A_1 and A_2 (whose genotypic values are $-a$ and $+a$, respectively, and which in the diploid case would be the alternative homozygotes A_{11} and A_{22}) largely overlap (see Fig. 1).

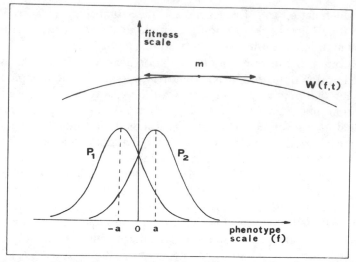

Figure 1.   The model studied in the present chapter. For explanations see text.

We assume that the fitness of an individual with phenotypic value f is determined by a function $W(f, t)$, which is defined as follows:

$$W(f, t) = \exp\left[\frac{-[f - m(t)]^2}{\sigma_m^2}\right] \tag{3}$$

m denotes the optimal phenotype, which is assumed to be attained at $f = m$ and fluctuates at random over time; σ_m^2, which is a measure of the inverse of the "strength" of selection, is assumed to be very large in comparison with the range of phenotypic values.

The fluctuations of the optimum are driven by the equation

$$m(t) = m_0 + \xi(t) \tag{4}$$

where $\xi(t)$ is a random process, the properties of which will be discussed later.

The environmental deviation acts on the genotypic values $-a$ and $+a$ as a Gaussian noise, so that the phenotypic values of the individuals of the two genotypes A_1 and A_2 are normally distributed around $-a$ and $+a$:

$$P_1(f) = \frac{1}{\sqrt{2\pi}\,\sigma_1} \exp\left[\frac{-(f+a)^2}{2\sigma_1^2}\right] \tag{5}$$

$$P_2(f) = \frac{1}{\sqrt{2\pi}\,\sigma_2} \exp\left[\frac{-(f-a)^2}{2\sigma_2^2}\right] \tag{6}$$

σ_1 and σ_2 are measures of the environmental effect which causes the observed spread of the phenotypic values.

The fitness of an allele at a given time can be computed by integrating the product of the distribution of the individuals carrying that allele with the fitness function at that time. Thus we obtain, from Eqs. (3), (5), and (6).

$$w_1(t) = \frac{1}{\sqrt{2\pi}\,\sigma_1} \int_{-\infty}^{+\infty} \exp\left[\frac{-[f-m(t)]^2}{2\sigma_m^2} + \frac{-(f+a)^2}{2\sigma_1^2}\right] df \tag{7}$$

$$w_2(t) = \frac{1}{\sqrt{2\pi}\,\sigma_2} \int_{-\infty}^{+\infty} \exp\left[\frac{-[f-m(t)]^2}{2\sigma_m^2} + \frac{-(f-a)^2}{2\sigma_2^2}\right] df \tag{8}$$

It is important to note at this point that two different time scales are present in Eqs. (7) and (8). The former, whose time unit can be called τ_m, is explicit and measures the speed at which the optimum fluctuates; the latter, whose time unit can be called τ_f, does not appear in Eqs. (7) and (8), but it determines the speed of the process resulting in the distributions (5) and (6). Because the environmental variability, which causes the spread of the phenotypes, can work during the whole of ontogenetic development, τ_f is to be considered of the order of the generation time.

In principle, three possibilities are to be examined:

1. $\tau_m \ll \tau_f$. Under this condition the fitness of a given phenotype is undetermined (because it changes too quickly), and the model loses all meaning, since the operation of integrating the products of the functions (3) by (5) and (6) cannot be carried out. Also, from a biological point of view, it does not make sense that the individual fitness changes into a value completely new almost instantaneously.

2. $\tau_m \sim \tau_f$. In this case the fluctuation of the optimum can be regarded as a white noise, that is, the fitness of a genotype is uncorrelated over generations; this situation can have a (little) biological importance, but it is still insoluble by RMT because the system is strongly inertial and we do not have

sufficient "microscopic" information to adopt either the Itô or the Stratonovich prescription.

3. $\tau_m \gg \tau_f$. In a sense we are obliged to think in terms of this condition, which is precisely the assumption justifying Eqs. (7) and (8). In this case the inertia disappears, because the action of the environment on the individual development is almost instantaneous in comparison with the fluctuation of the optimum, and the external noise $\xi(t)$ becomes a non-white noise, that is, the fitness of a genotype is autocorrelated over generations.

Note that if the environmental deviation is excluded from the model, then the lack of inertia arises even in case 2, and there are no theoretical differences between cases 2 and 3.

We are now in a position to apply the arguments of Chapter X to these considerations. Thus we can conclude that the Fokker-Planck equation to be associated with the evolution equation

$$\dot{x} = x(1-x)\left[w_1(t) - w_2(t)\right] \tag{9}$$

has the Stratonovich form. Gillespie[3] was right.

To express this Stratonovich form explicitly as a function of the parameters of our model, let us replace Eq. (3) with

$$w(f, t) = \kappa \frac{1 - [f - m(t)]^2}{\sigma_m^2} \tag{10}$$

This is justified by the assumption that σ_m is much larger than σ_1 and σ_2. Thus the selection of coefficient of Eq. (9), which is defined as

$$s(t) = w_1(t) - w_2(t) \tag{11}$$

becomes, from Eqs. (7) and (8),

$$s(t) = \kappa\left[\sigma_2^2 - \sigma_1^2 - 4am(t)\right] \tag{12}$$

By using the AEP (see Chapter II) we are led naturally to the Stratonovich form [to be associated with Eq. (9)]:

$$\frac{\partial}{\partial t}\sigma(x; t) = \left\{\alpha\frac{\partial}{\partial x}g(x) + \beta\frac{\partial}{\partial x}g(x)\frac{\partial}{\partial x}g(x)\right\}\sigma(x; t) \tag{13}$$

where $g(x) = x(1-x)$ and

$$\alpha = \kappa\left(\sigma_1^2 - \sigma_2^2 + 4am_0\right) \tag{14}$$

and

$$\beta = 16a^2\kappa^2 \int_0^\infty \langle \xi(0)\xi(t)\rangle \, dt \tag{15}$$

α can be regarded as a parameter measuring the "asymmetry" of the noise, that is, directional selection in favor of a given allele; β is a measure of the "intensity" of the noise. Assuming $\xi(t)$ to be an Uhlenbeck-Ornstein Gaussian noise with correlation functions

$$\langle \xi(t)\rangle = 0 \tag{16}$$

$$\langle \xi(0)\xi(t)\rangle = e^{-\gamma t}\langle \xi^2\rangle_{eq} \tag{17}$$

β turns out to be given by

$$\beta = 16a^2\kappa^2 \langle \xi^2\rangle_{eq}\frac{1}{\gamma} \tag{18}$$

Note that if Eq. (17) holds, then the time τ_m is defined by $\tau_m = \gamma^{-1}$.

III. THE TIME NEEDED TO REACH A NEW DISTRIBUTION OF GENE FREQUENCIES AND THEIR EQUILIBRIUM DENSITY

Let us make the following change in the time scale:

$$t \to t' = Ft$$

This allows us to write Eq. (13) as follows:

$$\frac{\partial}{\partial t}\sigma(x;t) = \Gamma\sigma(x;t) \equiv \left\{ \frac{\alpha}{F}\frac{\partial}{\partial x}g(x) + \frac{\beta}{F}\frac{\partial}{\partial x}g(x)\frac{\partial}{\partial x} \right\}\sigma(x;t) \tag{19}$$

where $g(x) = x(1-x)$. From now on, for simplicity, the rescaled time t' will be denoted again by the symbol t.

To determine the relaxation properties of our system it is necessary to define a dynamical property which is independent of the initial distribution of gene frequency, for example, the correlation function

$$\langle x(0)x(t)\rangle_{eq} = \int x\left\{ e^{\Gamma^\dagger t}x \right\}\sigma_{eq}(x)\, dx \tag{20}$$

where $\sigma_{eq}(x)$ is the equilibrium distribution of x and $\Gamma^\dagger$ is the operator adjoint to that defined in Eq. (19). Its actual form depends on the value assigned to F. Let us now consider the time-dependent function

$$\Phi_C(t) = \frac{\langle x(0)x(t)\rangle_{eq} - \langle x\rangle^2_{eq}}{\langle x^2\rangle_{eq} - \langle x\rangle^2_{eq}} \tag{21}$$

Its Laplace transform can be evaluated via a continued fraction expansion (see Chapters I, III, and IV) and is denoted by $\hat{\Phi}_C(z)$. Note that $\Phi_C(t)$ is a function that decays (in general nonmonotonically) from the initial value $\Phi_C(0) = 1$ to zero as $t \to \infty$. If this decay can be approximated by an exponential function, the time

$$T = \hat{\Phi}_C(0) \tag{22}$$

elapses before the decay function reaches e^{-1}. We assume T, Eq. (22), as our definition of the relaxation time.

We shall refer to the following special cases of Eq. (19): (i) $F = \beta$ and (ii) $F = \alpha$, which lead respectively to

$$\frac{\partial}{\partial t}\sigma(x;t) = \Gamma_A\sigma(x;t)$$

$$\equiv \left[A\frac{\partial}{\partial x}g(x) + \frac{\partial}{\partial x}g(x)\frac{\partial}{\partial x}g(x) \right]\xi(x;t) \tag{23}$$

where $A = \alpha/\beta$, and

$$\frac{\partial}{\partial t}\sigma(x;t) = \Gamma_Q\sigma(x;t)$$

$$\equiv \left[\frac{\partial}{\partial x}g(x) + Q\frac{\partial}{\partial x}g(x)\frac{\partial}{\partial x}g(x) \right]\sigma(x;t) \tag{24}$$

where $Q = \beta/\alpha$.

The first scheme allows us to explore the region where the value of asymmetry is vanishingly small, while keeping finite the intensity of noise; the second enables us to study the case of vanishingly small noise at finite values of asymmetry. Figure 2 shows T as a function of A and Q.

The results of Fig. 2a correspond exactly to intuition: given a certain amount of optimum fluctuation, any increase in the asymmetry parameter would cause an increase in the rate at which the favorite allele gains frequency in the population, so that the relaxation time decreases asymptotically. The results of Fig. 2b, on the other hand, are rather surprising. Given a certain

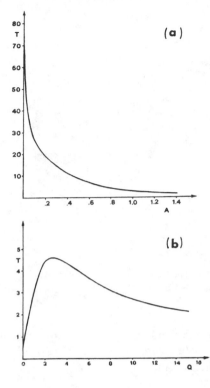

Figure 2. The relaxation time T as a function of the parameters $A(=\alpha/\beta)$ and $Q(=\beta/\alpha)$.

amount of directional selection, the increase in the fluctuation intensity has the effect of *increasing* the relaxation time in the region where the intensity is small in comparison with the asymmetry (i.e., when $\beta/\alpha < 2.5$). This means that the fluctuations of the optimum can make the system *more* stable. However, this happens just before getting to a threshold, after which the relaxation time starts to decrease asymptotically. It is noteworthy that the decay of gene frequency can be slowed down by up to 10 times, owing to random fluctuation, in comparison with the purely deterministic condition ($Q = 0$). This relaxation-time increase is difficult to explain intuitively, but from a biological point of view it is important, because it is brought about by conditions which can be thought of reasonably as being present in natural ecosystems.

Anyway, we shall try to explain the result of Fig. 2 in terms of potentials and noise intensity. If we use the identity

$$\frac{\partial}{\partial x}g(x)\frac{\partial}{\partial x}g(x) = \frac{\partial^2}{\partial x^2}g^2(x) - \frac{\partial}{\partial x}g(x)g'(x) \qquad (25)$$

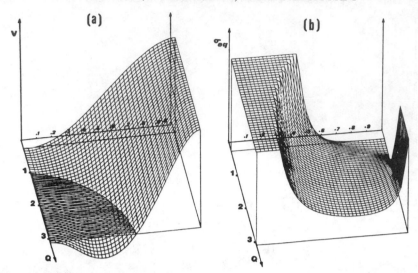

Figure 3.   (*a*) The effective potential as a function of gene frequency x and the parameter Q. (*b*) The corresponding equilibrium distributions (which are artificially truncated).

We note that the second term on the right-hand side is a deterministic term of the same form as the first term on the right-hand side of Eq. (24). Both terms contribute to defining a sort of "effective potential." Thus we obtain

$$\frac{\partial}{\partial t}\sigma(x;t) = \left[\frac{\partial}{\partial x}\left(\frac{\partial V}{\partial x}\right) + \frac{\partial^2}{\partial x^2}D(x)\right]\sigma(x;t) \qquad (26)$$

where V, the "effective potential," is

$$V = -\frac{x^3}{3} + \frac{x^2}{2} - \frac{Q}{2}x^2(1-x)^2 \qquad (27)$$

and D, the intensity of the stochastic force, is

$$D = Qx^2(1-x)^2 \qquad (28)$$

Note that the stochastic force has, for any value of $Q \neq 0$, a maximum at $x = 1/2$ and vanishes at $x = 1$ and $x = 0$. The potential, on the contrary, shows a more complex behavior as a function of x and Q. Its surface is shown in Fig. 3*a*. Let us discuss its properties together with the corresponding stationary distribution of gene frequency, which from Eq. (24) is found to be

$$\sigma_{eq} = \frac{(1-x)^{1/Q-1}}{x^{1/Q+1}} \qquad (29)$$

and is shown in Fig. 3*b*. The potential and the equilibrium distribution as functions of A are not shown because they are easily extrapolated from Fig. 3 by taking into account that $A = 0$ corresponds to $Q = \infty$ and vice versa.

It can be seen that the potential is always positive and has a monotonic dependence on x when $Q < 1$ (or $A > 1$). In this case the asymmetry parameter is the driving force and the equilibrium distribution is simply unimodal. At the point $Q = 1$ ($A = 1$) there is a transition in the nature of both the potential and the equilibrium density. As Q increases from 1, a well begins to appear in the potential, which establishes an absolute minimum at $x \neq 0$, while the equilibrium distribution becomes bimodal.

It is worthwhile remarking that the standard picture borrowed from classical mechanics, according to which the maximum population density is attained at the potential minimum, does not apply to the present case. In fact, the classical view is applicable to systems in which the stochastic forces are additive, while in our model the noise is multiplicative, that is, dependent on the variable of interest. This means that the standard behavior can be recovered only when the stochastic force is not the leading agent driving the evolution of the system. In our case this happens when $Q \ll 1$ or $A \gg 1$. When the stochastic force is the prevailing agent, it acts as a flux of energy from outside into the system. This works against the direction imposed by the potential (as discussed in depth in Chapter X). It causes the appearance of significant asymptotic frequencies for the unfavorite allele. However, there is so little probability mass at intermediate values of gene frequency that the population can be considered effectively monomorphic. Note that the transition point of Fig. 3 (which is at $Q = 1$) does not correspond to the maximum relation time of Fig. 2b. In fact, this maximum is at the value of Q (2.7) at which the function of potential is negative for $x < .5$ and positive for $x > .5$. This suggests that a relationship exists between the inverted trend of relaxation time and the effective potential. We are investigating this possibility.

IV. CONCLUDING REMARKS

The difficulty of choosing on a rational basis an algorithm for the calculus of the density function associated with Eq. (2) led Guess and Gillespie[32] to develop an elegant system to solve the discrete formula for the change of gene frequency in diploid populations under stochastic selection [i.e., $\Delta_t x = x(1 - x)s_t/2(1 + xs_t)$]. They obtained[4] three different limiting continuous models by changing the value of the "autocorrelation of the environment": respectively, the cases of weakly, moderately, and strongly autocorrelated environments. In our work the autocorrelation of the environment, that is, the non-whiteness of the optimum fluctuation, is measured by the parameter $1/\gamma$ of Eqs. (16) and (17). From this point of view, three time scales are involved in our model. The first two are defined in Section II and give, respectively, the spreading time of the phenotypic values, which we here assume equal to the generation time τ_g, and the speed of the optimum fluctuation measured by $1/\gamma$. The third time scale is given by T, the relaxation time studied in Section III.

This paper is based on the assumptions

$$1 = \tau_g \ll \frac{1}{\gamma} \ll T \tag{30}$$

The condition $\tau_g \ll 1/\gamma$ is discussed in Section II. If $\gamma \to \infty$, the model loses meaning, and this is the ultimate reason for adopting the Stratonovich form; however, a further element of caution emerges by considering the presence of environmental deviation: It introduces some kind of "inertia" into the system, thus invalidating the Stratonovich form coming from the application of the AEP (see Chapter X). In a sense we must assume that the friction affecting the "velocity" corresponding to the variable x is very large, and this always happens when both the conditions of Eq. (30) are met.

On the basis of AEP alone we can tackle the problem where $1/\gamma$ is not much smaller than T. It is possible, indeed, to evaluate in a fairly systematic way correction of orders higher than $1/\gamma$. However, we are in a position to assert that the basic structure of Eq. (13) would be left unchanged, so that no relevant changes from the picture of Section III may be expected.

In conclusion, we may say that in the absence of environmental deviation the results of Section III are valid for environments whose autocorrelation ranges through all the possible reasonable values; if we take into account the environmental deviation, this has the interesting effect of making tractable by RMT only those cases in which the fitnesses are autocorrelated in time. Under such conditions our treatment demonstrates that the rate of extinction of genetic variability depends on two parameters, respectively measuring the noise intensity and the asymmetry (i.e., fundamentally the range of the optimum fluctuation and the long-term difference in mean fitness). Taking into account asymmetry only, we obtained the anticipated result that any increase in this factor diminishes the time of quasi-fixation of alleles; the increase in the intensity of noise has, in contrast, a more complex effect, producing an increase of quasi-fixation time in the region where the noise is weak compared to the asymmetry.

Our main conclusion is that random fluctuations of the environment can slow down the extinction of genetic variability by up to 10 times in comparison with the absence of fluctuation. The slowing down is brought about when the range of the optimum fluctuation has about the same value as the long-term difference in mean fitness.

Acknowledgments

We thank Dr. S. Jayakar for reading the manuscript and for his helpful comments and criticism.

References

1. S. Wright, *Proc. Nat. Acad. Sci. U.S.A.*, **31**, 382 (1945).

2. M. Kimura, *Genetics*, **39**, 280 (1954).

3. J. H. Gillespie, *Genet. Res.*, **21**, 115 (1973).

4. J. H. Gillespie and H. A. Guess, *Amer. Natur.*, **112**, 897 (1978).

5. W. Horstemke and M. Malek-Mansour, *Z. Phys.*, **B24**, 307 (1976).

6. W. Horstemke and R. Lefever, *Phys. Lett.*, **64A**, 19 (1977).

7. L. Arnold, W. Horstemke, and R. Lefever, *Z. Phys.*, **B29**, 367 (1978).

8. K. Kitahara, W. Horstemke, and R. Lefever, *Phys. Lett.*, **70A**, 377 (1979).

9. R. Lefever and W. Horstemke, *Bull. Math. Biol.*, **41**, 469 (1979).

10. R. Lefever and W. Horstemke, *Proc. Nat. Acad. Sci. U.S.A.*, **76**, 2490 (1979).

11. J. Smythe, F. Mass, P. V. E. McClintock, and D. Clarkson, *Phys. Lett.*, **974**, 95 (1983).

12. J. Smythe, F. Mass, and P. V. E. McClintock, *Phys. Rev. Lett.*, **51**, 1062 (1983).

13. H. Fujisaka and S. Grossman, *Z. Phys.*, **43**, 69 (1981).

14. S. Faetti, P. Grigolini, and F. Marchesoni, *Z. Phys.*, **B47**, 353 (1982).

15. J. M. Sancho, M. San Miguel, S. L. Katz, and J. D. Gunton, *Phys. Rev.*, **26A**, 1589 (1982).

16. S. Faetti, C. Festa, L. Fronzoni, P. Grigolini, F. Marchesoni, and V. Palleschi, *Phys. Lett.*, **49A**, 25 (1983).

17. R. Graham and A. Schenzle, *Phys. Rev.*, **26A**, 1676 (1982).

18. M. Suzuki, K. Kaneko, and F. Sasagawa, *Prog. Theor. Phys.*, **65**, 828 (1981).

19. Y. Homada, *Prog. Theor. Phys.*, **65**, 850 (1981).

20. M. Suzuki, *Adv. Chem. Phys.*, **46**, 195 (1980).

21. P. Glansdorff and I. Prigogine, *Thermodynamic Theory of Structure, Stability and Fluctuations*, Wiley, New York, 1971.

22. G. Nicolis and I. Prigogine, *Self-Organization in Non-Equilibrium Systems*, Wiley, New York, 1977

23. H. Haken, *Synergetics*, Springer, Berlin, 1977.

24. J. H. Gillespie, *Theor. Pop. Biol.*, **3**, 241 (1972).

25. S. Wright, *J. Genet.*, **30**, 243 (1935).

26. K. Kojima, *Proc. Nat. Acad. Sci. U.S.A.*, **45**, 989 (1959).

27. R. C. Lewontin, *Genetics*, **50**, 757 (1964).

28. B. D. H. Latter, *Genetics*, **66**, 165 (1970).

29. M. G. Bulmer, *Genet. Res.*, **19**, 17 (1972).

30. R. Lande, *Genet. Res.*, **26**, 221 (1976).

31. M. Kimura, *Proc. Nat. Acad. Sci. U.S.*, **54**, 731 (1981).

32. H. A. Guess and J. H. Gillespie, *J. Appl. Probability*, **14**, 58 (1977).

33. E. Mayr and W. B. Provine, eds., *The Evolutionary Synthesis*, Harvard Univ. Press, Cambridge, MA, 1980.

34. J. B. S. Haldane and S. D. Jayakar, *J. Genet.*, **58**, 237 (1962).

35. E. R. Dempster, *Cold Spring Harbor Symp. Quant. Biol.*, **20**, 25 (1955).

36. J. Felsenstein, *Ann. Rev. Genet.*, **10**, 253 (1976).

37. J. Maynard-Smith and R. Hoekstra, *Genet. Res.*, **35**, 45 (1980).

XII

STOCHASTIC PROCESSES IN ASTROPHYSICS: STELLAR FORMATION AND GALACTIC EVOLUTION

F. FERRINI, F. MARCHESONI, and S. N. SHORE

CONTENTS

I. INTRODUCTION

The application of stochastic methods to astrophysical problems has a long and colorful history. Perhaps the first area of investigation, in analogy with the successes of nineteenth century statistical mechanics, was that of stellar dynamics. The sidereal universe was treated as a gas of massive bodies in a phase space whose natural coordinates referred to the galactic plane and center. This treatment, initiated by Schwarzschild[1] and Eddington[2] was capped with the review by Chandrasekhar.[3] It included the elucidation of the velocity distribution function, the variation of the velocity dispersion with galactocentric position and age (later explained in part by Spitzer and Schwarzschild,[4]) and the discovery of the rich field of statistical stellar dynamics (see, e.g., reviews by Kurth,[5] Mihalas and Routly,[6] Oort,[7] and Mihalas and Binney,[8] and references therein). At about the same time, the methods were applied to stochastic line broadening in atomic systems and eventually to stellar atmospheres (see Mihalas,[9] Griem,[10] and Chandrasekhar[3]). The application of such methods to stable fluctua-

493

tions in the brightness of the Milky Way and the statistical mechanics of gravitational encounters between stars has also been very productive.[11,12] Turbulence theory has been applied to dynamo models (see review by Parker[13]) and to propagation of cosmic rays through the galaxy by diffusive motion in both energy and spacetime (see Ginzburg and Ptuskin[14] for review).

Since these areas have been covered extensively in the literature,[15-20] we shall not add to the already groaning mass of tomes on the subject with this survey. Instead, we shall concentrate on those processes which have in the past decade been brought to bear on the problem of galactic evolution, star formation, and global problems of the large-scale distribution of the galactic population of the universe. Most of these methods fall into categories of broad generality and can in effect be labeled by the methods employed in their investigation rather than the area being studied. We shall therefore proceed by separating the problems according to the method rather than the topic being investigated. This is done for several reasons. The field of stochastic phenomena in astrophysics in particular and physical science in general has taken on the appearance of a growth industry since the introduction of computer methods about 15 years ago. The time for diffusion of techniques between fields has been decreasing, and still there is some lack of communication between theorists in different disciplines concerning the similarity of approaches. Since, for example, population ecology has spawned many of the nonlinear methods used in the modeling of the chemical history of the galaxy, and since percolation techniques can be used for any form of lattice-dominated phase transition from QCD to galactic structure, we feel that this separation by method will assist the reader by allowing for easy comparison between techniques and setting in the different areas of astrophysical investigation. Having presented our philosophical justification, then, we should now present our basic categories of methods:

Coagulation Phenomena. Agglomeration and fragmentation calculations which bear direct kinship to the Fokker-Planck treatments but also include discussion of expectation of *N*-body systems for which the initial distribution function and dynamics cannot be or are not treated in the continuum limit.

Percolation. Application of local interactions to the problem of generating long-range order, including phase transitions and morphogenesis, to large-scale discrete models.

Langevin Systems. Nonlinear deterministic and stochastic representations of interacting states or populations with and without consideration of spatial variance.

Fokker-Planck Equations. Fully stochastic realizations of the Langevin systems and the Monte Carlo simulations to which they give rise.

II. AGGLOMERATION PHENOMENA (COAGULATION EQUATION APPLICATIONS)

One recent development in astrophysical stochastic processes has been the widespread use of coagulation calculations for both nucleation phenomena and dust formation and processes that relate to the distribution function for masses and mass ratios in forming stellar systems. The use of the coagulation equation for the study of star and stellar system formation in particular has been quite recent and warrants a review.

A. Solar System Formation

Perhaps the first studies to employ agglomeration were those related to the formation of the solar system. The use of the coagulation equation, essentially a macroscopic version of the master equation for systems which can be treated as being controlled by one independent variable and time, has been of some importance in recent simulations of the process of star formation. Employed for some time in the study of nucleation,[21-28] the results were first used extensively by Safronov and his collaborators[29] in the modeling of planetary formation in the solar nebula.

Various stages can be identified by the main physical process acting in the evolution of low-mass protosolar nebula models.

In the initial phase a disk was formed by the dust grains settling to the central plane. The original disk structure can be specified by the application of some rather basic hydrodynamical constraints. Generally it is assumed that the disk is dominated by turbulent heating and is not self-gravitating.[30,31] The grains can grow up to 1 cm, due to the condensation they undergo in the cooling nebula. The turbulence can prevent the disk from becoming dense enough to suffer the gravitational instability which could fragment it in higher mass pieces.[32] The mechanism that allows the growth of the grains up to dimensions of the order of 1 meter or more is coagulation.

The grains will proceed to collide with each other and then both fragment and stick.[29,33-35] The growth of the individual grains, or fragments, can then be represented by a Smoluchowski equation of the form

$$\frac{\partial N(\mu, t)}{\partial t} = \frac{1}{2} \int_0^\mu \alpha(\mu, \mu') N(\mu', t) N(\mu - \mu', t) \, d\mu'$$

$$- \int_0^\mu N(\mu) N(\mu') \alpha(\mu, \mu') \, d\mu' \qquad (2.1)$$

where $\alpha(\mu, \mu')$ is the sticking probability, μ is the mass of the particle, and $N(\mu, t)$ is the number of particles of mass μ at time t. In the case of particles in the solar nebula, several arguments have been adduced to supply some phenomenological form for α. These are outlined by Nakagawa et al.,[34] Safronov and Ruzmaikina,[36] and Wetherill.[37] Simple simulations can be carried out with such systems, assuming that the collision rules are well specified, and the results show that the formation of large-scale agglomerated bodies can proceed quite easily. The approximation made in the treatment by Nakagawa et al.[34] and Morfill[35] is that of a continuous medium, in which case the equation for the evolution of the size distribution is a diffusion equation. Such an equation, the limit of the full master equation for the system, can be solved numerically for the distribution of surface density (not particle sizes explicitly) as a function of heliocentric distance. A mass spectrum can be calculated as a function of distance by introducing the agglomeration conditions explicitly and then solving for the variation of an initial particle spectrum. The evolution of that distribution can then be followed as a function of time.

Within this new higher density medium, it is nonetheless possible for gravitational clumping to occur[29,38,39] and for bodies of sizes of the order of 1 km to form.

The subsequent evolution of this swarm of planetesimals is governed by their relative gravitational interactions and collisions. With the exception of a thermodynamical model based on phase transition,[40,41] this problem has been treated numerically[42-45]; the approach is to follow explicitly the evolution of the system of planetesimals as an N-body problem. The individual particles are labeled with the appropriate orbital parameters, and the equations of motion for a large number of planetesimals are then integrated. The distribution of the particles initially simulates the structure of the protosolar disk, and the subsequent evolution of the protoplanets can be followed allowing for the same kind of collisional dynamics first assumed by Dole[46] and Isaacman and Sagan[47] in their models for synthetic solar systems.

When a fragment grows to 1000 km, it can capture the residual gas in the nebula, and this is the main process of growth together with the rare collisions with residual planetesimals.

The asteroid belt, after decades of disregard, has recently been the object of various theoretical works, even for its aspect of planetesimal system. Its dynamical structure is intriguing for the presence of chaotic regions in the distribution of asteroids in phase space,[48-50] revealing the intrinsic nondeterministic nature of the N-body nonintegrable gravitational problem and the possibility of testing Kolmogorov-Arnold-Moder theory.[51] The collisional evolution, as determined by several dynamical models,[52-56] based on the coagulation equation and other statistical treatments such as Monte Carlo calculations, affects the physical properties of these objects. Detailed observations, such as photometric light curves, can test the prediction capabilities of the theory.

B.   Star Formation—Initial Mass Function

The basic data for stochastic simulations of galaxies and their constituent populations and metallicity evolution is the initial mass function (IMF), which represents the mass distribution with which stars are presumed to form. Its derivation from the observed distribution of luminosity among field stars (refs. 57 and 58 and references therein) and from star clusters[59,60] involves many detailed corrections for both stellar evolution and abundance variations among the observed population. The methods for achieving the IMF from the observed distribution are most thoroughly outlined by Miller and Scalo[60] but can be stated briefly, since they also relate to an accurate testing of various proposed stochastic methods. It should first be noted that the problems encountered for stellar distributions are quite similar to those with which studies of galaxies and their intrinsic properties have to deal.

Given the observed distribution of stellar masses among both cluster and field stars, the problem is to correct for evolutionary and metallicity effects. The actual mass is rarely measured; instead, one observes the bolometric brightness and surface temperature or color. The first correction which must then be made is to determine the age of the star and correct for the variation in these parameters with time. If the star is in a cluster, then it is possible to at least get a handle on the appropriate age for all the members and to use theoretical interior models to trace all the stars back in time and surface parameters to the stage of hydrogen core burning, the so-called zero-age main sequence (ZAMS). The population of stars along this part of the evolutionary history of the cluster is presumably the population with which the cluster formed. The same is true for field stars if the relative populations of the various main sequence groups (which are ordered by temperature or spectral type) can be determined from a complete survey (that is, complete down to some limiting apparent brightness with appropriate statistical corrections for the fainter, inaccessible part of the sample). The rate of star formation can then be assumed, and the IMF obtained from modeling the population statistics.

When this is done, some parametric form for the mass spectrum has to be assumed. The initial approximation, that of a power law, is referred to as Salpeter mass function, following Salpeter.[59] This approximation, of course, cannot apply over the entire range of possible masses, since the lower masses produce divergence in the total population. It is usual to specify three parameters: the upper and lower mass cutoffs and the exponent. While not useful in a fundamental way for explicating the origin of the mass spectrum, it is a convenient parametrization for models of star formation and the populations of external galaxies.

If we imagine that the stellar population has been formed continuously over a period of time, but that the distribution function for the formation of new stars is stable (in a stochastic sense), then the IMF should be reflected in the more evolved members of the sample population. That is, the mean age of stars should be younger for the most massive, and the relative number of massive to low mass stars should be stable. The primary complication to this would be, and in fact is, that the stars of the upper main sequence do lose a considerable amount of mass during post-hydrogen core exhaustion evolution, and therefore can repopulate the lower mass tail of the IMF. However, this phenomenon is not really under control in the theoretical work of tracing back the sample to the ZAMS under the constraint of a history of star formation, since the algorithms for the computation of the evolution of these massive, high luminosity stars dominated by mass loss are not well understood.[61-63] For the lower mass stars, specifically for those less than 10 $M_\odot$, the situation is a good deal better. These stars evolve essentially conservatively until the extreme red giant stage, which represents only a few stars in any sample and therefore can be ignored.

The average star encountered in the field will be evolved, and therefore corrections for the expansion of the envelope and consequent increase in the luminosity must be applied, and changes in the metallicity (gradient over both space and time) must also be applied to the lower mass stars.

It is still necessary to make assumptions, many of them ad hoc, concerning the time dependence of the star formation rate (see Section IV for the Langevin nonlinear treatment which allows us to circumvent this problem as well) and then evolve a chosen population forward in time to obtain the final IMF.

While the ideal way of proceeding would appear to be a fully stochastic simulation of the population, including the effects of mass loss and rotation, all of which are chosen from distribution functions which are known at present, this has only been done for a few cluster models by Elmegreen and Mathieu.[64]

It would be most useful to apply this to the field population in general. In addition, models are currently being studied which allow for the formation of stars of different masses by using different reaction channels in the Langevin systems of the following sections in order to see whether the IMF is a stable stochastic function of time. If it changes, the star formation rate, the metallicity evolution of the disk, and the IMF variations become inexorably linked and impossible to separate.[65]

Broadly speaking, the treatment of fragmentation of molecular clouds falls into two categories in its astrophysical guises: analytic approximations to the coagulation equation and numerical simulations of accumulations among fragments or of fragmentation during collisions or collapse of clouds. We shall review these together, since the analytic treatment often precedes the introduction of numerical methods in a large variety of different contexts.

However, since we are dealing in this review primarily with problems related to star formation and galactic evolution, we shall ignore the work that has been done on dust formation and nucleation of classical (chemical) systems. These have been extensively reviewed by Abraham,[21] Burton,[26] and Draine and Salpeter[25] for problems of astrophysical interest. We shall only refer to this literature for analogies which may be of some aid in establishing new directions for work on megascopic systems like interstellar clouds.

The idea that fragmentation is the source for stars in clouds was first expressed most comprehensively by Hoyle,[66] who argued that the observed distribution of galactic masses and the constituent stellar masses can be represented as the result of a hierarchical fragmentation. The dominant mass in this case is the Jeans mass, the size of a self-gravitating blob which will be critically stable against collapse if it remains isothermal. This mass is given by

$$M_J = 1.86 \left(\frac{R}{G} \right)^{3/2} \frac{T^{3/2}}{\rho^{1/2}} \tag{2.2}$$

where R is the perfect gas constant, G is the gravitational constant, T is the temperature of the cloud, and ρ is the mean density. Hoyle's consideration was elaborated on by Hunter,[67] who followed the collapse of a sphere through the critical stage, and by Mestel,[68] who argued that magnetic fields can play a critical role in supporting such a self-gravitating system against collapse. Indeed, if there is any charge present in the cloud, the magnetic field will actually dominate the stability of the system.[13] The scale of the critical mass is then the most probable mass to form at any stage in the collapse of a cloud.[69,70] Armed with this scale, it is possible to compute whether observed clouds in the intestellar medium are stable against fragmentation. The simple answer is that the average molecular cloud contains many thousands of Jeans masses and so should be unstable to spontaneous star formation. The support of such systems has alternatively been ascribed to magnetic fields[71] and turbulence.[72,73]

It is such considerations that have given rise to the application of analytic treatments of the coagulation equation. The first such treatment for clouds is that of Nakano,[74] who assumed that all fragments initially have the Jeans mass. Taking several forms for the agglomeration coefficient, which was assumed to be a simple mass-dependent quantity (or constant), the coagulation equation was numerically integrated. For $\alpha_a(\mu, \mu')$ constant or $\alpha_b(\mu, \mu') = \mu^{1/3} + \mu'^{1/3}$, the resultant mass distribution is peaked between $\log m$ of 1 and 2, while for $\alpha_d(\mu, \mu') = \mu + \mu'$ and $\alpha_c(\mu, \mu') = \alpha_b(\mu, \mu')^2$ the distribution is a monotonically decreasing function of mass. In general, coefficients α_a and α_b lead to something like a log-normal distribution, while α_d tends toward a power law. Case α_c is intermediate.

An elaboration on this procedure was provided by Normal and Silk,[75] who treat the growth of T Tauri stars in turbulent clouds by the coagulation equation and provide an analytic solution for the evolution for a simple power-law input. The turbulence of the cloud derives from the turning on of the T Tauri stage, which through stellar wind stirring maintains the turbulence required to support the clouds against collapse. The key feature of their treatment is a concentration on the low-mass stars, which are near the Jeans mass for the cloud and consequently the most probable stars to form.[76] This treatment, however, has several shortcomings. It ignores the infall of material, which may be of some importance in the evolution of the system (especially if there is any stimulated formation of fragments or agglomeration process), and it also ignores the decay of individual fragments. The latter may be due either to collapse and the subsequent formation of a star (if the mass is greater than the Jeans mass and the fragment is not supported by either turbulence or magnetic field) or the evaporation by stars which may already have formed. This is a relevant problem for the evolution of molecular clouds, since recent observations[77] have shown that x-ray emission by T Tauri stars which have been formed previously in the cloud may alter the subsequent evolution of the fragments. Therefore, one of the pressing problems in the theory of fragmentation and evolution of the cores of molecular clouds is the inclusion of the effects of previously present stars on the dynamics and thermal-ionization structure of the medium. While Norman and Silk[75] do treat feedback processes in a simple way, there is still lacking a general theory of the detailed effects of the appearance of stars on the internal structure and evolution of molecular clouds.

A simple treatment of fragmentation which includes the hierarchical model is due to Larson.[78] Assuming successive, but random, bifurcations of a sample of collapsing fragments, he derives a binomial distribution for the mass spectrum:

$$f(n, m) = \left(\frac{1}{2}\right)^n \binom{n}{m} \tag{2.3}$$

where there are n stages of fragmentation and each fragment has a mass fraction 2^{-m}. If the probability of fragmentation is p, then this generalizes to

$$f(n, m; p) = p^m (1-p)^{n-m} \binom{n}{m} \tag{2.4}$$

which for $n \to N$ becomes a Gaussian distribution for the fragments:

$$f(N, m; p) \simeq [2\pi p(1-p)N]^{-1/2} \exp - \left[\frac{(m - Np)^2}{2p(1-p)N}\right] \tag{2.5}$$

Since each division is essentially a Poisson-type trial (that is, there is a probability p at each step which is independent of the previous history of the fragment), there is a random normal distribution of the fragments with the mean value being Np and the dispersion being $2p(1 - p)N$. The fluctuation is therefore of the order of $N^{1/2}$, as expected from such a Poisson trial. There are no terms related to the agglomeration of these fragments, which will tend to cause the distribution to evolve away from the Gaussian form. Larson points out that such a fragmentation scheme will ultimately lead, in its full realization, to a fractal distribution of masses (see ref. 80) which is characteristic of turbulent distribution (relevant also for the properties of molecular clouds[73]). It should be noted that the IMF to which Larson[78] compares his data does not agree with that derived by Miller and Scalo,[60] and his function is quite different from the results of the coagulation calculations. The treatment of statistical fragmentation has recently been discussed by Zinnecker[79] and Elmegreen[76] for cluster and binary system formation. Returning to the picture advocated by Hoyle, they treat fragmentation as a multiplicative stochastic process in which the probability of fission of a fragment at any step is Markovian, that is, independent of the history of the fragment. In this fashion, they do not explicitly include the dynamics of the collapse phase. As in the case of the broken-stick distribution known from ecological models (see ref. 81 for a review), they obtain a log-normal mass spectrum which is in good agreement with the initial mass function obtained for galactic stars by Miller and Scalo[60] and which asymptotically provides the power law spectrum suggested by Salpeter.[59] Elmegreen and Mathieu[64] have generalized this treatment to the formation of star clusters by arguing that the formation of massive stars shuts off the process of stable fragmentation-star formation within the cloud. Therefore, only in the most massive systems, where there is a likelihood of forming OB stars, will there be a short formation time. Elmegreen[76] argues that statistically long formation times (of the order 10^9 years for the typical system) may characterize the formation of clusters. The resultant mass spectrum should be dominated by the low-mass stars. In addition, the long formation period may give rise to an intrinsic spread in the metallicity and time-dependent properties of the stars.

A novel approach to the fragmentation problem is due to Ferrini et al.,[82] who describe the internal cloud dynamics by means of a nonlinear Lagrangian in a scalar field, whose modulus square represents the density distribution inside the cloud. By solving analytically the nonlinear resulting Klein-Gordon equation, they calculate the fragmentation spectrum. Although the model is not explicitly self-gravitating, the Jeans mass enters into the calculation as a local instability criterium and in the normalization constant, and therefore the gravitational triggering and dynamics of the collapse are implicitly included in the model. The model is essentially a log-normal distribution, where the time scale for the propagation of perturbation in the nonlinear field is smaller than the local free fall time, and therefore agrees well with the Miller and Scalo[60] fits for the field stars. In addition, the IMF is of the form suggested by Elmegreen and Zinnecker for the mass spectrum resulting from multiplicative stochastic processes, as discussed above. Moreover, since the Jeans mass of the cloud depends on the possible presence of magnetic fields, the expected cloud fragmentation spectrum can be derived under very general astrophysical assumptions.

The formation of multiple star systems has long plagued the theory of star formation. Usually termed the angular momentum problem, it is the product of the fact that molecular clouds rotate. Assume that the rotation rate for a 1 $M_\odot$ (solar mass) cloud is 10^{-15} s^{-1}, simply the galactic differential rotation rate for a 1 parsec cloud. If the angular momentum is conserved during the collapse, the resultant velocity of the finally formed single star would be enough to prevent stability (the rotational distortion of the final star will be such that an equatorial cusp is inevitable). Calculations using three-dimensional fluid dynamical codes[87-93] show that the collapse of rotating clouds will certainly form multiple systems.

We refer the interested reader to the paper by Bossi et al.,[94] where this problem is discussed in detail in its parallel aspects of star formation and planetary system formation.

III. PERCOLATION MODELS

There are two primary treatments of percolation in astrophysics, one connected with galactic structure and the other, far less well developed or understood, with magnetic dynamos. We shall concentrate on the first (see the reviews by Broadbeut and Hammersley[95] and Hammersley[96]). The morphological structure of disk galaxies has been a problem of long standing since the introduction of the Hubble classification scheme.[97] This taxonomy uses both the nuclear spheroid (the so-called bulge-to-disk ratio) and the openness of the spiral arms in the disk to provide a classification. The deterministic basis of this scheme, that there is a sequential evolution between the taxa as originally described by Jeans,[98] has been shown not to apply, but the system has survived because numerous other properties of galaxies appear to be well correlated with the Hubble type. A fully hydrodynamic (in effect deterministic) description is provided by the density wave theory of Lin and Shu[99] and its nonlinear elaboration by Roberts.[100] The theory is well reviewed by Toomre,[101] but we shall discuss it briefly, since it can be shown to fit into the stochastic scheme.

Disk systems composed of stars and presumably gas and which are self-gravitating are intrinsically unstable to the growth of a spectrum of longitudinal density waves. The stellar population acts as a compressible fluid whose number density perturbation causes the gravitational potential to vary with azimuth and radius in the disk. This in turn produces accelerations which support the density wave, and in principle the system might be self-sustaining. The resultant structure of the disk would be spiral in appearance, since the shear of the system will produce spiral longitudinal perturbations, and the pattern speed through the system should be fixed by the local condition of shearing given by the epicyclic frequency. If the wave is sufficiently strong, nonlinear hydrodynamical models show that a standing shock will develop at the spiral arms, resulting in compression of the disk's gaseous component and the possible driving of either star formation or cloud formation. Although this should in principle also feed into the distribution function for the stellar population, such an effect has not yet been included with sufficient generality in the (analytic) models to comment on this reaction with the basic density wave predictions. Toomre[101] and Zang[102] have both shown that the system is unstable to winding and that the spiral pattern is indeed not stable on long (many-rotation) time scales. It is in this context that the percolation models have been introduced.

The idea that star formation can have an induced component which depends on the available stellar population in the disk was first realized by Elmegreen and Lada[103] and Herbst and Assousa,[104] and at about the same time by Mueller and Arnett[105] for stellar systems. We shall discuss the effect of such an assumption on models of star formation in local models in Section IV; now we can show that this picture carries over quite nicely into the global models. However, in these pictures, since there has been little analytic work on the large-scale structure (with the exception of Shore[106] and Fujimoto and Ikeuchi[107]), we must confine the discussion to the more abstract aspects of percolation theory on differentially rotating planes and then discuss the reinterpretation of the results in light of the modeling that has been done to date.

The simulations of the stochastic models, for the study of the morphology of star-forming disks as a function of the probability of stimulated star formation and the rotational velocity of the system, have been performed by several groups, notably Gerola and Seiden[108,109]; Schulman, Seiden and Gerola[110,111]; Comins[112]; Freedman and Madore[113]; Statler et al.[114]; and Feitzinger et al.[115]

The basic principle of all these models is that star formation can be viewed as a percolation phenomenon, with the metastructure of the system resulting from the short-range interaction between neighboring cells. The stimulus of Conway's life game has proven of great importance in this field; the system is governed by the rules that the activity of a cell is controlled by the neighboring population of active cells and that the propagation of influence between two cells can occur with or without time delay. This is augmented with the "dynamical" condition that the disk on which the percolation occurs can be treated as differentially rotating. Unlike the Conway game, this means that there is a global structuring and replenishment of active sites. In addition, it implies that the percolation, if not taking place on a fixed grid, will have a variable critical probability from place to place in the disk. This latter problem is circumvented by assuming that the rotation curve for the disk is flat; that is, the rotational frequency varies as the inverse of the radius. Such an expedient not only fixes the number of cells that are needed to treat the evolution of the disk, it also provides for a global percolation parameter. The models show that the peak of the rotation curve, that is, the rate at which the cells are dynamically forced to communicate, is the basic percolation parameter. The primary success of the models has been the reproduction of stable spiral structure. The propagation of the spiral pattern occurs at the percolation velocity, determined only by the maximum of the rotational frequency. The rate of replenishment of a cell's gas is also determined by this velocity, since the differential rotation is scaled to this speed, and therefore the robustness of the global star formation is dominated by the maximum of the rotational velocity.

Predictions of the color, metallicity, and star formation gradients across such disks can be compared directly with observations[116–118] for a few galaxies, and the rates of star formation with time can be derived. One of the most remarkable observations, which is in effect similar to that observed in chemical systems, is that modes of coherent oscillations are possible[115] which have the appearance of ring galaxies. The implicitly nonlinear stochastic models do not allow one to follow the details of star formation as such but rather present a morphology which can be classified in the same fashion as those of galaxies. This classification has been stressed by Shore and Comins[119] and also by the originators of the various systems.[97,120] In the absence of detailed information about the rotation curve and distribution of stellar constituents, the only information available for a galaxy is global in nature—its morphology, total gas content, integrated luminosity, and integrated spectral type. Ultimately, the goal of the SSPSF or density wave models must, it seems fair to say, be to provide some causal connections between the gross features of these systems and the detailed "microphysics" by which they come into being.

IV. LANGEVIN SYSTEMS AND THE FOKKER-PLANCK EQUATION

Langevin systems of coupled nonlinear equations have been used recently in the modeling of galactic evolution, in the framework of the so-called one-zone model (OZM). The OZM is a model in which a certain limited region of a galaxy is considered and its content in gas, stars, and, eventually, clouds is studied by neglecting the variations of the basic galactic properties across the region[121,122]; it can allow for exchange of matter with the external medium.

The astrophysical problem of justifying on theoretical grounds the morphology of galaxies (spiral and elliptical, with their different content in stars and gas), their chemical evolution (initial rapid enrichment of metals, i.e., any element heavier than hydrogen and helium), and, finally, the attempt to trace a classification based on different physical aspects of the evolution, has been tackled by employing the approach of cooperative systems. In these models a scenario is proposed where the large-scale dynamics are related to the local microscopic interactions. At the same time a macroscopic description (e.g., the interplay of various phases, the metallicity) is derived by means of few (stochastic) variables.

The mathematical structure of the models is their unifying background: systems of nonlinear coupled differential equations with eventually nonlocal terms. Approximate analytic solutions have been calculated for linearized or reduced models, and their asymptotic behaviors have been determined, while various numerical simulations have been performed for the complete models. The structure of the fixed points and their values and stability have been analyzed, and some preliminary correspondence between fixed points and morphological classes of galaxies is evident—for example, the parallelism between low and high gas content with elliptical and spiral galaxies, respectively.

Typical is the oscillating behavior of the solutions, the astrophysical meaning of this phenomenon being straightforward: bursts in the star formation rate have been observed in young galaxies, and the color distribution of older galaxies is again evidence of nonmonotonic star formation history. The burst time scale can be calculated from the parameters of the models. At the same time, the local models are particularly suited to describing irregular galaxies, while the nonlocal models reproduce the large-scale pattern of spiral galaxies. Finally, the chemical evolution of galaxies can be reproduced with great care. In conclusion, the common matrix of modeling is the synergetic behavior of the system: A few variables would describe the evolution, while microphysics intervenes in the processes which determine the values of the parameters.

Shore[123] first introduced some Langevin models in which the effects of induced star formation on the evolution of a galaxy are investigated. In Table I we summarize all the models described in the literature; we will analyze only the main features and refer to the original papers for a more detailed discussion. Let us call $s(t)$ and $g(t)$ the mass fraction of stars and gas, re-

TABLE I
Langevin Systems Presented in the Literature for the Modeling
of Galactic Evolution

Model	Comments	Ref.
1. $s(t) + g(t) = 1$ $\dot{s}(t) = as(t)g(t) = as(t)[1 - s(t)]$	The meaning of s, g, and c is in the text as well as that of the various rates. Model 4 includes the effects of lag between generations of stars.	123
2. $\dot{s}(t) = -rs(t) + as(t)g(t)$ $\dot{g}(t) = -dg(t) + f$		
3. $\dot{s}(t) = -rs(t) + as(t)c(t) + bg(t)^n$ $\dot{c}(t) = -dc(t) - a's(t)c(t) + eg(t)^m$ $\dot{g}(t) = rs(t) + f - eg(t)^m - bg(t)^n$		
4. $\dot{s}(t) = -\gamma s(t) + \beta s(t)g(t - \Delta t)$ $\dot{g}(t) = \gamma' s(t - \Delta t) + \mu - \beta s(t - \Delta t)g(t)$		
5. $\dot{s}(r,t) = -ds(r,t) + as(r,t)g(r,t) + \eta \nabla^2 s(r,t)$ $g(r,t) = 1 - s(r,t)$	A diffusion (nonlocal) term and a rotation term are introduced.	106
6. $\left(\dfrac{\partial}{\partial t} - \eta \nabla^2 \right) s(\mathbf{r},t) = -ds(\mathbf{r},t) + as(\mathbf{r},t)[1 - s(\mathbf{r},t)]$ $\quad - \Omega \dfrac{\partial}{\partial \phi} s(\mathbf{r},t)$ $g(\mathbf{r},t) = 1 - s(\mathbf{r},t)$		
7. $\dot{s}(t) = -rs(t) + as(t)g(t)$ $\dot{g}(t) = -dg(t) + r's(t) - hs(t)^n + f_0 + F(t)$	The model contains spontaneous star formation and stochastic infall.	125
8. $\dot{s}(t) = -rs(t) + as(t)g(t)$ $\dot{g}(t) = -dg(t) - 1s(t)^k + f_0 + F(t)$	Detailed numerical simulations are performed on these models.	127
9. $\dot{s}(t) = -rs(t) + as(t)c(t) + bg(t)^n$ $\dot{c}(t) = -dc(t) - a's(t)c(t) + eg(t)^m$ $\dot{g}(t) = rs(t) - eg(t)^m - bg(t)^n + f_0 + F(t)$		
0. $\dot{s}(t) = -rs(t) + as(t)c(t)$ $\dot{c}(t) = -dc(t) + rs(t) - a's(t)c(t) + f_0 + F(t)$		

506

TABLE I (*Continued*)

Model	Comments	Re
11. $\dfrac{\partial}{\partial t} X(t) = aX(t) - X(t)[bY(t) - cY(t)^2 + dY(t)^3]$ $\left[\dfrac{\partial}{\partial t} Y(t) = -\alpha Y(t) + X(t)[\beta + \gamma Y(t) - \delta Y(t)^2]\right.$ $+ D\nabla^2 Y(t)$	Model 11, like 12 and 13, is a model for the interstellar medium. X is the density of the interstellar gas, Y the density of interstellar diffusive energy.	107
12. $\dfrac{\partial}{\partial t} X(r,t) + v\dfrac{\partial}{\partial r} X(r,t)$ $= aX(r,t) - X(r,t)[bY(r,t) - cY(r,t)^2 + dY(r,t)^3]$ $\dfrac{\partial}{\partial t} Y(r,t) + v\dfrac{\partial}{\partial r} Y(r,t)$ $= -\alpha Y(r,t) + X(r,t)[\beta + \gamma Y(r,t) - \delta Y(r,t)^2]$ $+ D\nabla^2 Y(r,t)$		140
13. $\dfrac{dX_c}{dt} = AX_w - BX_c X_h^2 + D_c\nabla^2 X_c + \Omega(r)\dfrac{\partial X_c}{\partial\theta}$ $\dfrac{dX_h}{dt} = BX_c X_h^2 - X_h X_w + D_h\nabla^2 X_h + \Omega(r)\dfrac{\partial X_h}{\partial\theta}$ $\dfrac{dX_w}{dt} = X_h X_w - AX_w + D_w\nabla^2 X_w + \Omega(r)\dfrac{\partial X_w}{\partial\theta}$	X_c refers to cold clouds, X_w to warm gas, and X_h to hot gas.	138

spectively. A simple model assumes the rate of star formation to be determined by the rate of depletion of interstellar gas as follows:

$$\dot{s}(t) = -rs(t) + as(t)g(t)$$
$$\dot{g}(t) = -dg(t) + f \tag{4.1}$$

where f, the rate of infall of halo material, is assumed constant, r is the rate of return of stellar material to the gas phase, a is the rate of induced star formation, and d is the rate of gas consumption. A more realistic model can be obtained from the generalized one-zone three-phase model consisting of diffuse gas, clouds, and stars:

$$\dot{s}(t) = -rs(t) + as(t)c(t) + bg(t)^n$$
$$\dot{c}(t) = -dc(t) - a's(t)c(t) + eg(t)^m \tag{4.2}$$
$$\dot{g}(t) = rs(t) + f - eg(t)^m - bg(t)^n$$

Now b is the modified Schmidt[124] rate (spontaneous star formation), a' is the induced rate of cloud consumption, d is the rate of cloud destruction by background sources, and e is the rate of formation of clouds out of diffuse gas.

The nonlinear one-zone models can be generalized by the introduction of explicitly stochastic terms for any of the rates. The easiest, and physically most interesting, term to introduce is that of time-variable infall. In this case, Ferrini et al.[125] have shown that there are analytic solutions possible for the simple two-component gas model which agree well with both the equilibrium behavior predicted by the linearized models and the deterministic systems.

Consider the system

$$\dot{s}(t) = -rs(t) + as(t)g(t)$$
$$\dot{g}(t) = -dg(t) + r's(t) - hs(t)^n + f + F(t) \tag{4.3}$$

where all the variables have the same meaning as for the deterministic systems, but the additional term $F(t)$ is assumed to be a random variable with the property

$$\langle F(t)F(t') \rangle = 2D\delta(t-t') \tag{4.4}$$

This is the approximation of a Wiener variable, which renders the system (4.3) a coupled Langevin system. The correlation function for F is therefore of the character of a diffusion coefficient, and the system can be solved by standard Lagrangian methods.

Assuming that the change in the gas fraction is negligible, Ferrini et al.[125] solve the system by putting it into the form

$$\dot{s}(t) = \frac{a}{d}(f - f_0)s(t) + \frac{ar'}{d}s(t)^2 \frac{ha}{d}s(t)^{n+1} + \frac{a}{d}s(t)F(t) \tag{4.5}$$

where f_0 is defined to be rd/a. This is, then, an equation associated with a potential function $V(s)$, which is characteristic of systems with multiplicative noise and which is of the form

$$\frac{d}{a}V(s) = -\frac{1}{2}(f - f_0)s^2 - \frac{r'}{3}s^3 + \frac{h}{n+2}s^{n+2} \tag{4.6}$$

It should be noted that if n is of the order 2, this is a cusp manifold (which is one which will show bifurcation behavior of the form discussed by Shore and Comins[119]). The analysis of this system presented by Ferrini et al.[125] shows that this is indeed a system with multiple equilibrium states. Having this form for the potential, there exists a Fokker-Planck equation for the evolution of the probability distribution function for the stellar population of the system, which is one that yields the expectation for the number of stars as a function of time. The evolution equation for this system is

$$\frac{\partial}{\partial t} P(s; t) = -\frac{a}{d} \frac{\partial}{\partial s} \left[(f - f_0)s - hs^{n+1} + \frac{a}{d} Ds \right] P(s; t)$$

$$+ \frac{aD}{d} \frac{\partial^2}{\partial s^2} \left[s^2 P(s; t) \right] \tag{4.7}$$

which gives

$$\langle s \rangle = \left(\frac{dh}{aDn} \right)^{-1/n} \frac{\Gamma \left[(f - f_0)d/aDn + 1/n \right]}{\Gamma \left[(f - f_0) d/aDn \right]} \tag{4.8}$$

with $\sigma^2(s) = aD/dh$, $f > f_0$. As expected, the variance is linearly dependent on D, while the mean stellar fraction varies as D (noting that the case $n = 2$ is then essentially that of a Poisson limit evolution).

A similar treatment in which the rates themselves can all be treated as random variables has been given by Shore.[106] Here the treatment by Bartlett[126] is used, which allows for the evolution of the coupled system under the explicit assumption that $g = 1 - s$. This will not be true in the case of infall but well approximates the evolution of the closed system. The Fokker-Planck equation in this case results from the interpretation of the coupled one-zone evolution equations as both being Langevin equations. The infall characteristics are not specified, nor is it necessary to state at the outset what the variance of the rate coefficients is like. The equation is of the form

$$\frac{\partial}{\partial t} P(s; t) = as(1 - s) \frac{\partial^2}{\partial s^2} P(s; t) - d(1 - s) \frac{\partial}{\partial s} P(s; t) \tag{4.9}$$

which has as a solution

$$P(s; t) = P_0 e^{\lambda t} \left[{}_2F_1(\alpha, \beta_+, \gamma; s) + A_2 F_1(\alpha, \beta_-, \gamma; s) \right] \tag{4.10}$$

where

$$\gamma = -\frac{d}{a} \qquad -(\alpha + \beta) = 1 + \frac{d}{a} \qquad \alpha\beta = \frac{\lambda}{a} \tag{4.11}$$

and $\beta_\pm$ are the roots of $\beta^2 + (1 + d/a)\beta + \lambda/a = 0$. Again, this is a solution which shows multiple roots and the bifurcation character one is led to expect from the deterministic models.

A numerical treatment of these equations has been provided recently both by Ferrini and Marchesoni[127] and by Ferrini et al.[128] The analysis proceeds as follows.

The system (4.3) is coupled to the metallicity equation

$$\frac{d}{dt}(Zg) = -(1 - R)Z\dot{s} + P_Z\dot{s} + Z_f f \qquad (4.12)$$

where Z_f is the metallicity of the infalling material, P_Z is the rate of stellar production of Z, and R is the rate of return of material to the gas phase. This is the same equation explored in Tinsley[122] and Shore.[123] The rate of star formation ψ has been replaced by the specific rate from the nonlinear models. We can then couple this directly to the evolution equations for the system and solve for the stellar and gas contribution explicitly, without the usual assumptions of previous work about the rate of star formation with time (see, e.g., Audouze and Tinsley,[129] Pagel and Edmunds,[130] Twarog,[131] and Miller and Scalo[60] for cases in which the rate of star formation must be assumed in the analysis of field star metallicity). The distribution function for the metallicity has been given by Ferrini et al.,[125] and the explicit deterministic evolution of the system has been solved by Shore.[123] A version of the system with deterministic (that is, constant) infall of halo material has been presented previously by Lynden-Bell[132] and agrees with the more detailed solution by Shore.[123]

In previous exploration of these systems, the analytic solutions were sought which would describe the asymptotic stability of such a system. We now drop the assumptions required for the linearized treatment and examine the full system. We begin by assuming that the infall can be a stochastic variable in time. Such behavior is expected for a galaxy in a cluster, for example, for which collisional stripping or infall might be occurring, or for any region of the galaxy through which stars are randomly passing. In such a case as this last one, the rate coefficients should also be random functions. For the moment, we will examine only the metallicity evolution of the system. We have computed a set of models for reasonable values of the parameters with stochastic infall. The values chosen for the parameters are

$$a = a' = 0.10 \qquad h = 0.05 \qquad n = 1.84*$$
$$r = r' = 0.10 \qquad f = 0.04$$

and $Z_f = 10^{-3}$. It was also assumed that the infall had no metallicity dispersion but that it had a mean value of f, with a dispersion ± 0.04. For com-

*Sanduleak.[133]

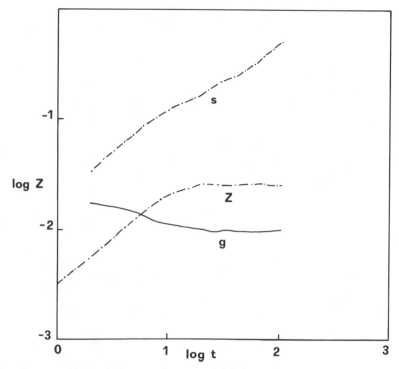

Figure 1. *s, g, Z* vs. *t*, in the case of constant infall *f*. *s* and *g* arc in arbitrary units; unit of time is 10^8 years.

parison, we have also computed a fully deterministic model with the same parameters. In all cases, the initial ratio g_0/s_0 is 10^3 (essentially all gas) and $Z_0 = 10^{-3}$. The results are shown in Fig. 1. The stellar fraction rises quickly, with an asymptotically linear form, while the gas fraction decreases slowly and thereafter remains fixed. The metal abundance of the system saturates, even while the stars continue to increase. There is little difference in the evolution, in this particular system, of the stellar and gas phases with the deterministic rates. However, for a critical value of $r = 0.08$, all other parameters kept fixed, the metallicity reaches an initial maximum and then slowly declines. This sort of behavior is reminiscent of Larson,[134] which shows that when the star formation rate is very large in comparison with the death rate, the metallicity actually peaks during the early evolution of the system.

We have also tried models in which the fluctuations increased in amplitude with time, and the metallicity also increased stochastically with time. Again the metallicity of the system saturates at approximately solar values,

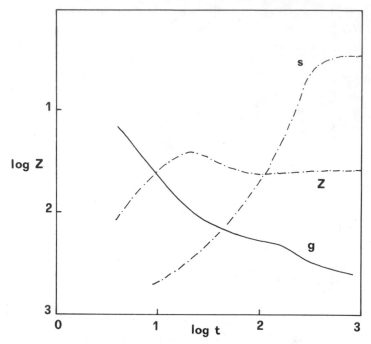

Figure 2. s, g, Z vs. t, in the case of exponentially decreasing infall.

although the stellar fraction behaves like the stochastic system described previously. Finally, in Fig. 2 we show the results for the evolution, assuming that the infall decreases like e^{-pt}, where p is a free parameter. This is the result of allowing f to fluctuate to zero during the course of the infall, although the rate can temporarily increase. Here, the stellar fraction saturates (decreasing ψ). Again, the metallicity changes with time in a fashion almost indistinguishable from the previous cases. In short, models of this sort seem to suggest that while the infall is necessary in order to explain the general evolution of the system, very different histories of this infall can produce essentially the same result in the final state. That our galaxy seems to have essentially constant metallicity, constant birth rate, and increasing star-to-gas ratio at about the same rate as the birth rate suggests that the appropriate description is one with either constant or stochastic infall at a rate comparable to the rate of induced star formation.

In order to further specify the nature of models, we should add that a three-level system is the only one completely appropriate for the modeling of a galaxy. This has also been discussed recently by Seiden,[135] who has labeled these phases "active" and "inactive." the active phase is the gas, since it is from this that the clouds and eventually the stars are formed. The inactive phase , or the clouds in our picture,[109] is a form of holding phase for the material.

It does not seem without interest to evaluate the linearized stochastic case, both from the standpoint of introducing the formalism and because there is some evidence that systems forming stars can be treated as self-regulated and therefore equilibrated systems.[75,136,137] Let us assume a simplified two-level system:

$$\dot{s}(t) = as(t)g(t) - rs(t)$$
$$\dot{g}(t) = r's(t) - a's(t)g(t) + f \tag{4.13}$$

so that we now have

$$\frac{d}{dt}\mathbf{X} = A\mathbf{X} + \mathbf{\Phi} \tag{4.14}$$

where

$$\mathbf{\Phi} = \begin{pmatrix} a_1 s_0 g_0 - r_1 s_0 \\ s_0 r'_1 - a'_0 g_0 + f \end{pmatrix} \tag{4.15}$$

and

$$A = \begin{pmatrix} a_0 g_0 - r_0 & a_0 s_0 \\ r'_0 - a'_0 g_0 & - a'_0 s_0 \end{pmatrix} \tag{4.16}$$

are the state operators for the system. Now consider the diagonalizing transformation T such that

$$T^{-1}\mathbf{\Phi} = \begin{pmatrix} \left(\dfrac{r_1}{r_0} - \dfrac{a'_1}{a_0} \right) r_0 s_0 + f \\[2ex] \left(\dfrac{a_1}{a_0} - \dfrac{r_1}{r_0} + \dfrac{a'_1}{a_0} - \dfrac{r_1}{r_0} \right) s_0 r_0 - f \end{pmatrix} \tag{4.17}$$

Equation (4.14) can now be integrated to yield

$$g(t) = c_0 e^{-a_0 s_0 t} + \int_0^t (R - A)_x r_0 s_0 \exp[-a_0 s_0 (t - x)] \, dx$$
$$+ \int_0^t \exp[a_0 s_0 (x - t)] f(x) \, dx \tag{4.18}$$

The important thing to note here, where we can take c_0 to be given by the initial conditions, is that the entire process can be viewed as a stochastic Itô equation, nowhere assuming anything about the differentiability of the infall parameter $f(t)$. The advantage of this approach for the generalization of the evolution equations to the stochastic regime in which we do not linearize the system is therefore clear; the infall (or outflow, we need not specify the sign of f) can be a stepwise continuous function of time, or even discontinuous. The rate of change of the gas fraction in the system will be determinable regardless.

The reason for dwelling on this point is that the evolution of any galactic system will be influenced by random processes occurring in the environment in which it finds itself. The chance encounter between two galaxies in a cluster will cause infall time-dependent but stochastic in character. The occurrence of supernovae in portions of the system will be random in time, and the input of energy to the interstellar medium and consequent alteration of the local conditions for star formation will also be inherently stochastic in nature. Thus, having in hand a qualitative formalism for the analysis of such effects may serve as a useful starting point for a discussion of further developments in the nonlinear theory. Furthermore, this will be useful in the explication of the stochastic metallicity evolution of the system, since it is the fluctuations of the stellar and gas fraction in any part of the system that will drive the alteration of the metal abundance of the disk. Provided the system is well mixed, which is fine for the laboratory but not necessarily for a spiral galaxy, there will be a predictable spread in the abundances derivable from the evolution equations for the system.

The extension of these models to two dimensions, a prerequisite for realistic models of spiral galaxies, can be accomplished by using stochastic methods for justification. A diffusion equation for the stellar population including birth and death terms was first asserted by Shore[106] and also recently employed by Nozakura and Ikeuchi.[138] It is possible, however, to derive this equation from first principles provided the spatial distribution for the stellar velocities has a random component as well as that due to the differential rotation of the galaxy.

Assume that the distribution of orbital eccentricities is a random function of space. Then, at any position in the galaxy, there will be stars on either inbound or outbound legs of their orbits, distributed about the mean motion at that galactocentric radius. The master equation for the stellar population of that region of the disk will be given, then, by

$$
\begin{aligned}
P_t(x,t) = {} & A_+ P(x-dx, t-dt) + A_- P(x+dx, t-dt) - B_+ P(x,t) \\
& - B_- P(x,t) - rP(x,t) + f\big(\big(P(x,t-dt)[1-P(x,t)] \\
& + P(x,t)[1-P(x,t-dt)]\big)\big)
\end{aligned}
\tag{4.19}
$$

where the coefficients $A_\pm$ and $B_\pm$ refer to the motion of stars through the region without changes in the composition of the stellar component of the system, and r and f are the death and stimulated birth rates. It is assumed that the gas fraction g is given by $1 - s$, as before. It is then simple to see that this equation reduces, in the continuum limit, to the diffusion equation

$$\frac{\partial s}{\partial t} = \eta \nabla^2 s - \Omega \frac{\partial s}{\partial \phi} + as(1 - s) - rs \tag{4.20}$$

which is given by Shore.[106] The key reason for the diffusion term appearing is that the master equation couples the population at $x - dx$ with that at $x + dx$, thereby giving rise to a second derivative. It is also the case that the diffusion is dominated by the tendency, due to the random distribution of orbital eccentricities, to appear like a small diffusive variation in the stellar population in the zone superimposed on the convective derivative due to differential rotation of the centroid. Clearly, this model can be generalized to three dimensions, depending upon the choice of the vertical structuring of the rotation law, since for the halo polulation the orbits form a more or less spherical distribution about the galactic plane. The two populations can, in fact, be coupled through the stellar distribution function $P(x, t)$, which can be broken into subcomponents (subpopulations) and which can then change the birth and death terms (since, for instance, there is no current formation of halo stars while these become supernovae or form planetary nebulae during their passage through the disk).

In the one-zone picture, the metallicity evolution can be solved using the coupled star-gas evolution equations, and the same is true for this case. If we assume that the terms in the metallicity function are only spatially dependent through the stellar population evolution equation, then it is possible to solve explicitly for the metallicity as a function of position in the galaxy.

One recent attempt at a phenomenological model for evolution of the metallicity of the galactic disk has been presented by White and Audouze.[139] Their picture assumes that gas is recycled, possibly nonconservatively, between the interstellar medium and stellar interiors, where nuclear processing alters the abundances and increases the metallicity. On being returned to the interstellar medium, the material is mixed so that subsequent generations of stars will draw from this polluted source of matter. They proceed as follows. Take f to be the fraction of the disk locked up in long-lived stars and g to be the remaining fraction in gas, which can be polluted by an amount ΔZ_1 in its heavy metal abundance (the species of element being considered will remain momentarily unspecified). The yield of the medium is defined to be

$$\tilde{Y}_1 = (f^{-1} - 1)g\Delta Z_1 \tag{4.21}$$

The probability that any parcel of gas will not have been incorporated into stars in N events and will finally find itself inside a long-lived star is

$$P(N) = f(1-f)^N \qquad (4.22)$$

while the probability that this parcel will have been enriched n times out of N will be the conditional probability

$$P(n|N) = \binom{N}{n} g^n (1-g)^{N-n} \qquad (4.23)$$

If the Poisson process of enrichment is assumed (the gas is completely mixed and randomly enriched), then, using the Chapman-Kolmogorov equation, we have

$$P(n) = \sum_{N=n}^{\infty} P(n|N) P(N) \qquad (4.24)$$

for the probability of n enrichments. Therefore, the metallicity increase is the expectation value

$$\langle Z_1 \rangle = \Delta Z_1 \sum n P(n) \qquad (4.25)$$

This is simply, for the gas-star model we have chosen,

$$\langle Z_1 \rangle = (a^{-1} - 1) \Delta Z_1$$
$$\sigma^2 = \langle Z_1^2 \rangle (1-a)^{-1} \qquad (4.26)$$

The rate of increase of metals is therefore dependent upon

$$a = \frac{f}{g + f - fg} \qquad (4.27)$$

which is the probability that the parcel will be in the gas phase at the time of sampling. We note that this system is assumed to be open, since in the case of $f = 1 - g$, we see that the system is unbounded if $g = 1$ and that therefore this is not a realistic model for the system. The inherent stochastic dispersion of the model results from the fact that the infalling material will not be contaminated until it is incorporated into the disk material and enriched from the star formation occurring in the disk.

Although the model has been elaborated by White and Audouze to study the variations among different elements, the essentials of the model remain unchanged. We have described it as phenomenological because it does not include the effects of the alteration of the conditions of the galactic disk on the star formation rate or the spectrum of mass or the effects of processing of the stellar component of the system. This absence of feedback essential to an understanding of the time development of the metallicity is also one of the basic features which we have covered with our models. This paper[139] is, however, noteworthy in being the first attempt to understand both the variation of metal abundance and the cosmic spread associated with star formation and enrichment processes in the disk. The conclusion of this paper is that infall is indeed necessary, as discussed by Lynden-Bell[132] and Shore,[106] and also by Ostriker and Thuan.[121] Its failure to consider the change in the star formation characteristics as a function of the change in the composition of the disk does, however, limit severely the use of this model to real galaxies.

V. MATHEMATICAL TREATMENT OF A SIMPLE ASTROPHYSICAL MODEL

In the foregoing sections we reviewed some stochastic models for stellar formation and galactic evolution widely employed in the literature. Here, we shall discuss explicitly a simplified astrophysical model where the stochastic processes introduced for mimicking the complexity of the relevant interactions are dealt with by recourse to the analytical tools of the previous chapters in this volume (notably Grigolini and Marchesoni, Chapter II).

Even though we focus on the detailed treatment of a single example, we make some preliminary remarks:

1. All the approaches mentioned above which adopt Langevin or the Fokker-Planck formalism are to be regarded as merely phenomenological in nature. The corresponding Langevin (or Fokker-Planck) equations cannot be derived from a proper global Hamiltonian descriptions, no matter which restrictions are imposed. That is mainly due to our scarce knowledge of the intimate structure of the systems under study. Generally our Langevin phenomenological equations are written down in two steps. First of all we must recognize and characterize the relevant constituents whose interplay is likely to determine the time evolution of the observed physical quantities (i.e., the choice of variables and deterministic terms). Thereafter, we try to account for the presence of the remainder by assuming that some parameters, first introduced as deterministic, undergo stochastic fluctuations with statistics that cannot be readily referred to the actual dynamics of the system.

2. This procedure, in spite of the possible criticism as arbitrary, is useful for testing the stability of more refined but purely dynamical models. Ferrini

et al.[125,127] showed that small fluctuations of the parameters of nonlinear models can affect dramatically both the time evolution and the stationary equilibrium state of the system. This effect can be explained in terms of basic mechanisms such as the so-called noise-induced phase transitions introduced in great detail by Faetti et al. (Chapter X). The problem of the exact Langevin equations lies outside the limits of our discussion.

3.   Most of the mathematical techniques reviewed in this volume provide subsidiary tools to the authors addressing the stochastic methods in astrophysics. The reduced model approach can be successfully employed for treating delay equations like those involved in the metallicity problem. In this case we can suggest a sensible criterion for adding new macrovariables to a simpler starting model.[128] On the other hand, an abiabatic elimination scheme (like the AEP in Chapter II) is of great use when we need to simplify a set of stochastic equations without losing relevant information. An example of this reduction technique for a specific astrophysical model is worked out in the following. Finally, the CFP of Grosso and Pastori Parravicini (Chapter III) is certainly a flexible numerical algorithm for explicitly computing the time dependence of related statistical quantities of astrophysical interest.

Let us choose as a starting point for a stochastic galactic evolutionary model the set of deterministic equations (4.3). Such a simplified version of a two-phase picture of the galactic medium assumes $s(t)$ and $g(t)$ as the relevant constituents of the system. If we imagine that the matter exchange with the galactic halo is a random process, the rate of infall f is now being given by $f + \eta(t)$. For simplicity we assume $\eta(t)$ to be a white-Gaussian noise with zero mean value and correlation

$$\langle \eta(t)\eta(0) \rangle = 2D\delta(t) \tag{5.1}$$

The two-phase stochastic model we address now reads

$$\dot{s}(t) = -rs(t) + as(t)g(t)$$
$$\dot{g}(t) = r's(t) - a's(t)g(t) + f + \eta(t) \tag{5.2}$$

Here r is the rate of star decay, r' is the rate of return of mass to gas, a is the rate of induced star formation, and a' is the rate of star breeding.

In spite of its simplicity, application of the AEP to the Langevin system (5.2) is not straightforward. Most notably we must slightly improve the procedure summarized in Chapter II so as to deal with systems like this one, where the distinction between fast and slowly relaxing variables is not clear-cut. Let us start by studying the fixed point structure of (5.2). In the absence of stochastic terms $[\eta(t) = 0]$ we find only one fixed point (s_0, g_0),

$$s_0 = \frac{fa}{ra' - ar'} \qquad g_0 = \frac{r}{d} \qquad (5.3)$$

with $s_0 > 0$ when $r/a > r'/a'$.

This inequality corresponds to the circumstance that stars burn gaseous matter at a positive rate. The energy production by stellar combustion is not accounted for by our model. It is noteworthy that if $s(t)$ is positive at a fixed time, it will be bounded in the positive axis for all times. We can prove this statement by noting that all time derivatives of s vanish when s vanishes. Indeed, if $s(t_0) = 0$ at $t = t_0$, then $\dot{s}(t_0) = 0$. From Eq. (5.2) we also write down a general expression for the nth derivative:

$$s^{(n+1)}(t) = a[s(t) + g(t)]^n \qquad n \geq 1 \qquad (5.4)$$

where the right-hand side is to be expanded formally as the nth power of a binomial, but the powers denote the derivative order. By induction, at $t = t_0$, $s^{(n)}(t_0) = 0$ for any n. This property holds in the presence of external fluctuations as well. Let us change variables as follows:

$$g \rightarrow G = g - g_0$$
$$s \rightarrow S = s - s_0 \qquad (5.5)$$
$$t \rightarrow \tau = at$$

The set of stochastic equations (5.2) can be rewritten as

$$\dot{S} = SG + s_0 G$$
$$\dot{G} = -\frac{f}{s_0 a} S - \frac{a'}{a} s_0 G - \frac{a'}{a} SG + \frac{\eta(t)}{a} \qquad (5.6)$$

On changing notation:

$$x = \frac{S}{s_0} \qquad v = G \qquad \gamma = \frac{a'}{a} s_0$$

Equations (5.6) show a form resembling the explanatory systems studied in Chapter II:

$$\dot{x} = xv + v$$
$$\dot{v} = -\frac{f}{a} x - \gamma v - \gamma xv + \frac{\eta(t)}{a} \qquad (5.7)$$

The corresponding Fokker-Planck equation reads

$$\frac{\partial}{\partial \tau}\rho(x,v;\tau) = \Gamma\rho(x,v;\tau)$$

$$= \left\{ -v\frac{\partial}{\partial x}x - v\frac{\partial}{\partial x} + \frac{f}{a}x\frac{\partial}{\partial v} + \gamma x\frac{\partial}{\partial v}v \right.$$

$$\left. + \gamma\frac{\partial}{\partial v}v + \frac{D}{a^2}\frac{\partial^2}{\partial v^2} \right\}\rho(x,v;\tau) \tag{5.8}$$

where $\rho(x,v;\tau)$ denotes the probability distribution of (x,v) at time τ.

In the following we apply the AEP in the case of large viscosity γ and small fluctuations in intensity D. We determine explicitly the range of parameter values for which our perturbation approach is valid. Our strategy now consists of considering $v(t)$ as a fast-relaxing variable and $x(t)$, which related to the star population, as the observable of interest. We notice that in the present case the usual prescription for writing down the Langevin equation corresponding to the first-order perturbation approximation is very suspect. Indeed, the Smoluchowski approximation is often obtained by putting $\dot{v} = 0$ in Eq. (5.7) so that we obtain

$$\dot{x} = -\frac{f}{a\gamma}x + \frac{\eta(t)}{a\gamma} \tag{5.9}$$

The corresponding Fokker-Planck equation is given by

$$\frac{\partial}{\partial \tau}P(x;\tau) = \frac{1}{\gamma}\frac{\partial}{\partial x}\left[\frac{f}{a}x + \langle v^2\rangle\frac{\partial}{\partial x}\right]P(x;\tau) \tag{5.10}$$

where $\langle v^2\rangle = D/a^2\gamma$, and $P(x;\tau)$ is the reduced probability distribution of the observable x. The equilibrium distribution is

$$\bar{P}(x) = \mathcal{N}\exp\left(-\frac{f}{a}\frac{x^2}{2\langle v^2\rangle}\right) \tag{5.11}$$

where $\mathcal{N}$ is a normalization constant. From Eq. (5.5) we know that

$$s(t) = s_0[x(t)-1] \tag{5.12}$$

It implies that even though the equilibrium distribution (5.11) is centered around $x = 0$ (i.e., $s = s_0$), negative values of $s(t)$ are allowed. This is in contrast with the exact result proved above.

The naive approximation on Eq. (5.9) fails because $v(t)$ cannot be considered such a fast relaxing variable with respect to $x(t)$ as to assume $\dot{v} = 0$ in Eq. (5.7). The presence of a mixing term $-\gamma xv$ would lead us to employ AEP with some caution. We can get rid of $v(t)$ as promised, but only when D can be considered small. The condition of a large γ imposed by Smoluchowski is no longer enough for treating our system.

In order to apply the AEP of Chapter II, we separate the Fokker-Planck operator Γ on Eq. (5.8) into an unperturbed part Γ_0 and a perturbation part Γ_1, so that

$$\Gamma = \Gamma_0 + \Gamma_1 \tag{5.13}$$

where

$$\Gamma_0 = \gamma \left[\frac{\partial}{\partial v} v + \langle v^2 \rangle \frac{\partial^2}{\partial v^2} \right] \tag{5.14}$$

and

$$\Gamma_1 = -v \frac{\partial}{\partial x} x - v \frac{\partial}{\partial x} + \frac{f}{a} x \frac{\partial}{\partial v} + \gamma x \frac{\partial}{\partial v} v \tag{5.15}$$

The three relevant parameters γ, f/a, and $\langle v^2 \rangle$ have been suitably defined above. The result of our perturbative approach is a Fokker-Planck equation for the reduced probability distribution $P(x; \tau)$ of the form

$$\frac{\partial}{\partial \tau} P(x; \tau) = \sum_{r=0}^{\infty} \Gamma_r P(x; \tau) \tag{5.16}$$

where Γ_r are the perturbation terms of the Fokker-Planck operator of rth order with respect to the perturbation parameter $1/\gamma$. However, such a counting rule is not reliable in the present case because the last term Γ_1 is proportional to γ.

If we proceed further, disregarding such a warning, we easily find the explicit expression for $\Gamma_1 + \Gamma_2$ ($\Gamma_0 = 0$):

$$\Gamma_1 + \Gamma_2 = \frac{1}{\gamma} \frac{\partial}{\partial x} \left\{ Q_1(x) + \langle v^2 \rangle \frac{\partial}{\partial x} Q_2(x) \right\} \tag{5.17}$$

with

$$Q_1(x) = \frac{f}{a} \left(x - \frac{x^3}{\gamma} \right) + 3 \langle v^2 \rangle x (1 + x) \tag{5.18}$$

and

$$Q_2(x) = 1 - 3x^2 + 2x^3 \qquad (5.19)$$

We have adopted the notation of Grigolini and Marchesoni (Chapter II). The diffusion coefficient $Q_2(x)$ exhibits the following properties:

1. $Q_2(-1) = 0$
2. $Q_2'(-1) = 0$ (prime denotes derivative after x)
3. $Q_2(1) < 0$.

According to Eq. (5.12), $x = -1$ corresponds to $s = 0$, and $x = 1$ to $s = 2s_0$. On the other hand, the (stationary) equilibrium distribution $\bar{P}(x)$ of a standard Fokker-Planck equation of the type (5.17) is given by

$$\bar{P}(x) = \frac{\mathcal{N}_c}{Q_2(x)} \exp\left[-\int_c^x \frac{Q_1(x')}{Q_2(x')} \, dx' \right] \qquad (5.20)$$

where $\mathcal{N}_c$ is a suitable normalization constant. We can easily check that $\bar{P}(x)$ vanishes in $x = -1$, as it should, but becomes meaningless (i.e., negative) around $x = 1$. Moreover, on calculating Γ_2 we would notice that the term $\gamma x(\partial/\partial v)v$ in Γ_1 is responsible for producing contributions proportional to $1/\gamma$, while those should be wholly accounted for by Γ_1. Such a mechanism works at higher perturbation order as well, so that our perturbation criterion has to be completely restated.

Let us assume that

$$\gamma \gg 1 \qquad \langle v^2 \rangle \ll \frac{f}{a} \qquad (5.21)$$

so that $Q_1(x)$ can be approximated by

$$Q_1(x) = \frac{f}{a} x \qquad (5.22)$$

Restrictions (5.21) are not enough to make all terms coming from Γ_r (with $r > 2$) negligible for determining $Q_2(x)$. On the contrary, we must sum all contributions to each Γ_r in order to pick up the terms proportional to $\langle v^2 \rangle / \gamma$. For large γ, that is, $(f/a)^{1/2}, \langle v^2 \rangle \ll \gamma$, the remainder is certainly comparatively small.

Section III of Chapter II has shown that Γ_r can be written as a sum of terms $\mathcal{D}_n^{(m,0)}$ with $m + n = r$ and their products $\mathcal{D}_{n_1}^{(m_1,0)} \cdots \mathcal{D}_{n_k}^{(m_k,0)}$ with $m_1 + \cdots + m_k + n_1 + \cdots + n_k = r$. Among these, only $\mathcal{D}_1^{(m,0)}$ can give rise to contributions proportional to $\langle v^2 \rangle / \gamma$; the others generate terms of order $1/\gamma^2$ or higher.

Let us now focus on any single pair $\mathscr{D}_1^{(2n,0)}$ and $\mathscr{D}_1^{(2n+1,0)}$, $n \geq 0$. Following the basic rules of our AEP (Chapter II, Section IV), we note that:

1. Each $\mathscr{D}_1^{(2n,0)}$ is the integral product of $2n+2$ Γ_1's.
2. Γ_1 consists of terms odd in v, except for the last one, which is the cause of our difficulties and is even in v.
3. The first Γ_1 in any product $\mathscr{D}_1^{(m,0)}$ contributes by means of only two odd components, $-v(\partial/\partial x)x - v(\partial/\partial x)$.
4. Since the global balance of the powers of v must be even and nonnegative, another Γ_1 at least contributes to the product through its odd components.
5. The other $2n$ Γ_1 in $\mathscr{D}_1^{(2n,0)}$ can contribute a factor $\gamma x(\partial/\partial v)v$ each. The formal perturbation order $2n+1$ is then decreased by at most $2n$: these are precisely the terms $1/\gamma$ which are to be resummed.
6. On counting the derivative order after x, we note that only the components of Γ_1 which are proportional to v (i.e., odd) contain an x derivative. Since we are interested in the contributions displaying only two such components (v), and factors like $\gamma x(\partial/\partial v)v$ do not affect the final power of v, we conclude that the corresponding terms arc order $\langle v^2 \rangle/\gamma$, contain a second-order x derivative, and therefore contribute to $Q_2(x)$.
7. We readily prove that the numerical factors coming from the internal integrals cancel out those obtained when moving $\partial/\partial v$ toward the left exactly.

Table II helps the reader visualize our resummation rules. From 7 we prove immediately that the contributions to $Q_2(x)$ from the $2n+1$ terms in part (b) of Table II are identical. Contributions from the pair $\mathscr{D}_1^{(2n,0)} + \mathscr{D}_1^{(2n+1,0)}$ can be reordered as in part (c). The signs are as in Table II. The operatorial part of the integral factors in square brackets in (c) can be rewritten as follows ($\partial = \partial/\partial x$):

$$(A) = \partial(1 - x - 2x^2)$$
$$(B) = 2\partial(1+x)x \qquad (5.23)$$

Since any Σ corresponds to the operator $\partial(1+x)$ and any π to a power of x, we finally obtain

$$Q_2(x) = \sum_{n=0}^{\infty} x^{2n}\left[(2n+1)Q_2^{(0)}(x) + 2nx(1+x)^2\right] \qquad (5.24)$$

where $Q_2^{(0)}(x)$ is given in Eq. (5.19). If we truncate the series on Eq. (5.24) for $n = 0$, we recover the naive approximation (5.17)–(5.19).

The series (5.24) can be easily resummed on employing the properties of the geometrical series:

$$Q_2(x) = \begin{cases} \infty & |x| > 1 \\ 1 & |x| \leq 1 \end{cases} \qquad (5.25)$$

The corresponding equilibrium distribution is determined by substituting Eqs. (5.22) and (5.25) into Eq. (5.20)

$$\bar{P}(x) = \begin{cases} 0 & |x| > 1 \\ \mathcal{N}'\exp\left(-\frac{a}{f}\frac{x^2}{2\langle v^2 \rangle}\right) & |x| \leq 1 \end{cases} \qquad (5.26)$$

After a very complicated elimination procedure, we recovered a Fokker-Planck equation (and the corresponding equilibrium distribution) which closely resembles the ingenuous approximation in Eq. (5.9). The stochastic observable now ranges within the interval $[-1,1]$ so that the normalization

TABLE II
Visualization of the Resummation Procedure

(a) Notation

$$\Sigma = v\frac{\partial}{\partial x}(1+x) \qquad \Pi = \gamma x\frac{\partial}{\partial v}v$$

(b) The counting rule for reckoning $D_1^{(2n,0)}$ terms contributing to $Q_2(x)$

$$\left.\begin{array}{c} \Sigma\ \Sigma\ \Pi\ \overset{2n}{\cdots}\ \Pi \\ \Sigma\ \Pi\ \Sigma\ \Pi\ \overset{2n-1}{\cdots}\ \Pi \\ \cdots\cdots\cdots\cdots \\ \Sigma\ \Pi\ \overset{2n}{\cdots}\ \Pi\ \Sigma \end{array}\right\}\ 2n+1\ \text{terms}\ \alpha\ \dfrac{\langle v^2 \rangle}{\gamma}$$

(c) Diagrammatic expression on $Q_2(x)$ (see text).

$$\left\{\begin{array}{l} \Sigma\left[\ \Sigma\ \right]\Pi\ \overset{2n}{\cdots}\ \Pi \\ \Sigma\left[-\Sigma\ \Pi\right]\Pi\ \overset{2n}{\cdots}\ \Pi \\ \Sigma\left[-\Pi\ \Sigma\right]\Pi\ \overset{2n}{\cdots}\ \Pi \end{array}\right\}\times(2n+1) + \left\{\begin{array}{l} \Sigma\left[\ \Sigma\ \ \Pi\right]\Pi\ \overset{2n}{\cdots}\ \Pi \\ \Sigma\left[\ \Pi\ \ \Sigma\right]\Pi\ \overset{2n}{\cdots}\ \Pi \end{array}\right\}\times 2n$$

(A)   (B)

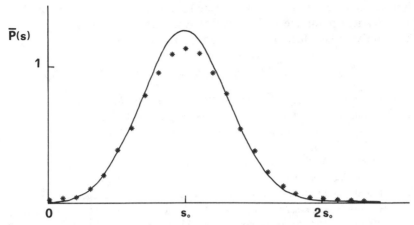

Figure 3. (Stationary) equilibrium distribution for $\bar{p}(s)$. (∗) refer to the result of our numerical simulation for $f/a = 1$, $\langle v^2 \rangle = 0.1$, $\gamma = 5$. The accuracy of our data is evaluated to be about 10%. The solid line represents our theoretical prediction.

constant $\mathcal{N}'$ in Eq. (5.26) is not to be mistaken with $\mathcal{N}$ in Eq. (5.11). This implies that the astrophysical quantity $s(t)$ assumes values between 0 and $2s_0$.

The upper bound $2s_0$ is to be regarded as an artifact of the perturbation criterion we adopted for calculating $Q_2(x)$ in Eq. (5.25). In order to check the reliability of our treatment, we carried out a numerical simulation of the stochastic system (5.2). A detailed description of the numerical algorithm is available elsewhere.[128] The comparison between the analytical expression for $\bar{P}(x)$ and the result of our simulation is illustrated in Fig. 3. We note that the agreement with our predictions is fairly close. The lower bound for $s(t)$ is correctly recovered, while a long tail lingers over the limiting value $2s_0$. Such a constraint is expected to disappear as we proceed further with our perturbation method.

We finally summarize the restrictions under which the procedure described above is reliable. Taking into account definition (5.3), we rewrite conditions (5.21) as follows:

$$0 < \left(\frac{r}{a} - \frac{r'}{a'} \right) \ll \frac{f}{a} \ll \left(\frac{f}{a} \right)^2 \frac{a^2}{D} \tag{5.27}$$

The conditions previously stated now read

$$0 < \left(\frac{r}{a} - \frac{r'}{a'} \right) \ll \left(\frac{f}{a} \right)^{1/2}, \qquad \left(\frac{f^2}{D} \right)^{1/2} \tag{5.28}$$

We notice that these inequalities can be satisfied for many different choices of the physical parameters. We recall that in our perturbation approach we defined only three effective parameters, f/a, γ, and $\langle v^2 \rangle$.

References

1. K. Schwarzschild, *Nachr. Kgl. Ges. Wiss. Gottingen*, **1907**, 614; **1908**, 191.
2. A. Eddington, *Mon. Not. Roy. Astron. Soc.* **75**, 366 (1914); **76**, 572 (1916).
3. S. Chandrasekhar, *Principles of Stellar Dynamics*, Dover, New York, 1942; *Rev. Mod. Phys.* **15**, 1 (1943).
4. L. Spitzer and M. Schwarzschild, *Astrophys. J.*, **114**, 385 (1951).
5. R. Kurth, *Introduction to Stellar Statistics*, Pergamon, New York, 1967.
6. D. Mihalas and F. Routly, *Galactic Astronomy*, Freeman, San Francisco, 1968.
7. J. H. Oort, in *Stars and Stellar Systems*, Vol. V, A. Blaauw and M. Schmidt, eds., University of Chicago Press, Chicago, 1965, p. 455.
8. D. Mihalas and J. Binney, *Galactic Astronomy*, Freeman, San Francisco, 1981.
9. D. Mihalas, *Stellar Atmospheres*, Freeman, San Francisco, 1979.
10. H. Griem, *Plasma Spectroscopy*, McGraw-Hill, New York, 1964.
11. A. T. Barucha-Reid, *Elements of the Theory of Markov Processes and Their Applications*, McGraw-Hill, New York, 1960.
12. S. Chandrasekhar and G. Munch, *Astrophys. J.*, **112**, 380 (1950).
13. E. N. Parker, *Cosmic Magnetic Fields*, Oxford Univ. Press, Oxford, 1979.
14. V. L. Ginzburg and V. S. Ptuskin, *Rev. Mod. Phys.* **48**, 161 (1976). G. Setti, G. Spada, and A. W. Wolfendale, eds. *Origin of Cosmic Rays*, *IAU Symposium 94*, Reidel, Dordrecht, 1981.
15. M. Lecar ed., *Gravitational N-Body Problem* (IAU Colloquium 10), Reidel, Dordrecht, 1972.
16. J. Moser, *Stable and Random Motions in Dynamical Systems* (*With Special Emphasis on Celestial Mechanics*), Princeton Univ. Press, Princeton, 1973.
17. E. Knobloch, *Vistas Astron.*, **24**, 39 (1980).
18. D. F. Gray and J. L. Linsky eds., *Stellar Turbulence* (IAU Colloquium 51), Springer Verlag, Heidelberg, 1980.
19. P. J. E. Pebbles, *The Large Scale Structure of the Universe*, Princeton Univ. Press, Princeton, 1980.
20. J. P. Eckmann, *Rev. Mod. Phys.*, **53**, 643 (1981).
21. F. F. Abraham, *Homogeneous Nucleation Theory*, Academic Press, New York, 1974.
22. B. Donn, N. C. Wickramasinghe, J. P. Hudson, and T. P. Stecher, *Astrophys. J.*, **153**, 451 (1968).
23. A. Ziabicki, *J. Chem. Phys.*, **48**, 4368 (1968).
24. G. H. Walker, *Astrophys. Lett.* **16**, 115 (1975).
25. B. T. Draine and E. E. Salpeter, *J. Chem. Phys.*, **67**, 2230 (1977).
26. J. J. Burton, in *Statistical Mechanics*, Part A. *Equilibrium Techniques*, B. J. Berne, ed., Plenum Press, New York, 1977, p. 195.
27. B. T. Draine, *Astrophys. Space Science*, **65**, 313 (1979).

28. J. S. Langer and A. J. Schwartz, *Phys. Rev.* **A21**, 948 (1980).

29. V. S. Safronov, *Evolution of the Protoplanetary Cloud and Formation of the Earth and the Planets*, Nauka, Moskow, 1969; NASA TT F-677.

30. N. I. Shakura and R. A. Sunayev, *Astron. Astrophys.*, **23**, 337 (1973).

31. J. E. Pringle, *Ann. Rev. Astron. Astrophys.*, **19**, 137 (1981).

32. S. J. Weidenschilling, *Icarus*, **44**, 172 (1980).

33. G. V. Pechernikova, V. S. Safronov, and E. V. Zvyagina, *Sov. Astron.*, **20**, 346 (1976).

34. Y. Nakagawa, C. Hayashi, and K. Nakazawa, *Icarus*, **54**, 361 (1983).

35. G. E. Morfill, *Birth and Infancy of Stars*, Les Houches, S. S. 1984.

36. V. S. Safronov and T. V. Ruzmaikina, in *Protostars and Planets*, T. Gehrels, ed., Univ. of Arizona Press, Tucson, p. 545 (1978).

37. G. W. Wetherill, in *Protostars and Planets*, T. Gehrels, ed., Univ. of Arizona Press, Tucson, p. 565 (1978).

38. P. G. Goldreich and W. R. Ward, *Astrophys. J.*, **183**, 1051 (1973).

39. A. Coradini, C. Federico, and G. Magni, *Astron. Astrophys.*, **98**, 173 (1981).

40. P. Farinella and P. Paolicchi, *Moon Planets*, **19**, 327 (1978).

41. P. Farinella, F. Ferrini, and P. Paolicchi, *Moon Planets*, **20**, 371 (1979); **21**, 405 (1979).

42. R. Greenberg, W. K. Hartmann, and C. R. Chapman, *Icarus*, **35**, 1 (1978).

43. G. W. Wetherill, *Ann. Rev. Astron. Astrophys.* **18**, 77 (1980).

44. S. I. Ipatov, *Sov. Astron.*, **25**, 617 (1981).

45. T. M. Eneev and N. N. Kuzlov, *Sov. Astron. Lett.*, **5**, 252 (1979).

46. S. H. Dole, *Icarus*, **13**, 494 (1970).

47. R. Isaacman and C. Sagan, *Icarus*, **31**, 510 (1977).

48. A. Milani and A. Nobili, *Astron. Astrophys.*, **133**, 231 (1984).

49. J. Wisdom, *Astron. J.*, **87**, 577 (1982); *Icarus*, **56**, 51 (1983).

50. V. Szebehely, ed., *Application of Modern Dynamics to Celestial Mechanics*, Reidel, Dordrecht, 1982.

51. C. Froëschle and H. Scholl, *Astron. Astrophys.*, **57**, 33 (1977); **72**, 246 (1979).

52. G. W. Wetherill, *J. Geophys. Res.*, **72**, 2429 (1967).

53. J. W. Dohnanyi, *J. Geophys. Res.*, **74**, 2531 (1969).

54. B. Hellyer, *Mon. Not. Roy. Astron. Soc.*, **148**, 383 (1970).

55. P. Farinella, P. Paolicchi, and V. Zappala, *Icarus*, **52**, 409 (1982).

56. B. Hellyer, *Mon. Not. Roy. Astron. Soc.*, **154**, 279 (1971).

57. P. J. van Rhijn, in *Galactic Structure*, A. Blaaw and M. Schmidt, eds., Univ. of Chicago Press, Chicago, 1965, p. 27.

58. A. G. D. Philip and A. Upgren, *The Luminosity Function* (IAU Colloq.), 1983.

59. E. E. Salpeter, *Astrophys. J.*, **121**, 161 (1955).

60. G. E. Miller and J. M. Scalo, *Astrophys. J. Suppl. S.*, **41**, 513 (1979).

61. C. Chiosi, in *Effects of Mass Loss on Stellar Evolution* (IAU Col. 59), C. Chiosi and R. Stalio, eds., Reidel, Dordrecht, 1981, p. 29.

62. D. C. Abbott, *Astrophys. J.*, **263**, 723 (1982).

63. S. N. Shore and N. Sanduleak, *Astrophys. J. Suppl. S.* **279**, 909 (1984).

64. B. G. Elmegreen and R. D. Mathieu, *Mon. Not. Roy. Astron. Soc.*, **203**, 305 (1983).

65. S. N. Shore and R. Duncan, (1985) in preparation.

66. F. Hoyle, *Astrophys. J.*, **118**, 513 (1953).

67. C. Hunter, *Astrophys. J.*, **136**, 594 (1962); **139**, 570 (1964).

68. L. Mestel, *Quart. J. Roy. Astron. Soc.*, **6**, 161, 265 (1965).

69. R. B. Larson, *Mon. Not. Roy. Astron. Soc.*, **184**, 69 (1978).

70. F. Palla, E. E. Salpeter, and S. Stahler, *Astrophys. J.*, **271**, 632 (1983).

71. T. C. Mouschovias, in *Protostars and Planets*, T. Gehrels, ed., Univ. of Arizona Press, Tucson, p. 209 (1978).

72. R. C. Fleck, *Astrophys. J.*, **246**, L151 (1981).

73. F. Ferrini, F. Marchesoni, and A. Vulpiani, *Phys. Lett.*, **A92**, 47 (1982).

74. T. Nakano, *Prog. Theor. Phys.*, **36**, 515 (1966).

75. C. Norman and J. Silk, *Astrophys. J.*, **238**, 158 (1980).

76. B. G. Elmegreen, *Mon. Not. Roy. Astron. Soc.*, **203**, 1011 (1983).

77. T. Montmerle, L. Koch-Miramond, E. Falagarone and J. E. Grindlay, *Astrophys. J.*, **269**, 182 (1983).

78. R. B. Larson, *Mon. Not. Roy. Astron. Soc.*, **161**, 133 (1973).

79. H. Zinnecker, *MPI-PAE Extraterr.*, **167** (1981).

80. B. B. Mandelbrot, *Fractals: Form, Chance and Dimension*, Freeman, San Francisco (1977).

81. R. M. May, *Nature*, **261**, 459 (1976).

82. F. Ferrini, F. Marchesoni, and A. Vulpiani, *Mon. Not. Roy. Astron. Soc.*, **202**, 1071 (1983).

83. L. B. Lucy, *Astron. J.*, **82**, 1013 (1977).

84. R. A. Gingold and J. J. Monaghan, *Mon. Not. Roy. Astron. Soc.*, **181**, 375 (1977).

85. R. B. Larson, *Mon. Not. Roy. Astron. Soc.*, **184**, 69 (1978).

86. D. Wood, *Mon. Not. Roy. Astron. Soc.*, **194**, 201 (1981).

87. M. L. Norman and J. R. Wilson, *Astrophys. J.*, **224**, 497 (1978).

88. P. Bodenheimer, *Astrophys. J.*, **224**, 488 (1978).

89. J. Buff, H. Gerola, and R. F. Stellingwerf, *Astron. J.*, **230**, 839 (1979).

90. A. P. Boss and P. Bodenheimer, *Astrophys. J.*, **234**, 289 (1979).

91. A. P. Boss, *Astrophys. J.*, **237**, 866 (1980).

92. P. Bodenheimer, D. C. Black, and J. E. Tohline, *Astrophys. J.*, **242**, 209 (1980).

93. N. A. Pumphrey and J. M. Scalo, *Astrophys. J.*, **269**, 531 (1983).

94. M. Bossi, F. Ferrini, and P. Paolicchi, *Mon. Not. Roy. Astron. Soc.*, (1985) in press.

95. S. R. Broadbent and J. M. Hammersley, *Proc. Camb. Phil. Soc.*, **53**, 629 (1957).

96. J. M. Hammersley, *Proc. Camb. Phil. Soc.*, **53**, 642 (1957).

97. E. P. Hubble, *Realm of the Nebulae*, Yale Univ. Press, New Haven, 1936.

98. J. H. Jeans, *Astronomy and Cosmogony*, Cambridge Univ. Press, Cambridge, 1929.

99. C. C. Lin and F. H. Shu, *Astrophys. J.*, **140**, 646 (1964).

100. W. W. Roberts, *Astrophys. J.*, **158**, 123 (1969).

101. A. Toomre, *Ann. Rev. Astron. Astrophys.*, **15**, 437 (1977).

102. T. A. Zang, Ph.D. thesis, MIT, Cambridge, 1976.

103. B. G. Elmegreen and C. D. Lada, *Astrophys. J.*, **214**, 725 (1977).

104.   W. Herbst and G. E. Assousa, *Astrophys. J.*, **217**, 473 (1977).

105.   M. W. Mueller and W. D. Arnett, *Astrophys. J.*, **210**, 670 (1976).

106.   S. N. Shore, *Astrophys. J.*, **265**, 202 (1983).

107.   M. Fujimoto and S. Ikeuchi, *Publ. Astron. Soc. Japan*, **36**, 319 (1984).

108.   H. Gerola and P. E. Seiden, *Astrophys. J.*, **223**, 129 (1978).

109.   P. E. Seiden and H. Gerola, *Astrophys. J.*, **233**, 56 (1979); *Fund. Cosm. Phys.*, **7**, 241 (1982).

110.   P. E. Seiden, L. S. Schulman, and H. Gerola, *Astrophys. J.*, **232**, 702 (1979).

111.   H. Gerola, P. E. Seiden, and L. S. Shulman, *Astrophys. J.*, **242**, 517 (1980).

112.   N. F. Comins, *Mon. Not. Roy. Astron. Soc.*, **194**, 169 (1981); *Astrophys. J.*, **266**, 543 (1983); **274**, 595 (1983).

113.   W. Freedman and B. F. Madore, *Astrophys. J.*, **265**, 140 (1983).

114.   T. Statler, N. F. Comins, and B. F. Smith, *Astrophys. J.*, **270**, 79 (1983).

115.   J. V. Feitzinger, A. E. Glassgold, H. Gerola, and P. E. Seiden, *Astron. Astrophys.*, **98**, 371 (1981).

116.   V. C. Rubin, W. K. Ford, and M. S. Roberts, *Astrophys. J.*, **230**, 35 (1979).

117.   V. C. Rubin, W. K. Ford, N. Thonnard, and D. Burstein, *Astrophys. J.*, **261**, 439 (1982).

118.   F. C. Bruhweiler, S. B. Parsons, and J. D. Wray, *Astrophys. J.*, **256**, L49 (1982).

119.   S. N. Shore and N. F. Comins, preprint, 1985.

120.   S. Van den Berg, *Ann. Rev. Astron. Astrop.*, **13**, 217 (1975); *J. Roy. Astron. Soc. Canada*, **69**, 105 (1975); *Astrophys. J.*, **206**, 883 (1976).

121.   J. P. Ostriker and T. X. Thuan, *Astrophys. J.*, **202**, 353 (1975).

122.   B. M. Tinsley, *Astrophys. J.*, **192**, 629 (1974); **216**, 548 (1977); *Fund. Cosm. Phys.*, **5**, 287 (1980).

123.   S. N. Shore, *Astrophys. J.*, **249**, 93 (1981).

124.   M. Schmidt, *Astrophys. J.*, **129**, 243 (1959).

125.   F. Ferrini, F. Marchesoni, and A. Vulpiani, *Mem. Soc. Astron. It.*, **54**, 217 (1983).

126.   M. S. Bartlett, *Introduction to Stochastic Processes*, Cambridge Univ. Press, Cambridge, 1956.

127.   F. Ferrini and F. Marchesoni, *Astrophys. J.*, **287**, 17 (1984).

128.   F. Ferrini, S. N. Shore, and F. Marchesoni, *Astron. Astrophys.*, (1985) in press.

129.   J. Audouze and B. M. Tinsley, *Ann. Rev. Astron. Astrophys.*, **14**, 43 (1976).

130.   B. E. J. Pagel and M. G. Edmunds, *Ann. Rev. Astron. Astrophys.*, **19**, 77 (1981).

131.   B. Twarog, *Astrophys. J.*, **242**, 242 (1980).

132.   D. Lynden-Bell, *Vistas Astron.*, **19**, 299 (1975).

133.   N. Sanduleak, *Astron. J.*, **74**, 47 (1969).

134.   R. B. Larson, *Mon. Not. Roy. Astron. Soc.*, **176**, 31 (1976).

135.   P. E. Seiden, *Astrophys. J.*, **266**, 555 (1983).

136.   J. Franco and D. P. Cox, *Astrophys. J.*, **273**, 243 (1983).

137.   J. Franco and S. N. Shore, *Astrophys. J.* **285**, 813 (1984).

138.   T. Nozakura and S. Ikeuchi, *Astrophys. J.* **279**, 40 (1984).

139.   S. S. White and J. Audouze, *Mon. Not. Roy. Astron. Soc.*, **203**, 603 (1983).

140.   M. Fujimoto, (1985) preprint.

AUTHOR INDEX

Numbers in parentheses are reference numbers and indicate that the author's work is referred to although his name is not mentioned in the text. Numbers in *italics* show the pages on which the complete references are listed.

Abbott, D. C., 498(62), *527*
Abbott, R. J., 202(40), *223*
Abou-Chacra, R., 177(54), *181*
Abragam, A., 321(3, 4), 322(3), *387*
Abraham, F. F., 495(21), 499(21), *526*
Abramowitz, M., 51(39), *80*, 230(3), *275*
Adelman, S. A., 4(17, 18), *27*, 156(27), *179*, 233(6), *275*, 424(70), *442*
Afsar, M. N., 302(54), *319*
Akhiezer, N. I., 81(3), *130*, 324(34), *388*
Anderson, J. E., 301(51), 317(51), *319*
Anderson, P. W., 137(9), 149(9), 172(45), 177(45, 54), *179, 180, 181*
Andretta, M., 188(30), *222*
Angell, C. A., 278(1, 3, 4), 296(4), 308(65), *318, 319*
Apfel, A., 295(38), *319*
Arnett, W. S., 503(105), *529*
Arnold, L., 478(7), *491*
Assousa, G. E., 503(104), *529*
Atkins, P. W., 334(48b, 49), 348(48b), *388*
Audouze, J., 510(129), 517(139), *529*

Bach, P., 308(65), *319*
Bagchi, B., 184(19, 20), 203(19, 20) *222*
Baker, G. A., Jr., 87(12), *131*
Balucani, U., 24(35), *27*, 208(47), 209(47), *223*, 239(7, 8), 245(7, 8), 260(7, 8), 267(7, 8), *275*
Balzarotti, A., 137(10), *179*
Baraff, G., 163(37), *180*
Baram, A., 325(43), 348(43), *388*
Baranova, N. B., 219(56), *223*
Barucha-Reid, A. T., 494(11), *526*
Bassani, F., 134(2), 135(2), 161(2), 163(35), *178, 180*
Beer, N., 161(30), *180*
Belch, A. C., 279(13), 280(19), 282(13), 293(13, 19, 34), *318, 319*

Bellissent-Funel, M. C., 317(84), *320*
Benoit, H., 183(1), 188(1), *222*
Berne, B. J., 25(37), *27*, 208(46), 217(38), *233*, 399(25), 430(25), *441*
Bernholc, J., 163(36), *180*
Bertolini, D., 301(52), 303(52), 317(83), *319, 320*
Beveridge, D. L., 281(22), 282(22), 283(22), *318*
Binney, J., 493(8), *526*
Bishop, M., 226(1), *275*
Bixon, M., 23(33), *27*, 105(24), *131*, 154(24), *179*
Black, D. C., 502(92), *528*
Bleaney, B., 358(50a), *388*
Blomberg, C., 398(19), 423(68), *441, 442*
Blum, L., 334(48d), 348(48d), *388*
Blumberg, R. L., 298(48), *319*
Blumstein, C., 108(29), 122(29), 124(29), *131*
Bochov, G. N., 2(10), *26*
Bodenheimer, P., 502(88, 90, 92), *528*
Boned, C., 316(79, 80), *320*
Boon, J. P., 25(37), *27*, 309(68), *320*
Boss, A. P., 502(90, 91), *528*
Bossi, M., 502(94), *528*
Bossis, G., 187(27), 189(27), *222*, 226(2a-c), 261(2a-c), 262(2a), 263(2a), 264(2a), 267(2a,b), *275*, 304(57), *319*
Bowman, J. M., 402(29), *441*
Brand, H., 3(12), *26*, 410(34), 428(34), 429(34), *441*, 446(3), 458(3), 470(3), *474*
Brent, R. P., 163(34), *180*
Brezinski, C., 81(5), 87(5), 114(5), *131*
Broadbent, S. R., 502(95), *528*
Brooks, C. L., 4(18), *27*
Brot, C., 187(27), 189(27), *222*, 226(2b), 261(2b), 266(2b), 267(2b), *275*, 304(57), *319*

531

SUBJECT INDEX

Activation:
  in chemical reaction, 62
  energy, 298
  entropy, 392
  time, 471
  volume, 392, 430
Adiabatic approximation, 135, 402
Adiabatic elimination procedure, 3, 29–80,
      188, 208
  applications, 65
  approaches to, comparison of, 77
  creation and destruction operator, 50
  deviation from, 470
  for diffusion-like equation, 70
  division of dynamical operator into
      perturbation and unperturbed parts,
      74
  and Fokker-Planck equation, 78
  history, 29
  and inertia, 470
  and Liouville operator, 61
  mechanical time scale, 65
  and molecular angular velocity, 70
  and Mori theory, 41
  projection operator, choosing best, 73
  slowing down prediction, 471
  stochastic normal modes, 57
  and synergetics, 34
  systematic version, 30
  and Zwanzig projection method, 34
    application of, 44
Agglomeration, 494
  for solar system formation study, 495
Algorithm, for ESR spectrum, 351
Alleles, 480
Alternating fields:
  circularly polarized, 219
  induced transients, 200
Ammonia, 402
Amplitude randomness, 202
Analog simulation, 453
  functions, 454
Anderson model, 172

Anderson's localization theory, 178
Angular coordinates, 246
Angular momentum, 199
Angular velocity:
  equilibrium distribution, 269
  and translation, 246
Angular velocity autocorrelation, oscillation
      envelopes, 205
Angular velocity decay, effects of excitation,
      272
Annihilation operator, 244
  definition, 249
Anticommutation operator, 339
Arrhenius factor, 416
Arrhenius fitting curve, 297
Asteroid, 496
Astrophysical stochastic models:
  adiabatic elimination procedure
      application, 518
  mathematical treatment, 517
  two-phase, 518
Astrophysical stochastic processes, 493
  coagulation calculations, 495
  initial mass function, 497
  percolation models, 502
  solar system formation, 495
  sticking probability, 495
Astrophysics:
  dust formation, 495
  nucleation phenomena, 495
Atmosphere, stellar, 493
Atomic energy levels, 136
Atomic orbitals, 135
Atom-solid collisions, 156
Augmented space techniques, 176
Autocorrelation function, 97
  angular velocity, 266
  dipole correlation, 290
  excited, decay of, 239
  first-order mixed, 198
  individual dipole, 266
  Kirkwood factor, 304
  Laplace transform, 98

543